A Level
Advancing Biology for OCR
Year 1 and AS

B

Series Editor
Fran Fuller

Authors
Michael Fisher
Dawn Parker
Jennifer Wakefield Warren

OXFORD
UNIVERSITY PRESS

OXFORD
UNIVERSITY PRESS

Great Clarendon Street, Oxford, OX2 6DP, United Kingdom

Oxford University Press is a department of the University of Oxford. It furthers the University's objective of excellence in research, scholarship, and education by publishing worldwide. Oxford is a registered trade mark of Oxford University Press in the UK and in certain other countries

First published in 2015

British Library Cataloguing in Publication Data
Data available

978-0-19-834097-3

10 9 8 7 6 5

Paper used in the production of this book is a natural, recyclable product made from wood grown in sustainable forests. The manufacturing process conforms to the environmental regulations of the country of origin.

Printed and bound by CPI Group (UK) Ltd, Croydon, CR0 4YY

This resource is endorsed by OCR for use with specification H022 AS Level GCE Biology B (Advancing Biology) and year 1 of H422 A Level GCE Biology B (Advancing Biology). In order to gain endorsement this resource has undergone an independent quality check. OCR has not paid for the production of this resource, nor does OCR receive any royalties from its sale. For more information about the endorsement process please visit the OCR website www.ocr.org.uk/

AS/A Level course structure

This book has been written to support students studying for OCR AS Biology B and for students in their first year of studying for OCR A Level Biology B. It covers the AS modules from the specification, the content of which will also be examined at A Level. There is an index at the back to help you find what you are looking for. If you are studying for OCR AS Biology B, you will only need to know the content in the blue box.

AS exam

Year 1 content

1 Development of practical skills in biology
2 Cells, chemicals for life, transport, and gas exchange (Chapters 1–7)
3 Cell division, development, and disease control (Chapters 8–15)

Year 2 content

4 Energy, reproduction, and populations (Chapters 16–22)
5 Genetics, control, and homeostasis (Chapters 23–31)

A level exam

A Level exams will cover content from Year 1 and Year 2 and will be at a higher demand. You will also carry out practical activities throughout your course.

Contents

How to use this book

This book contains many different features. Each feature is designed to support and develop the skills you will need for your examinations, as well as foster and stimulate your interest in biology.

Terms that you will need to be able to define and understand are highlighted by **bold text**.

Learning outcomes

- → At the beginning of each topic, there is a list of learning outcomes.
- → These are matched to the specification and allow you to monitor your progress.
- → A specification reference is also included.
  Specification reference: 2.1.3

Application features

These features contain important and interesting applications of biology in order to emphasise how scientists and engineers have used their scientific knowledge and understanding to develop new applications and technologies.

1 All application features have a question to link to material covered with the concept from the specification.

Study Tips

Study tips contain prompts to help you with your understanding and revision.

Extension features

These features contain material that is beyond the specification. They are designed to stretch and provide you with a broader knowledge and understanding and lead the way into the types of thinking and areas you might study in further education. As such, neither the detail nor the depth of questioning will be required for the examinations. But this book is about more than getting through the examinations.

1 Extension features also contain questions that link the off-specification material back to your course.

Synoptic link

These highlight the key areas where topics relate to each other. As you go through your course, knowing how to link different areas of biology together becomes increasingly important. Many exam questions, particularly at A Level, will require you to bring together your knowledge from different areas.

Summary Questions

1 These are short questions at the end of each topic.

2 They test your understanding of the topic and allow you to apply the knowledge and skills you have acquired.

3 The questions are ramped in order of difficulty.

Worked example:

Worked examples take you through a calculation step-by-step, you can then test your skills using the summary questions and chapter practice questions.

Module 1 gives an overview of the key practical skills you need to develop through your course. There is also a table which points you to which topics cover the practicals.

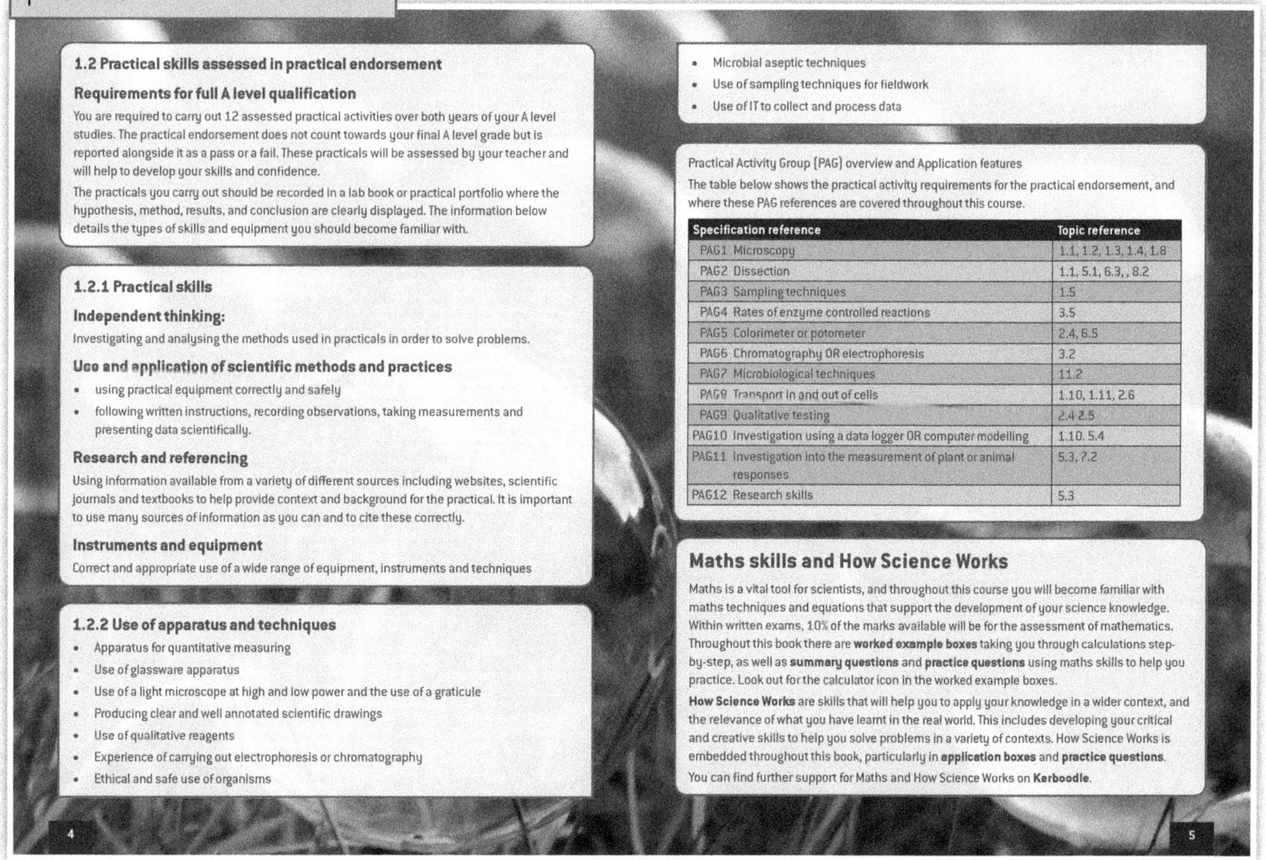

1.2 Practical skills assessed in practical endorsement

Requirements for full A level qualification

You are required to carry out 12 assessed practical activities over both years of your A level studies. The practical endorsement does not count towards your final A level grade but is reported alongside it as a pass or a fail. These practicals will be assessed by your teacher and will help to develop your skills and confidence.

The practicals you carry out should be recorded in a lab book or practical portfolio where the hypothesis, method, results, and conclusion are clearly displayed. The information below details the types of skills and equipment you should become familiar with.

1.2.1 Practical skills

Independent thinking:

Investigating and analysing the methods used in practicals in order to solve problems.

Use and application of scientific methods and practices

- using practical equipment correctly and safely
- following written instructions, recording observations, taking measurements and presenting data scientifically.

Research and referencing

Using information available from a variety of different sources including websites, scientific journals and textbooks to help provide context and background for the practical. It is important to use many sources of information as you can and to cite these correctly.

Instruments and equipment

Correct and appropriate use of a wide range of equipment, instruments and techniques

1.2.2 Use of apparatus and techniques

- Apparatus for quantitative measuring
- Use of glassware apparatus
- Use of a light microscope at high and low power and the use of a graticule
- Producing clear and well annotated scientific drawings
- Use of qualitative reagents
- Experience of carrying out electrophoresis or chromatography
- Ethical and safe use of organisms
- Microbial aseptic techniques
- Use of sampling techniques for fieldwork
- Use of IT to collect and process data

Practical Activity Group (PAG) overview and Application features

The table below shows the practical activity requirements for the practical endorsement, and where these PAG references are covered throughout this course.

Specification reference	Topic reference
PAG1 Microscopy	1.1, 1.2, 1.3, 1.4, 1.8
PAG2 Dissection	1.1, 5.1, 6.3, , 8.2
PAG3 Sampling techniques	1.5
PAG4 Rates of enzyme controlled reactions	3.5
PAG5 Colorimeter or potometer	2.4, 6.5
PAG6 Chromatography OR electrophoresis	3.2
PAG7 Microbiological techniques	11.2
PAG8 Transport in and out of cells	1.10, 1.11, 2.6
PAG9 Qualitative testing	2.4 2.5
PAG10 Investigation using a data logger OR computer modelling	1.10. 5.4
PAG11 Investigation into the measurement of plant or animal responses	5.3, 7.2
PAG12 Research skills	5.3

Maths skills and How Science Works

Maths is a vital tool for scientists, and throughout this course you will become familiar with maths techniques and equations that support the development of your science knowledge. Within written exams, 10% of the marks available will be for the assessment of mathematics. Throughout this book there are **worked example boxes** taking you through calculations step-by-step, as well as **summary questions** and **practice questions** using maths skills to help you practice. Look out for the calculator icon in the worked example boxes.

How Science Works are skills that will help you to apply your knowledge in a wider context, and the relevance of what you have learnt in the real world. This includes developing your critical and creative skills to help you solve problems in a variety of contexts. How Science Works is embedded throughout this book, particularly in **application boxes** and **practice questions**.

You can find further support for Maths and How Science Works on **Kerboodle**.

4

5

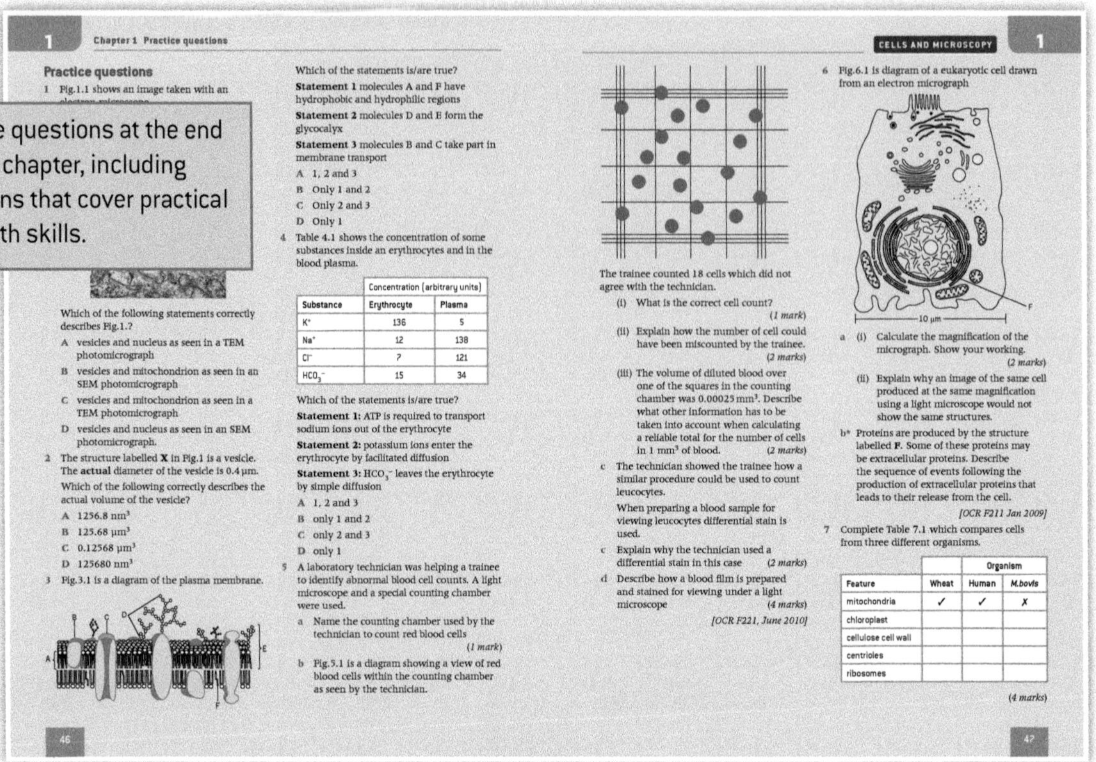

Practice questions at the end of each chapter, including questions that cover practical and math skills.

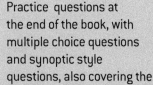

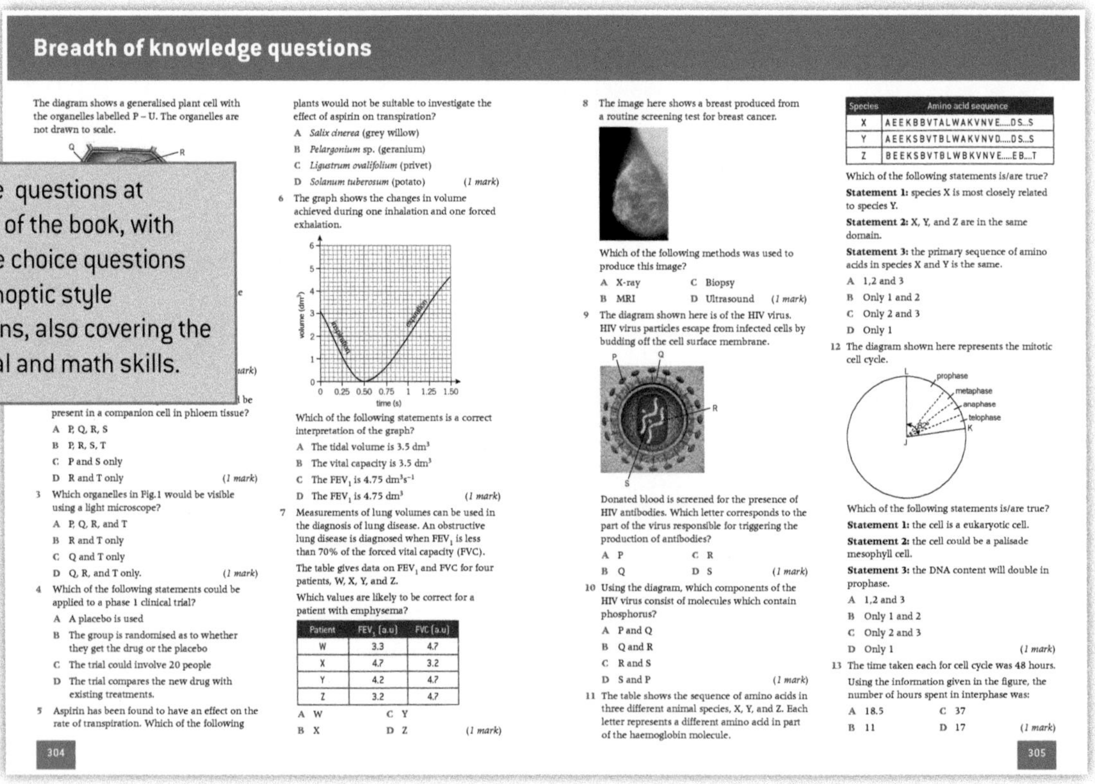

Practice questions at the end of the book, with multiple choice questions and synoptic style questions, also covering the practical and math skills.

Kerboodle

This book is supported by next generation Kerboodle, offering unrivalled digital support for independent study, differentiation, assessment, and the new practical endorsement.

If your school subscribes to Kerboodle, you will also find a wealth of additional resources to help you with your studies and with revision.

- Study guides
- Maths skills boosters and calculation worksheets
- On your marks activities to help you achieve your best
- Practicals and follow up activities to support the practical endorsement
- Interactive objective tests that give question-by-question feedback
- Animations and revision podcasts
- Self-assessment checklists

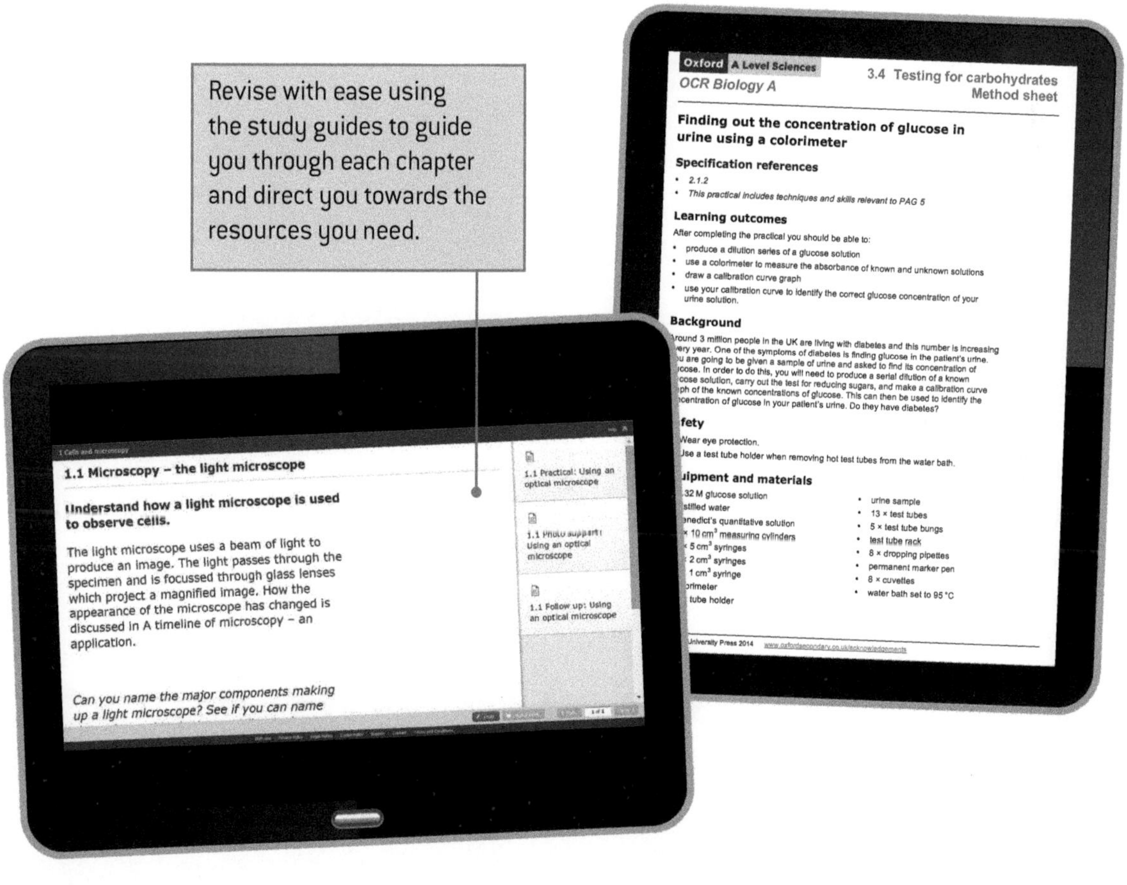

For teachers, Kerboodle also has plenty of further assessment resources, answers to the questions in the book, and a digital markbook along with full teacher support for practicals and the worksheets, which include suggestions on how to support and stretch students. All of the resources are pulled together into teacher guides that suggest a route through each chapter.

MODULE 1

Development of practical skills in biology

Developing your practical skills is a fundamental part of a complete education in science. Having a good grounding of practical skills will advance your understanding of Biology and help to prepare you for studying beyond A level. Biology is a dynamic subject in which our understanding constantly changes – largely as a result of developments in practical research.

You will carry out a number of practicals during both the AS and A level Biology course. You will be assessed on your practical skills in two different ways:

- Written examinations (AS and A level)
- Practical endorsement (A level only)

It is a good idea to keep a record of your practical work during your AS year of study, this will be very useful when you come to revise for your exams. This may also be the start of your A level course and you can use these practicals later as part of your practical endorsement. You can find more details of the practical endorsement from your teacher or from the specification.

Practical coverage throughout this book

Practical skills are covered in a number of ways in this book and in the supporting materials. By studying **Application boxes** and **Practice questions** in this student book, and by using the practical resources on **Kerboodle**, you will have many opportunities to learn about scientific method and how to carry out practical activities.

1.1 Practical skills assessed in written examinations

There are four key components of practical skill assessments. These are laid out here with a checklist to help you as you study.

1.1.1 Planning

- Experimental design
- Identification of variables
- Evaluation of experimental method

Skills checklist

- ☐ Forming a hypothesis
- ☐ Selecting suitable equipment
- ☐ Considering accuracy and precision
- ☐ Identifying dependent and independent variables
- ☐ Identifying variables which need to be controlled

1.1.2 Implementing

- Use of practical apparatus
- Appropriate units for measurement
- Presenting observations and data

Skills checklist

- [ ] Confidence using apparatus and techniques correctly
- [ ] Understanding SI units and prefixes
- [ ] Results table design
- [ ] Presenting data in the most suitable way:
  - Scatter graph
  - Line graph
  - Bar chart
  - Pie chart

1.1.3 Analysis

- Processing, analysing, and interpreting results
- Appropriate mathematical skills for data analysis
- Use of appropriate number of significant figures
- Plotting and interpreting graphs

Skills checklist

- [ ] Understanding results and using this to reach valid conclusions
- [ ] Using mathematical skills to process results
- [ ] Using significant figures correctly
- [ ] Plotting and interpreting graphs
  - labelling axes correctly
  - using appropriate scales
  - reading intercepts and gradients from graphs

1.1.4 Evaluation

- Evaluate results to draw conclusions
- Identify anomalies
- Explain limitations in method
- Precision and accuracy of measurements
- Uncertainties and errors
- Suggest improvements to help improve experimental design

Skills checklist

- [ ] Evaluate results to draw sound conclusions
- [ ] Understand and explain any limitations in the method, and make suggestions on how the method could be improved
- [ ] Understand accuracy of measurements and margins of error (including percentage error) and uncertainty in apparatus

1.2 Practical skills assessed in practical endorsement

Requirements for full A level qualification

You are required to carry out 12 assessed practical activities over both years of your A level studies. The practical endorsement does not count towards your final A level grade but is reported alongside it as a pass or a fail. These practicals will be assessed by your teacher and will help to develop your skills and confidence.

The practicals you carry out should be recorded in a lab book or practical portfolio where the hypothesis, method, results, and conclusion are clearly displayed. The information below details the types of skills and equipment you should become familiar with.

1.2.1 Practical skills

Independent thinking:

Investigating and analysing the methods used in practicals in order to solve problems.

Use and application of scientific methods and practices

- using practical equipment correctly and safely
- following written instructions, recording observations, taking measurements and presenting data scientifically.

Research and referencing

Using information available from a variety of different sources including websites, scientific journals and textbooks to help provide context and background for the practical. It is important to use many sources of information as you can and to cite these correctly.

Instruments and equipment

Correct and appropriate use of a wide range of equipment, instruments and techniques

1.2.2 Use of apparatus and techniques

- Apparatus for quantitative measuring
- Use of glassware apparatus
- Use of a light microscope at high and low power and the use of a graticule
- Producing clear and well annotated scientific drawings
- Use of qualitative reagents
- Experience of carrying out electrophoresis or chromatography
- Ethical and safe use of organisms

- Microbial aseptic techniques
- Use of sampling techniques for fieldwork
- Use of IT to collect and process data

Practical Activity Group (PAG) overview

The table below shows the practical activity requirements for the practical endorsement, and where these PAG references are covered throughout this course.

Specification reference		Topic reference
PAG1	Microscopy	1.1, 1.2, 1.3, 1.4, 1.8
PAG2	Dissection	1.1, 5.1, 6.3, 8.2
PAG3	Sampling techniques	1.5
PAG4	Rates of enzyme controlled reactions	3.5
PAG5	Colorimeter or potometer	2.4, 6.5
PAG6	Chromatography OR electrophoresis	3.2
PAG7	Microbiological techniques	11.2
PAG8	Transport in and out of cells	1.10, 1.11, 2.6
PAG9	Qualitative testing	2.4 2.5
PAG10	Investigation using a data logger OR computer modelling	1.10. 5.4
PAG11	Investigation into the measurement of plant or animal responses	5.3, 7.2
PAG12	Research skills	5.3

Maths skills and How Science Works

Maths is a vital tool for scientists, and throughout this course you will become familiar with maths techniques and equations that support the development of your science knowledge. Within written exams, 10% of the marks available will be for the assessment of mathematics. Throughout this book there are **worked example boxes** taking you through calculations step-by-step, as well as **summary questions** and **practice questions** using maths skills to help you practice. Look out for the calculator icon in the worked example boxes.

How Science Works are skills that will help you to apply your knowledge in a wider context, and the relevance of what you have learnt in the real world. This includes developing your critical and creative skills to help you solve problems in a variety of contexts. How Science Works is embedded throughout this book, particularly in **application boxes** and **practice questions**.

You can find further support for Maths and How Science Works on **Kerboodle**.

1 CELLS AND MICROSCOPY

1.1 Microscopy – the light microscope

Specification reference: 2.1.1

Learning outcomes

Demonstrate knowledge, understanding, and application of:

- → the importance of microscopy in cell theory
- → how to use a light microscope.

Study tip

Remember that increasing the magnification will continue to increase the size of an image but it does not always increase the resolution – so the image can become blurred even though it may be bigger.

Microscopy

Microscopy is the area of science that uses microscopes to view samples and objects that cannot be seen with the naked eye. In microscopy, the image of the object (the specimen being viewed under the microscope) is magnified.

Light microscopy

The most common type of microscope is the light (optical) microscope. It contains one or more lenses and has refractive glass to focus light into the eye. The **magnification** of a light microscope using visible light is up to 1500× with a **resolution** limit of 200 nm. Magnification is the number of times an image is enlarged compared to the actual object. Resolution is the ability to distinguish between two separate points. The resolution of a microscope can be improved by using radiation with a shorter wavelength, for example, ultraviolet light or a beam of electrons.

Light passes from a bulb under the stage and passes through a condenser lens and up through the specimen. The beam of light is then focused through the **objective lens** and through the eye piece lens. Microscopes have different objective lenses that can be selected to achieve different magnifications. These are usually ×4, ×10, and ×40. The eyepiece lens also magnifies the image, usually ×10.

The total magnification for any specimen can be determined by the equation:

Total magnification = eye piece magnification × objective magnification

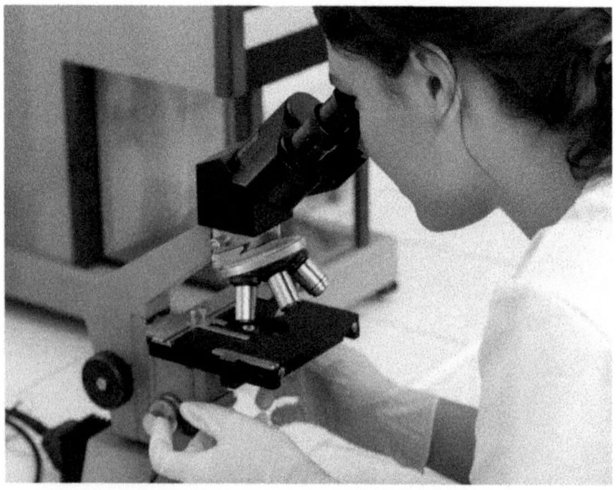

▲ **Figure 1** *Scientist using a light microscope*

An advantage of the light microscope is that it enables observation of a wide range of specimens, including living organisms for example, *Euglena* and *Daphnia*. Smear preparations and thin sections of tissue can also be studied. Light microscopes are widely used in education as they are easy to use, relatively inexpensive, do not require specialist training, and can be transported easily. A disadvantage, however, is that due to their limited resolution most internal cellular structures cannot be seen.

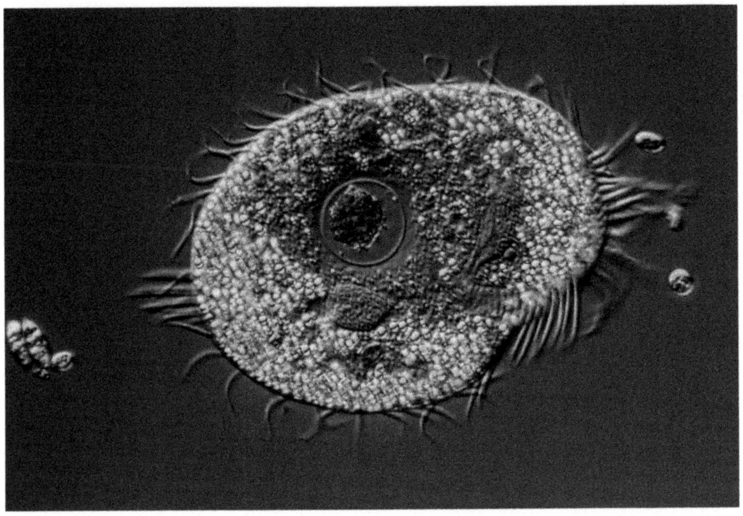

▲ **Figure 2** *Light micrograph of a ciliate protozoan. This single-celled organism is recognisable by the hair-like cilia protruding from the cell body, which it uses for feeding and locomotion, ×300 magnification*

Stains are often used to make different parts of the specimen stand out more clearly, as many biological specimens are colourless and almost transparent. Stains usually colour just a particular part of a cell, for example, iodine solution colours starch grains blue-black. Some stains, such as methylene blue and iodine solution, can be added directly to living cells. Other stains need the specimen to be fixed by adding acetic acid or alcohol to make the proteins and nucleic acids insoluble, hence fixing them in position. This process kills the cells and the stains can be added before or after the fixing process.

Worked example: Calculating magnification

The magnification of an object can be calculated using the equation:

$$\text{magnification} = \frac{\text{size of image}}{\text{size of object}}$$

This formula can be rearranged to calculate the actual size of the object from an image from the magnification:

$$\text{size of object} = \frac{\text{size of image}}{\text{magnification}}$$

The international unit for length is the metre (m). However, this is an inappropriate unit of measurement to use when recording the sizes of cells and organelles viewed down a microscope. Instead the micrometre is used (μm). One millimetre (mm) is divided into 1000 equal divisions, each one is 1μm, i.e., 1μm = 1 × 10^{-6}m. Most animal cells are between 40 and 100μm long.

On some occasions it is necessary to use an even smaller unit of scale. Each μm can be divided into 1000 equal divisions; each division is 1 nanometre (nm). 1nm = 1 × 10^{-9}m. Cell surface membranes are usually around 7nm wide and ribosomes are around 20nm in diameter.

Example 1

If the size of the image is 3cm but the actual specimen is 45μm calculate the magnification to the nearest whole number.

Step 1 Convert the size of the image from cm into the units of mm.
3 × 10 = 30mm

Step 2 Convert the size of the image into μm.
30 × 1000 = 30000μm

Step 3 Calculate the magnification using the formula:

$$\text{magnification} = \frac{\text{image size}}{\text{object size}}$$

$$= \frac{30\,000}{45} = 667 \text{ (to the nearest whole number)}$$

Example 2

If a nucleus measures 120mm on a diagram with a magnification of ×15000, what is the actual size of the nucleus in μm?

1 Convert the size of the nucleus from mm into the units of μm = 120 × 1000 = 120000μm

2 Calculate the actual size using the formula:

$$\text{object size} = \frac{\text{image size}}{\text{magnification}} = \frac{120\,000}{15\,000} = 8\,\mu\text{m}$$

Study tip

Remember:

1000μm = 1mm (i.e., to convert mm to μm, multiply by 1000)

1000000nm = 1mm

Study tip

Make sure you look carefully at the number of decimal places or significant figures needed in your answer – the answer could be 666.7 (to one decimal place), 667 (to the nearest whole number) or 700 (to one significant figure).

Remember when stating a magnification that there are no units as it is a ratio.

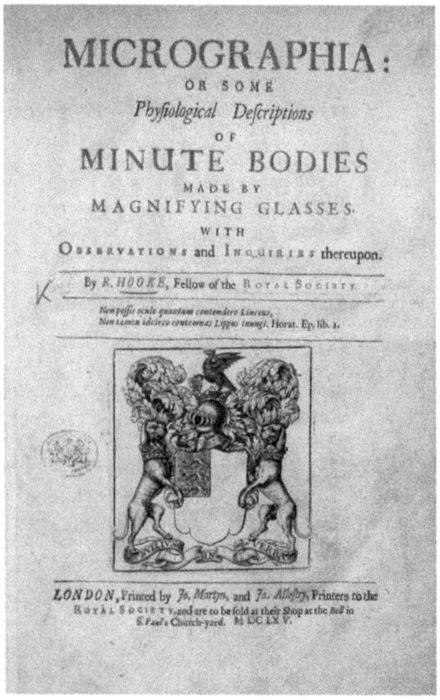
MICROGRAPHIA: OR SOME Physiological Descriptions OF MINUTE BODIES MADE BY MAGNIFYING GLASSES. WITH OBSERVATIONS and INQUIRIES thereupon.

By R. HOOKE, Fellow of the ROYAL SOCIETY.

LONDON, Printed by Jo. Martyn, and Ja. Allestry, Printers to the ROYAL SOCIETY, and are to be sold at their Shop at the Bell in S. Paul's Church-yard. M DC LX V.

▲ Figure 3 *Title page of* Micrographia: or, some physical descriptions of minute bodies made by magnifying glasses, with observations and enquiries thereupon *by the English scientist Robert Hooke (1635–1703)*

Development of the cell theory

Microscopes were invented in the 17th century and as they developed, so did our knowledge of cell structure. In 1665 the scientist Robert Hooke used microscopes to make detailed observations of a variety of specimens. He published his observations in a book called *Micrographia*. To Hooke the plant cells resembled cells where monks lived, and he used this analogy to coin the term 'cell'. In the mid-19th century Matthias Schleiden and Theodor Schwann along with many other scientists such as Rudolf Virchow developed what is known today as the cell theory.

Today the current cell theory states that:

- The cell is the basic unit of all life forms.
- All living organisms are made up of one cell (unicellular) or many cells (multicellular).

- Metabolic processes take place inside cells.
- New cells are derived from existing cells.
- Cells possess the genetic material of an organism and this can be passed from parent cells to new daughter cells.
- A cell is the smallest unit of an organism capable of surviving independently.

Summary questions

1 State why the specimen must be very thin when using a light microscope. *(1 mark)*

2 a The actual width of the ciliate protozoan shown in Figure 2 is 170 μm. Calculate the magnification of Figure 2. Show your working. *(2 marks)*

b At a magnification of ×14 000 a structure appears 8 mm long. Calculate the actual length of the specimen in μm. Give your answer in μm to 1 decimal place. *(2 marks)*

3 Some animal cells are 50 μm in diameter. If a cell is observed under a microscope that can magnify the specimen by 1500, calculate the diameter of the cell on the image in mm. *(2 marks)*

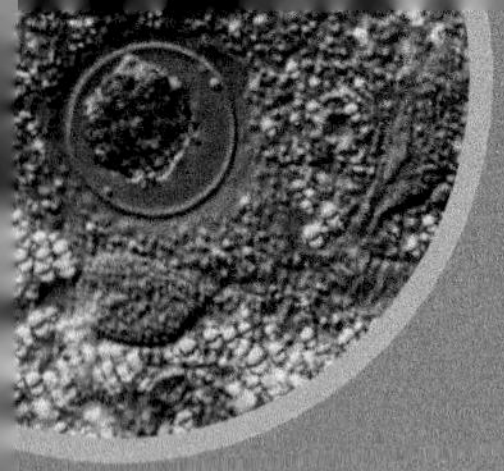

1.2 Microscopy – the electron microscope

Specification reference: 2.1.1

Learning outcomes

Demonstrate knowledge, understanding, and application of:

→ the advantages and disadvantages of an electron microscope

→ the different types of electron microscope.

Electron microscopy

The **electron microscope** (EM) is a more complex type of microscope, which creates an image using a beam of electrons as the form of radiation. In 1931 the first magnified image using a beam of electrons was produced by German engineers Max Knoll and Ernst Ruska. Ruska built the first electron microscope prototype in 1933. This basic prototype had a resolving power of 50 nm. In recognition for his work Ruska was awarded the Nobel Prize for his work 50 years later even though the prototype was not really fit for practical use.

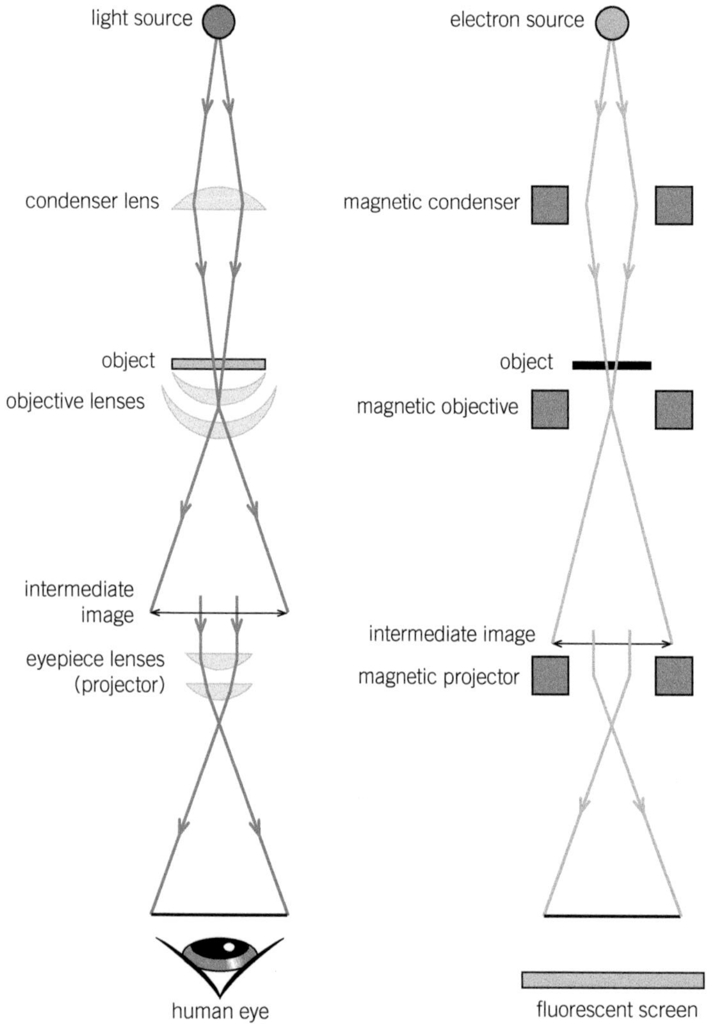

▲ **Figure 1** *The difference in radiation pathways in a light and electron microscope*

Using a beam of electrons which have a shorter wavelength results in a much greater resolving power (0.1 nm). This is a significant advantage over the light microscope as higher magnifications can be achieved enabling smaller objects and finer detail to be seen. Disadvantages of electron microscopes are that they are large and expensive, require trained personnel to use them, and require a specially designed room to be used in. The electrons are negatively charged so the beam of electrons can be focused using electromagnets, which make up the condenser.

A disadvantage of an EM is that the specimens for examination must be placed into a near vacuum, as air molecules absorb the electrons. This means the specimens must be dehydrated and dead. A complex staining process is also needed to prepare specimens. As the human eye is not able to detect or respond to electrons, the electrons are focused on to a fluorescent screen, which emits visible light where the electrons hit. The visible image on the screen is called a photomicrograph. It is also possible that artefacts may occur in the photomicrograph as a result of the preparation technique, that is, structures may appear in the final image that are not actually part of the natural specimen.

The final images produced by EMs are always black, grey, and white. However, colour can be added to images by specialist computer programs to produce false-colour electron micrographs. There are two main types of EM – transmission and scanning electron microscopes.

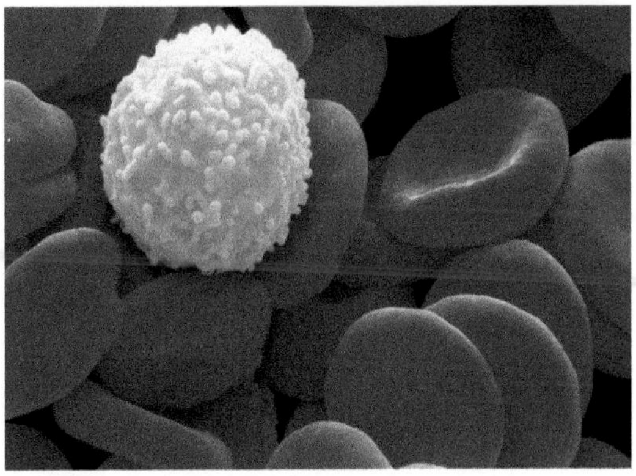

▲ Figure 2 *False-colour scanning electron micrograph of erythrocytes and a white blood cell, ×3 500 magnification*

Transmission electron microscope (TEM)

Specimens to be observed using a transmission electron microscope (TEM) must be a lot thinner than those to be observed using light microscopes because electrons cannot penetrate materials as well as light rays can. Heavy metals are used to stain the specimen for the TEM as the atoms of these heavy metals have large, positively charged nuclei that scatter the electrons. These scattered electrons do not hit the fluorescent screen and so leave a dark area in the image. The structures in the cell that take up the stain appear as dark images. Images produced by the TEM are two-dimensional and are usually black and white.

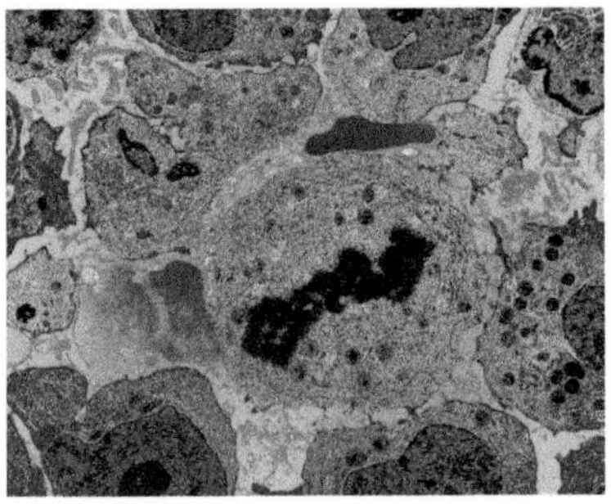

▲ Figure 3 *Coloured transmission electron micrograph of malignant plasma cells (a type of leucocyte) in a tumour. The large, irregular nuclei (dark blue) in these cells indicate they are actively dividing, ×1 580 magnification*

Scanning electron microscope (SEM)

Scanning electron microscopes (SEMs) work differently to the TEM as the electrons do not pass through the actual specimen. Instead they are reflected off its surface. The beam of electrons is passed backwards and forwards over the surface of the specimen in a regular pattern. The pattern of the scattered electrons reflects the contours of the specimen and this information is used to produce a 3-D image using computer analysis.

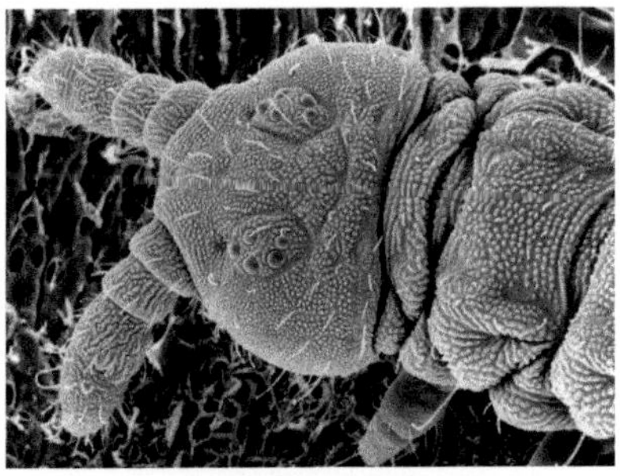

▲ Figure 4 *Scanning electron micrograph of a springtail. Springtails are the most primitive insects, dating back 300 million years in the fossil record, ~×700 magnification*

Confocal scanning microscope

Confocal laser scanning microscopy (CLSM) was developed in the 1980s. It can be used to obtain high resolution images and 3-D reconstructions. Its main advantage is that it has the ability to produce focused images of thick specimens at various depths. This process is known as optical sectioning:

- A point source of light is directed on to the object plane.
- A point of emitted fluorescence light or reflected light from the object is then directed through a detector pinhole, enhanced using a photomultiplier, and then displayed on a computer screen as a pixel.
- By scanning the object point by point and line by line, very thin and blur-free optical sections are recorded one pixel at a time.
- A series of optical sections can be combined to form an image stack, which enables 3-D images to be produced.

Fluorescent markers can also be used for the detection of biological objects. This technique is commonly used in the fields of cell biology, genetics, microbiology, and developmental biology.

Study tip

Make sure you practise looking at SEM and TEM photomicrographs and can accurately identify cell organelles.

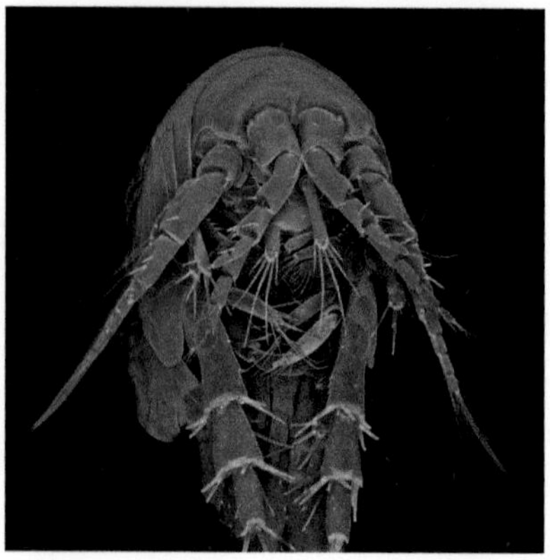

Figure 5 *Laser scanning confocal micrograph of a juvenile amphipod viewed from the front. Amphipods are a large order of crustaceans. This specimen is roughly 1.5 mm long*

Summary questions

1 Explain why the resolving power of a SEM is greater than that of a light microscope. *(1 mark)*

2 Complete the table to compare the different types of microscope. *(14 marks)*

Feature	Light microscope	Electron microscopes	
		Scanning electron microscope	Transmission electron microscope
Radiation source			
Method to focus radiation			
Maximum resolution			
Maximum magnification			
Biological specimens			
Relative thickness of specimen			
Example of a stain that can be used			
Image obtained			
Size of microscope			
Preparation of the specimen			
Relative cost			
Affected by magnetic fields			
Risk of distortion of material			

3 Tissues need heavy metal stains as the molecules from which the tissues are composed consist of 'light elements'. Read Topic 1.9, Cell membranes, and name one molecule found in all cell membranes. List the 'light elements' this molecule contains. *(2 marks)*

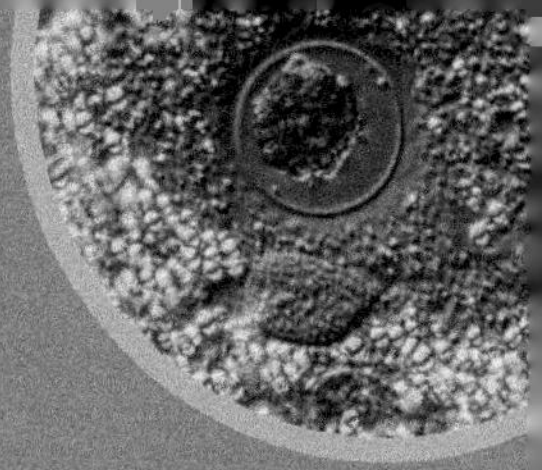

1.3 Differential staining and blood smears

Specification reference: 2.1.1

Differential staining of plant cells

Biological specimens can be examined in a living or preserved form. Using a preserved form enables the specimen to be cut into sections and to be treated with a variety of stains, revealing different structures within the tissues and cells.

Specimens may need to be sectioned (Figure 1) before being prepared for examination under the microscope.

Sectioning is just one of the stages involved in preparing a tissue for examination under the microscope:

- The stages in the preparation of a temporary slide are – fixation, staining, and mounting.
- The stages in the preparation of a permanent slide are – fixation, dehydration, clearing, embedding, sectioning, staining, and mounting.

The preparation of temporary slides of tissues for light microscopy is generally rapid and simple. The tissue can be sectioned before fixation (preservation and sterilisation) using 70% alcohol. Temporary stains can be set up by placing the tissue on a clean glass side and adding a few drops of the stain. The specimen is then covered using a thin glass coverslip to exclude dust and air and to protect the high-power objective lens on the microscope. If the specimen begins to dry out or if prolonged examination is likely to be needed then the specimen can be mounted in glycerine after staining.

Differential staining can be carried out on a variety of plant tissues. A differential stain makes some structures appear darker or different in colour from other structures. Some examples of permanent and temporary stains suitable for plant tissue are shown in Table 1.

▼ **Table 1** *Some permanent (P) and temporary (T) stains suitable for plant tissue*

Stain	Final colour	Suitable for
Eosin (P)	Red	Cellulose
Toluidine (P)	Blue Purple	Lignin Cellulose
Iodine–KI (T)	Blue-black	Starch

For example, in vascular tissue in plants, toluidine stains lignified cell walls blue and cellulose cell walls purple, allowing xylem vessels to be distinguished from other cells.

Learning outcomes

Demonstrate knowledge, understanding, and application of:

→ how to differentially stain plant cells

→ how to differentially stain blood cells

→ how to prepare and examine blood smears.

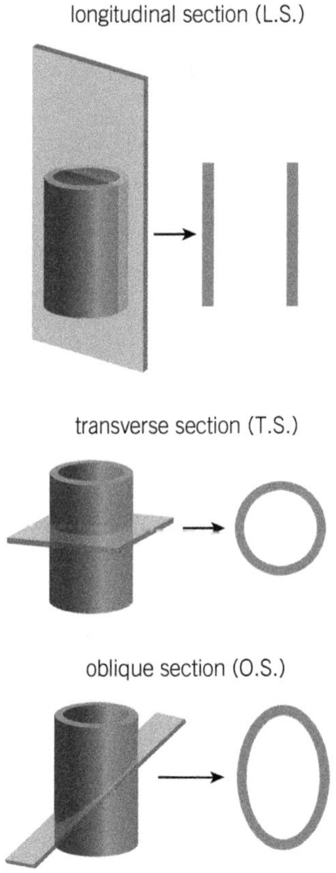

▲ **Figure 1** *The three different types of section that can be cut through a plant stem*

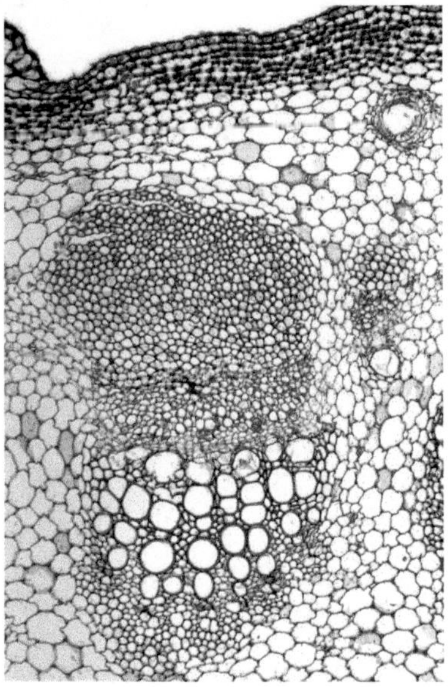

▲ Figure 2 *Light micrograph of a section through a young stem from a sunflower (*Helianthus annuus*), showing the many vascular bundles. The phloem is stained dark blue and the xylem is stained red ×18 magnification*

Differential staining of blood cells

Blood samples contain two main types of cells, **erythrocytes** (red blood cells) and leucocytes (white blood cells). Erythrocytes can be clearly seen under a microscope but to be able to distinguish between the different types of leucocyte (monocytes, neutrophils, and lymphocytes) they need to be stained to show the differences in the shapes of their nuclei. Structures within cells vary in the type and/or quantity of stain that they take up.

Romanowsky stains are a group of stains used by pathologists:

- Leishman's stain – the blood is allowed to dry first and then it is fixed with methanol. The slide is flooded, left for 2 minutes then diluted with water and left for a further 5–7 minutes. The slide is then washed until it appears pale pink to the naked eye.
- Wright's stain – this is widely used for performing differential white blood cell counts, which are routinely ordered when infections are suspected.

The staining process can be automated in a laboratory to allow many slides to be prepared at the same time.

Preparing and examining blood smears

Some blood diagnostic tests involve observing the appearance of the blood. To do this a small sample of the blood is spread onto a microscope slide, usually by a machine, to make a blood film. It is important to get the film on the microscope slide correct:

- If it is too thick, the individual cells cannot be seen.
- If it is too thin, the small number of cells present may not be representative.

Blood smears can also be prepared by hand:

- A small drop of blood is placed at the end of a sterile, clean, and dry microscope slide.
- A spreader (another sterile microscope slide) is held at a 30° angle and is pushed along the slide to spread the blood back over the slide (i.e., form a smear).
- The microscope slide is immediately labelled with the patient's details (sometimes a bar code is used).
- The slide is allowed to dry in the air to enable the cells to stick to the slide.
- A fixative is used to preserve the cells.

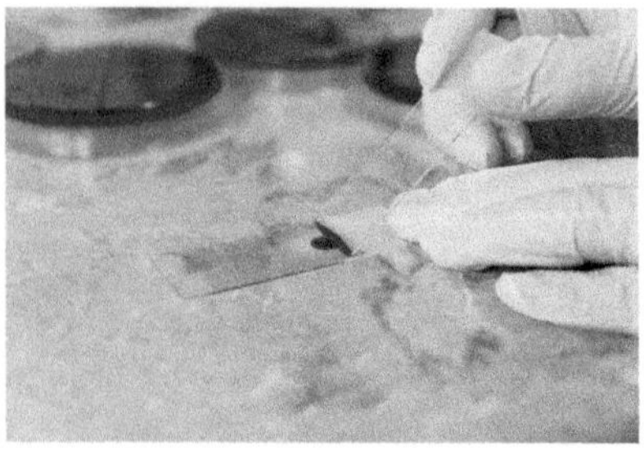

▲ Figure 3 *Preparing a blood smear. A drop of blood has been placed on one end of a microscope slide and a spreader slide is being used to smear the blood across the slide*

Trained pathologists viewing the slide under a microscope can then identify what cells are present or identify pathogens, such as blood parasites. Examination of stained blood films is important in the diagnosis of many diseases and disorders. The shapes, sizes, and relative numbers of the different kinds of cells present can indicate the state of the patient's health.

Summary questions

1 a Suggest why latex-free gloves are worn when preparing blood smears. *(2 marks)*

b Suggest why the microscope slide must be free from grease. *(1 mark)*

2 a Suggest what is meant by the term cryosectioning and when it might be used. *(2 marks)*

b Suggest a possible disadvantage of using this form of sectioning. *(1 mark)*

3 Calculate the area of the stem in the transverse and oblique sections in Figure 1 using the following formulae.

In the transverse section: $A = \pi r^2$, where r = radius of the section.
In the oblique section: $A = \pi ab$, where a and b are the lengths shown on the following diagram.

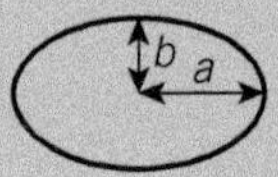

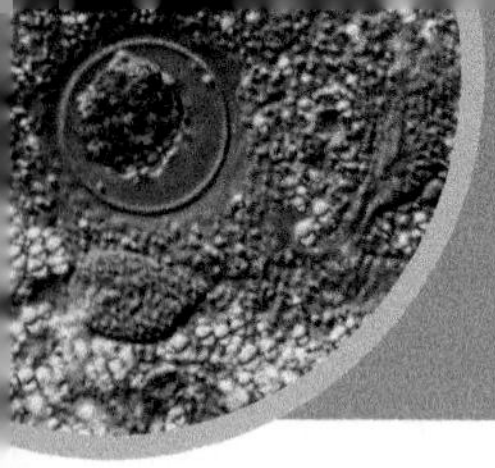

1.4 Cells of the blood

Specification reference: 2.1.1

Learning outcomes

Demonstrate knowledge, understanding, and application of:

→ the structure of animal cells as illustrated by blood cells

→ studying cells under a light microscope and producing biological drawings.

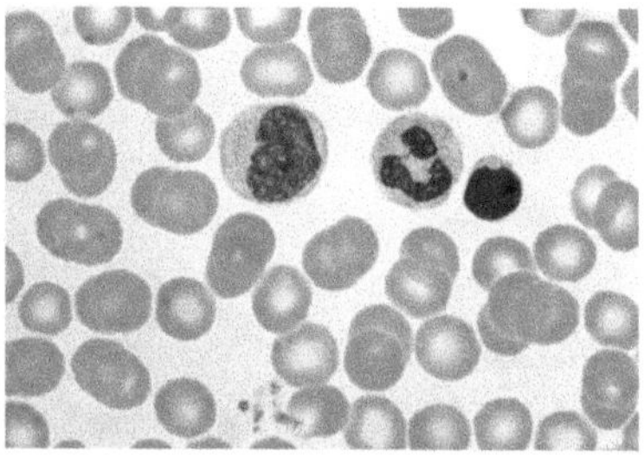

▲ **Figure 1** *Light micrograph showing a human blood smear. Erythrocytes (red blood cells) are pink. Three types of leucocytes (purple) can also be seen (from left to right they are a monocyte, a neutrophil and a lymphocyte), ×550 magnification*

Synoptic link

You will find out more about the structure and function of the circulatory system in Topic 6.1, The transport system in mammals.

Blood

A person with a body mass of 70 kg will usually have around 5.5 litres of blood. In vertebrates blood is made of:

- approximately 53.3% plasma, which is mostly water and proteins, glucose, mineral ions, **hormones**, and carbon dioxide
- approximately 45% red blood cells
- approximately 0.7% white blood cells
- approximately 1% platelets.

Functions of the blood

The blood performs many vital functions within the body – made possible by the different types of blood cells carrying out different functions due to their specific adaptations. Some of these functions include:

- delivery of oxygen to tissues
- delivery of nutrients to tissues
- removal of waste from tissues
- immunological protection
- clotting
- transport of cell signalling molecules, for example, hormones
- acting as a buffer to regulate body pH (acidity)
- distribution of heat to regulate core body temperature.

Erythrocytes

The most common type of blood cell are erythrocytes. Their function is to deliver oxygen from the lungs to the body and carrying carbon dioxide from the tissues to the lungs. Red blood cells develop in the bone marrow and circulate in the blood stream for about 100–120 days.

Erythrocytes have a diameter of approximately 6.2–8.2 μm and are approximately 2–2.5 μm thick at their thickest point. Red blood cells contain haemoglobin, an iron-containing protein that can bind reversibly with oxygen to form oxyhaemoglobin. In humans, mature red blood cells are flexible and oval biconcave discs. They lack a cell nucleus and most organelles, enabling them to:

- have a large surface area to volume ratio
- become biconcave – enabling them to be flexible and squeeze through the capillaries.

New erythrocytes are produced in the bone marrow from erythropoietic stem cells. This process is called erythropoiesis and is stimulated by a hormone produced by the kidney called erythropoietin (EPO). Erythropoiesis takes approximately seven days. Immature red blood cells are known as reticulocytes, and as well as mature erythrocytes they do not have a nucleus. Reticulocytes can be identified in a blood smear when they are stained using methylene blue.

Thrombocytes

Thrombocytes are also known as platelets. They are approximately 2–3 μm in diameter and are biconvex discs of cytoplasm fragments surrounded by a cell surface membrane. They are produced in the bone marrow and do not have a nucleus. On a stained blood smear platelets appear as dark purple spots. Their main function is to contribute to blood clotting and clot formation. Platelets live for about 6–7 days.

> **Synoptic link**
>
> You will find out more about the principle and process of blood clotting in Topic 3.6, Blood clotting and enzymes.
>
> You will need to understand the roles of leucocytes in the immune system. This is covered in Topic 12.1, The immune system.

Leucocytes

Leucocytes (white blood cells) play an important role in the immune system. There are at least five different types of leucocytes. They are larger in size than erythrocytes but they are less abundant. They can be divided into two categories:

- Granulocytes – granulocytes contain granules (usually lysosomes) in their cytoplasm, which all stain differently. Neutrophils are one example of granulocytes.
- Agranulocytes – these have an absence of granules in their cytoplasm. Some examples include lymphocytes, **monocytes**, and **macrophages**.

Neutrophils

Neutrophils help the body to defend against fungal or bacterial infections. On a blood smear, a multi-lobed nucleus can be seen. Neutrophils engulf and break down bacteria in a process known as phagocytosis. They are present in large amounts in the pus of wounds. These cells are not able to renew their lysosomes and therefore die after breaking down a few pathogens (usually after about five days). These cells have the ability to leave the capillaries by squeezing through the fenestrations in the capillary wall. This is made possible by having a lobed nucleus, which gives the neutrophil greater flexibility.

Lymphocytes

Lymphocytes are characterised by their large, deeply stained nucleus which is surrounded by a relatively small amount of cytoplasm. Lymphocytes can be categorised into two groups:

- **B lymphocytes** – these produce immunoglobulins, also known as antibodies.
- **T lymphocytes** – there are several different types of T lymphocyte, examples include, helper T cells (produce cytokines and help co-ordinate the immune response), cytotoxic T cells (bind to antigens on virus infected cells or tumour cells and destroy them), and natural killer cells.

Despite their different functions both B and T cells look the same on blood smears.

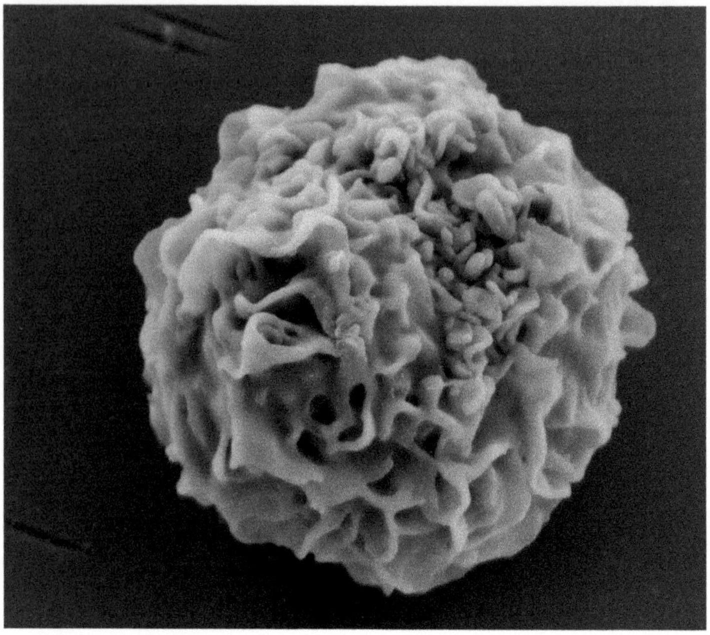

▶ Figure 2 *Scanning electron micrograph of a monocyte white blood cell (magnification × 12 000)*

Monocytes

These are the largest kind of leucocyte and, like neutrophils, they also carry out phagocytosis but live for much longer. Eventually, monocytes leave the bloodstream and differentiate into tissue macrophages, which remove dead cell debris and attack microorganisms. Macrophages are found in the lungs, lymph nodes, and liver. Unlike neutrophils, monocytes are able to replace their lysosomes. They have the kidney-bean shaped nucleus and are typically agranulated (lacking in granules) so the cytoplasm appears clear.

Studying cells under the light microscope

Almost all cells are too small to be seen with the naked eye, so a microscope is needed to study them. The light microscope can be used in schools to study cells of the blood, and drawings of cells used to record the structure of these cells. When making biological drawings of cells you should:

- Complete all drawings in pencil – a sharp 2H pencil will give the best lines but HB is also acceptable.
- Use a single, clear continuous line (no ragged lines) – join up lines to form a continuous structure.
- Accurately record exactly what is observed – do not rely on memory.
- Add correct labels – these are just the names of structures that are visible. Label lines drawn with a ruler.
- Add annotations – details such as size and colour.
- Add the total magnification used (Topic 1.1) for a reminder of how to calculate total magnification.

There must be no shading or colouring of any part of the drawing.

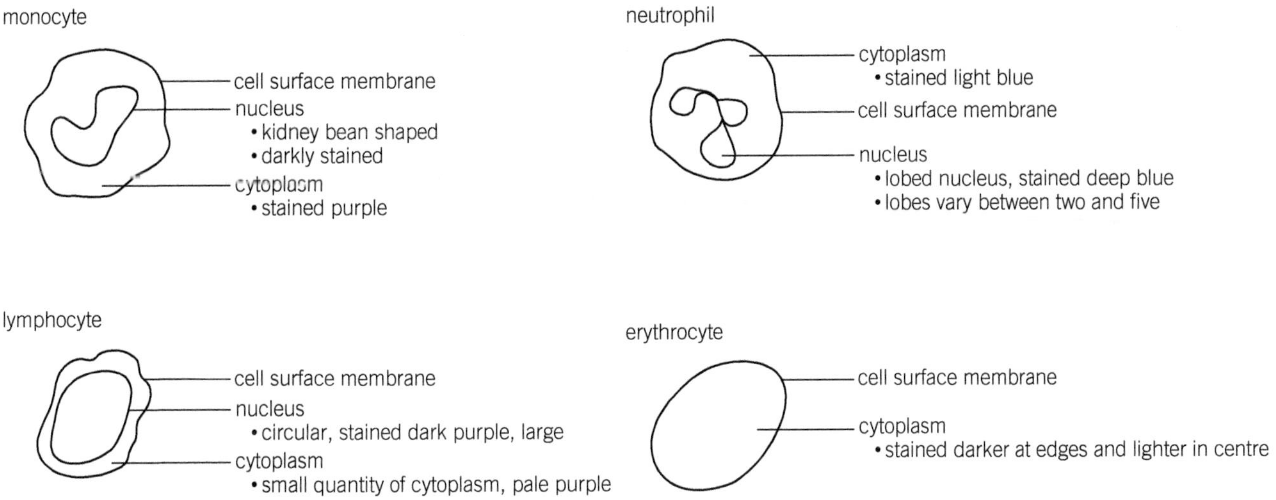

▲ **Figure 3** *Examples of drawings of blood cells viewed under the light microscope*

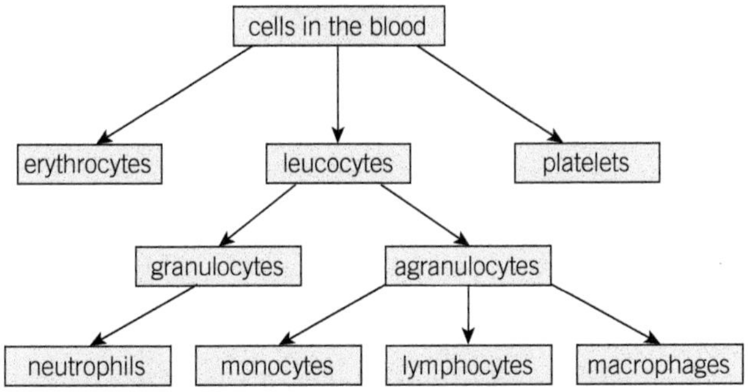

▲ **Figure 4** *Summary of blood cells and the relationships between them*

Summary questions

1 State one waste product carried by erythrocytes. *(1 mark)*

2 Copy and complete the table to compare the functions of three components of blood using ticks and crosses. *(3 marks)*

Function	Erythrocyte	Platelet	Leucocyte
Contains enzymes involved in clotting			
Forms antibodies			
Carries out phagocytosis			

3 **a** Suggest why red blood cells are unable to divide and have limited repair capabilities. *(1 mark)*

b Suggest why the lack of a nucleus means viruses do not target mammalian red blood cells. *(2 marks)*

c Using the information in Figure 2 calculate the maximum width of the monocyte in micrometres (μm) to:

i one decimal place *(1 mark)*

ii the nearest whole number *(1 mark)*

iii one significant figure. *(1 mark)*

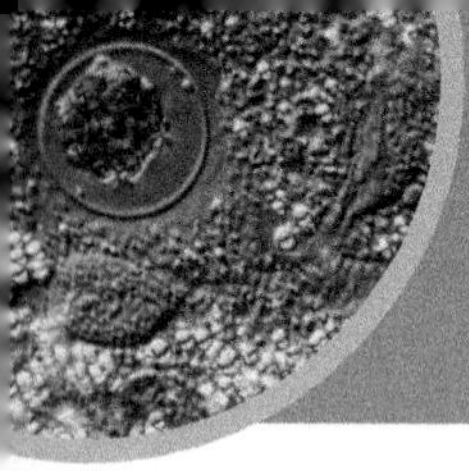

1.5 Counting cells

Specification reference: 2.1.1

Learning outcomes

Demonstrate knowledge, understanding, and application of:

→ how to count cells using a haemocytometer

→ how to use flow cytometry to count cells.

Methods of counting cells

Haemocytometer

A haemocytometer is a specialised microscope slide made of thickened glass which is used to count cells, it was invented by Louis-Charles Malassez. In the centre of the slide there is a platform which has grooves surrounding it. There is also a grid etched onto the platform, the squares in the grid measure 0.2 × 0.2 mm and some squares in the centre of the grid have triple-lines. The depth of each square is exactly 0.1 mm when a cover slip is placed over the platform. It is therefore possible to count the number of cells in a specific volume and calculate the concentration of cells in the overall sample.

It is essential when viewing a sample of blood that:

- it is thoroughly mixed before a sample is taken to ensure the sample is representative
- an appropriate dilution is used so that the number of cells can be counted accurately
- the process is repeated so that a mean can be calculated.

Counting cells using a haemocytometer

When the haemocytometer is observed under the microscope, sometimes cells lie on top of the triple lines around the edge of the square. If all the visible cells are counted, then this will generate a cell count that is too high. To stop this, a rule is applied called the North-West rule:

- if a cell lies on the middle of the triple line on the north or west grid it is counted
- if a cell lies on the middle of the triple line of the south or east grid it is not counted.

Worked example: Calculating the number of cells in a sample

The diagram shows a haemocytometer slide containing a yeast solution that has been diluted 200 times. The triple-lined square has a volume of 0.2 (width) × 0.2 (height) × 0.1(depth) mm. Use this information to calculate the number of yeast cells in 1 mm^3 of the original solution.

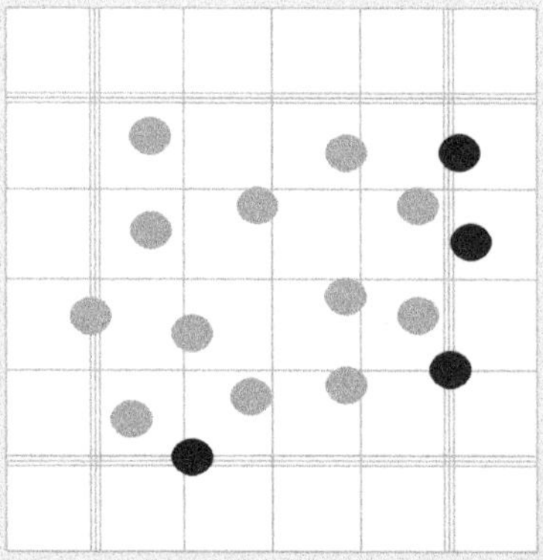

Step 1 Using the North-West rule there are 12 yeast cells in the diluted sample – 11 within the triple-lined square and 1 on the west line that is also counted.

Step 2 This means there are 12 yeast cells in 0.1 × 0.2 × 0.2 mm^3 = 0.004 mm^3

Step 3 Therefore the number of yeast cells in 1 $mm^3 = \frac{1}{0.004} \times 12 = 3000$ yeast cells

Step 4 However the sample was diluted 200 times. Therefore the number of yeast cells in 1 mm^3 of the original sample = 3000 × 200 = 600 000 yeast cells.

In real life the mean from at least five squares is used.

Flow cytometry

Flow cytometry is an electronic counting apparatus that uses a laserbeam passed over a stream of blood to count cells. It allows analysis of the physical and chemical characteristics of the blood, simultaneously counting up to thousands of particles per second. It is a technique commonly used in the diagnosis of blood cancers.

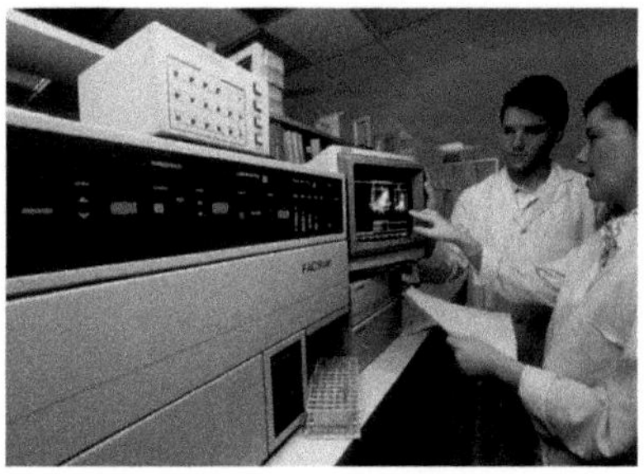

▲ **Figure 1** *Technicians use a flow cytometer to analyse a sample of human blood cells*

There are five main parts:

- a flow cell – a liquid stream which carries cells in single file through the light beam so they can be counted accurately
- a measuring system
- a detector
- an amplification system
- a computer with appropriate software that can analyse the signals.

Specific antibodies can be 'tagged' (attached) to different fluorochromes. These tagged antibodies recognise and target specific antigens inside cells or on the surface of cell membranes. Fluorochromes can also be attached to a chemical that binds to a specific part of the DNA, cell membrane, or other cellular structure. Each fluorochrome has its own peak of excitation and emission wavelength. Lamps or lasers with different wavelengths are used to excite the fluorochromes depending on the detectors available. This light or laser causes the tagged cells to fluoresce so that they can be counted by a detector. The specific light scattering and fluorescent characteristics of each cell as they pass the laser beams is used for counting and sorting.

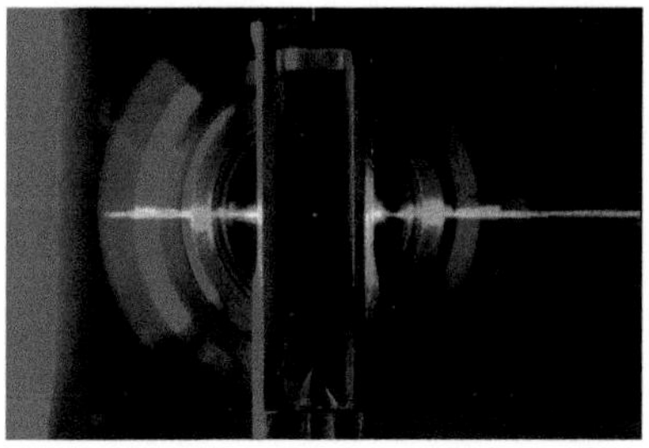

▲ **Figure 2** *A flow cytometry laser beam*

Summary questions

1 Suggest two reasons why the first and second counts from a haemocytometer slide may be different. *(2 marks)*

2 Explain how a serial dilution could be carried out on a yeast cell culture to obtain a range of dilutions between 10^{-1} and 10^{-4}. *(5 marks)*

3 Other than blood cell and pathogen counts, suggest another use for the haemocytometer. *(1 mark)*

1.6 Cell ultrastructure

Specification reference: 2.1.1

Learning outcomes

Demonstrate knowledge, understanding, and application of:

- → the ultrastructure of typical plant and animal eukaryotic cells
- → the ultrastructure of a typical prokaryotic cell
- → the similarities and differences between the structure of plant, animal, and prokaryotic cells.

Cells

Each multicellular organism is made up of hundreds to billions of cells. Each cell type has a specific internal structure that makes it efficient for its role. The internal structure of a cell can be divided up into areas specialised for certain functions. This is called **compartmentalisation**. The compartments are usually surrounded by a membrane.

Structure of a leucocyte

Leucocytes are **eukaryotic** cells, meaning they possess a distinct nucleus and membrane-bound organelles. Using an electron microscope it is possible to study the structure of the organelles.

All organelles within an animal cell are contained by the **cell surface membrane**. The structure and role of the cell surface membrane is discussed in detail in Topics 1.9 to 1.11.

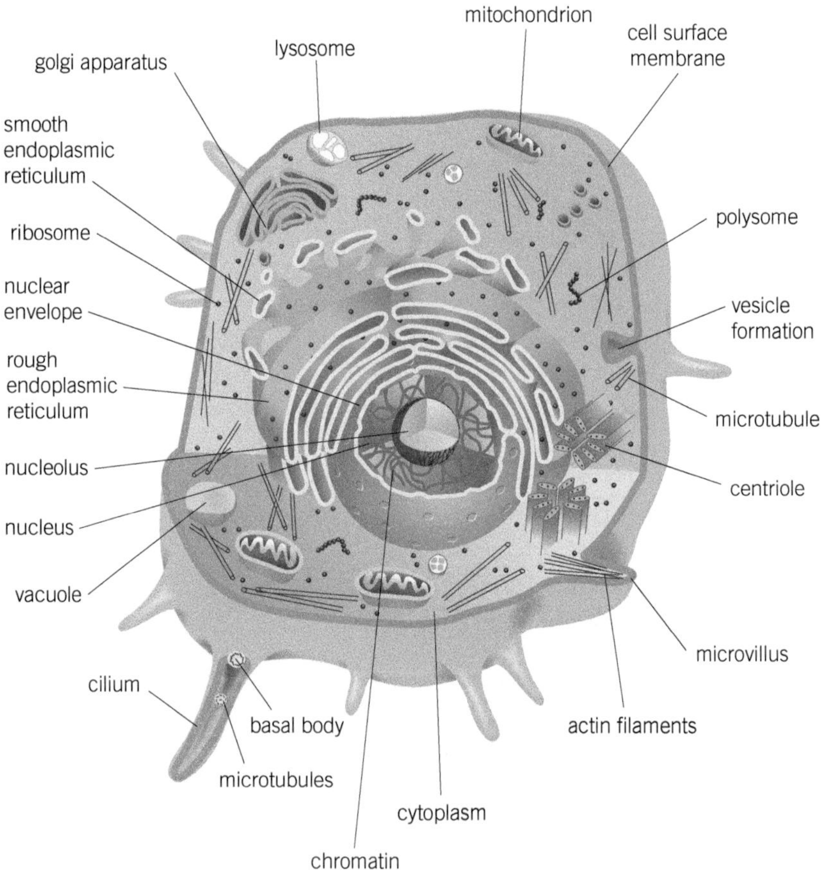

▲ **Figure 1** *A generalised animal cell*

Nucleus and nucleolus

The nucleus is the most important structure of a eukaryotic cell as it contains the genetic, hereditary material DNA and controls the cell's activities. The nucleus is a large, spherical organelle that is 10–20 µm in diameter. It is surrounded by a double membrane, the nuclear envelope. The outer membrane is continuous with the rough **endoplasmic reticulum** (RER). The nuclear envelope controls the entry and exit of materials to and from the nucleus. There are typically ~3000 nuclear pores, each about 40–100 nm in diameter in the envelope. These pores allow the passage of large molecules out of the nucleus, for example, messenger RNA (mRNA). The liquid within the nucleus is called the nucleoplasm and contains the chromatin (DNA) and the nucleolus – a small spherical structure that produces ribosomal RNA (rRNA) and assembles the **ribosomes**.

The function of the nucleus is to:

- control the activities of the cell via the production of mRNA
- contain the genetic material in the form of DNA coiled around proteins to form linear chromosomes
- manufacture rRNA and ribosomes.

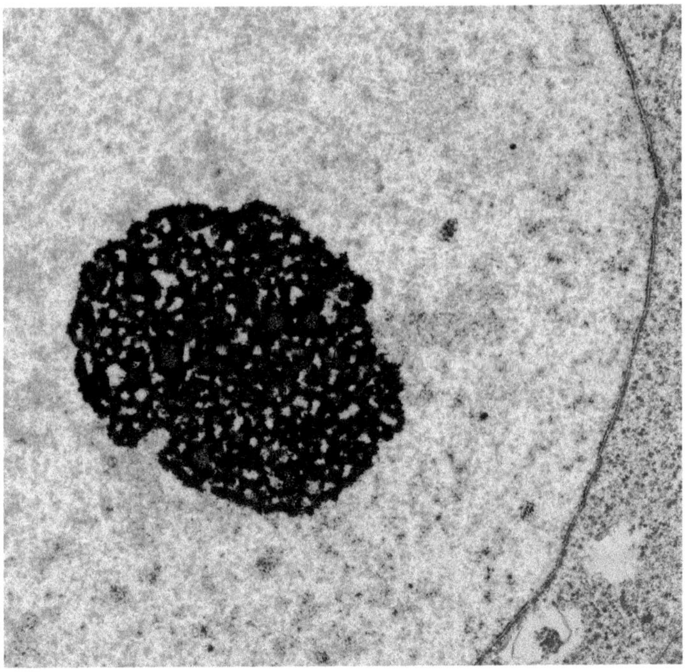

▲ **Figure 2** *Transmission electron micrograph of a cancer cell nucleolus. The nucleolus is the most active part of the nucleus. Here the nucleolus is abnormally large, as cancer cells are unusually active cells. ×4000 magnification*

Mitochondria

Mitochondria are rod-shaped organelles typically between 1 and 10 µm in length. They are surrounded by the mitochondrial envelope – the outer membrane controls materials entering and exiting the mitochondrion and the inner membrane is folded to form the cristae.

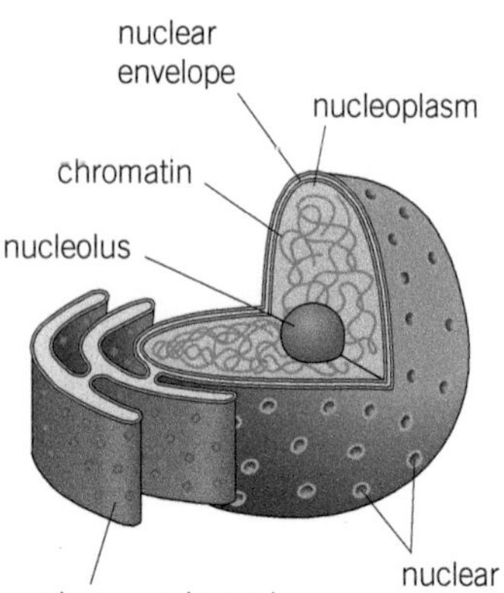

▲ Figure 3 *The relationship between the nucleus, nucleolus, and endoplasmic reticulum*

The cristae provide a large surface area for enzymes involved in the synthesis of ATP to be attached to. The liquid inside the cristae, the matrix, contains proteins (including enzymes for the breakdown of carbohydrates), lipids, and small amounts of circular DNA.

The function of the mitochondria is to:

- carry out the later stages of aerobic respiration
- produce ATP, an energy-carrier molecule.

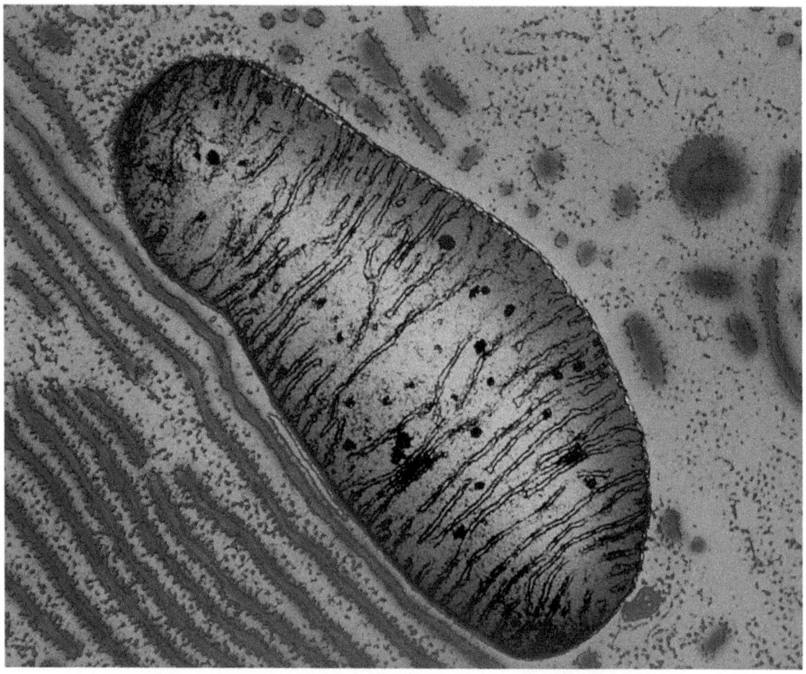

▲ Figure 4 *Coloured transmission electron micrograph of a single mitochondrion (blue) in a human pancreas cell (×64 000 magnification)*

Endoplasmic reticulum

Endoplasmic reticulum (ER) is a series of flattened membrane-bound sheets that spread through the cytoplasm. The membrane-bound sacs are called cisternae. Two different forms of ER exist:

- Smooth endoplasmic reticulum (SER) – these cisternae are often tubular in appearance. The function of the SER is to synthesise, store, and transport both lipids and carbohydrates.
- Rough endoplasmic reticulum (RER) – has ribosomes attached to the outer surfaces of the cisternae. The RER is continuous with the outer nuclear membrane. The function of the RER is to provide a large surface area for protein and glycoprotein synthesis and to provide a pathway for transporting proteins and other materials through the cell.

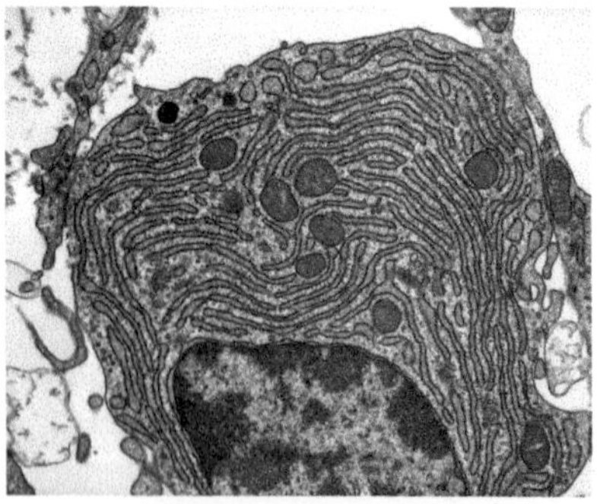

▲ Figure 5 *Transmission electron micrograph of a section through a cell, showing membranes of rough endoplasmic reticulum (dark lines). The flattened stacks of membranes are studded with ribosomes (small dots), ×6000 magnification*

Ribosomes

Ribosomes are *not* membrane bound but instead are cytoplasmic granules made from rRNA and protein. They are found free in the cytoplasm or associated with RER. Ribosomes in eukaryotic cells are 80S ribosomes and they are typically 25–30 nm in diameter with a

1:1 ratio of rRNA to proteins. Each ribosome consists of a small (40S) and large (60S) subunit. The symbol 'S' in these two measurements is a Svedberg unit. Svedberg units are a measure of the sedimentation rate of particles, for example, the rate at which an organelle reaches the bottom of a test tube during centrifugation.

The role of ribosomes is to join amino acids in a specific order, this order is dictated by messenger RNA (mRNA) molecules. The small ribosomal subunit reads the RNA and the large subunit joins amino acids together to form a polypeptide chain.

Golgi apparatus

Leucocytes, like almost all eukaryotic cells, contain **Golgi apparatus**. Like SER in appearance, it is made from a stack of membrane-bound sacs (cisternae) and small hollow spherical vesicles. A typical mammalian Golgi body will contain between 40 and 100 stacks, with between three to six cisternae per stack. Proteins and lipids that are made by the ER are passed to the Golgi apparatus. Proteins are modified by the Golgi apparatus by the addition of non-protein portions, for example, the addition of oligosaccharides or 'glycans' to form a **glycoprotein** or by the addition of phosphates (phosphorylation). The Golgi apparatus also labels the proteins to enable them to be accurately sorted and transported to their correct destinations. After modification, the proteins and lipids are packaged into vesicles, which bud off from the ends of the Golgi apparatus for transport around the cell.

The functions of the Golgi apparatus are to:

- assemble polypeptides into proteins
- pack proteins and carbohydrates into vesicles for secretion by **exocytosis**.

Vesicles

Vesicles are small, fluid containing, membrane-bound organelles. They are used in processes such as secretion (exocytosis), uptake (**endocytosis** and phagocytosis), and transport of materials in the cytoplasm. They are also involved in transport within the cell, cell metabolism, and enzyme storage.

Lysosomes

Lysosomes are spherical membrane-bound vesicles (sacs) that are produced by the Golgi apparatus and contain up to 50 different types of enzymes (usually proteases and lipases). They are typically 1.0 μm in diameter and have an internal pH of approximately 5. The function of the lysosomes is to digest unwanted materials in the cytoplasm. They have an important role in phagocytosis where they fuse with phagocytic vesicles so the enzymes can digest engulfed pathogens (Figure 6).

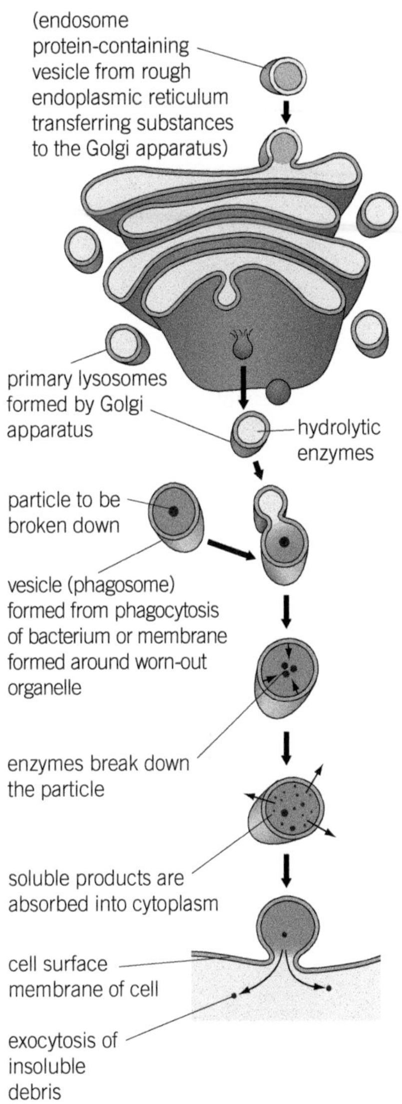

▲ Figure 6 *The relationship between the Golgi and lysosome activity*

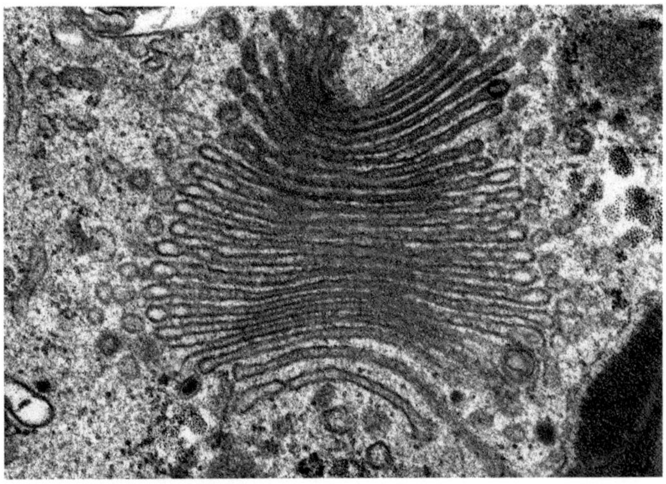

▲ Figure 7 *False colour transmission electron micrograph of a Golgi apparatus (orange), ×12 500 magnification*

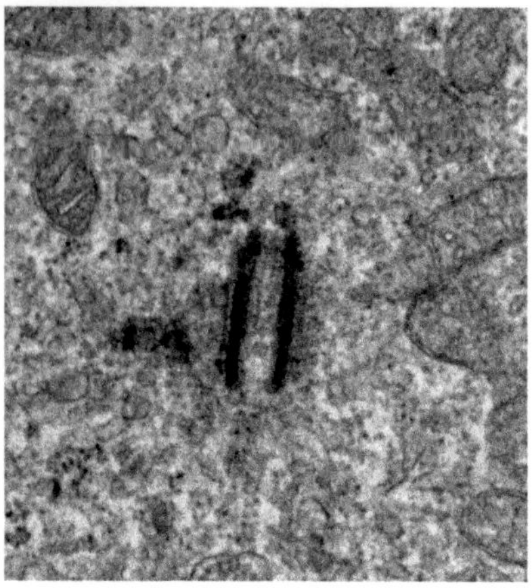

▲ **Figure 8** *Transmission electron micrograph of centrioles (dark lines at centre), approximately × 80 000 magnification*

Centrioles

A **centriole** is a cylindrical cell structure that is found in most eukaryotic cells. It is composed mainly of a protein called tubulin organised into microtubules. Centrioles are short cylinders with a 9 + 0 pattern of microtubules, meaning there are nine outer microtubules and no central microtubules. Centrioles are involved in organising spindle fibres which are involved in cell division. Centrioles are found in pairs surrounded by a membrane, forming a centrosome, which is associated with the nuclear membrane. Before a cell divides, the centrosome duplicates so each centrosome will contain one mother (original) centriole and one new centriole, arranged at right angles to each other. During **mitosis** the centrosomes migrate to opposite poles so that each daughter cell receives one centrosome.

Structure of a palisade mesophyll cell

In leaves, the main site of photosynthetic tissue is the palisade mesophyll cells. Palisade mesophyll cells are also eukaryotic cells and contain common organelles with animal cells. However, like other typical plant cells they also contain some additional organelles.

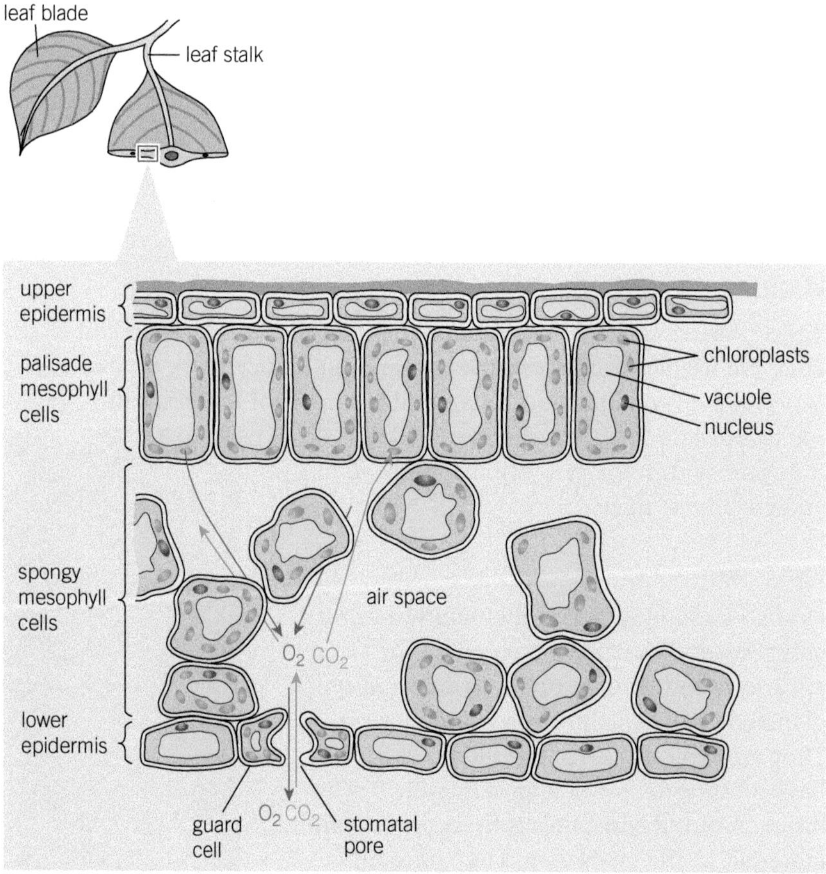

▲ **Figure 9** *Vertical section through a dicotyledonous leaf*

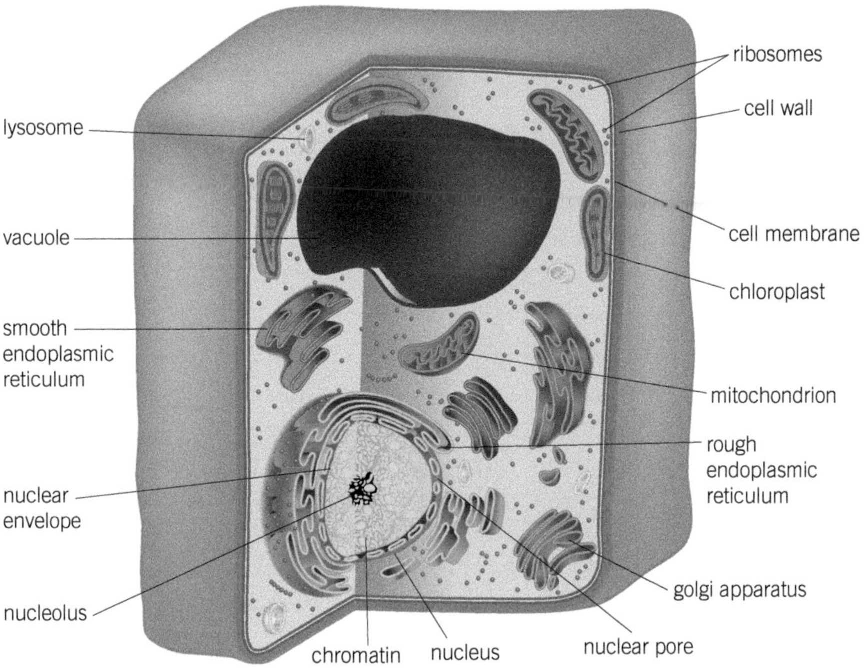

▲ Figure 10 *A generalised plant cell*

Chloroplasts

Chloroplasts vary in size and shape but are usually disc-shaped, 2–10 μm long and 1 μm in diameter. They are surrounded by a double membrane, the chloroplast envelope. All chloroplasts have at least three membrane systems:

- The outer chloroplast membrane – this outer layer of the envelope allows small molecules and ions to easily diffuse across but not large proteins.
- The inner chloroplast membrane – this is in contact with the fluid-filled centre of the chloroplast (the stroma) and controls the passage of materials in and out of the stroma. It is also the site of fatty acids, lipids, and carotenoids synthesis. The inner membrane is continuous with the thylakoid system.
- The thylakoid system – these are suspended in the stroma. They are made from a collection of membrane-bound sacs called thylakoids. Chlorophyll molecules and other photosynthetic pigments are found embedded in the thylakoid membranes.

The stroma is where the second stage of the process of photosynthesis occurs (the light independent stage). The stroma contains DNA, chloroplast ribosomes (70 s), starch granules, and many proteins.

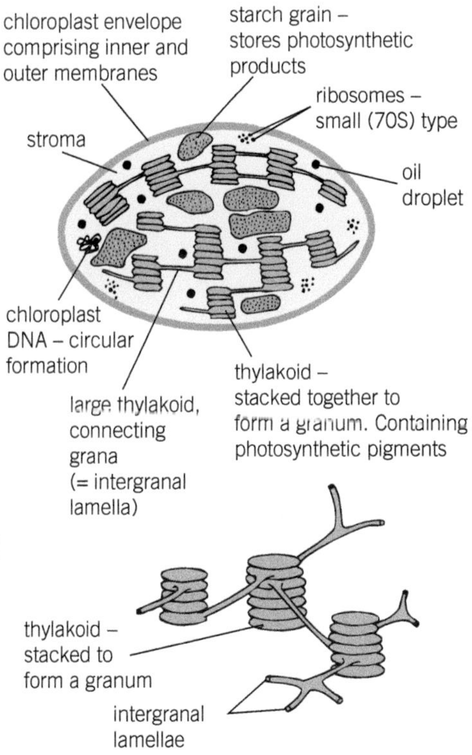

▲ Figure 11 *The structure of a chloroplast*

Large vacuole

Vacuoles are large vesicles that contain mainly water. The vacuole contents are separated from the cytoplasm by a membrane called the tonoplast. Salts, sugars, and organic acids are dissolved in the water. Mature plant cells usually have a large, central permanent vacuole, which plays an important role in cell turgidity and overall plant support.

Cell wall

All plant cells possess a cell wall made from the polysaccharide **cellulose**. In the cell wall, cellulose microfibrils are held together by hydrogen bonds. The microfibrils form cross links, which gives the cell wall strength. This provides the cell with mechanical strength and prevents it from bursting when water enters by osmosis. It also allows water to pass along the microfibrils to aid the passage of water through the plant.

Structure of *Escherichia coli*

All bacteria such as *E. coli* are described as **prokaryotic cells** because they have no true nucleus or nuclear envelope. Prokaryotic cells are usually smaller than eukaryotic cells, with volumes being 1000 or 10 000 times less than a typical animal or plant cell, respectively. Like eukaryotic cells, prokaryotic cells have features which are unique to them.

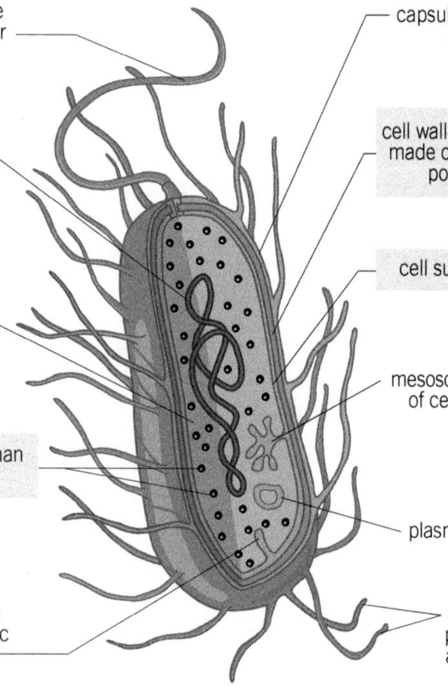

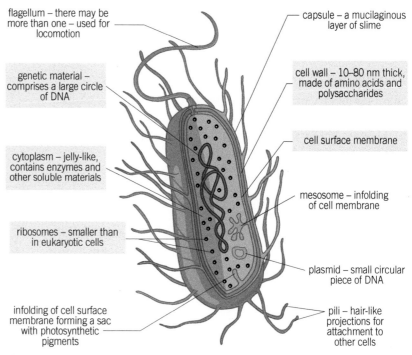

▲ **Figure 12** *A generalised prokaryotic cell, features in blue boxes are also common to eukaryotic cells*

Circular DNA

Circular DNA contains the genetic information for replication and protein synthesis.

Plasmids

Plasmids are very small circular pieces of DNA, physically separate from the main DNA within the cell. Plasmids can replicate independently of the main DNA and carry genes that increase the survival of the bacteria, for example, the genes to produce enzymes that break down antibiotics.

Mesosome

The **mesosome** is formed by invagination of the cell surface membrane. This has been proposed as a possible site of respiration but more recent studies suggest that such structures might be artefacts produced during the preparation of specimens for electron microscopy.

Pili

Pili are hair-like structures made of protein, which extend through the cell wall. They help bacteria to stick (adhere) to each other or to other surfaces. They are not found in all bacteria.

Flagella

Some prokaryotic cells have a **flagellum** (plural – flagella) which is used for movement. Bacterial flagella rotate due to the presence of a rotating bearing at the base of the flagellum (the smallest motor known!). For example, *Helicobacter pylori*, which causes stomach ulcers, uses multiple flagella to propel itself.

▼ Table 1 *Comparison of different cells*

Feature	Prokaryotic cells	Eukaryotic cells	
		Animal cells	Plant cells
cytoskeleton	absent	present	present
nucleus	No true nucleus DNA found naked in the cytoplasm	distinct nucleus surrounded by nuclear envelope DNA associated with histone proteins	distinct nucleus surrounded by nuclear envelope DNA associated with histone proteins
nucleolus	absent	present	present
DNA	circular strands	linear molecules, found arranged as chromosomes	linear molecules, found arranged as chromosomes
organelles	no membrane-bound organelles present	membrane-bound organelles present	membrane-bound organelles
chloroplasts	none present but may contain photosynthetic pigments	absent	present
ribosomes	small 70S	larger 80S	larger 80S
cilia and flagella	sometimes present – but different to those in animal cells	frequently present – usually a 9 + 2 arrangement	rarely present
lysosomes	no	yes	no – but vacuole carries out the same function
endoplasmic reticulum	absent	present	present
cell wall	present – made of peptidoglycan (murein)	absent	present – made of cellulose or pectin
vacuole	absent	if present – small and temporary	large, central, and permanent
storage molecule	glycogen granules and lipid droplets	glycogen granules	starch grains

Summary questions

1 Suggest why the cytoplasm of a B lymphocyte will have a larger proportion of each of the following structures compared to a monocyte:

a RER *(2 marks)*

b Golgi apparatus *(2 marks)*

2 a If the nucleus of a cell is measured to have a radius of 7.5 μm and is assumed to be spherical, calculate the volume of the nucleus to the nearest whole μm. (Hint: the volume of a sphere can be calculated using the formula volume = $\frac{4}{3}\pi r^3$) *(3 marks)*

b If the cell in part b was calculated to have a total volume of 20 500 μm^3, calculate the percentage of the cell that is made up of the nucleus. Give your answer to one decimal place. *(2 marks)*

3 Discuss the validity of the statement 'All plant cells have chloroplasts.' *(4 marks)*

1.7 Interrelationship of organelles

Specification reference: 2.1.1

Division of labour

Gastrin is a peptide hormone that is made in tissues of the stomach. It causes the production and secretion of gastric acid. Gastric acid is composed of 0.5% hydrochloric acid (HCl), potassium chloride (KCl), and sodium chloride (NaCl). Gastric acid plays a vital role in activating digestive enzymes and making ingested proteins unravel so that digestive enzymes can hydrolyse them.

Gastrin is produced by the G-cells of the duodenum and stomach. It is secreted into the bloodstream. The production of gastrin demonstrates division of labour within the G-cell and shows the relationship between many cell organelles.

Learning outcomes

Demonstrate knowledge, understanding, and application of:

→ the relationship between different organelles involved in the production and secretion of proteins.

Sequence of protein synthesis

All proteins require the involvement of many organelles in their production, in a specific order and sequence:

1. The gene for the production of the protein is located in the nucleus.
2. The genetic information for the production of the specific protein is copied into mRNA by the process of transcription.
3. The mRNA leaves the nucleoplasm in the nucleus via the nuclear pore and enters the cytoplasm.
4. The mRNA attaches to the ribosome (which may be free in the cytoplasm or attached to endoplasmic reticulum, RER).
5. The ribosome reads the instructions carried by the RNA nucleotides in the mRNA and translates them into a polypeptide chain.
6. The assembled protein in the RER is pinched off in a vesicle and is transported to the Golgi apparatus.
7. The Golgi apparatus packages the protein.
8. The Golgi vesicle is moved to the cell surface membrane with the aid of the cytoskeleton, where the protein gastrin is secreted out of the cell.

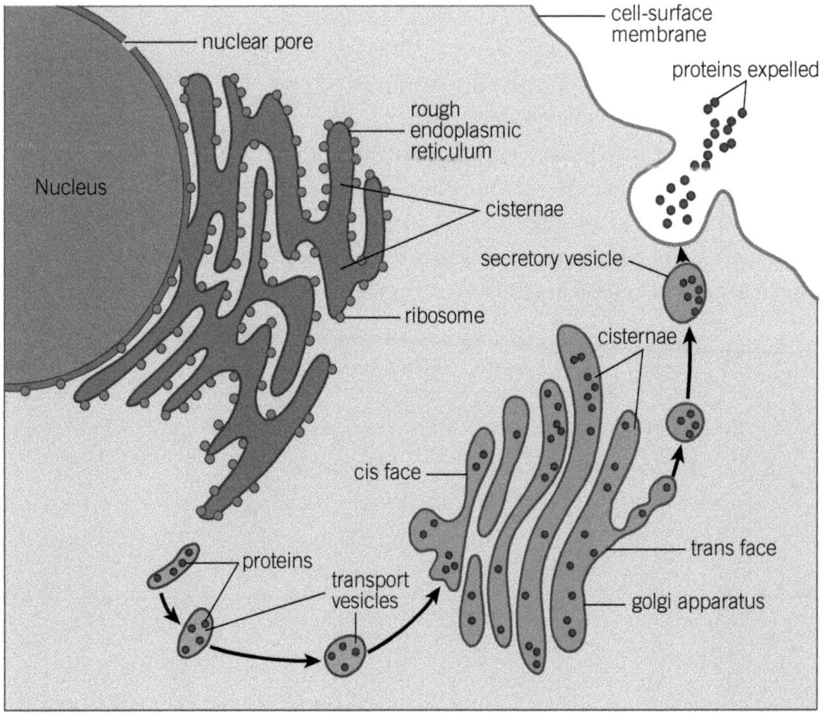

▲ Figure 1 *Protein synthesis*

Cytoskeleton

Protein synthesis is carried out in the cytoplasm and involves the movement of different organelles and chemical substances. Cell organelles are moved through the cytoplasm by the cytoskeleton. In eukaryotic cells there are three main kinds of cytoskeletal filaments:

- microfilaments – these are made from actin and are the thinnest filaments.
- intermediate filaments – these are 10 nm in diameter and are more stable than microfilaments. They are made of proteins such as keratin.
- **microtubules** – these are 23 nm in diameter and made from tubulin (a specialised protein) arranged in hollow cylinders.

These three different types of filaments maintain the cell's structure and shape and play important roles in intracellular transport and mitosis. Molecules called motor proteins provide the driving force to move organelles and chemical substances along these filaments.

Motor proteins

The cytoskeleton also includes hundreds of associated proteins like molecular motors, crosslinkers, and nucleation-promoting factors. One example of a motor protein is the muscle protein myosin which powers the contraction of muscle fibres in animals. Motor proteins are essential in actively transporting proteins, membrane-enclosed organelles, and vesicles through the cytoplasm to their appropriate locations in the cell.

Motor proteins move by binding to a cytoskeletal filament (microtubules or actin filaments) and using energy derived from the hydrolysis of *ATP* to move along the filament. The proteins cycle between being bound strongly to the filament and being unbound. The process occurs in a cycle of filament binding, conformational change, filament release, conformational relaxation, and filament rebinding. The motor protein and its associated organelle or molecule, move one step at a time along the filament (each 'step' is usually only a few nanometers).

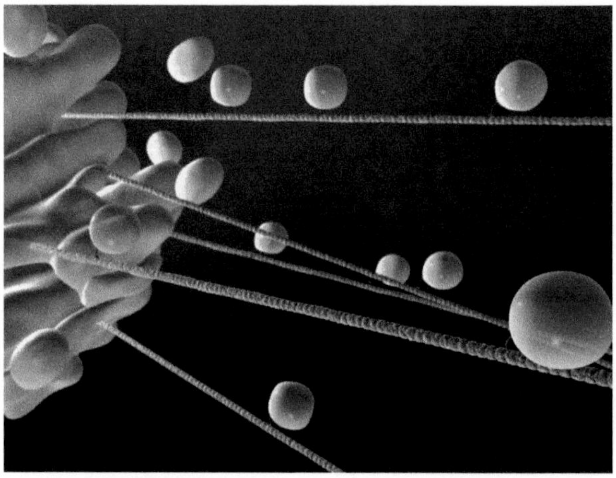

▲ **Figure 2** *Computer generated image of vesicles (spheres) being transported from a Golgi body (blue, left) around the cell by microtubules (string-like). The vesicles are attached to the microtubules by motor proteins (small orange string structures), which 'walk' along the microtubule*

1 a Describe the energy transfer carried out by motor proteins
  b Other eukaryotic cells also use motor proteins to enable them to carry out their function. State the name of the cell and how the motor proteins are used within the cell.

Summary questions

1 Suggest why the term 'mechanochemical cycles' is often used when describing the action of motor proteins. *(3 marks)*

2 Describe three roles of the cytoskeleton in a eukaryotic cell. *(3 marks)*

3 A pulse-chase experiment can be carried out to track the pathway of a radioactive molecule (e.g., an amino acid) through a cell. At the start of the experiment a large dose of radioactively labelled amino acid is added to a cell culture. The cells will use these radioactive amino acids to synthesise new proteins. The chase phase begins when a large sample of a non-radioactive version of the same amino acid is added. The procedure allows the radioactively labelled protein to be tracked through the cell.

a Data from a pulse-chase experiment are shown in the graph. The four lines show the relative levels of radioactivity found in four cell organelles – endosomes, Golgi apparatus, lysosomes, and RER. Using your knowledge about the roles of organelles in protein synthesis, state which line on the graph represents each organelle. *(4 marks)*

b Suggest why each line peaks at the same level of radioactivity. *(1 mark)*

c Suggest what type of tissue could be used to carry out this investigation and what protein would need to be tracked. *(2 marks)*

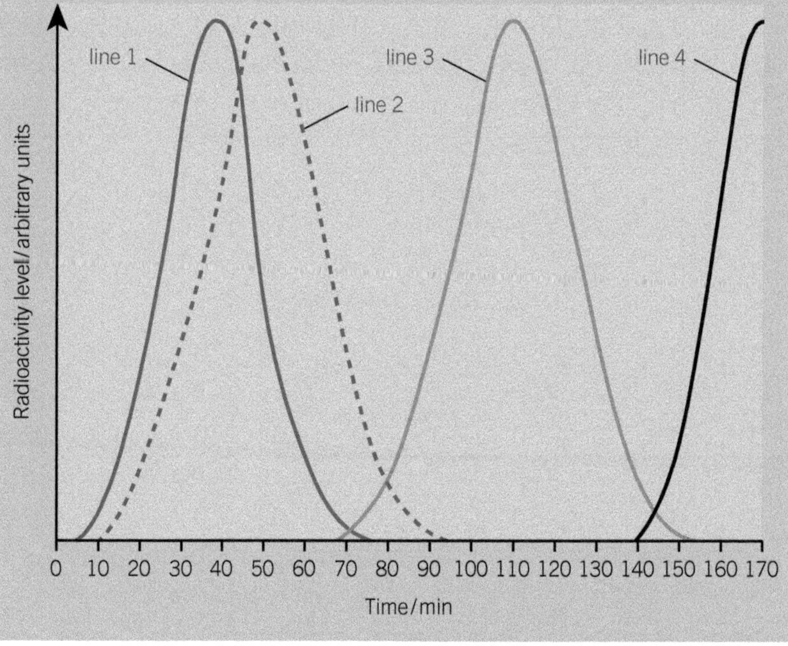

1.8 Measuring cells and organelles

Specification reference: 2.1.1

Learning outcomes

Demonstrate knowledge, understanding, and application of:

→ how to use a stage micrometer and eyepiece graticule to measure cells and organelles

→ how to calculate the area of the field of view.

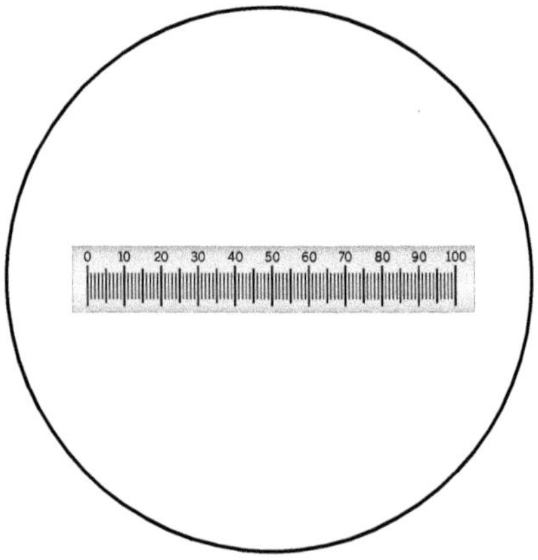

▲ **Figure 1** *An eyepiece graticule*

Measuring cells

Any ocular microscope (viewed using the human eye) needs to be calibrated to enable the viewer to measure accurately the specimen being observed. To do this two pieces of specialist equipment are used:

- a stage micrometer
- an eyepiece graticule.

Eyepiece graticule

The **eyepiece graticule** is a transparent disc fitted into the eyepiece of the microscope. These can be fitted to existing eyepieces or eyepieces can be purchased with graticules already in place. The disc is marked with a fine grid or a graduated line. The absolute size of the grid or gradiations is not important as this is what will be calibrated.

Stage micrometer

The **stage micrometer** is a specialised microscope slide with a finely divided scale marked on it. The scale is used to calibrate the eyepiece graticule. The scale has to be accurately produced to give reference dimensions, which makes stage micrometers much more expensive than eyepiece graticules. The stage micrometer is usually 1 cm long and subdivided into ten divisions. Each of these divisions is therefore 1 mm. Each division is then normally divided into ten divisions of 0.1 mm or 100 µm each.

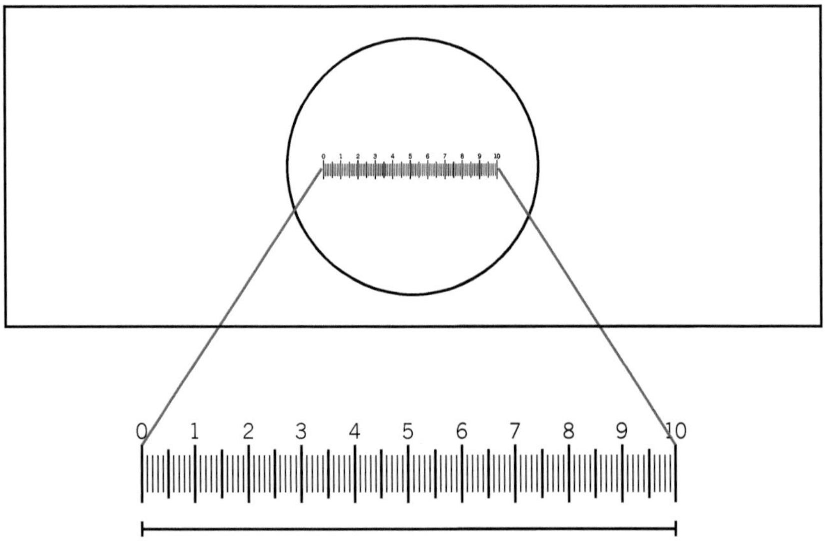

▲ **Figure 2** *A stage micrometer*

Carrying out a calibration

When you carry out a calibration, you have to calibrate each objective lens separately. This will result in different calibration factors for each objective. You start with the lowest power objective on the microscope, aligning the scale on the stage micrometer with the scale of the eyepiece graticule. Then you take a reading from the scales. You use these readings to calculate the calibration factor for the objective lens you are using. The following worked example shows how to calibrate the graticule for the ×40 objective lens.

Worked example: Performing a calibration

12 divisions on the eyepiece graticule equal 25 divisions on the stage micrometer. For the particular stage micrometer being used 100 divisions = 1 mm. Use this information to calculate the length of one eyepiece division. You may need to revise the orders of magnitude of units for linear dimensions, which is covered in Topic 1.1.

Step 1 Each micrometer division = $\frac{1}{100}$ = 0.01 mm = 10 µm

Step 2 12 eyepiece divisions = 25 × 10 µm = 250 µm

Step 3 One eyepiece division = $\frac{250}{12}$ = 20.8 µm

If you now observed a specimen and found the length of the specimen measured 16 eyepiece divisions, calculate its length.

The length of the specimen would be 16 × 20.8 = 332.8 µm.

Study tip

When asked to carry out a calculation, make sure you look carefully at the instructions:

Do you have to show your working?

Does your answer have to be given to a set number of decimal places or significant figures?

Does your answer make sense? Always double check!

Field of view

The **field of view** refers to the circular area visible down the microscope. The stage micrometer or calibrated graticule can be used to measure the radius of the field of view and, using this value, the area of the field of view can be calculated.

Worked example: Calculating the area of the field of view

The diameter of a field of view is measured as 0.45 mm. Use this to calculate the area of the field of view.

Step 1 The diameter in this case is 0.45 mm and must be divided by 2 to give the radius: $\frac{0.45}{2}$ = 0.225 mm

Step 2 The area of a circle = πr^2, where r is the radius. In this case area = $\pi \times 0.225^2$ = 0.159 mm^2

Summary questions

1 The field of view through a microscope has a radius of 1 mm. Calculate the area of the field of view to one decimal place. *(2 marks)*

2 Using the information from Figure 4 in Topic 1.6 calculate the maximum dimension of the mitochondrion in µm to one decimal place. *(2 marks)*

3 If a 1 cm scale bar was added to Figure 4 in Topic 1.6, what actual length would this scale bar be representing? Give you answer in µm. *(2 marks)*

Specification reference: 2.1.1

Learning outcomes

Demonstrate knowledge, understanding, and application of:

- → the components of cell membranes
- → the fluid mosaic model of membrane structure
- → the different roles of cell membranes.

Cell membranes

Cell membranes exist:

- around the outside of all cells (the cell surface membrane) where they control the entry and exit of molecules in cells
- within eukaryotic cells surrounding organelles.

Within eukaryotic cells, membranes:

- control the entry and exit of molecules in organelles
- isolate organelles to enable them to carry out specific chemical reactions
- provide an internal transport system
- concentrate enzymes and substrates to increase efficiency
- isolate enzymes that can cause cellular damage
- maintain internal specific conditions at the optimum level, for example, pH (acidity)
- can be moved around the cell to where they are needed for other processes
- provide surfaces for chemical reactions.

They exist as single membranes or as double membranes forming an envelope (e.g., the nuclear envelope, chloroplast envelope, and mitochondrial envelope). Membranes within cells and at the surface of cells share a common chemical composition and structure.

Components of cell membranes

Phospholipids

The main components of all cell membranes are phospholipid molecules. Each phospholipid molecule has two distinct parts – a phosphate head and hydrocarbon tail.

Phospholipids are made of two fatty acids joined to a polar head. This head always contains a phosphate group and a glycerol molecule and usually another polar group. The two fatty acids are joined to the glycerol molecule by two separate condensation reactions to form two covalent bonds called ester bonds.

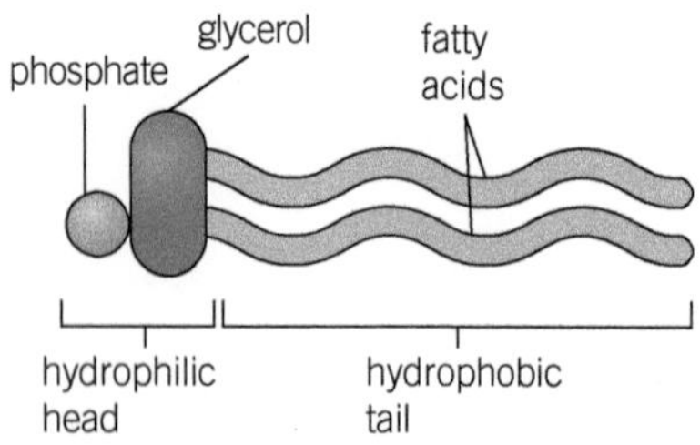

▲ Figure 1 *Simplified structure of a phospholipid*

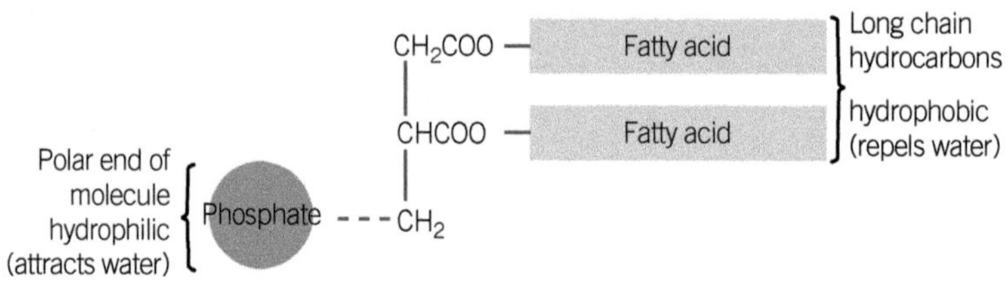

▲ Figure 2 *The structure of a phospholipid*

A fatty acid is a carboxylic acid with a long hydrocarbon tail. They are important sources of energy in the body. Fatty acids vary in:

- the types of bonds between carbon atoms – fatty acids that have carbon–carbon double bonds are known as unsaturated fatty acids, while fatty acids without double bonds are known as saturated fatty acids
- the number of carbon atoms in the hydrocarbon tail – most fatty acids which occur naturally have an even number of carbon atoms in the hydrocarbon tail, from 4–28.

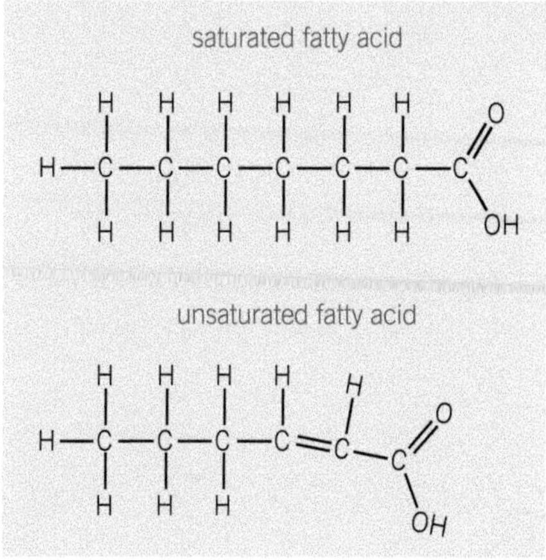

▲ **Figure 3** *The two types of fatty acid*

Each phospholipid molecule therefore has a region which is **hydrophilic** (phosphate head) and one which is **hydrophobic** (hydrocarbon tail). When phospholipids are mixed with water they will spontaneously form micelles – spheres where the phosphate heads are on the outside next to the water and the hydrocarbon tails all point inwards away from the water. If more phospholipids are added they form a double layer called a phospholipid bilayer which is usually around 7 nm thick.

The phospholipid bilayer:

- allows lipid-soluble molecules to cross the membrane
- prevents water-soluble molecules crossing the membrane
- allows the membrane to be flexible
- allows the membrane to be stable.

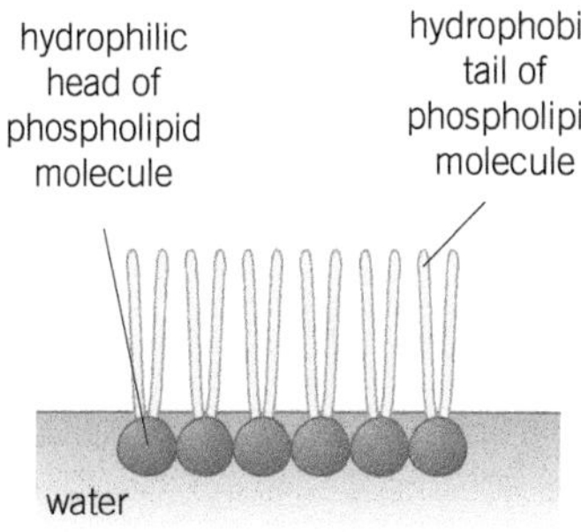

a *Monolayer*

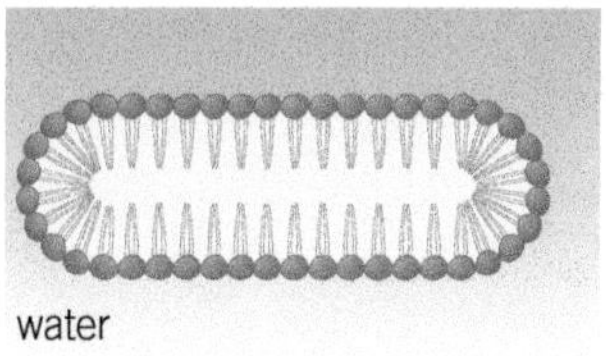

b *Bilayered sheet*

▲ **Figure 4** *Phospholipid arrangements in water*

Some phospholipids are modified and have a carbohydrate portion added to them to form glycolipids.

Triglycerides are the most common form of lipid in the body, but they are not found in cell membranes. They are entirely hydrophobic and often called 'neutral fats'. They usually associate with each other or other hydrophobic molecules to form spherical bodies or large droplets inside cells. They are insoluble in cytosol or other aqueous solutions – however they will dissolve in organic solvents.

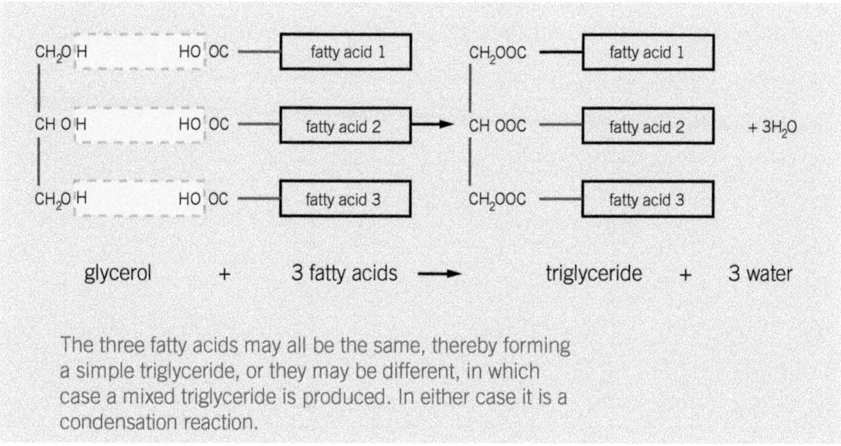

▲ **Figure 6** *The formation of a triglyceride*

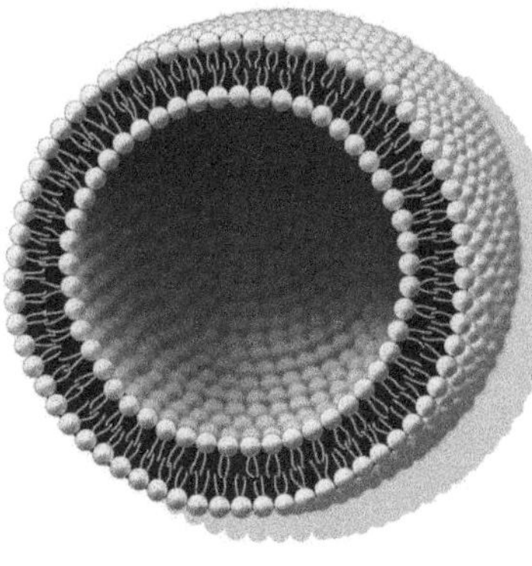

▲ **Figure 5** *Phospholipid bilayer forming part of the cell surface membrane*

Proteins

Proteins are found floating within the phospholipid bilayer in two different ways:

- Integral proteins occur within the bilayer and are sometimes called intrinsic proteins.
- Peripheral proteins are found at the edges of the bilayer and are sometimes called extrinsic proteins.

Membrane proteins carry out a variety of functions:

- provide the membrane with structural support
- carry water-soluble molecules across the membrane
- form ion channels to allow active transport
- help cells adhere to each other
- act as receptors for hormones and neurotransmitters
- act as antigens allowing cells to recognise one another.

Some extrinsic proteins on the outer cell surface membrane are modified and have a carbohydrate portion added to them to form a glycoprotein. The carbohydrate molecules on glycoproteins and glycolipids collectively form a region called the glycocalyx. This has a role in cell to cell adhesion.

Cholesterol

Cholesterol is a lipid-like substance (steroid) and can be found distributed throughout the cell surface and other membranes. Most of the cholesterol molecule is hydrophobic so it is attracted to the hydrophobic hydrocarbon tails in the centre of the membrane. However, one end of the cholesterol is hydrophilic and is attracted to the phosphate heads of the phospholipids. The amount of cholesterol in cell membranes varies. Cholesterol provides strength to the cell membrane, which is important in the absence of a cell wall, and plays an important role in regulating the fluidity of the membrane. It also prevents water and water-soluble ions leaking out of the cell and reduces the lateral movement of phospholipids within a monolayer.

Fluid mosaic model

All the components of the cell membrane are arranged in the fluid mosaic model. It is described as fluid because the individual phospholipid molecules can move laterally within a monolayer or flip between monolayers. It is described as a mosaic as the proteins floating in the phospholipid bilayer vary in shape, size, and structure making a pattern.

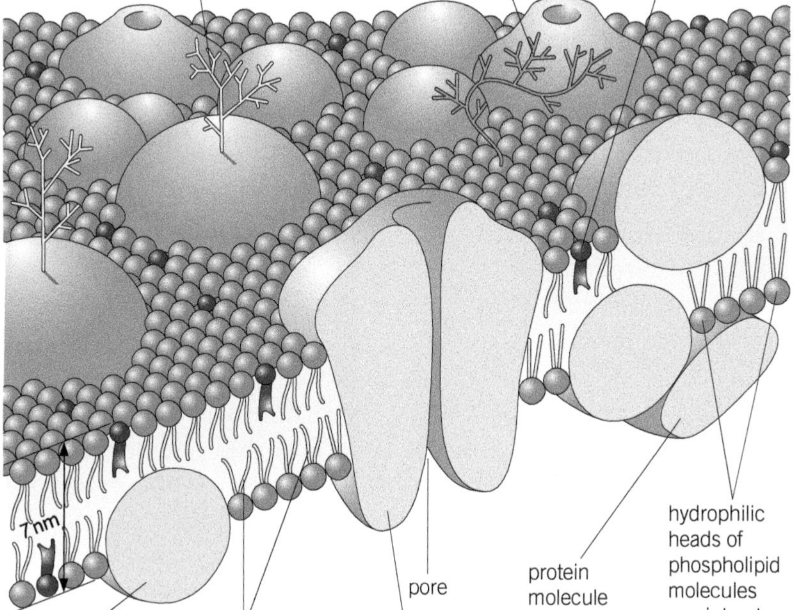

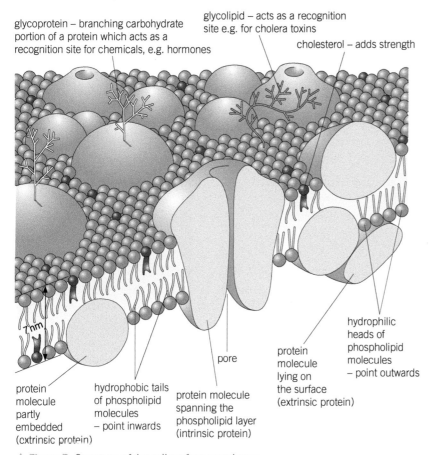

▲ **Figure 7** *Structure of the cell surface membrane*

Study tip

When answering questions on phospholipids check carefully what the question is asking for – is it the *structure* of a phospholipid molecule or the *properties* of the molecule that is required?

Summary questions

1 State which part of the cell surface membrane will allow the passage of:
  a a molecule which is soluble in lipids (*1 mark*)
  b a mineral ion. (*1 mark*)

2 Explain what would happen to plasma membranes that had:
  a too much cholesterol (*2 marks*)
  b too little cholesterol. (*2 marks*)

3 Some cells have different proteins on their 'upper' surface compared to their 'bottom' surface (e.g., cells lining the proximal convoluted tubule in the kidney).
  a Explain why the fluidity of the cell surface membrane would be a problem for these cells. (*1 mark*)
  b Suggest how this could be overcome. (*2 marks*)

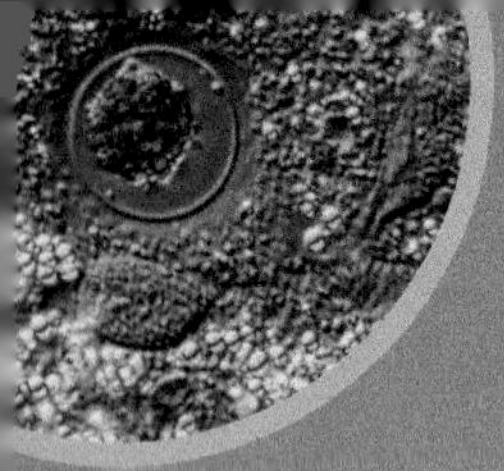

1.10 Passive movement across cell membranes

Specification reference: 2.1.1

Learning outcomes

Demonstrate knowledge, understanding, and application of:

- → how molecules enter and leave cells without the need for energy input
- → the process of diffusion
- → the process of facilitated diffusion.

Passive transport mechanisms

All molecules have kinetic energy and move randomly in liquids and gases. Molecules in a given space will collide as they move and, as a result, the path of their movement changes direction. If the same number of molecules is confined in a smaller space then the molecules will collide more frequently with each other. Movement of molecules that happens as a result of kinetic energy alone, with no additional energy input, is described as a passive process.

Simple diffusion

Simple **diffusion** is the net movement of molecules from an area of high concentration to an area of low concentration down the concentration gradient. The process of diffusion can be summarised in the following steps:

1. The ten particles on the left hand side of the container move randomly as they possess kinetic energy. They will collide with each other and the side of the container. As there are no molecules on the right hand side of the container, movement only occurs in one direction and since there is a large concentration gradient diffusion is rapid.
2. The molecules are beginning to spread out but remain in motion. Movement is now in both directions with some molecules moving from the left to the right and vice versa. As there is still a higher concentration on the left hand side of the container, it is more likely that a molecule will move from the left hand side to the right hand side. As the concentration is smaller than in diagram 1, the rate of diffusion is slower.
3. The molecules have now moved so that they are equally spaced out within the container. There is therefore no concentration gradient within the container. However, molecules still possess kinetic energy and therefore are still moving randomly but no diffusion occurs.
4. The molecules are still evenly spread out and there is still no concentration gradient but the actual position of the molecules has changed. This is called a dynamic equilibrium. The distribution of the molecules remains even but individual molecules are continuously changing their position. There is no net movement of molecules.

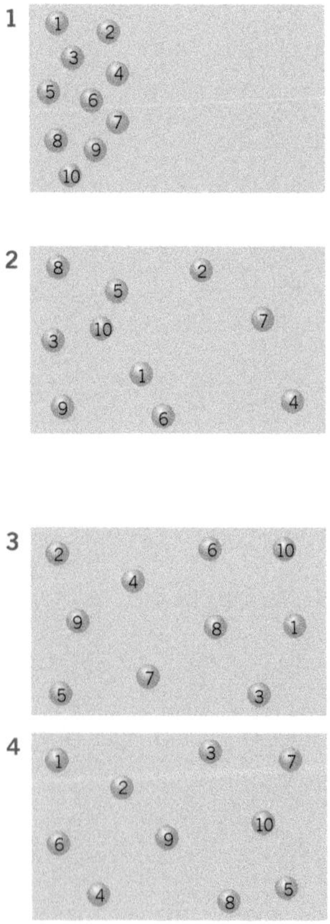

▲ **Figure 1** *Diffusion*

Factors affecting the rate of diffusion across cell membranes

These include the:

- concentration gradient – the steeper the gradient the faster the rate of diffusion

- surface area for diffusion – the greater the surface area the faster the rate of diffusion
- diffusion distance – the distance over which the molecules have to diffuse will affect the rate of diffusion – the shorter the distance the faster the rate of diffusion
- size of the molecule – the smaller the molecule the faster the rate of diffusion
- composition of the cell membrane – the greater number of protein channels in the membrane, the faster the diffusion of hydrophilic molecules.

Study tip

Remember the diffusion rate is the number of molecules that move across a cell membrane in a given period of time. So if there is an increase in the diffusion rate either the molecules are moving faster *OR* more molecules are crossing in the same given period of time for a different reason.

Facilitated diffusion

Molecules that are water-soluble are not able to cross the cell membrane using simple diffusion, even if there is a concentration gradient. These molecules must be transported using facilitated diffusion. In facilitated diffusion specialised proteins are used to carry molecules from one side of the membrane to the other:

- Channel proteins – these form selective 'pores' in the membrane – aqueous 'tunnels.' through the hydrophobic lipid bilayer. This means ions can pass through. Some of these channels can be gated, meaning they can be opened or closed depending on the presence of a potential difference (voltage) across the membrane or by the presence of another molecule (the ligand).

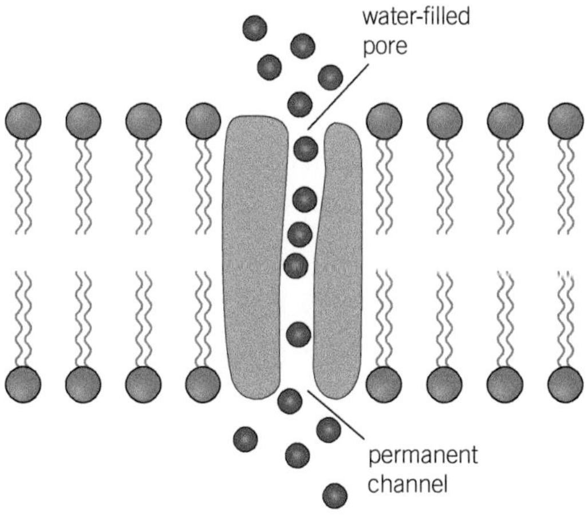

▲ Figure 2 *Facilitated diffusion using channel proteins*

- Carrier proteins – these are specialised proteins that have a specific shape for the molecule that they transport. When the molecule binds to the protein it undergoes a conformational change (allostery) and passes the molecule to the other side of the membrane. An example of this form of facilitated diffusion is the uptake of glucose and amino acids in the small intestine.

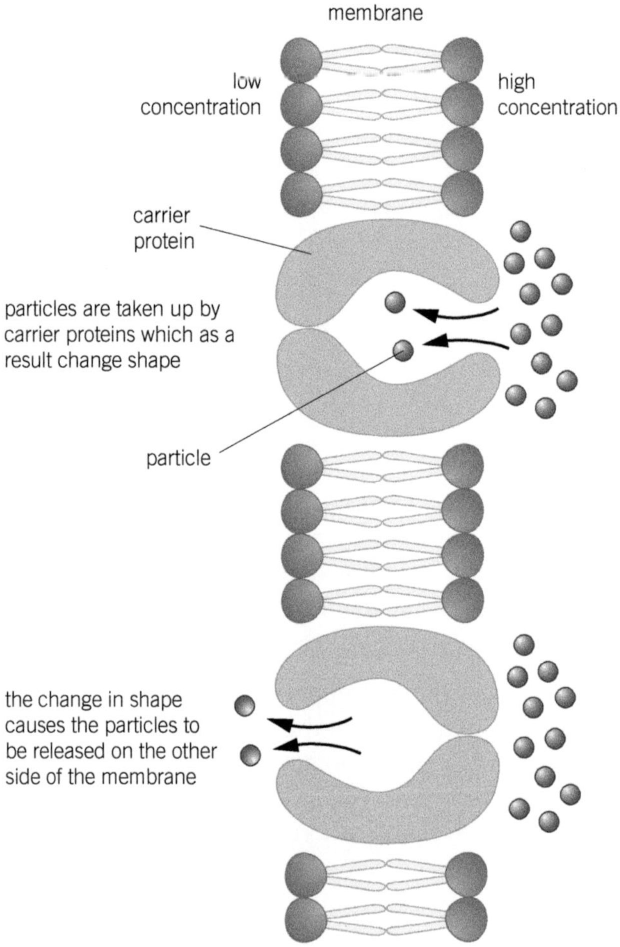

▲ **Figure 3** *Facilitated diffusion using carrier proteins*

Different cell membranes possess different carrier and channel proteins. This gives different cells and organelles control over which molecules can pass in or out across the membrane.

Summary questions

1 Give two examples where diffusion occurs in:
  a humans (*2 marks*)
  b plants. (*2 marks*)

2 Explain why glucose molecules do not simply diffuse through the phospholipid bilayer. (*2 marks*)

3 Suggest how the permeability of cell surface membranes to glucose could be varied. (*2 marks*)

1.11 Active movement across cell membranes

Specification reference: 2.1.1

Active transport mechanisms

In some situations, the molecules that need to be transported across the cell membrane have to be transported against the concentration gradient, that is, from an area of low concentration to an area of high concentration. For this to occur additional energy has to be added as the kinetic energy of the molecules is insufficient. This extra energy is supplied by the **hydrolysis** of ATP produced during cellular respiration. The mechanisms that require the input of ATP are called active processes. There are two main types of active transport mechanism.

Active transport

Embedded in the membrane are specialised **carrier proteins** that span the whole membrane. Each of these carrier proteins is specific for one or a few molecules. The molecule binds to a complementary binding site on the carrier protein. ATP also binds to a separate binding site on the carrier molecule. The energy that is released from the hydrolysis of ATP is used to change the structure of the carrier

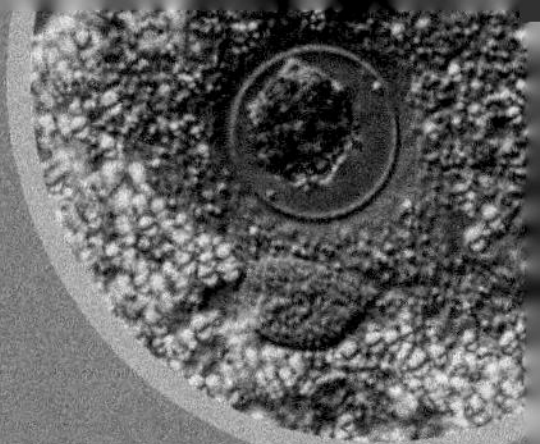

Learning outcomes

Demonstrate knowledge, understanding, and application of:

→ how molecules enter and leave cells with the need for energy input

→ the process of active transport

→ the process of bulk transport: endocytosis and exocytosis.

Outside of cell
Low concentration of molecule, e.g., glucose
membrane

Inside of cell
High concentration of molecule, e.g., glucose

a Carrier protein takes up molecules from outside plasma membrane.

molecule, e.g., glucose

carrier protein spanning membrane

b Glucose molecules bind to carrier protein and ATP attaches to the plasma membrane protein on the inside of the cell.

c Binding of glucose molecules to carrier protein causes the protein to change shape so that access for the glucose molecules is open to the inside of the plasma membrane but closed to the outside.

d This new shape of the protein no longer binds the glucose molecules and so they are released to the inside of the plasma membrane with the aid of energy released from the breakdown of ATP to ADP + P_i

e The release of the glucose molecules causes the protein to go back to its original shape and so it is available to take up more glucose molecules from the outside.

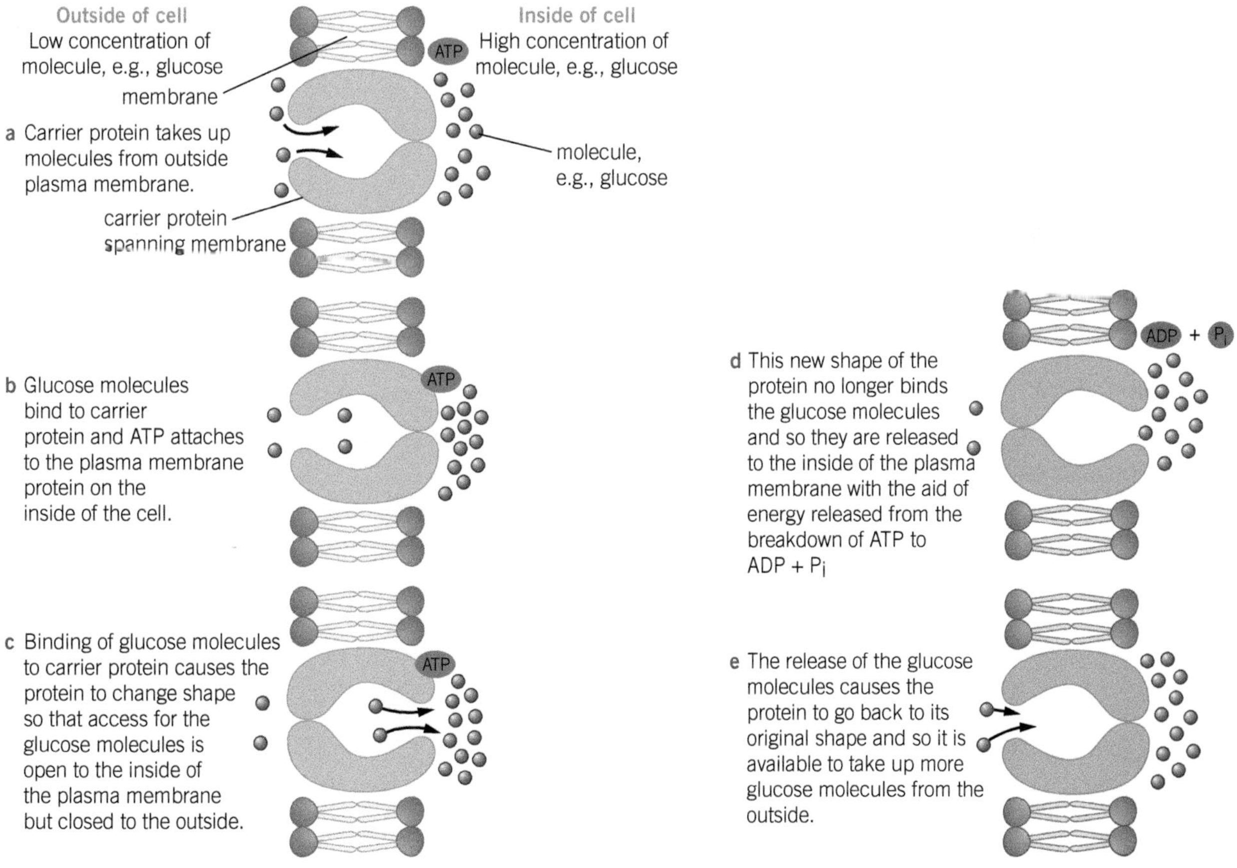

▲ Figure 1 *Active transport*

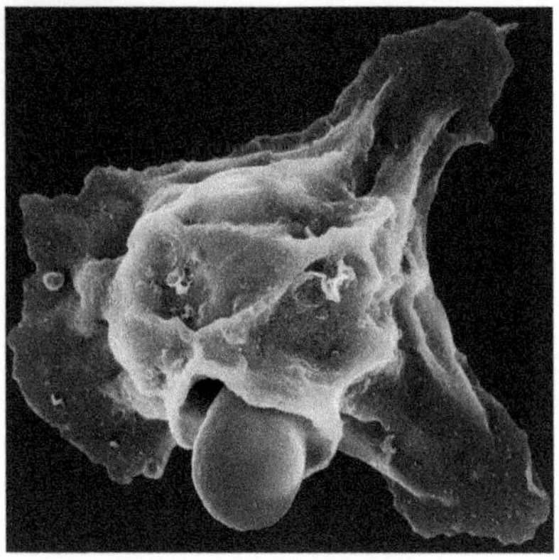

▲ **Figure 2** *Coloured scanning electron micrograph of a neutrophil white blood cell engulfing and destroying a fungal hypha (round, lower centre), approximately × 4000 magnification*

Study tip

Remember different prefixes are used to indicate the direction or material being transported:

endo – into

exo – out of

pino – liquid material

phago – solid material

For example, *endophagocytosis* is the movement of large quantities of solid material into a cell

protein, which moves the molecule from one side of the membrane to the other against the concentration gradient. The carrier protein has undergone a conformational change. Once the molecule is released, the carrier protein changes back to its original shape. As only specific molecules can be transported by each type of carrier protein the process is said to be selective.

Examples of active transport include:

- the uptake of iodine from blood by cells in the thyroid gland
- the uptake of mineral ions such as magnesium ions (Mg^{2+}) into the root hair cells of plants.

Bulk transport

Some substances need to be transported from one side of the cell surface membrane to the other in large quantities. This is not possible using active transport. Instead a different process called bulk transport is used. There are two types of bulk transport:

- **endocytosis** – the bulk movement of material into the cell
- **exocytosis** – the bulk movement of material out of the cell.

Endocytosis

Endocytosis is the process used to take large quantities of materials into the cell. The cell surface membrane folds around the materials outside the cell. The ingested material is trapped in a membrane-bound vesicle or vacuole. The two main types of endocytosis are:

- pinocytosis – cells engulf liquid material
- phagocytosis – cells engulf solid material.

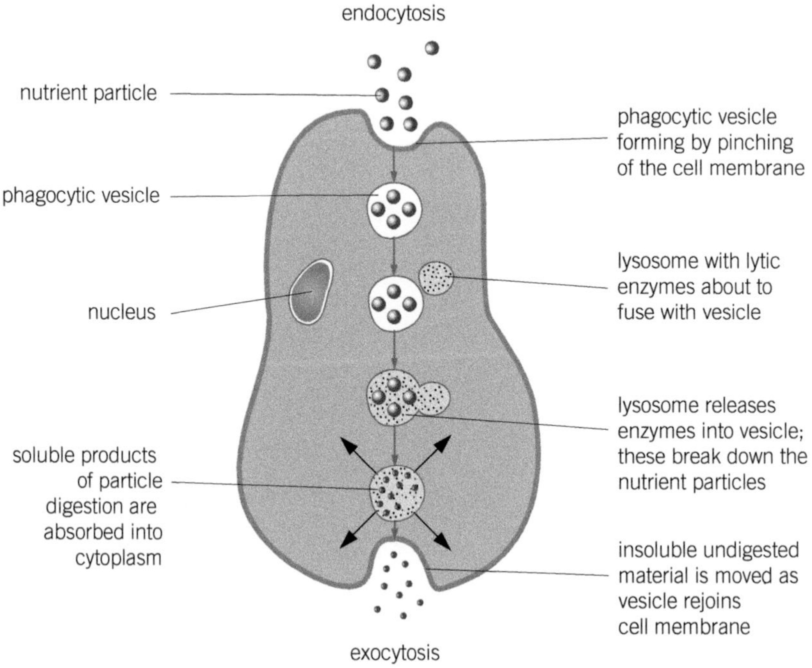

▲ **Figure 3** *Endocytosis and exocytosis*

Excocytosis

Exocytosis is the process used by cells to remove large quantities of materials from the cell. It is the opposite of endocytosis. Exocytosis releases hormones from the cells that produce them (e.g., gastrin from G-cells of the stomach and duodenum, insulin from β-cells of the pancreas).

Summary questions

1 State an example of exocytosis being used in:
  a secretion *(1 mark)*
  b excretion. *(1 mark)*

2 Copy and complete the table to compare active transport and facilitated diffusion. *(8 marks)*

	Facilitated diffusion	Active transport
move molecules down a concentration gradient		
move molecules against a concentration gradient		
involves intrinsic proteins		
involves carrier proteins		
carrier proteins possess at least two binding sites		
carrier protein has a complementary binding site for the molecule to be transported		
requires ATP		
involves allostery		

3 a State two ways that the rate of active transport can be increased. Explain in detail how each method works. *(6 marks)*
  b Explain why active transport allows the accumulation of molecules in an area but facilitated diffusion does not. *(2 marks)*

Practice questions

1 The image shows an image taken with an electron microscope.

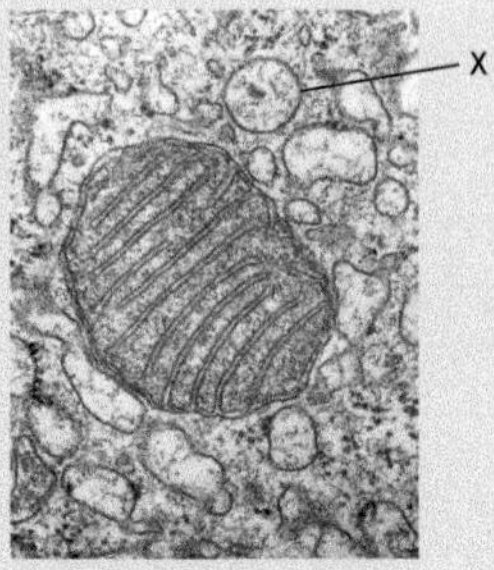

Which of the following statements correctly describes the image?

A vesicles and nucleus as seen in a TEM photomicrograph

B vesicles and mitochondrion as seen in an SEM photomicrograph

C vesicles and mitochondrion as seen in a TEM photomicrograph

D vesicles and nucleus as seen in an SEM photomicrograph *(1 mark)*

2 The structure labelled **X** in the image is a vesicle. The **actual** diameter of the vesicle is 0.4 μm.

Which of the following correctly describes the actual volume of the vesicle?

A 1256.8 nm^3

B 125.68 $μm^3$

C 0.12568 $μm^3$

D 125680 nm^3 *(1 mark)*

3 The image here shows a diagram of the plasma membrane.

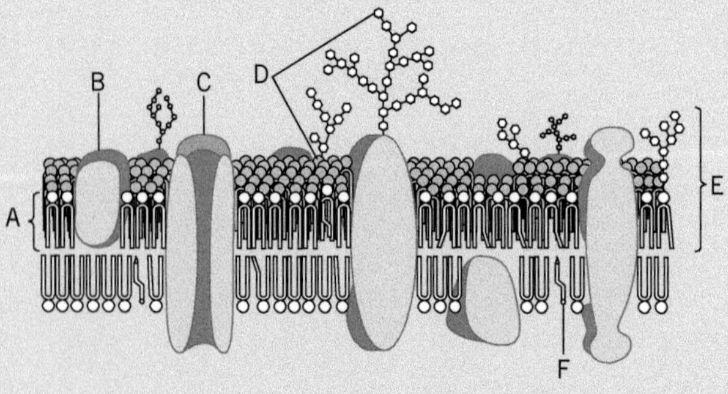

Which of the statements is/are true?

Statement 1: molecules A and F have hydrophobic and hydrophilic regions.

Statement 2: molecules D and E form the glycocalyx.

Statement 3: molecules B and C take part in membrane transport.

A 1, 2 and 3

B Only 1 and 2

C Only 2 and 3

D Only 1 *(1 mark)*

4 The table shows the concentration of some substances inside an erythrocyte and in the blood plasma.

	Concentration (arbitrary units)	
Substance	**Erythrocyte**	**Plasma**
K^+	136	5
Na^+	12	138
Cl^-	7	121
HCO_3^-	15	34

Which of the statements is/are true?

Statement 1: ATP is required to transport sodium ions out of the erythrocyte.

Statement 2: potassium ions enter the erythrocyte by facilitated diffusion.

Statement 3: HCO_3^- leaves the erythrocyte by simple diffusion.

A 1, 2 and 3

B Only 1 and 2

C Only 2 and 3

D Only 1 *(1 mark)*

5 A laboratory technician was helping a trainee to identify abnormal blood cell counts. A light microscope and a special counting chamber were used.

a Name the counting chamber used by the technician to count red blood cells *(1 mark)*

b The diagram shows a view of red blood cells within the counting chamber as seen by the technician.

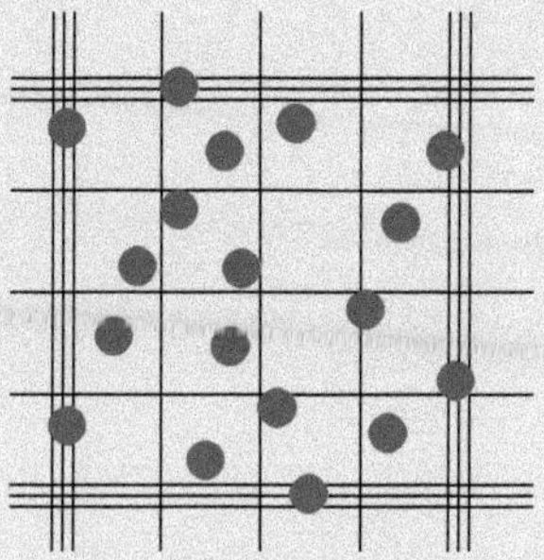

The trainee counted 18 cells which did not agree with the technician.

(i) What is the correct cell count? (*1 mark*)

(ii) Explain how the number of cell could have been miscounted by the trainee. (*2 marks*)

(iii) The volume of diluted blood over one of the squares in the counting chamber was 0.00025 mm³. Describe what other information has to be taken into account when calculating a reliable total for the number of cells in 1 mm³ of blood. (*2 marks*)

c The technician showed the trainee how a similar procedure could be used to count leucocytes.

When preparing a blood sample for viewing leucocytes differential stain is used.

Explain why the technician used a differential stain in this case (*2 marks*)

d Describe how a blood film is prepared and stained for viewing under a light microscope (*4 marks*)

[OCR F221, June 2010]

6 The diagram shows a eukaryotic cell drawn from an electron micrograph.

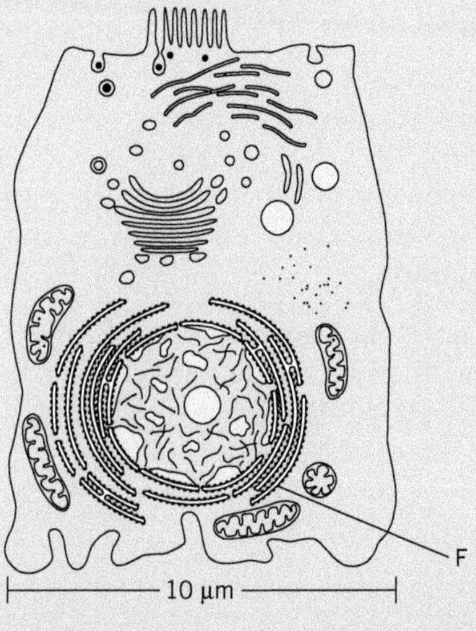

a (i) Calculate the magnification of the micrograph. Show your working. (*2 marks*)

(ii) Explain why an image of the same cell produced at the same magnification using a light microscope would not show the same structures (*2 marks*)

b* Proteins are produced by the structure labelled **F**. Some of these proteins may be extracellular proteins. Describe the sequence of events following the production of extracellular proteins that leads to their release from the cell. (*6 marks*)

[OCR F211 Jan 2009]

7 All living cells have a cell surface membrane. In 1972, Singer and Nicholson proposed a model for the arrangement of molecules within cell surface membranes.

The diagram represents the structure of a phospholipid molecule which is found in cell surface membranes.

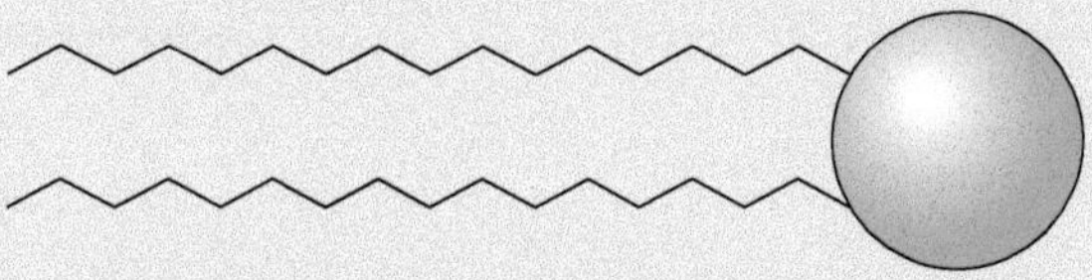

a Explain how the properties of phospholipid molecules are related to their arrangement within cell surface membranes. (*2 marks*)

b One piece of evidence used by Singer and Nicholson in their model was provided by the scientists Frye and Edidin.

Frye and Edidin fused together human and mouse cells to form a dikaryon.

- Antigens on the two types of cell were labelled with fluorescent markers by exposing the cells to antibodies.
- Mouse and human antibodies each carried a different coloured marker.
- The distribution of fluorescent markers was observed in the dikaryon.

Using the Singer Nicholson model, predict the distribution of the fluorescent markers in the dikaryon and justify your prediction. (*2 marks*)

2 THE IMPORTANCE OF WATER

2.1 The properties of water

Specification references: 2.1.2

Learning outcomes

Demonstrate knowledge, understanding, and application of:

- → the role of water in maintaining life
- → the importance of hydrogen bonds
- → the properties of water.

Water – a special molecule

All life forms depend on water. Water is the major constituent of most organisms. Two thirds of this water is contained inside cells and the remainder is found in extracellular biofluids, such as plasma in animals and phloem sap in plants.

Most biochemical reactions take place in water so without it there would be no life on planet Earth. Water has some unique properties that make it a very important biological molecule.

The structure of water

Water is a **polar** molecule made of an oxygen atom and two hydrogen atoms. The oxygen atom has a slightly negative charge and the hydrogen atoms have a slight positive charge. The water molecule is a bent molecule (rather than linear) and so the two hydrogen atoms are on one side of the molecule and form the positive pole whilst the oxygen atom is on the other, forming the negative pole.

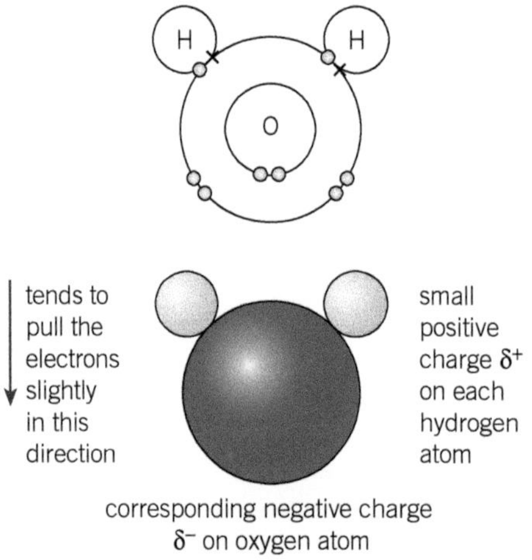

▲ **Figure 1** *A molecular model of a molecule of water, containing oxygen (red) and hydrogen (white)*

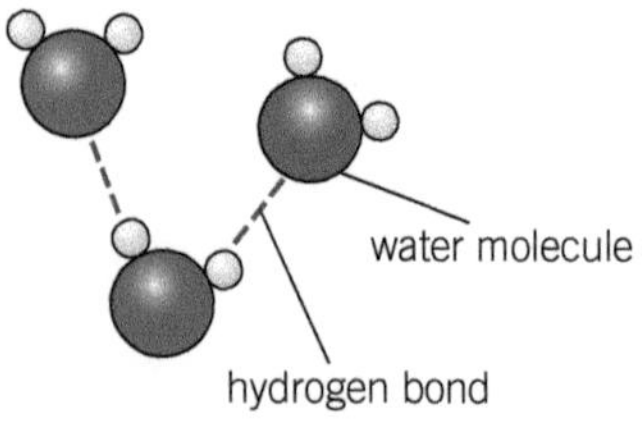

▲ **Figure 2** *Hydrogen bonds (shown by the dotted lines) are weak attractions found between atoms with slight negative charges and slight positive charges*

Water molecules are attracted to each other as the slight positive charges are attracted to the slight negative charges. These forces of attraction are **hydrogen bonds**.

Strictly speaking a hydrogen bond is a weak intermolecular force rather than an actual bond. Individually they are very weak but if there are a lot of hydrogen bonds in a given volume they are collectively quite strong.

Properties of water

Water has many different properties that make it unique (e.g, water has a high boiling point, water molecules are cohesive and adhesive). Some of these properties are essential for maintaining life.

Water is a solvent

Water molecules are attracted to other polar and charged particles, which makes water a good **solvent** for substances with these properties. The water molecules form a 'shell' around charged ions and other molecules that possess a slight charge on their surface. This prevents the ions and molecules from clumping together and so causes the ions and other molecules to dissolve. Non-polar molecules cannot dissolve in water and therefore cannot be transported in the plasma. Instead they have to be combined with molecules that are soluble, for example, cholesterol is insoluble in water so it is attached to lipoproteins to enable it to be transported in the plasma.

This property is important:

- for transporting substances around the body in the plasma, for example, glucose, insulin, and lymph
- in removing metabolic waste, for example, urea in urine
- in allowing chemical reactions to take place inside cells, for example, respiration, protein synthesis.

Ions dissolved in water within the body are called **electrolytes**.

Water has a high specific heat capacity

Hydrogen bonds restrict the movement of water molecules, which increases the amount of energy needed to break the hydrogen bonds. A large amount of energy is needed to make bodies of water change temperature. Water must lose a large amount of energy to cool down. This means the temperature of water bodies is relatively stable in comparison to air and land. This makes water an ideal habitat for aquatic organisms.

This property is also important as it prevents our internal body temperature changing quickly as a result of changes in the environment.

Water has a high latent heat of vaporisation

Water molecules need a high amount of heat to enable a molecule to separate from other molecules in a liquid to become a vapour. As it takes a large amount of heat to break the hydrogen bonds, this means that when water evaporates it has a cooling effect. This property of water is important since it allows our bodies to lose heat through sweating.

Water has cohesive and adhesive properties

Water is viscous as a result of its cohesive and adhesive properties. This property of water is important as water can be used as a lubricant in the form of pleural fluid (to minimise friction between the lungs and

Synoptic link

You will find out more about transport of water via xylem vessels in Topic 6.3, Structure of vascular tissue in plants.

thoracic rib cage) and mucus (e.g., to allow the passage of faeces down the colon). Cohesion and adhesion of water molecules is also critical to transport in xylem vessels in plants.

Oral rehydration solutions

The water contained within our bodies and the balance of ions dissolved in it can be seriously affected by diseases. It is estimated that Cholera (a bacterial disease) causes around 120 000 deaths per year. The bacteria which causes cholera (*Vibrio cholerae*) is transmitted between infected and non-infected people through the ingestion of water or food contaminated with faeces containing the bacteria.

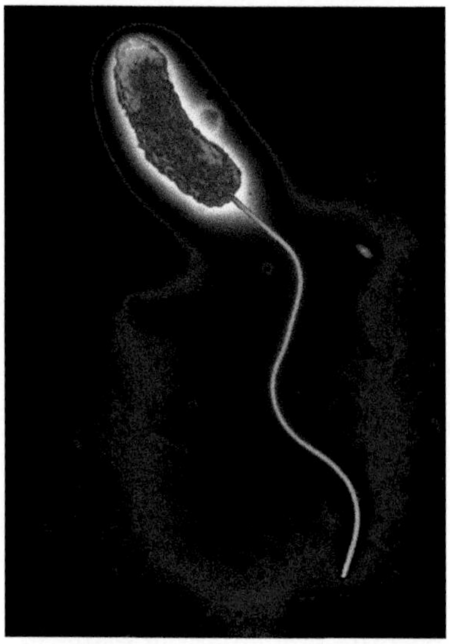

▲ **Figure 3** *Coloured transmission electron micrograph of a single* Vibrio cholerae *bacterium, ×10 000 magnification*

The cholera toxin is composed of six protein subunits – a single copy of the A subunit and five copies of the B subunit. The B subunit binds to the surface of target cells – enterocytes. Enterocytes are simple columnar epithelial cells found in the small intestines, colon, and appendix, which possess projections called microvilli on their surface. Once bound to the enterocyte, the entire cholera toxin complex enters the cell by endocytosis.

The toxin is transported to the Golgi apparatus, where the A subunit is then unfolded. The A subunit causes an increase in adenylate cyclase enzyme activity, which increases the intracellular concentration of the molecule cAMP to more than 100 times the normal level. This leads to the phosphorylation of the cystic fibrosis transmembrane conductance regulator (CFTR) chloride channel proteins. As a consequence, chloride ions leave the cell resulting in the secretion of H_2O, Na^+, K^+, and HCO_3^- into the intestinal lumen. The absorption of Na^+ and water into enterocytes is also decreased. The combined effect is the rapid loss of fluids from the intestine. In serious cases up to two litres per hour can be lost, leading to severe dehydration and other symptoms such as diarrhoea.

The symptoms of cholera can lead to low electrolyte concentrations in the blood within hours. This is called hyponatraemia and it can be life threatening. It is therefore vital that the sodium ions and water are replaced. However, the use of drinking water alone is not fully effective because water is lost from the cytoplasm of cells and is not being properly absorbed in the large intestine. Also, drinking water does not replace the ions that have been lost from the epithelial cells of the intestine.

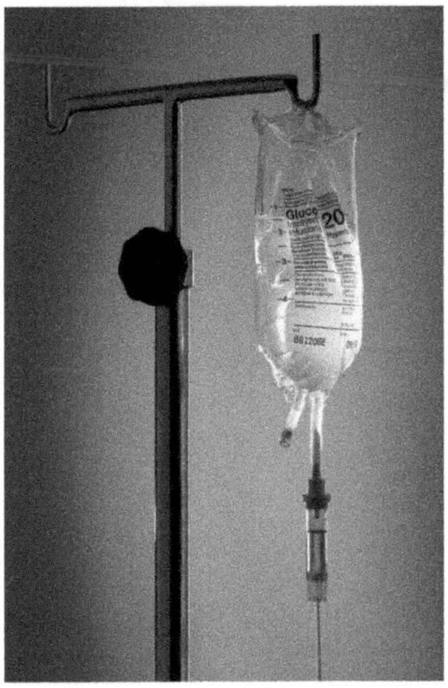

▲ **Figure 4** *A bag containing rehydration solution*

Oral rehydration solutions contain water, sodium chloride, glucose, potassium, and other electrolytes, such as chloride and citrate ions. Commercially available preparations can be bought over the counter but it is also possible to prepare a homemade solution. This can be made from eight level teaspoons of sugar and one level teaspoon of table salt dissolved in one litre of boiled drinking water.

1. Suggest why ingesting the bacterium *Vibrio cholerae* does not always result in the person developing cholera.
2. Using the information in Figure 3 suggest how the bacterium is able to pass from the stomach to the small intestine.
3. Suggest why enterocytes possess microvilli.
4. Suggest why, in countries where cholera is common, mothers are encouraged to breast feed their babies.
5. Suggest why the following are added to oral rehydration solutions:
   a. glucose
   b. sodium chloride
   c. potassium ions.
6. In Rwanda, a charity supplied the sports drink Gatorade to refugees after civil war broke out in April 2004 and was accused of making refugees with diarrhoea worse. The president of the charity, AmeriCares, defended the decision by saying, 'We stand by our decision to ship Gatorade to Rwandan refugees. In the absence of potable water, Gatorade, with its electrolytes and water, saved countless lives in a true triage situation.' Suggest why sports drinks such as Gatorade might not be suitable to rehydrate people with diarrhoea.

Synoptic link

You will learn more about how pathogens cause disease in Topic 11.1, Communicable diseases.

Summary questions

1. Explain why water is described as a polar molecule. *(1 mark)*
2. Is the interaction between water molecules and the wall of the xylem an example of cohesion or adhesion? Explain your answer. *(2 marks)*
3. Suggest why the protein collagen has a very high number of hydrogen bonds within its structure. *(2 marks)*

2.2 Mammalian and plant biofluids

Specification references: 2.1.2

Learning outcomes

Demonstrate knowledge, understanding, and application of:

→ the importance of water as:
- a major constituent of cells and body fluids
- the essential transport medium in plants and animals.

→ different types of mammalian body fluid:
- blood plasma
- serum
- tissue fluid
- lymph
- urine.

The importance of water

Water is the main constituent of every living cell, accounting for approximately 90% of the cytosol in some cells. The term cytosol refers to the liquid component of the cytoplasm in an intact cell, excluding any part that is contained within organelles. The results of research examining how water affects cell functions using brine shrimp showed that a 20% reduction in the water content of cells can significantly inhibit cell metabolism. As the cell continues to lose more water metabolic activity continues to decrease, and stops completely when the water content has been reduced by more than 70%.

Mammalian body fluids

Liquids produced within the body are called biofluids (or body fluids). These include fluids that are secreted or excreted from the body, for example – bile, semen, breast milk, amniotic fluid, and cerebrospinal fluid.

Body fluids can be placed into two groups:

- Intracellular fluids (e.g., cytosol) which are those that are found inside cells
- Extracellular fluids (e.g., **plasma**, **tissue fluid**) which are those that are found outside cells.

▼ **Table 1** *The amount and volume of body fluids found in the human body*

Body fluid	Amount of body weight (%)	Volume (litres)
total body fluid	60.0	42.0
intracellular fluid	40.0	28.0
extracellular fluid (ECF)	20.0	14.0
interstitial fluid (tissue fluid)	66.7 (of ECF)	9.4
plasma	33.3 (of ECF)	4.6

The main three body fluids, in terms of volume, are blood plasma, tissue fluid, and **lymph**.

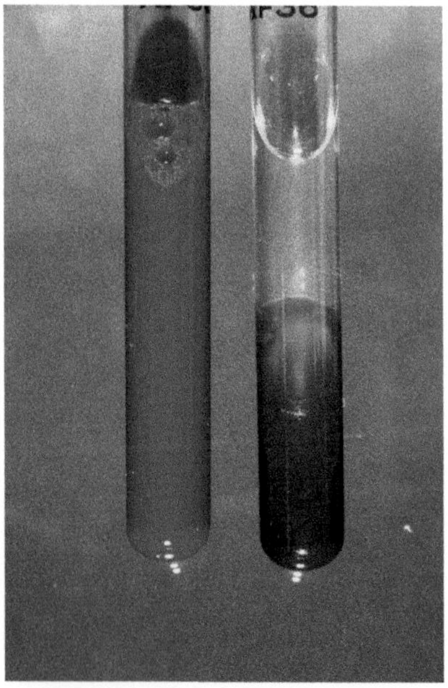

▲ **Figure 1** *Test tubes containing whole blood (left) and plasma-separated blood (right)*

Blood plasma

Approximately 55% of the blood is a liquid – the remaining 45% is made up of cells. The pale yellow liquid is called plasma. Plasma contains:

- plasma proteins (e.g., albumin, antibodies, clotting factors)
- absorbed nutrients (e.g., glucose, amino acids, fatty acids)

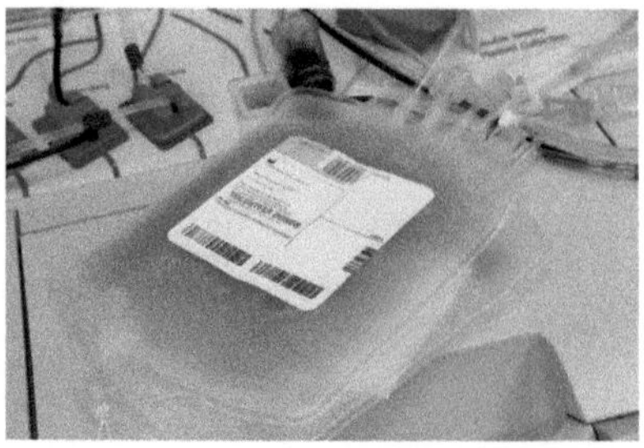

▲ **Figure 2** *A transfusion bag containing fresh, frozen blood plasma*

- excretory waste products (e.g., carbon dioxide, urea)
- hormones (e.g., insulin, adrenaline)
- electrolytes (e.g., sodium ions, chloride ions)
- heat.

Tissue fluid

Tissue fluid is formed as a result of hydrostatic pressure (the force created by the heart pumping blood from the left ventricle around the body in the arteries, arterioles, and capillaries) acting at the ateriole end of the capillaries. This pressure forces fluid out of the plasma into the spaces between the cells. The capillary walls are permeable to most components of the blood except most blood cells and large plasma proteins.

Synoptic link

You will be looking at the mechanism of tissue fluid formation in more detail in Topic 6.2, Blood pressure and tissue fluid.

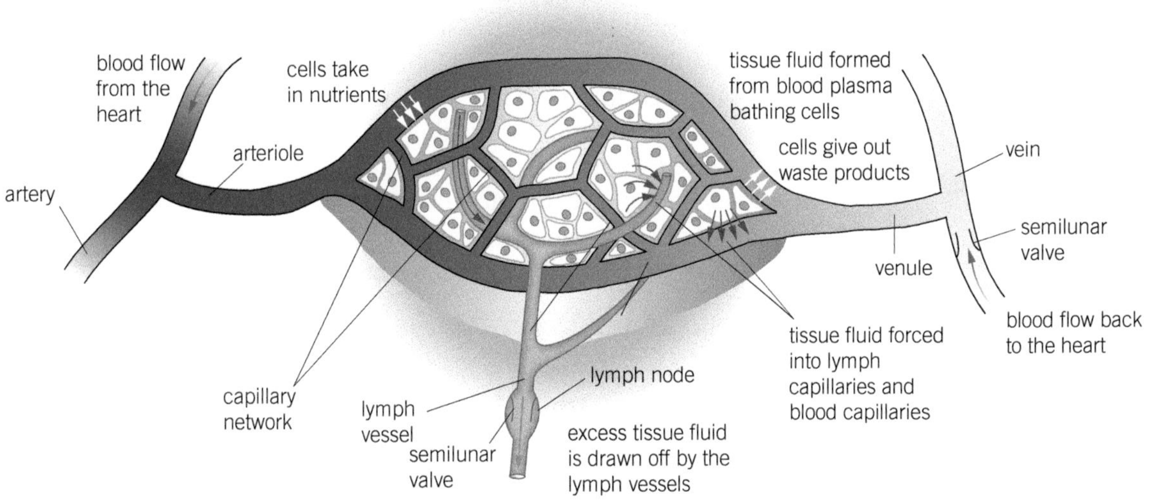

▲ **Figure 3** *Formation and return of tissue fluid*

The formation of tissue fluid enables capillaries to exchange substances with every cell in the body. Tissue fluid returns to the bloodstream via the capillaries and the lymphatic system.

Lymph

Approximately one tenth of the tissue fluid drains into blunt-ended lymphatic capillaries. These drain into larger lymph vessels that contain one-way semilunar valves. Once inside the lymph vessels the fluid is called lymph. Lymph drains back into the bloodstream in the subclavian veins (just under the collar bone). Here the lymph is mixed with the blood before joining the large vena cava vein and entering the heart. Lymph is moved by the contraction of skeletal muscles and the hydrostatic pressure of the tissue fluid that has left the capillaries.

Lymph nodes are found at intervals along the lymph vessels. These are the site of lymphocyte multiplication and storage. These white blood cells play a vital role in the immune system by producing antibodies.

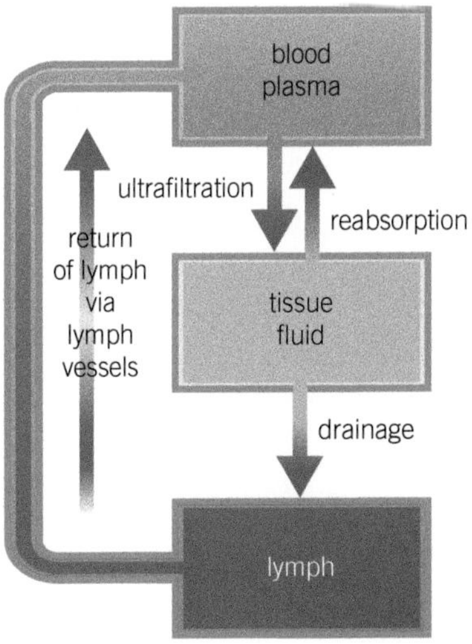

▲ **Figure 4** *The relationship between plasma, tissue fluid and lymph*

Other mammalian body fluids

Urine

Many metabolic reactions produce waste products. Some of these products have a high nitrogenous content and must be removed from the bloodstream. In humans, soluble waste products and excess water, sugars, and ions are excreted mainly by the urinary system and through perspiration (sweat).

The urinary system consists of two kidneys each with a ureter leading to the bladder and a urethra. Urine, the end result of the processes of filtration and reabsorption, typically contains a high concentration of urea and other toxic substances.

Serum

Blood plasma which has had the clotting factors removed from it is called **serum**. This body fluid contains electrolytes, antibodies, antigens, hormones, and other soluble proteins that are not involved in the coagulation (clotting) of blood. To obtain serum, whole blood is allowed to clot by leaving it at room temperature for 15–30 minutes. The clots and cells are then removed by centrifuging it for approximately 10 minutes. Following centrifugation the serum is immediately removed and stored at 2–8 °C. Serum is used in blood typing and also in several diagnostic tests, includng testing for immunoglobulin G (IgG) *Toxoplasma* antibodies.

Fluid retention in the body

Fluid retention in the body is known medically as Oedema (or dropsy). The build-up of fluid causes the affected tissues to become swollen. Oedema most commonly occurs in the feet and ankles (known as peripheral oedema). Oedema can occur in many areas of the body, for example, the brain (cerebral oedema), the lungs (pulmonary oedema), and the eyes (macular oedema).

◀ **Figure 5** *This child's distended abdomen is a sign of malnutrition and lack of food, a condition known as kwashiorkor*

Kwashiorkor

Kwashiorkor is caused by severe protein-energy malnutrition. It is characterised by bilateral pitting oedema (affecting both sides of the body) in the lower legs and feet, which as it progresses becomes more generalised to the arms, hands, and face. It is classified as follows:

- Mild – affects both feet
- Moderate – affects both feet plus lower legs, hands, or lower arms
- Severe – generalised oedema affecting both legs, arms, hands, feet, and face

1 In the form of malnutrition called protein-energy malnutrition (PEM), protein in the diet is used as an energy source rather than as a source of amino acids. Explain why PEM results in oedema.

Synoptic link

Hint: you may need to read Topic 3.1, Amino acids – the building blocks of proteins, if you are having difficulty answering the question in the extension box.

Biofluids in plants

Biofluids in plants include intracellular fluids such as the cytosol or cell sap, the vacuole contents, and the contents of specialised cells such as phloem sieve tube elements, and xylem vessels. Fluid is present in and between plant cell walls. The fluid in the xylem is very dilute in comparison to phloem sap and cell cytoplasm. Unlike phloem sap, the contents of the xylem do not contain sucrose but otherwise the composition is similar and consists of ions and some organic molecules including plant hormones such as auxins.

Synoptic link

The main force which draws water up from the roots through the stem of the plant is the evaporation of water vapour from the underside of the leaves through pores called stomata. This process is called transpiration. You will find out more about it in Topic 6.4, Transport of water in plants.

Summary questions

1 a Describe the function of the three main body fluids. *(3 marks)*
b State three substances delivered to cells and one substance removed from cells by the tissue fluid. *(3 marks)*

2 A man has a total blood volume of five litres. Calculate his volume of:
a intracellular fluid *(1 mark)*
b tissue fluid *(1 mark)*
c plasma. *(1 mark)*

Show your working and give your answers to one decimal place.

3 Compare the composition of the three main body fluids by copying and completing the table. *(5 marks)*

Component	Blood	Blood plasma	Tissue fluid	Lymph
Cells present				
Location				
Moved by				
Direction of flow				
Main components				

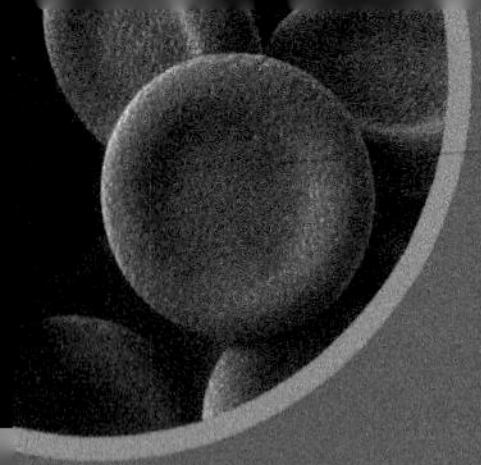

2.3 Carbohydrates in cells and biofluids

Specification references: 2.1.2

Learning outcomes

Demonstrate knowledge, understanding, and application of:

- → the concept of organic molecules
- → roles of carbohydrates
- → structure of monosaccharides
- → formation and breakdown of disaccharides
- → structure and function of polysaccharides.

Synoptic link

You will find out more about antigens in Topic 12.1, The immune system.

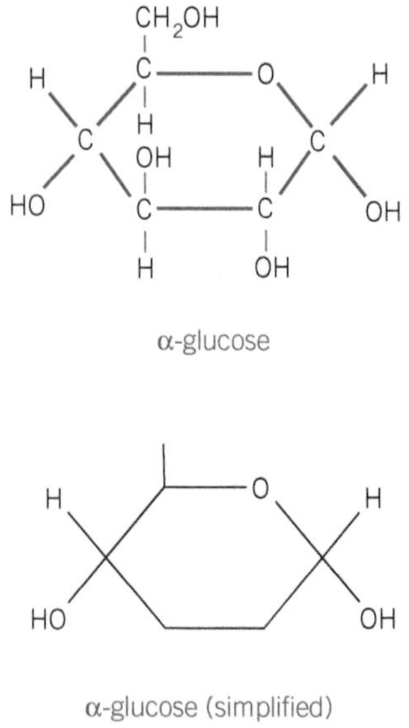

▲ **Figure 1** *The molecular arrangement of a glucose molecule*

Organic molecules

Living organisms have four main types of carbon compound: carbohydrates, proteins, lipids, and nucleic acids. Each of these types of compound is made up mainly of carbon, hydrogen, and oxygen, and a relatively small number of other elements in some cases. In addition, many biological molecules are **polymers** consisting of repeating units (monomers) joined in a number of ways.

Carbohydrates

Glucose is an example of a simple carbohydrate. Other more complex carbohydrates include starch, glycogen, and cellulose. Carbohydrates have many different roles in the body including:

- energy source (e.g., glucose for respiration)
- energy store (e.g., glycogen in liver and muscle cells and starch in plant cells)
- cell markers (e.g., attached to proteins to form glycoproteins which act as surface antigens).

Simple carbohydrates

Simple sugars are often referred to as monosaccharides. Glucose is an example of a monosaccharide. Most monosaccharides are small, soluble molecules that are sweet tasting. They have the general formula $C_x(H_2O)_y$.

Glucose is a hexose sugar as it contains six carbon atoms. All hexose sugars have the same formula $C_6H_{12}O_6$ but the atoms can be arranged in different ways. In Figure 1 the structure of α-glucose is shown.

Joining monosaccharides

Monosaccharides can be joined together to form **disaccharides**. Different disaccharides are made from different monosaccharides:

α-glucose + α-glucose = maltose

α-glucose + galactose = lactose

α-glucose + fructose = sucrose

When two monosaccharides are joined together a molecule of water is removed during the reaction and the two molecules are joined to give a single molecule joined by a covalent bond. This type of reaction is called a **condensation reaction**. The condensation reaction involves the removal of an −OH from one molecule and an −H from the other molecule. A covalent bond is formed between the two monosaccharides called a **glycosidic bond**.

The reaction can be reversed by the addition of a water molecule to split the glycosidic bond and break the disaccharide into two monosaccharides. This is called a hydrolysis reaction.

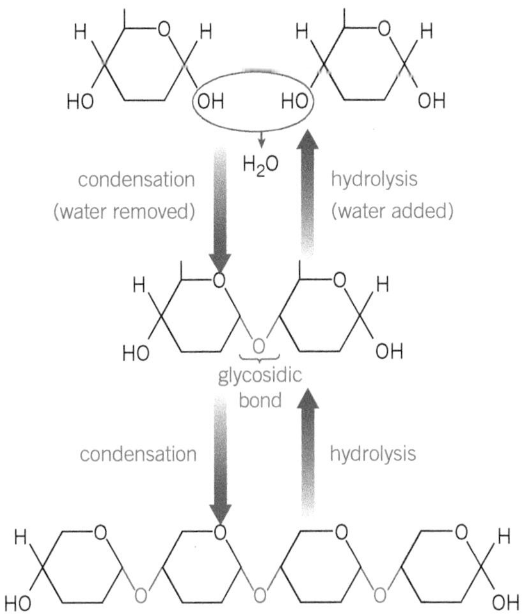

▲ **Figure 2** *Condensation and hydrolysis reactions between monosaccharides and disaccharides*

Condensation and hydrolysis reactions are vital in the formation and breakdown of compounds in cellular metabolism. The sum of all the biochemical reactions in the cell is collectively referred to as cellular metabolism. It includes all the reactions involved in synthesising and breaking down macromolecules, and those involved in generating small precursor molecules, such as amino acids for cellular requirements. The sequence of biochemical reactions which take place in metabolism are known as pathways.

Disaccharides

Disaccharides are more suitable for transport and storage than monosaccharides. The disaccharide sucrose is the form in which carbohydrates are transported in the **phloem** tubes of plants. It is also a storage carbohydrate in many plants, including sugar cane and sugar beet from which we obtain sugar.

Lactose is a disaccharide that is found in milk. It is formed from the condensation of α-glucose and galactose, resulting in a β-glycosidic bond.

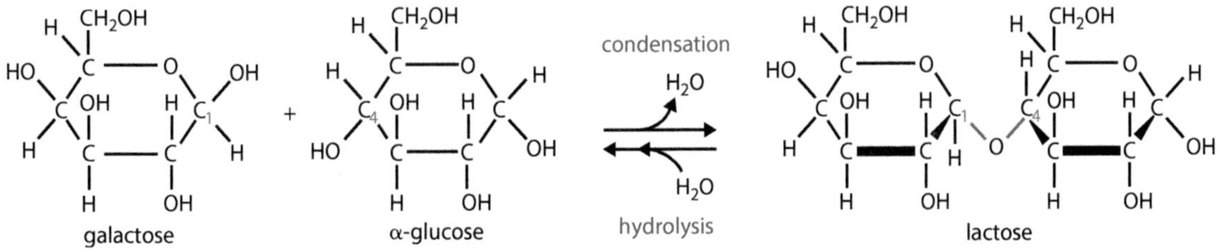

▲ **Figure 3** *The formation and breakdown of lactose,* $C_6H_{12}O_6 + C_6H_{12}O_6 \rightleftharpoons C_{12}H_{22}O_{11} + H_2O$

Study tip

Some tips for remembering the details of condensation and hydrolysis are:

Condensation – the process of joining molecules (think of condensing two objects into one) with the removal (release) of water.

Hydrolysis – 'hydro' means water and 'lysis' is to split – hence a hydrolysis reaction uses the addition of water to split a molecule by breaking a bond.

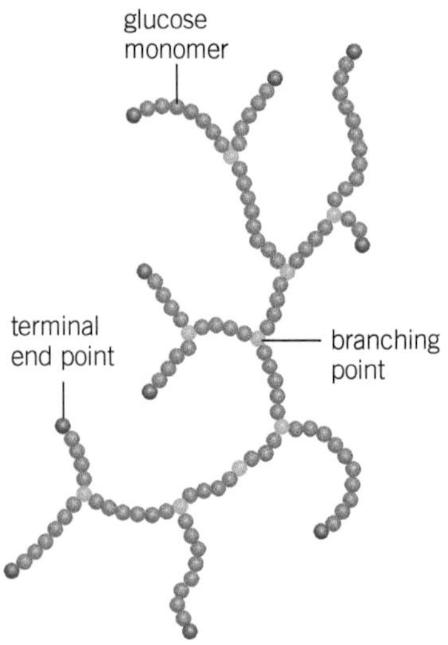

▲ **Figure 4** *Glycogen is made from many glucose molecules (green) linked by one of two types of glycosidic bonds. The less common of the two bonds, α-1,6-glycosidic bonds (shown in blue), cause the branching in the structure of the molecule*

Milk is rich in lactose and is a vital energy source, particularly for nursing infants. Our intestinal villi secrete the enzyme called lactase (β-D-galactosidase) to digest (hydrolyse) it. This enzyme cleaves the lactose molecule into its two monosaccharides, glucose and galactose, which can be absorbed.

Polysaccharides

Polysaccharides are polymers that are formed from many repeated condensation reactions joining many monosaccharides (**monomers**) together. These long chains of monosaccharides are held together by glycosidic bonds and can be branched or unbranched. They are large, insoluble molecules that are ideal storage molecules.

Glycogen is a storage polysaccharide found in animal cells (specifically the liver and skeletal muscle cells), fungal cells, and many prokaryotes but not in plant cells. Glycogen is a highly branched molecule, which means it has many terminal end points. This means that a large number of single glucose molecules can be hydrolysed (one from each terminal end point) at any one moment. During prolonged exercise, blood glucose levels can fall so liver glycogen is quickly hydrolysed to glucose to raise the glucose levels back up again.

Starch is a polysaccharide made in plant cells. Starch is made through the joining together of α-glucose molecules by condensation reactions. There are two forms of starch: amylose and amylopectin, which are compared in the table below. Both types of starch are hydrophilic but they are too large to be soluble in water, which makes them useful as energy storage compounds. Starch is used to store glucose (and hence energy) in seeds and storage organs such as tubers. Starch is also used as a temporary energy store in leaves when glucose is being made faster by photosynthesis than it can be used or moved to other parts of the plant.

▼ **Table 1** *A comparison of amylose and amylopectin*

		Amylose	Amylopectin
Similarities	monomer	α-glucose	
	type of reaction used to join monomers	condensation	
	type of bond formed between monomers	covalent	
	name of bond formed between monomers	glycosidic	
Differences	number of glucose monomers	300 to 3000	2000 to 200 000
	shape	linear	highly branched Branching takes place with α(1 → 6) bonds occurring every 24 to 30 glucose units
	properties in water	insoluble	soluble
	ease of digestion	less easily digested	easily digested
	space	more compact	less compact – making it easier to hydrolyse
	component of starch	approximately 30%	approximately 70%

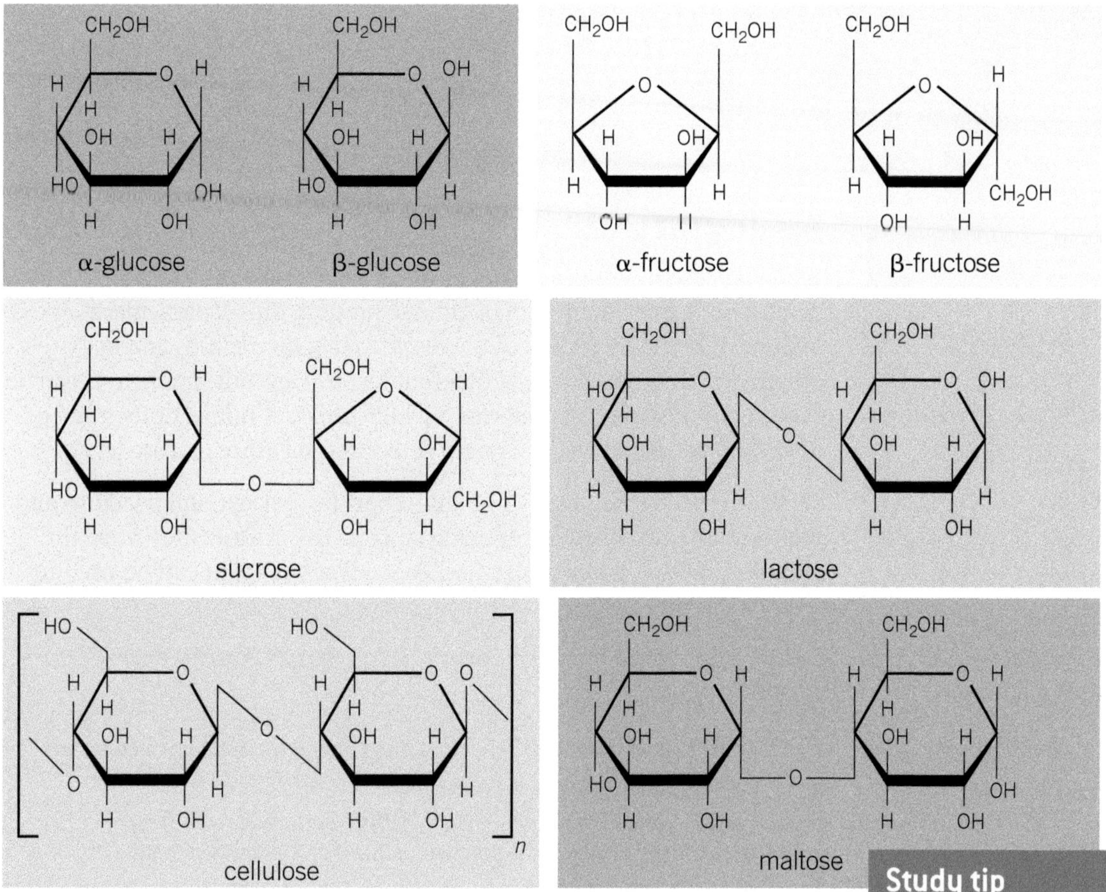

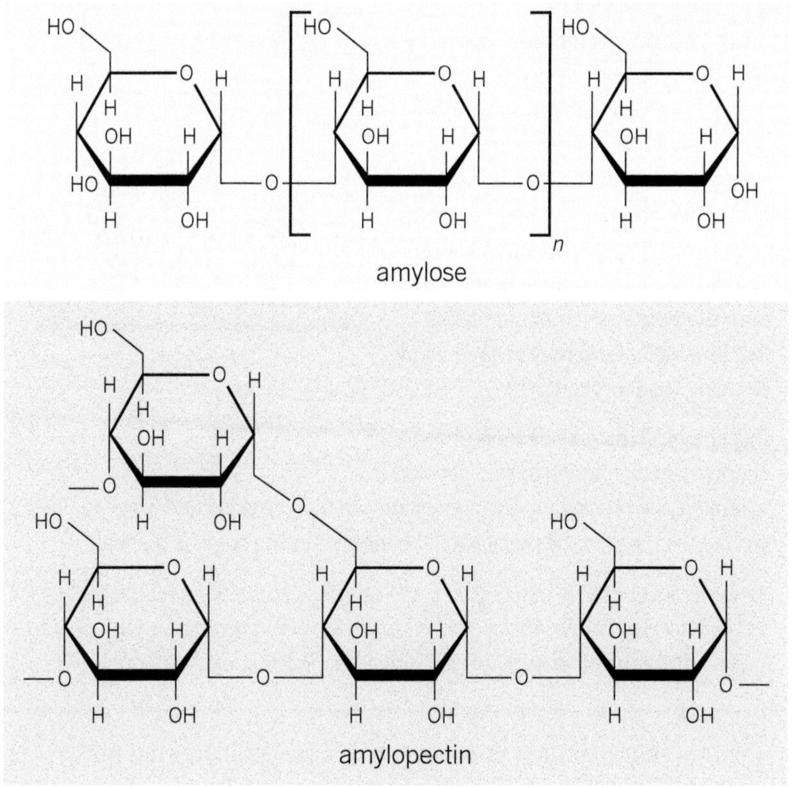

▲ **Figure 5** *The structures of different carbohydrates*

Study tip

Make sure that you can draw the ring structure of α-glucose correctly. The hydroxyl group on ^1C is below the ring. Remember: ABBA = alpha below, beta above (when considering the position of the —OH group on ^1C).

Summary questions

1 State two examples of monosaccharides other than glucose. *(2 marks)*

2 Maltose is a disaccharide made from the condensation of two α-glucose molecules.
   a Draw a diagram to show how this occurs. *(3 marks)*
   b Explain why a condensation reaction is described as an anabolic reaction. *(2 marks)*

3 Explain how the properties of glycogen make it an ideal storage molecule. *(6 marks)*

2.4 Detecting and measuring carbohydrates

Specification references: 2.1.2

Learning outcomes

Demonstrate knowledge, understanding, and application of:

→ different techniques to detect and measure glucose levels:
- Benedict's test
- biosensors
- colorimeters
- the importance of measuring blood glucose
- how to detect and measure starch levels.

Testing for glucose

Blood glucose monitoring is a way of testing blood and urine to determine the concentration of glucose present. This is particularly important in the treatment of the disease diabetes mellitus and in pregnancy. Different tests use different technology, but simple tests can be carried out in school laboratories, by individuals at home, or by trained medical practitioners to accurately detect and measure glucose levels.

All monosaccharides and some disaccharides, for example, lactose and maltose, are described as reducing sugars. This means they have the ability to reduce other molecules. **Benedict's** reagent can be used to test for the presence of a **reducing sugar**.

Benedict's test

1. Add an equal volume of Benedict's reagent and the solution to be tested for the presence of reducing sugars.
2. Heat in a boiling water bath for 5–10 minutes.
3. Observe the colour of the precipitate formed to estimate the quantity of reducing sugar present.

▲ **Figure 1** *Water baths with two test tubes containing glucose and Benedict's solution. The water bath on the right is warmer and the colour change is faster*

The reducing sugar reduces copper(II) ions in the Benedict's reagent to copper(I), which then forms a brick red copper(I) oxide precipitate. The colour of the solution changes from green through to yellow, orange, brown, and finally brick red according to the quantity of reducing sugar present. This is an example of a semi-quantitative test as it indicates a relative amount of reducing sugar present.

This chemical test can be made more quantitative by filtering the suspension to remove the precipitate. The precipitate can then be dried and weighed. The greater the mass of the precipitate, the more reducing sugar is present.

A Benedict's test carried out on a urine sample indicated a high level of reducing sugar present. A follow-up blood test showed normal concentrations of blood glucose. Suggest a possible explanation.

Some disaccharides such as sucrose (the sugar transported in the phloem in plants) are **non-reducing sugars**. There is no direct test to determine if a non-reducing sugar is present. First a reducing sugar test is carried out. If this gives a negative result *then* it can be tested for the presence of a non-reducing sugar.

Testing for a non-reducing sugar

A solution that has tested negative for the presence of a reducing sugar should then be tested for the presence of a non-reducing sugar.

1 First hydrolyse the sugar with an equal volume of dilute hydrochloric acid in a gently boiling water bath for five minutes.
2 Slowly add sodium hydrogencarbonate solution to neutralise the acid. Test the solution with pH paper to determine when the solution has reached neutral pH.
3 Re-test the solution in a water bath with Benedict's reagent in a boiling water bath for five minutes. If a non-reducing sugar is present it will now turn yellow/orange/brown/brick red.

Why is it essential to neutralise the acid prior to adding the Benedict's solution?

Study tip

Remember that Benedict's test does not just detect the presence of glucose – it will give a positive result in the presence of any reducing sugar.

Study tip

A reducing sugar can donate electrons to Cu^{2+} ions from Cu^{+} ions. The more reducing sugar there is present, the greater the number of electrons that can be donated and the more Cu^{2+} that can be reduced. Remember OILRIG – Oxidation Is Lost, Reduction Is Gain

Biosensors

A **biosensor** is an analytical device that is used to detect the presence of a chemical by combining a biological component. If the substance binds, a colour change or electrical change is produced.

Clinistrips

Commercially available test strips can be used to test glucose levels, such as Clinistix or Diastix. These are clinical sticks for the specific detection of glucose. They contain the enzyme glucose oxidase, which has been attached to the paper pad at the end of the test strip. This is an example of a biosensor-based test for glucose. Any glucose in the urine is oxidised to gluconic acid and hydrogen peroxide. The enzyme peroxidase is also present on the pad. It breaks down the hydrogen peroxide to oxygen and water. The oxygen then oxidises a colour dye on the test pad. The intensity of the colour change on the pad reflects the amount of glucose present in the urine.

Synoptic link

You will find out more about enzymes in Topic 3.4, The structure of enzymes.

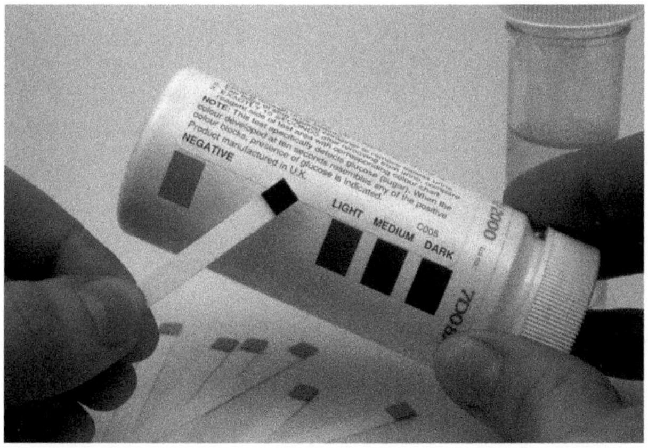

▲ **Figure 2** *The Clinistix test. The dark colouration on the reagent strip indicates a positive result. Shown on the jar, a light pink colour reaction indicates a negative (healthy) result. In diabetic patients, levels of glucose in the urine and blood are abnormally raised. This is because there is insufficient production of the hormone insulin, which allows the body to store and metabolise glucose*

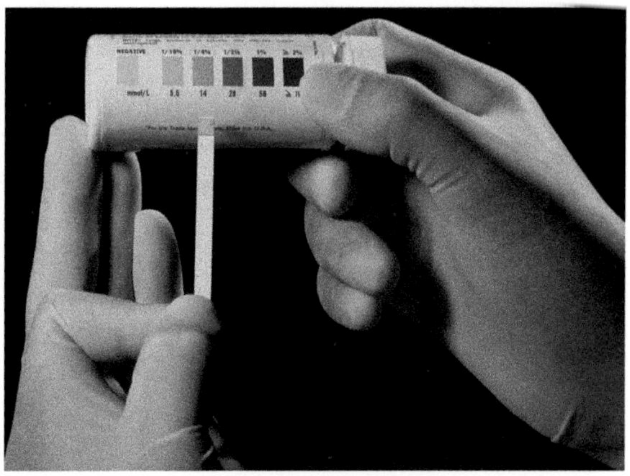

▲ **Figure 3** *Diastix test for glucose conducted on the urine of a diabetic patient. A negative result for glucose is seen. The patient's blood glucose is well controlled as no glucose is excreted in the urine. The levels of glucose in the urine help to establish how severe the diabetes is, and the type of treatment that is necessary*

Glucometers

Test strips can be impregnated with the immobilised enzyme glucose dehydrogenase. The enzyme converts glucose in blood to gluconolactone and produces a small current which is detected by an electrode (transducer) on the test strip. The current is read by a meter that produces a reading for blood glucose concentration in approximately 25–30 seconds. Normal blood glucose levels are between 3.89 and 5.83 mmol dm^{-3}.

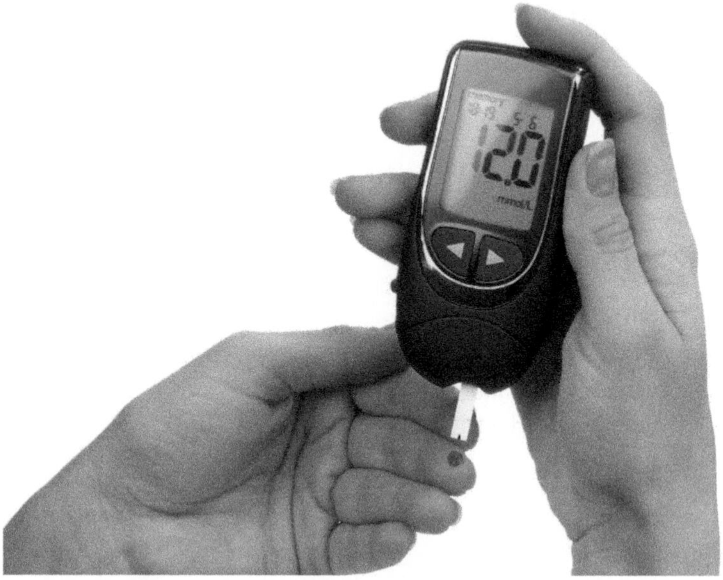

▲ **Figure 4** *A blood glucose measurement being performed by a diabetic. A drop of blood has been pricked from a finger and smeared on the test strip. The blood is analysed by the glucometer and the blood glucose level is displayed on the digital screen. In this case, the reading is 12.0 mmol dm^{-3}. This is a very high raised level of glucose*

Colorimeters

A **colorimeter** is a device that shines a beam of light through a solution and then measures the absorbance or transmission of particular wavelengths of light. A photoelectric cell detects the light that has passed through the solution and produces a reading to show either how much light has been absorbed by the solution or what percentage of the light has been transmitted through the solution.

Using a colorimeter

1. The colorimeter is first zeroed by placing a cuvette containing an appropriate 'blank' sample, for example, water.
2. A colour filter is used to give more accurate results, for example, if the colorimeter is being used to test for the presence of reducing sugar with Benedict's reagent then a red filter is used.
3. A **calibration curve** is determined. A Benedict's test is carried out on a series of solutions with known glucose concentrations, called glucose standards. The precipitate is filtered out of the solution and a colorimeter

reading is obtained for each filtrate. The readings are then plotted on a graph to show transmission (or absorption) against glucose concentration.

4 The solution of unknown glucose concentration is then tested with Benedict's reagent, filtered and the filtrate read using a colorimeter.

5 The transmission or absorption value is then used to find the concentration of glucose from the calibration curve, as shown in Figure 6.

This technique is known as an assay and it gives quantitative results.

▲ **Figure 5** *A student using a colorimeter to measure the hue of a sample of copper sulfate ($CuSO_4$)*

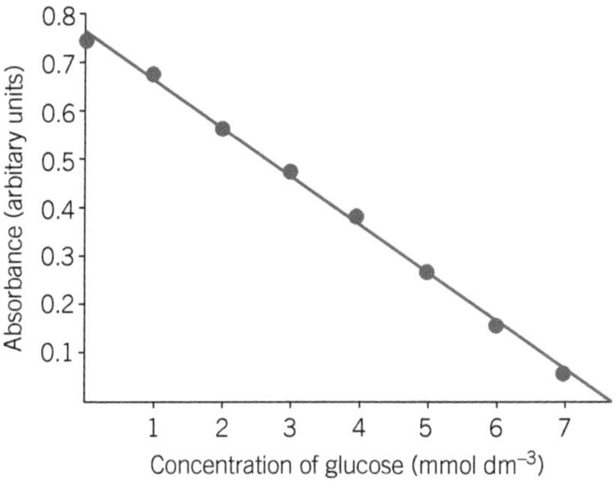

▲ **Figure 6** *A colorimeter calibration curve*

Why is it important to use the correct filter?

Testing for starch

Starch is easily detected by the addition of an iodine–KI reagent to a sample in a test tube or spotting tile at room temperature. In the absence of starch the iodine–KI reagent remains yellow-brown, whereas in the presence of starch it turns blue-black. Amylose will react to form a deep blue-black whereas amylopectin will produce a red-purple colour.

Iodine molecules can easily fit into the hole in the middle of the amylose helix and this produces the blue-black colour of the iodine–starch complex. This is a **qualitative** test as the results only indicate the presence or absence of starch, not the amount of starch present. Colorimetry using standard starch solutions to generate a calibration curve can be used to determine a quantitative result.

Summary questions

1 Suggest why it is not possible to distinguish between very high concentrations of reducing sugar in different samples even if their absolute concentrations are different. *(2 marks)*

2 Suggest three advantages of using the Diastix test strips (Figure 3) rather than the Clinistix strips (Figure 2). *(3 marks)*

3 a Explain why it is important that the cuvette used in a colorimeter is transparent and handled carefully. *(2 marks)*

b Sketch a colorimeter calibration curve for glucose showing:
i absorption
ii transmission. *(2 marks)*

2.5 Detecting and measuring proteins

Specification references: 2.1.2

Learning outcomes

Demonstrate knowledge, understanding, and application of:

- → why it is necessary to test for the presence of proteins
- → the biuret test
- → how proteins can be detected and measured in the blood.

▲ **Figure 1** *A urine sample*

Synoptic link

You will learn more about proteins, including their structure and functions in Topic 3.3, Protein structure – haemoglobin.

Proteins in body fluids

Protein in the urine can be an indication of kidney disease. Urine which has an abnormally high level of protein is called proteinuria. Research has shown that the level of kidney damage can be determined by both the level of proteinuria and the type of proteinuria, for example, if the urinary proteins are albumin only (a condition known as albuminuria), or other if other proteins are present as well. Small levels of proteinuria or albuminuria can increase the risk of developing heart and blood vessel disease.

Small proteins can be filtered through the kidney's glomerulus and some proteins are also produced in the genitourinary tract. In a healthy kidney total protein excretion is not normally higher than 150 mg in a 24 hour period or 10 mg per 100 ml in any one urine sample. If the level is higher than 150 mg in a 24 hour period the patient is diagnosed with proteinuria. If the levels rise higher than 3.5 g in a 24 hour period this is severe and the patient is diagnosed with nephrotic syndrome.

The biuret test

The biuret test can be used to detect the presence of peptide bonds and therefore the presence of proteins in a solution.

1. Add a sample of the solution to be tested to the same volume of sodium hydroxide.
2. Add a few drops of dilute (0.05%) copper sulfate solution and mix gently.
3. A purple colour indicates the presence of protein in the solution.

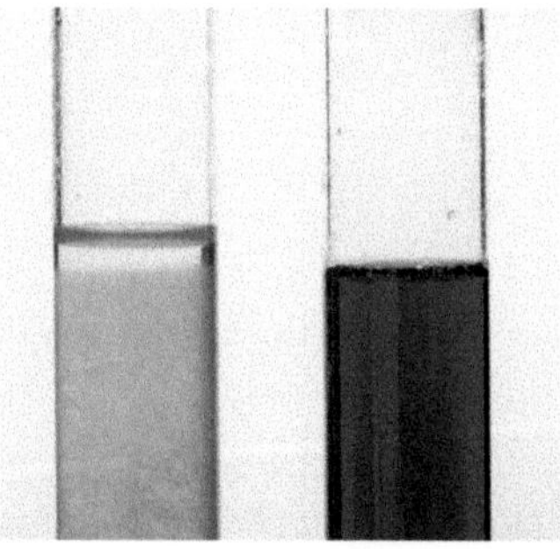

▲ **Figure 2** *Test tubes containing negative (left) and positive (right) tests for proteins using biuret reagent. The biuret reagent comprises sodium hydroxide and copper(II) sulfate. In the alkaline conditions provided by the sodium hydroxide, the peptide bonds form a complex with the copper(II) ions, giving the solution a deep violet colour. The more purple the solution, the more protein is present*

Suggest what colour you would expect to see if the sample contains amino acids instead of proteins.

Urine test strips

Chemical test strips can be used to detect the presence of proteins in urine. The test for proteins on a urine test strip relies on a visible colorimetric reaction (i.e., proteins alter the colour of some chemical dyes). In some test strips the protein test pad contains the pH indicator tetrabromophenol blue and a buffer to maintain the pH at a constant level. The indicator is yellow in the absence of proteins, but as the protein concentration increases the indicator colour changes through various shades of green until it becomes a dark blue.

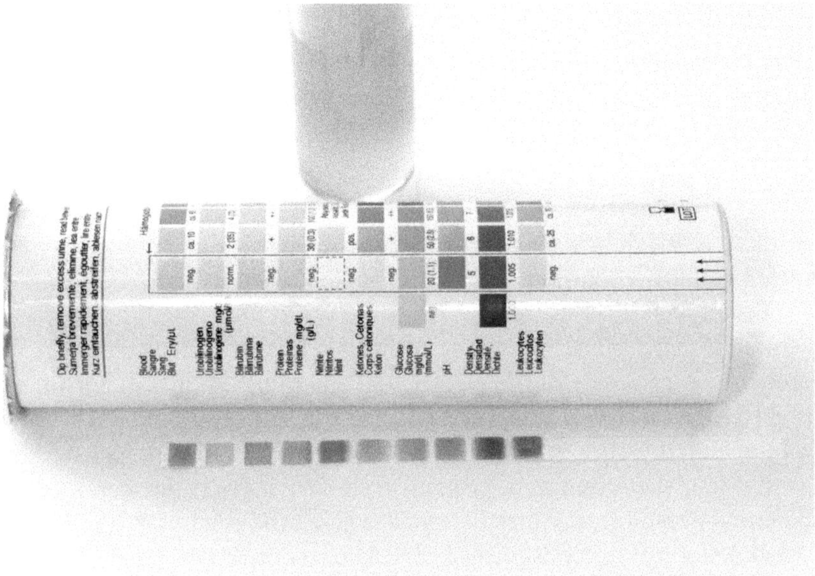

▲ **Figure 3** *Dipstick test for urine to detect the presence of protein as well as other substances such as blood, glucose, and nitrates*

Synoptic link

You will learn more about pH in Topic 3.5, Factors affecting enzyme controlled reactions.

Summary questions

1 Explain why *and* how the biuret reaction can be used to assess the concentration of proteins. *(4 marks)*

2 Suggest why only low concentrations of the main serum protein albumin are normally found in urine. *(2 marks)*

3 Suggest why very alkaline urine can affect the accuracy of the protein test strips. *(2 marks)*

2.6 Osmosis in cells

Specification references: 2.1.2

Learning outcomes

Demonstrate knowledge, understanding, and application of:

- → the process of osmosis
- → osmosis in animal cells
- → osmosis in plant cells
- → the effect of solutes and electrolytes on the water potential of body fluids.

Osmosis

Osmosis is a specialised form of diffusion. Water molecules are present on either side of cell surface membranes as the cytoplasm and the extracellular fluid are mainly water. Water molecules are small and move randomly and even though they are polar they can pass through tiny temporary gaps that appear between the phospholipid tails due to their small size.

Osmosis is defined as the passage of water molecules from an area of higher water potential to an area of lower water potential through a partially permeable membrane.

Water potential

A solute is any substance that can be dissolved in a solvent (e.g., water) to form a solution. The concentration of solute molecules in a solution will affect the tendency for the water molecules to move across the cell surface membrane. However, it is also affected by the pressure on each side of the membrane. The term **water potential** is used to describe the combined effect of the 'concentration of the solute' and 'pressure in a solution'. The presence of solutes such as sugars or the presence of proteins (colloids) changes the water potential of biofluids. Water potential is given the symbol Ψ (the Greek letter psi) and it is measured in kilopascals (kPa).

There are some key points to remember about water potential:

- Under standard state temperature (25 °C) and pressure (100 kPa), pure water is defined as having a water potential of zero.
- The addition of a solute to pure water will lower the water potential. This is because the solute molecules impede the movement of water molecules, reducing the number of collisions the water molecules make with the membrane. Consequently, the solution exerts less pressure on the membrane and has a lower water potential.
- The water potential of a solution is always less than zero, that is, the water potential of a solution will have a negative value.
- The greater the amount of solute added to the solution, the more negative the water potential will become.
- Water will always move from an area of higher water potential (less negative) to an area of lower water potential (more negative).

The process of osmosis

In Figure 1, the solution on the left hand side of the membrane has a higher water potential because there is a lower number of solutes present in the solution compared to the solution on the right hand side of the membrane. Both the solute and water molecules will possess kinetic energy and therefore move randomly. The partially permeable membrane allows the water molecules (solvent) to cross it but not the solute molecules. The water molecules will therefore move by osmosis from the left hand side of the membrane to the right hand side of the membrane down the water potential gradient. When the water potentials on both sides of the membrane are equal, a dynamic equilibrium is established and there is no net movement of water molecules.

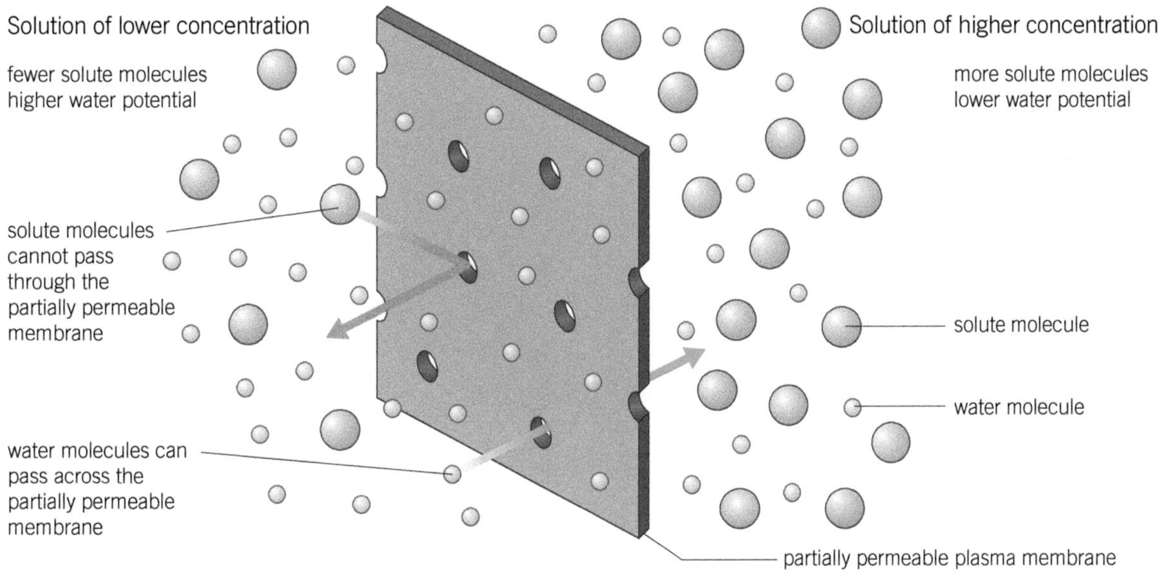

▲ **Figure 1** *Osmosis*

Osmosis in animal cells

Animal cells such as erythrocytes have a variety of solutes dissolved in their cytoplasm. The cell surface membrane is very thin and flexible but it is not able to withstand high pressure. This means when an animal cell is placed in a solution of pure water, water will move down the water potential gradient from the solution into the cytoplasm of the cell. This exerts a higher pressure on the cell surface membrane, which ruptures. In the context of erythrocytes this is known as haemolysis, but is more generally known as cytolysis. In some cases, such as when storing organs before transplantation, it is essential this cell and tissue damage does not occur. The organ is therefore kept in a solution that has the same water potential as the cells.

When an animal cell is placed in a solution of concentrated solute, water will still move down the water potential gradient, but this time it will move from the cytoplasm of the animal cell out of the cell into the surrounding external solution. This reduces the pressure on the cell surface membrane and the cell shrinks and shrivels. The cell is said to be crenated.

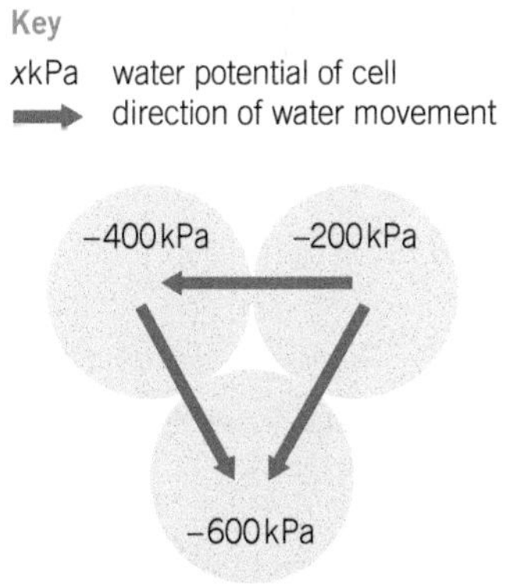

Water moves from higher water potential to lower water potential. The highest water potential is zero.

▲ **Figure 2** *Movement of water between cells down a water potential gradient*

▼ **Table 1** *Summary of osmosis in an animal cell*

Water potential (ψ) of external solution compared to cell solution	higher (less negative)	equal	lower (more negative)
Net movement of water	enters cell	neither enters nor leaves	leaves cell
State of cell	swells and bursts	no change	shrinks
Appearance of an example animal cell	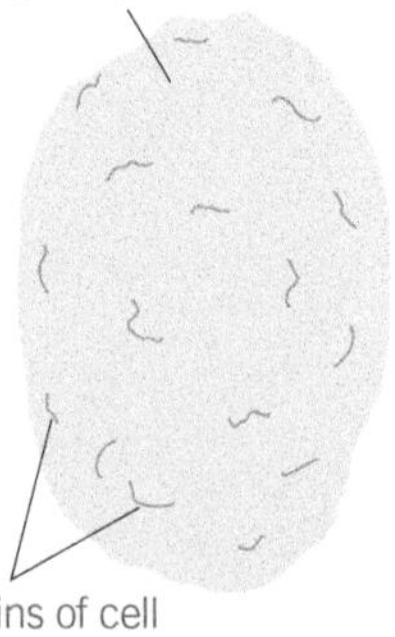	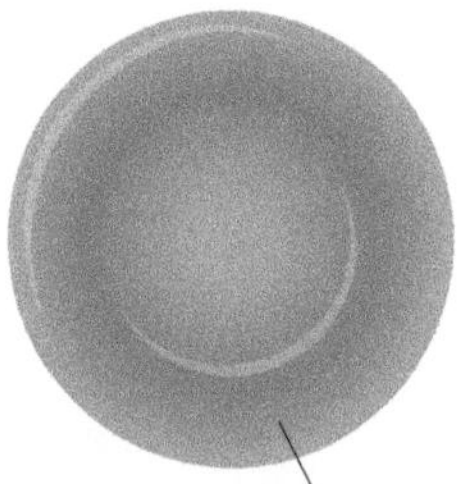	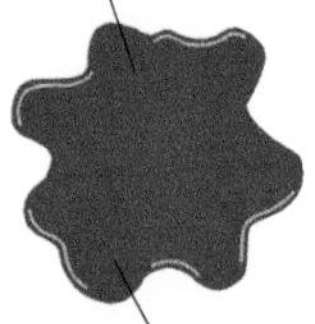

> **Synoptic link**
>
> You will need to understand the general structure of animal and plant cells. This is covered in Topic 1.6, Cell ultrastructure.

Water potential of the blood

The water potential of the blood is affected by solutes (e.g., plasma proteins, glucose) and electrolytes (e.g., H^+, K^+, Na^+, Cl^- and HCO_3^- ions). As the concentration of solutes or electrolytes increases in cells, the water potential of the plasma will decrease, which would cause water to leave the cells. The water potential of the blood must therefore be regulated to prevent loss or gain of water from cells.

Osmosis in plant cells

The process of osmosis in plants cells is affected by three structures unique to plant cells:

- the central vacuole – this contains salts, sugars, and organic acids in the solution
- the protoplast – this is made up from the cell surface membrane, cytoplasm, and tonoplast (the membrane around the vacuole)
- the cell wall – this is made from the polysaccharide cellulose and is tough and inelastic.

Plant cells do not rupture when they are placed in distilled water. As water enters the cell, the volume of the central vacuole and cytoplasm both increase. This causes the protoplast to exert a pressure on the cell wall. However, the strong cellulose cell wall can resist the pressure exerted on it and prevent further entry of water molecules. In this situation the plant cell is described as being turgid.

If the same plant cell is placed in a solution with a lower water potential, water will leave the cell by osmosis. This causes the volume of the cytoplasm and central vacuole to decrease. As water continues to leave the cell there is a point when the protoplast no longer presses on the cell wall. This stage is called incipient plasmolysis. As water continues to leave the cell and the protoplasm shrinks further the protoplast pulls away from the cell wall and the cell is described as plasmolysed.

▼ **Table 2** *Summary of osmosis in a plant cell*

Water potential (ψ) of external solution compared to cell solution	higher (less negative)	equal	lower (more negative)
Net movement of water	enters cell	neither enters nor leaves	leaves cell
Protoplast	swells	no change	shrinks
State of cell	turgid	incipient plasmolysis	plasmolysed
	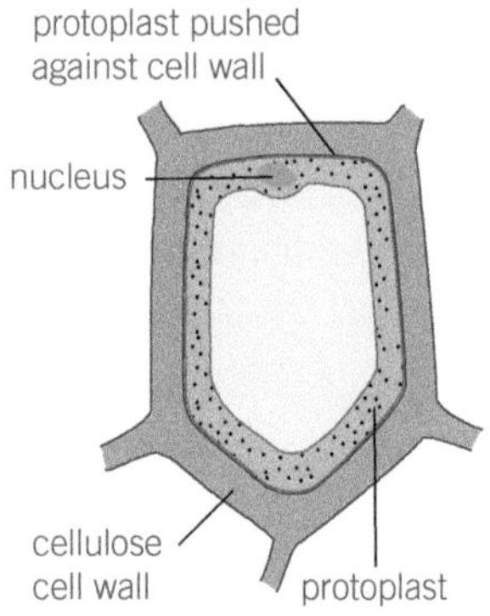	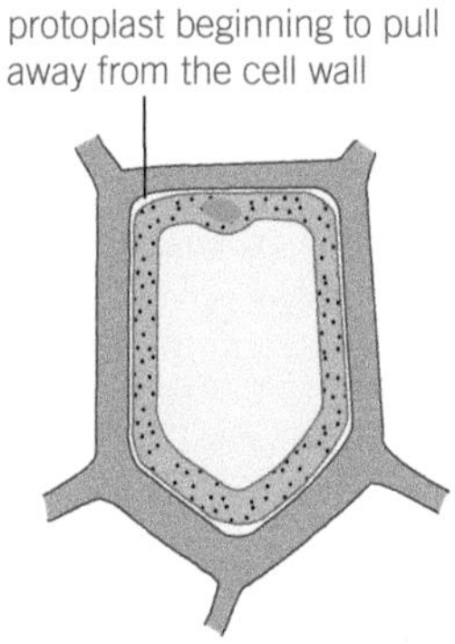	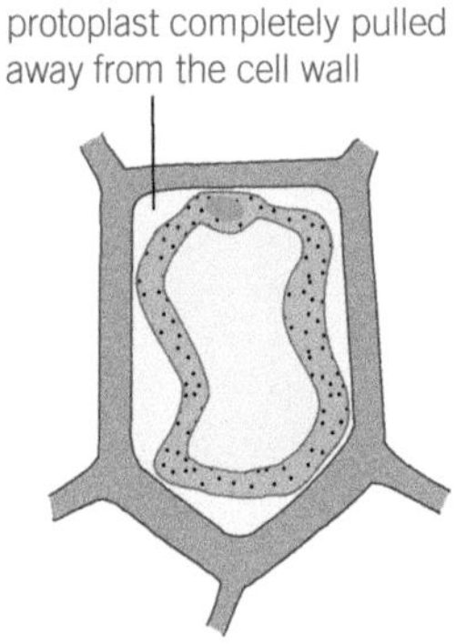

Summary questions

1 If an animal cell with a water potential of −700 kPa was placed in solutions with the following water potentials, what would happen to the cell in each case?

a −100 kPa

b −1500 kPa

c pure water

d −700 kPa *(4 marks)*

2 Some cells possess specialised protein channels called aquaporins, which allow water molecules to pass through the channel. State how the presence of aquaporins in the cell surface membrane affects the permeability of the cell surface membrane. *(1 mark)*

3 Explain why some plants wilt if the soil they are growing in becomes contaminated by seawater. *(3 marks)*

Study tip

Remember to use the term water potential when talking about osmosis. Do not use the term water concentration.

Remember diffusion is the movement of any molecule, but osmosis ONLY applies to the movement of water molecules.

Synoptic link

Osmosis is important in the process of transpiration, which you will study in detail in Topic 6.4, Transport of water in plants.

Practice questions

1 Which of the following are monomers containing carbon, nitrogen and phosphate?

A amino acids

B nucleic acids

C nucleotides

D monosaccharides *(1 mark)*

2 The diagram shows the ring form of two monosaccharides X and Y

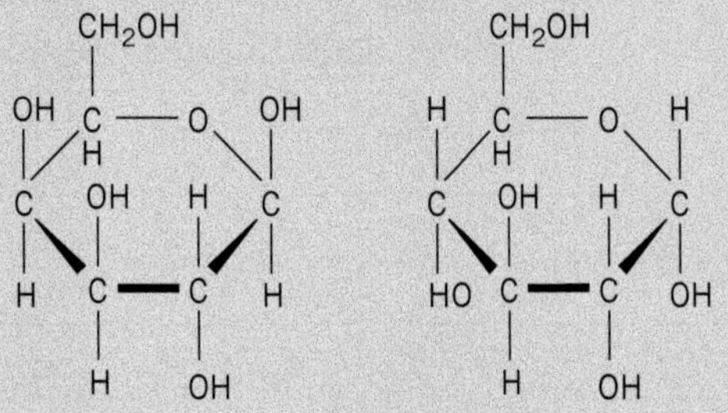

X Y

Which of the following statements is correct?

A condensation reactions between X and Y results in the formation of lactose.

B condensation reactions between X and Y result in the formation of sucrose.

C condensation reactions between X and alpha glucose results in the formation of maltose.

D condensation reactions between Y and alpha glucose results in the formation of lactose. *(1 mark)*

3 The diagram shows diagrams of four cells that have been placed in different solutions.

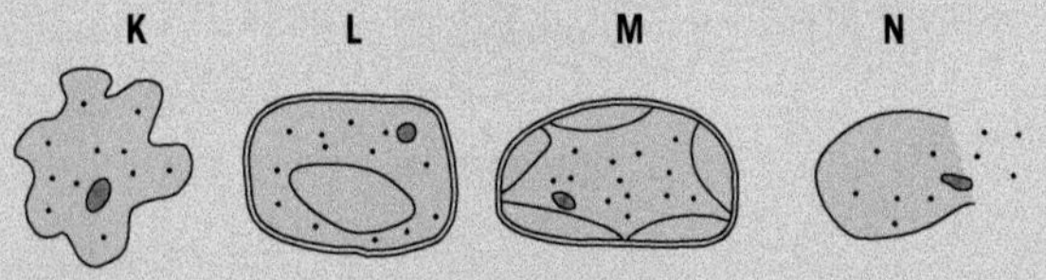

Which of the following statements is/are true?

Statement 1: Cells K and M have plasmolysed.

Statement 2: Cells K and M have been placed in a concentrated sucrose solution.

Statement 3: N and L have been placed in distilled water.

A 1, 2 and 3

B Only 1 and 2

C Only 2 and 3

D Only 1 *(1 mark)*

4 Some biochemical tests were carried out on a plant extract which was known to contain **starch and protein**.

Which of the following statements is/are true?

Statement 1: if iodine solution is added to the extract, a blue-black colour would be observed.

Statement 2: if biuret solution was added to the extract a violet colour would be observed.

Statement 3: if the extract is boiled with acid and then neutralised, a Benedicts test would result in a brick red colour.

A 1, 2 and 3

B Only 1 and 2

C Only 2 and 3

D Only 1 *(1 mark)*

5 A student followed a procedure to find the concentration of reducing sugars in a fruit juice. The first part of the method used was as follows:

- A range of glucose solutions of different concentrations was made up, starting with a 20 g dm^{-3} glucose solution.
- Each solution was boiled with excess Benedict's solution.
- When there was no further change in colour, the liquid was cooled and filtered.
- The absorbance of the liquid was measured with a colorimeter. (A colorimeter measures the amount of light that is absorbed by a coloured solution.)

The student's results are shown in the graph

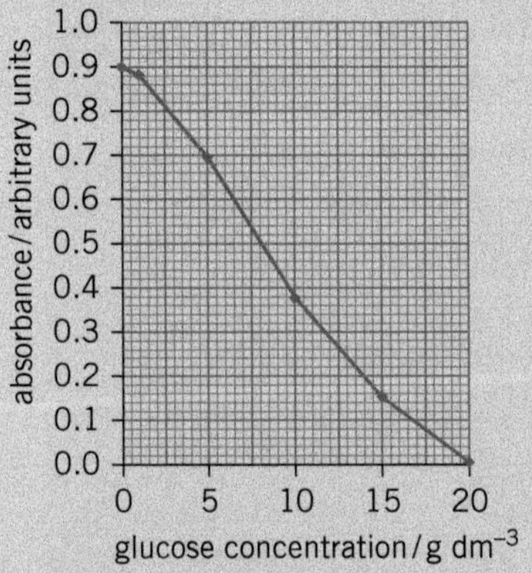

a (i) State two precautions that the student should have taken during the procedure to ensure that the results give a valid comparison between different sucrose solutions. *(2 marks)*

(ii) In the second part of the method, the student tested the fruit juice. The absorbance reading obtained was 0.6 arbitrary units.

Use the graph to determine the reducing sugar concentration of the fruit juice. (*1 mark*)

(iii) This procedure does not test for non-reducing sugars such as sucrose. How should the procedure be altered to determine the concentration of **non-reducing sugar** in the fruit juice? (*3 marks*)

[OCR 2801 June 2006]

6 The diagram shows a representation of part of a carbohydrate molecule called agarose.

One of the subunits of agarose is a sugar called galactose.

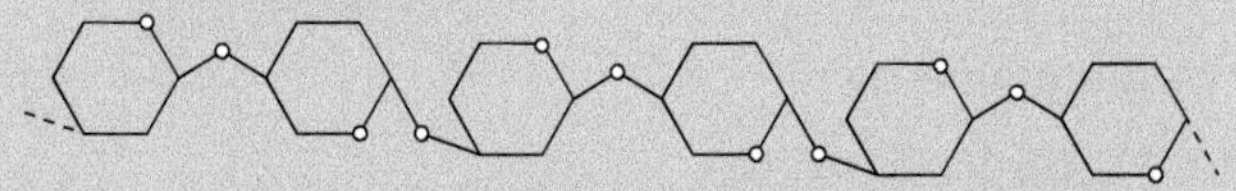

a (i) Identify the type of carbohydrate molecule of which the carbohydrate agarose is an example (*1 mark*)

(ii) Starch contains a carbohydrate called amylose. Amylose does not contain galactose.

Using the information in the diagram, identify **one** similarity and **one further** difference in structure between agarose and amylose. (*2 marks*)

b A student wished to demonstrate experimentally that bacteria cannot break down agarose. The student used a culture of *E.coli* bacteria which had been grown in a solution containing starch.

Two tubes, **A** and **B** were set up as follows:

Tube A: contained 0.1 cm^3 of the *E.coli* culture and 5 cm^3 of a nutrient solution in which agarose was the only carbohydrate.

Tube B: contained 0.1 cm^3 distilled water and 5 cm^3 of a nutrient solution in which agarose was the only carbohydrate.

Both tubes were incubated at 30 °C for 2 hours.

A sample from each tube was then tested for the presence of a reducing sugar.

The results are shown in the table.

Source of sample	Conclusion from test
Tube A	Very small amount of reducing sugar present
Tube B	No reducing sugar present

(i) Explain the purpose of tube B (*2 marks*)

(ii) The student wrote the following conclusion

My experiment showed that bacteria must be able to break down agarose. This is because reducing sugar was present in tube A.

Suggest an alternative explanation for the presence of reducing sugar in tube **A** that is **not** consistent with the student's conclusion (*1 mark*)

(iii) Suggest **two** ways in which the **validity** of the experiment could be improved. (*2 marks*)

c The student did not have access to a colorimeter.

*Describe how the student could carry out a chemical test for a reducing sugar and suggest how he could estimate the amount of reducing sugar in the sample from tube **A**. (*6 marks*)

7 Water is essential transport medium in plants and animals. The polar nature of the water molecule is related to the properties of water.

a Using the water molecule as an example, explain what is meant by a polar molecule (*2 marks*)

b The table shows how the density of water changes at different temperature.

Temperature (°C)	Density ($g\,cm^{-3}$)
30	0.9957
20	0.9982
10	0.9997
4	1.0000
0 (solid)	0.9150

(i) Calculate the percentage increase in the density of water as it cools from 30 to 20 °C Give your answer to 1 significant figure. (*2 marks*)

(ii) Suggest what difference an increase in the density of water would make to the supply of minerals to plant leaves. Explain your suggestion (*2 marks*)

3 PROTEINS AND ENZYMES

3.1 Amino acids – the building blocks of proteins

Specification reference: 2.1.3

Learning outcomes

Demonstrate knowledge, understanding, and application of:

- → the basic structure of an amino acid
- → properties of amino acids
- → different types of amino acids
- → how to join amino acids.

Synoptic link

You briefly learnt about actin in the context of motor proteins in Topic 1.7, Interrelationship of organelles. You will learn how genes are used to make proteins in Topics 4.6 and 4.7, Protein synthesis – transcription and translation.

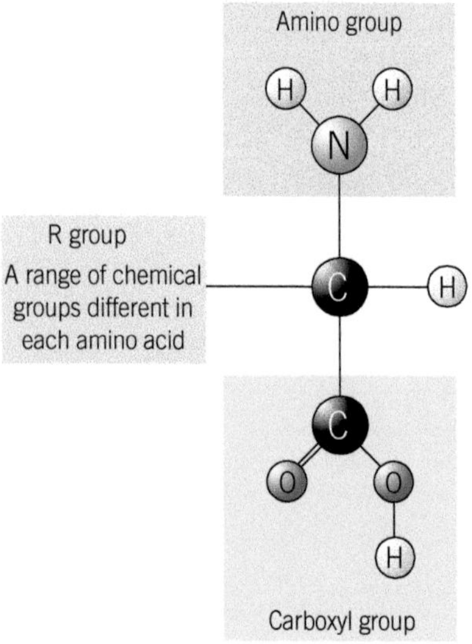

▲ **Figure 1** *The general structure of an amino acid*

Proteins

Cardiac muscle contains two important muscle proteins called actin and myosin. Without these contractile proteins the heart would not function. Defects in the ACTC1 gene (actin, alpha, cardiac muscle 1 gene) can lead to disorders such as:

- familial dilated cardiomyopathy
- familial hypertrophic cardiomyopathy
- familial restrictive cardiomyopathy.

You will study later how genes carry the instructions to produce functional proteins. Here we will look at protein structure and function.

Functions of proteins

Proteins have many functions including:

- enzymes (e.g., urease)
- antibodies (e.g., immunoglobulin A)
- cell membranes (e.g., channel proteins)
- structural roles (e.g., keratin)
- hormones (e.g., insulin, prolactin)
- transport (e.g., haemoglobin).

Structure of an amino acid

Proteins are polymers made from the repeating subunits of **amino acids** (monomers). There are 20 naturally occurring amino acids, although about 100 have been identified to date.

Each of the 20 naturally occurring amino acids has the same general structure. In the centre of the molecule is a carbon atom with four different chemical groups attached to it:

- an amino group ($-NH_2$) – this group has basic properties
- a carboxyl group ($-COOH$) – this group has acidic properties
- a hydrogen atom ($-H$)
- The R group – this varies between each of the 20 different amino acids and makes each amino acid unique. The simplest R group is that of the amino acid glycine where the R group is a hydrogen atom.

Properties of an amino acid

In solution, such as in plasma or the cytoplasm, amino acids can act in different ways:

- As a base – by accepting protons (H^+) to the amino group end of the molecule. This forms a positively charged amino acid ion.
- As an acid – by donating protons from the carboxyl group. This forms a negatively charged amino acid ion.

The amino acid can therefore exist as either positive or negative ions. Molecules that can act as bases or acids are described as amphoteric. This is important for the role of some proteins, such as haemoglobin, which act as buffers. This means the pH of the blood can be kept in a narrow range.

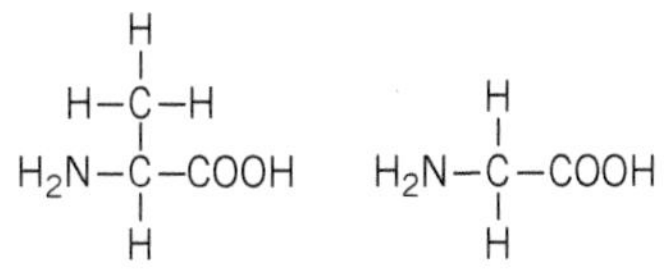

▲ **Figure 2** *The structure of two amino acids*

Synoptic link

Hydrolysis and condensation reactions are covered in Topic 2.3, Carbohydrates in cells and biofluids.

Joining amino acids

In the same way that two monosaccharides can be joined together using a condensation reaction, so two amino acids can be joined to form a dipeptide.

Water is removed during the chemical reaction – an —OH group is removed from the carboxyl group of one amino acid and a —H atom is removed from the adjacent amino acid. The two amino acids are linked by the formation of a covalent bond, called a **peptide bond**.

A peptide bond can also be broken by the addition of a water molecule during a hydrolysis reaction.

Study tip

A buffer solution is one that resists the tendency to change its pH even when small amounts of acid or alkali are added to it.

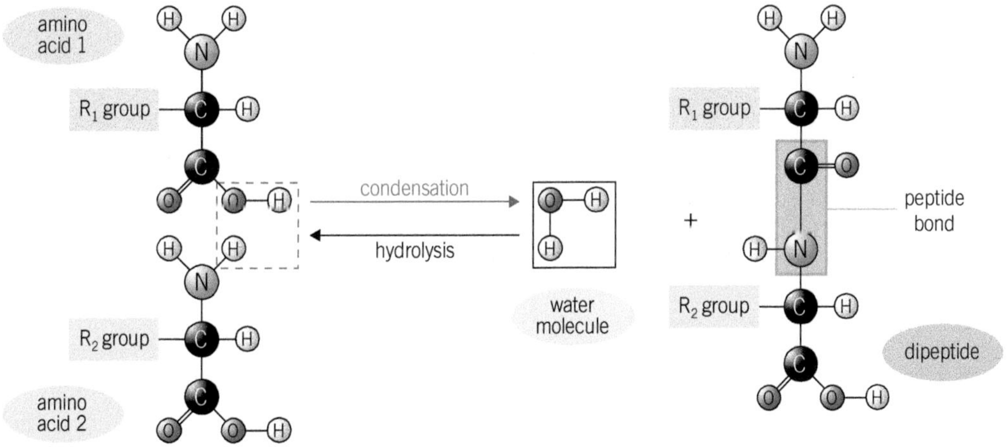

▲ **Figure 3** *The formation and splitting of a peptide bond*

This process can be repeated many times using many condensation reactions. This is called polymerisation and produces a long chain of many amino acids called a **polypeptide chain**.

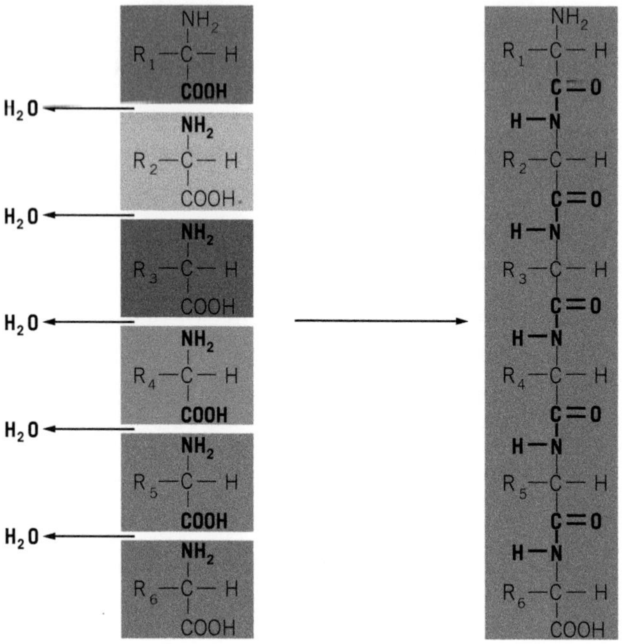

amino acid molecules (where R_1, R_2, R_3 etc. represent any of the 20 or so groups found in naturally occurring amino acids)

polypeptide (part of)

▲ **Figure 4** *The formation of a polypeptide chain*

Summary questions

1 Name two elements other than carbon, hydrogen, and oxygen that could be present in an R group. *(1 mark)*

2 Why is it important that the pH of the cytoplasm of erythrocytes is regulated? *(2 marks)*

3 The amino acid glycine has an R group consisting of a single hydrogen atom. Using an annotated diagram, show how a dipeptide made from two glycine molecules can be hydrolysed. *(4 marks)*

3.2 Separation and identification of amino acids

Specification reference: 2.1.3

Chromatography

Chromatography is a common technique that can be used to separate chemical substances according to their solubility. There are different forms of chromatography which use different mediums such as paper, silica gel, or columns. This is called the stationary phase. The mobile phase is the liquid or gas that is used to separate the different chemical substances (i.e., liquid or gas chromatography).

Amino acids can be separated using paper chromatography. All methods are based on the rate at which different chemicals migrate when they are in contact with both the stationary and mobile phases. Different chemicals naturally have different properties such as their molecular weight, polarity, and structure. These different properties mean that each molecule has a different relative affinity for each phase.

Learning outcomes

Demonstrate knowledge, understanding, and application of:

→ techniques used to separate amino acids

→ techniques used to identify amino acids.

Separating amino acids using chromatography

1. Very small volumes of solutions containing amino acids are applied using a narrow capillary tube (this process is sometimes called 'spotting') at the bottom of a rectangular piece of filter paper on a starting line (drawn on in graphite pencil). Each spot should be 2–3 mm in diameter. Label each spot (with pencil and below the starting line) to indicate its identity. Note – avoid getting fingerprints on the chromatography paper.
2. Allow the spots to dry by waving the paper in the air or using a heat lamp or hair dryer.
3. The paper is rolled into a cylinder and placed in a beaker that contains a few millilitres of the liquid mobile phase (e.g., a solution containing n-propanol, water and ammonia known as the eluting solvent). The solvent will wet the paper, but the sample spots should not be immersed. The paper should not touch the walls of the beaker.
4. The solution (sometimes called the eluting solvent) rises up the paper due to capillary action.
5. As the solvent travels up the paper it comes into contact with the amino acid 'spots'. The higher the affinity the amino acid has for the solvent compared to the stationary phase the further the amino acid will travel with the solvent – it will travel unimpeded by the stationary phase (paper). If the amino acid, however, has a greater affinity for the stationary phase than it does for the solvent then the amino acid will travel slower than the solvent as it adheres to the stationary phase.
6. When the solvent reaches 2 cm from the top of the filter paper, the paper is removed from the beaker and allowed to dry. A pencil mark is made to indicate how far the eluting solvent travelled. The separated amino acids are not visible at this stage.

7 To make the amino acids visible the paper is sprayed with a chemical called ninhydrin. Ninhydrin reacts with any amino acids present to form a blue-violet compound. At this point, any amino acids that are present can clearly be seen with one spot appearing for each amino acid present in the sample. The spots that appear furthest from the starting line are the amino acids which have travelled further as they have a greater affinity for the mobile phase/solvent.

8 The amino acids can be identified by calculating the retardation factor (R_f **value**) for each separated amino acid. This is the distance travelled by the amino acid divided by the distance travelled by the solvent. The measurements must be made from the line where the samples were applied onto the stationary phase (the origin) and the centre of the spot. R_f values can vary between zero (where the amino acid does not move from the starting line) to one (where the amino acid moves the same distance as the solvent). In Figure 1, d_A is the distance travelled by amino acid A, d_B is the distance travelled by amino acid B, and d_{solv} is the distance travelled by the solvent in the same period of time.

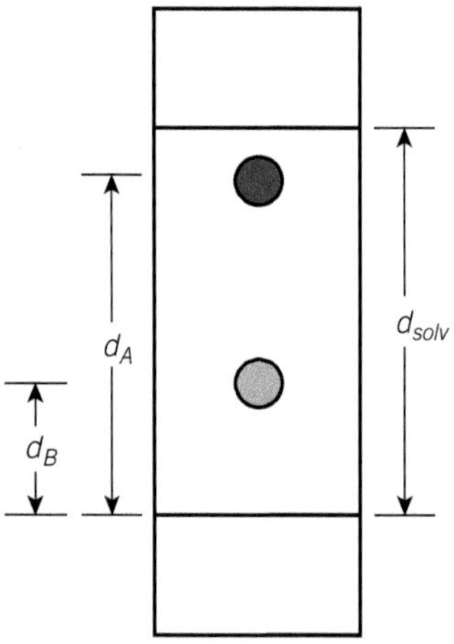

▲ **Figure 1** *Measuring d_A, d_B and d_{solv}*

What two factors determine the Rf value of an amino acid?

The R_f values for these two amino acids can be calculated by:

$$R_f(A) = \frac{d_A}{d_{solv}}$$

$$R_f(B) = \frac{d_B}{d_{solv}}$$

Although R_f values are not completely reproducible, they are reasonably good guides for identifying various amino acids. Paper chromatography is most effective for the identification of unknown substances when known samples are run on the same paper chromatograph with unknowns.

Why do different amino acids have different R_f values?

Each amino acid contains an amino group, $-NH_2$, and a carboxylic acid group, $-COOH$. The 20 different amino acids have a different side chain (R group).

In glycine, the simplest amino acid, R is a hydrogen atom. Eight amino acids have R groups that consist of carbon atoms with attached hydrogen atoms, for example, valine for which R is $-CH(CH_3)CH_3$. These nonpolar hydrocarbon side chains are hydrophobic, which lowers the water solubility of these amino acids.

Six amino acids have polar but neutral R groups that increase water solubility, for example, serine has the R group $-CH_2OH$. In two amino acids, glutamic acid and aspartic acid, the side chains carry carboxylic acid groups and three amino acids have basic R groups. Both acidic and basic R groups increase water solubility, though the solubility will be pH dependent. This means that the rate of migration of an amino acid will depend on the pH of the mobile phase, and this will vary from amino acid to amino acid.

Many proteins contain hundreds of amino acids. When a protein is heated in the presence of acid or base, it is hydrolysed, the peptide bonds are broken, and amino acids are released. The hydrolysed mixture can be studied using paper chromatography to determine the amino acids present in the protein.

Summary questions

1 State two amino acids which would have:
   a a high R_f value *(2 marks)*
   b a low R_f value. *(2 marks)*

2 a Explain why chromatography paper is marked with pencil and not pen. *(3 marks)*
   b Explain why you are warned not to get fingerprints on chromatography paper. *(3 marks)*

3 Suggest safety considerations that must be taken into account when carrying out paper chromatography. *(3 marks)*

3.3 Protein structure – haemoglobin

Specification reference: 2.1.3

Learning outcomes

Demonstrate knowledge, understanding, and application of:

- → how polypeptide chains are folded
- → the four different levels of protein structure
- → the structure of haemoglobin.

Haemoglobin

Haemoglobin is a complex protein that is essential for the transport of 98% of the oxygen – the rest is carried in solution. It is also found in the root nodules of leguminous plants, as leghaemoglobin, where its ability to bind oxygen creates the conditions for nitrogen fixation. In the body, haemoglobin enables blood to carry 70 times more oxygen than if the oxygen were simply dissolved in the plasma. Healthy humans have about 15 g of haemoglobin per decilitre of blood and this can bind with 200 ml of oxygen per litre. Haemoglobin is one of two respiratory pigments, the other being myoglobin which stores oxygen in muscle cells

Proteins may consist of a single polypeptide chain (e.g., lysozyme, the enzyme which destroys bacteria and is found in secretions such as nasal mucus and tears), or consist of a number of polypeptide chains (e.g., haemoglobin). There are four levels of structure found in polypeptides and proteins. The primary structure of a polypeptide or protein determines its secondary, tertiary, and quaternary structures.

Haemoglobin has a structure that consists of four polypeptide chains. There are two different types of polypeptide chain in each haemoglobin molecule:

- alpha polypeptide chain – has 141 amino acids, coded for by a gene on chromosome 16
- beta chain polypeptide – has 146 amino acids, coded for by a gene on chromosome 11.

In each haemoglobin molecule there are two alpha and two beta chains. Each of these polypeptide chains is associated with a haem group (prosthetic group).

Primary structure

The **primary structure** of a polypeptide and proteins is the sequence of amino acids in the polypeptide chain. The amino acids are linked by peptide bonds (Topic 3.1, Amino acids – the building blocks of proteins). A change in a single amino acid in the primary sequence can lead to a change in the shape and structure of the protein, which may make it non-functional. For example, a single amino acid substitution of glycine in the protein collagen can result in brittle bone disease.

Secondary structure

The primary structure of a polypeptide chain can be coiled or folded to form two different arrangements:

- α-helix – this causes the polypeptide chain to coil into a tubular shape
- β-pleated sheet – this causes the polypeptide chain to fold back on itself in flat parallel sheets.

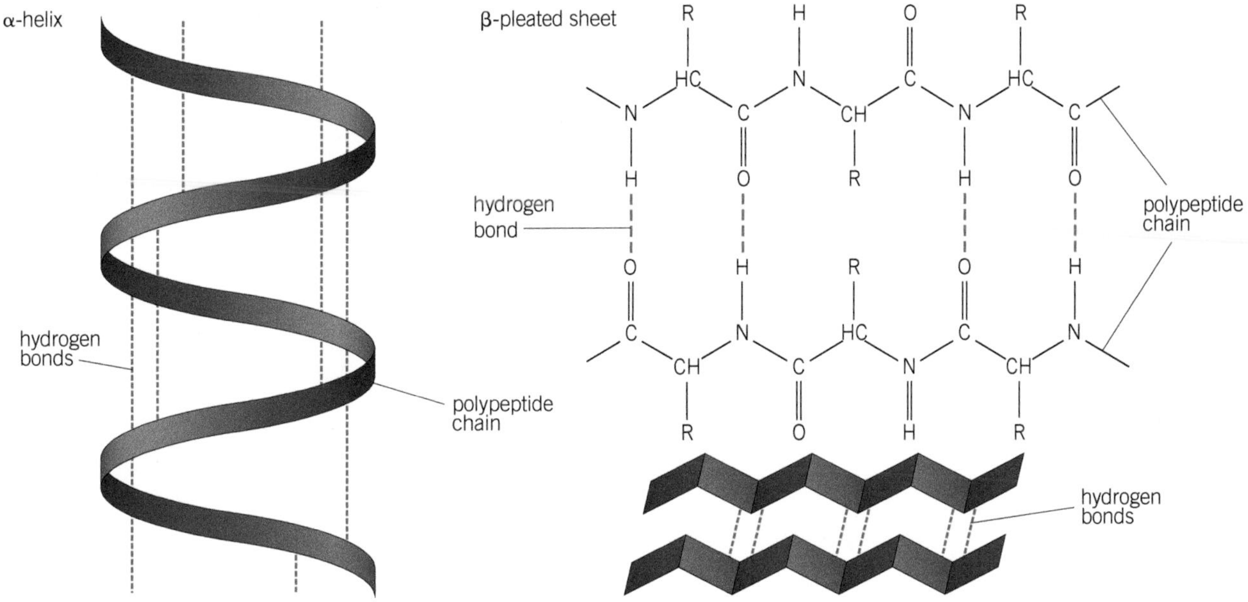

▲ **Figure 1** *The secondary structure of proteins*

These arrangements are held in place by hydrogen bonds. The –NH group of one amino acid has a slight overall positive charge, which is attracted to the negative charge on the oxygen of the –C═O group.

The **secondary structure** is determined by the properties of the R groups of the amino acids in the primary structure, but does not involve bonds between these groups.

Tertiary structure

The secondary structure of a protein can be further folded or coiled to form the **tertiary structure** to produce a more complex three-dimensional structure. To form this level of structure, four different types of interactions, which all occur between R groups, hold the polypeptide chain in place:

- Disulfide bonds – these are found between two cysteine amino acids. They are very strong covalent bonds that form between a sulfur atom in each R group of the cysteine amino acid.
- Ionic bonds – these are weaker than disulfide bonds and are formed between the carboxyl and amino groups on R groups. These bonds are easily broken by extreme pH and high temperatures.
- Hydrogen bonds – these are weaker than ionic bonds and are formed between the O atoms of the –C═O group and the H atoms of the –NH group. Each hydrogen bond is very weak but the combined effect of many hydrogen bonds is strong. These are easily broken by changes in pH and high temperature.

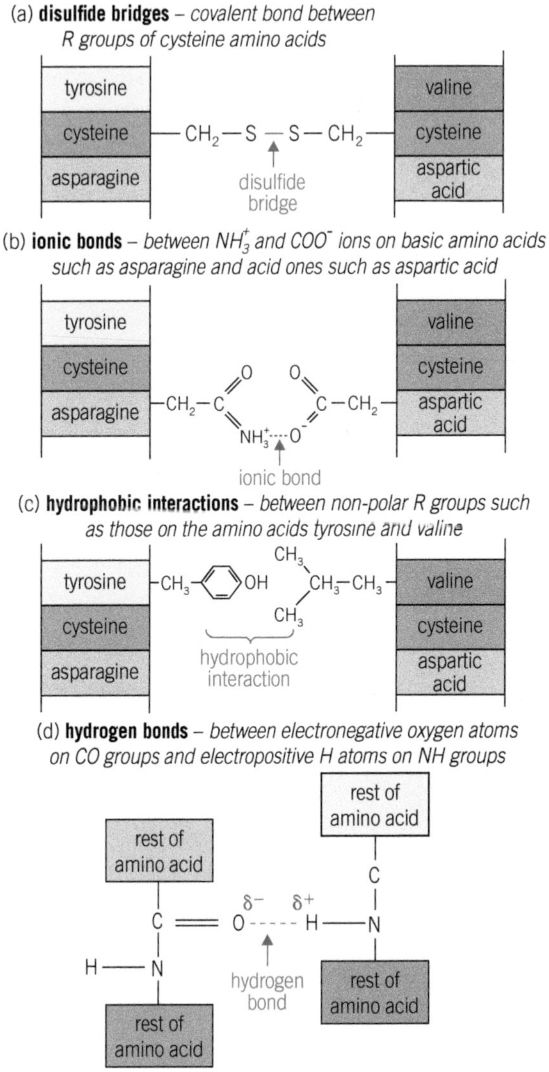

▲ **Figure 2** *Bonds and interactions found in the tertiary structure of proteins*

- Hydrophobic interactions – these are formed by non-polar R groups in amino acids, which are hydrophobic (repel water). These R groups are arranged so that they are positioned in the centre of the protein away from biofluids, such as the cytoplasm, plasma, etc.

Quaternary structure

Haemoglobin and many other proteins are composed of more than one polypeptide chain. Proteins which consist of more than one polypeptide chain joined together are said to have a **quaternary structure**. In some cases the functional protein may also require the addition of a non-protein (prosthetic) groups.

▼ **Table 1**

Example	Number of polypeptides	Prosthetic groups	Detail
carbonic anhydrase	1	Zinc ion	An enzyme that catalyses the conversion of carbon dioxide into hydrogen carbonate ions (or the reverse) in the cell.
integrin	2	none	Membrane protein used to make connections between structures inside and outside of cells.
collagen	3	none	Structural protein found in tendons, ligaments, and blood vessel walls.
haemoglobin	4	four haem groups – each containing an Fe^{2+} ion	Respiratory pigment that transports oxygen.

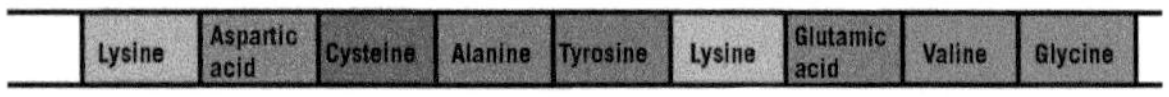

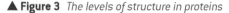

a The primary structure of a protein is the sequence of amino acids found in its polypeptide chains. This sequence determines its properties and shape. Following the elucidation of the amino acid sequence of the hormone insulin by Frederick Sanger in 1954, the primary structure of many other proteins is now known.

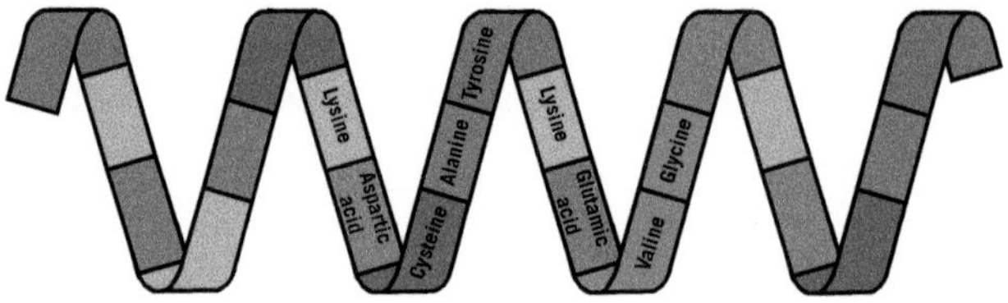

b The secondary structure is the shape which the polypeptide chain forms as a result of hydrogen bonding. This is most often a spiral known as the α-helix, although other configurations occur.

c The tertiary structure is due to the bending and twisting of the polypeptide helix into a compact structure. All four types of bond–disulfide, ionic hydrophobic and hydrogen–contribute to the maintenance of the tertiary structure.

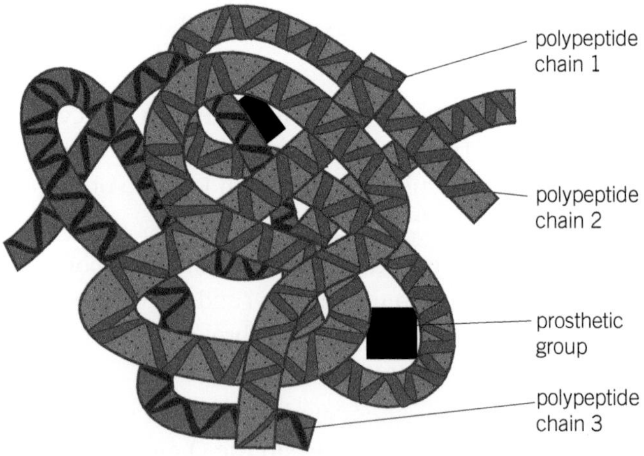

d The quaternary structure arises from the combination of a number of different polypeptide chains and associated non-protein (prosthetic) groups into a large, complex protein molecule.

▲ **Figure 3** *The levels of structure in proteins*

Assembling a haemoglobin molecule

Haemoglobin molecules are made when the four polypeptide chains, which are made on the ribosomes inside the cell that will become an erythrocyte, are combined with the four haem groups. This modification occurs in the endoplasmic reticulum and Golgi apparatus. When the haemoglobin content of an erythrocyte reaches approximately 30%, the nucleus, endoplasmic reticulum, Golgi apparatus, and mitochondria are broken down. This creates space for the haemoglobin molecules to be stored and aids the formation of the biconcave disc, which is essential to enable the erythrocyte to squeeze through the capillaries.

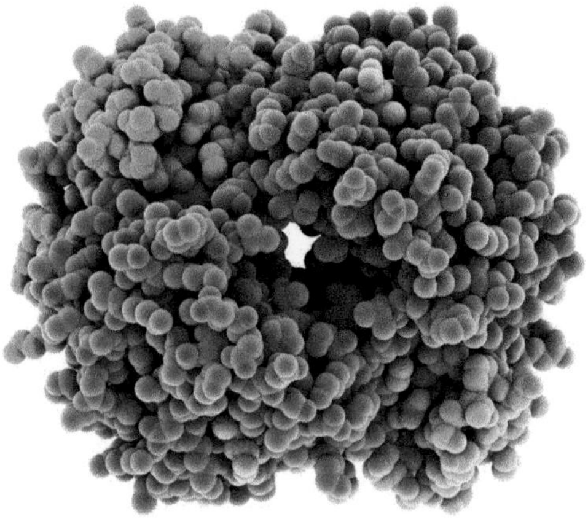

▲ **Figure 4** *Structure of a haemoglobin molecule – it consists of four globin proteins (polypeptide chains – orange, green, blue, and purple). Each globin protein is wrapped around a haem group*

Synoptic link

You will need to understand the roles of organelles in relation to protein synthesis. This is covered in Topic 1.7, Interrelationship of organelles.

Study tip

Primary structure has one type of bond – the peptide bond.

Secondary structure has two types of bond – the peptide bond and hydrogen bonds.

Tertiary structure has an additional *three* types of bond – disulfide, ionic and hydrophobic interactions. The bonds are all between R groups.

Summary questions

1 What is represented by the yellow structures of the haemoglobin molecule in Figure 4? *(1 mark)*

2 Copy and complete the table using ticks and crosses to compare the different bonds found in different levels of protein folding. *(3 marks)*

Level of folding	Disulfide	Hydrogen	Hydrophobic	Ionic	Peptide
Primary					
Secondary					
Tertiary					
Quarternary structure					

3 Compare the *structure* of catalase (an enzyme) and haemoglobin. (Hint – you can read about enzymes in Topic 3.4, The structure of enzymes) *(6 marks)*

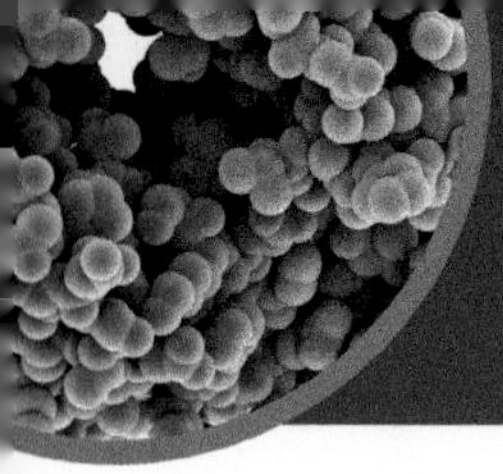

3.4 The structure of enzymes

Specification reference: 2.1.3

Learning outcomes

Demonstrate knowledge, understanding, and application of:

- → the structure of the enzyme catalase
- → how enzymes speed up chemical reactions
- → how enzymes work
- → the properties of enzymes.

▲ **Figure 1** *A molecular model of human catalase. This enzyme catalyses the breakdown of hydrogen peroxide to water and oxygen. Hydrogen peroxide is a highly toxic by-product of a number of normal cellular processes*

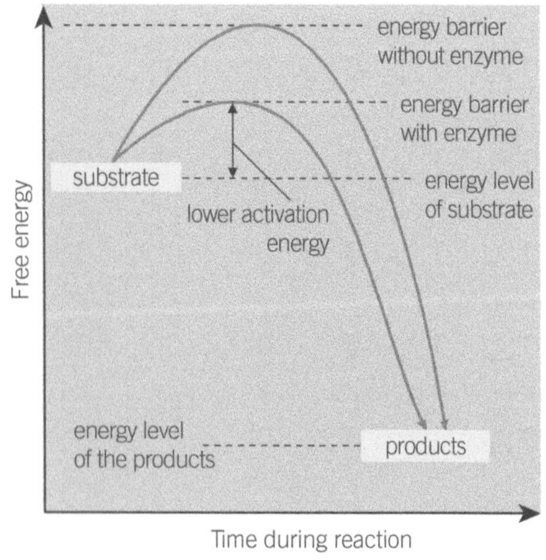

▲ **Figure 2** *How enzymes lower the activation energy and speed up the rate of chemical reactions*

The enzyme catalase

All cells require oxygen to enable aerobic respiration to release energy to carry out cell activities. However, oxygen is a reactive molecule that can cause serious problems if it is not carefully controlled. Inside cells, electrons are continually moved from site to site by carrier molecules. If one of these carrier molecules is exposed to an oxygen molecule, the electron may be accidentally transferred to it. This converts oxygen into highly reactive molecules such as superoxide radicals and hydrogen peroxide, which can attack the sulfur atoms and metal ions in proteins and mutate DNA. To deal with these situations, cells produce a variety of antioxidant **enzymes**. One of these antioxidant enzymes is catalase.

Catalase is an unusual enzyme as it is made from four polypeptide chains (subunits). Each subunit has its own **active site** that catalyses the decomposition of hydrogen peroxide. Catalase is an efficient enzyme that can decompose millions of hydrogen peroxide molecules per second. The active site is a cleft or depression on the surface of the enzyme. Each subunit also contains a prosthetic group, an iron ion, at the centre of a disc-shaped haem group. The importance of the enzyme catalase is shown by its prevalence, ranging from about 0.1% of the total protein in an *Escherichia coli* cell to approximately 25% of the protein in a hepatocyte.

How enzymes speed up reactions

All enzymes are globular proteins that act as biological **catalysts**, that is, they speed up the rate of a chemical reaction without being permanently altered themselves. This means that enzymes can be reused constantly. The chemical reactions would still take place without the presence of an enzyme but they would take place at a slower rate.

To convert a substrate to a product, the substrate molecules usually need to be given some extra energy. In living cells, this cannot be achieved by simply heating the molecules up to a higher temperature to make them collide with more energy. Endotherms are animals that are capable of the internal generation of heat. Reactions in endotherm cells normally take place at 37 °C and at this temperature most molecules do not possess enough energy to get over the energy barrier for the chemical reaction to occur. The minimum amount of energy needed to start the reaction (i.e., get over the energy barrier) and convert the substrate to the product(s) is called the **activation energy**. The reason that enzymes speed up the rate of reaction, sometimes by a factor of several millions, is because they lower this activation energy so allowing reactions to occur at 37 °C.

The lowering of the activation energy by the enzyme catalase means that the substrate (hydrogen peroxide) does not need to be raised to such a high energy level and can be converted into the products (water and oxygen) more easily at normal body temperatures.

The structure of catalase

Like all enzymes, catalase has a specific structure that enables it to carry out its function. Most enzymes are made from hundreds to thousands of amino acids, but it is usually only between three and 12 of these amino acids that form the precise part of the structure that is functional, the active site. The R groups of the three to 12 amino acids in the active site form temporary bonds with the substrate(s) and bring them closer together, lowering the activation energy. The remaining amino acids in the polypeptide change are responsible for determining the secondary, tertiary (and in some cases, quaternary) structure of the enzyme.

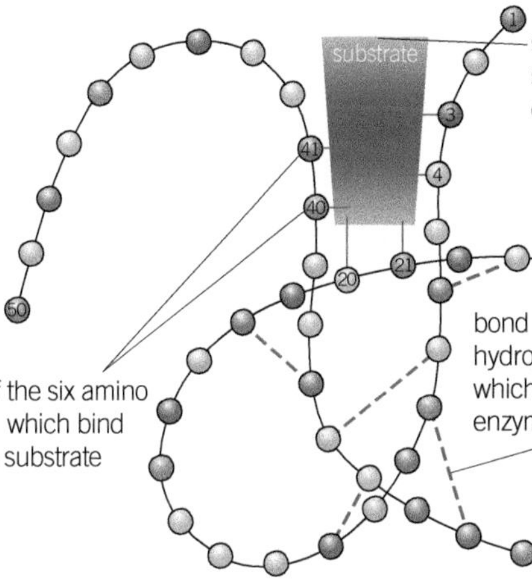

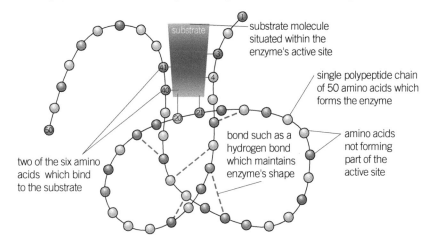

◀ **Figure 3** *An enzyme-substrate complex showing six of the 50 amino acids that form the active site*

Reducing the activation energy of a chemical reaction allows the enzyme to provide a *different pathway* for the reaction to follow.

Enzyme activity

Enzyme activity can be summarised into four stages:

1. The substrate(s) collides successfully with the active site, with sufficient energy and in the correct orientation.
2. The substrate(s) bind to the active site of the enzyme to form an enzyme-substrate complex (ESC).
3. Once the substrates are bound to the active site the substrate(s) are converted to the product(s).
4. The product(s) are released from the active site of the enzyme, leaving it vacant for another substrate(s) to bind again.

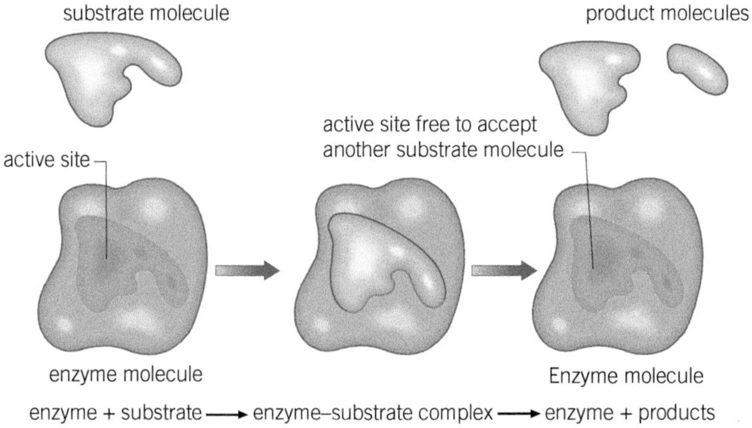

◀ **Figure 4** *The lock and key theory of enzyme activity*

Study tip

Remember the shape of the *substrate* is *complementary* to the shape of the active site. Do not use the term same as instead of complementary.

Specificity

All enzymes work on specific substrates. Only substrate(s) of a particular shape will fit into the active site of an enzyme. This is one significant difference between enzymes and non-biological catalysts.

Reversibility

Some chemical reactions are reversible:

$$A + B \rightleftharpoons C + D$$

The reaction can proceed in either direction but if the conditions change then the direction of the reaction may change. Enzymes such as carbonic anhydrase catalyse both the forward and reverse reactions.

In respiring tissues: $CO_2 + H_2O \xrightarrow{\text{carbonic anhydrase}} H_2CO_3$

In the lungs: $H_2CO_3 \xrightarrow{\text{carbonic anhydrase}} CO_2 + H_2O$

Fast acting

Enzymes can convert many molecules of a substrate to a product in a given period of time. The enzyme is said to have a high turnover number. This varies from several millions of molecules per minute as in the case of catalase, to a few hundred per minute for slow acting enzymes.

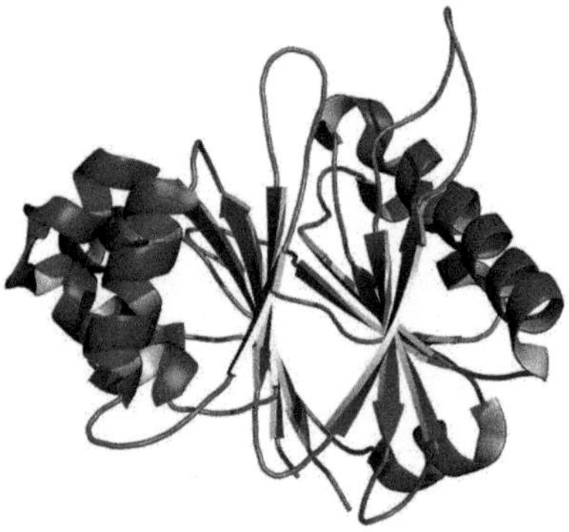

▲ **Figure 5** *Recombinant DNase I*

Summary questions

1 Explain how a successful collision between a substrate and an enzyme can occur. *(5 marks)*

2 Figure 5 shows the structure of the enzyme recombinant DNase I, which is used to treat cystic fibrosis.

a List three structural features that are common to all enzymes. *(3 marks)*

b State what is represented by:

i the red areas of the molecule *(1 mark)*

ii the blue areas of the molecule. *(2 marks)*

3 The diagram below shows metabolic pathway in an epithelial cell.

$$A \xrightarrow{E1} B \xrightarrow{E2} C \xrightarrow{E3} D$$

Each stage of the pathway is controlled by a specific enzyme (E1, E2, or E3).

a Suggest and explain what would happen to the concentration of molecule **D** if the concentration of **A** was reduced. *(2 marks)*

b Suggest and explain what would happen to the concentration of molecule **D** if the concentration of enzyme E1 was increased. *(4 marks)*

c Explain why enzyme E3 will not be able to convert molecule **B** to molecule **C**. *(3 marks)*

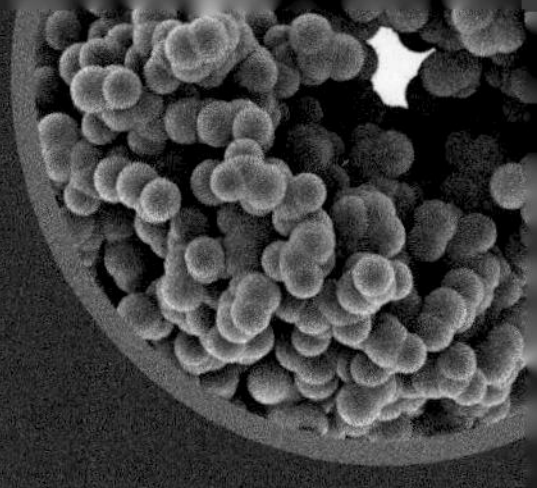

3.5 Factors affecting enzyme-controlled reactions

Specification reference: 2.1.3

Learning outcomes

Demonstrate knowledge, understanding, and application of:

→ how different factors affect the rate of enzyme-controlled reactions:

- temperature
- substrate concentration
- enzyme concentration
- pH
- inhibitors
- cofactors.

Temperature

An increase in temperature will have two key effects on any enzyme driven reaction.

- As the temperature increases (for example from 0 °C to 40 °C) the kinetic energy possessed by each molecule (substrates and enzymes) will increase. This means the random motion of all the molecules will be faster. The faster the molecules move, the greater the chance of a collision and the greater the probability of a successful collision. This will cause the rate of reaction to increase.
- However, as the temperature increases further (higher than 40 °C) it causes more vibrations to occur between the atoms within each molecule. These increased vibrations cause the weak hydrogen bonds to break in the enzyme structure. As a result, the enzyme loses its three-dimensional structure and shape. This distortion can cause the shape of the active site to change so that the substrate(s) is no longer complementary to it. The enzyme is described as **denatured** and no enzyme-subtrate complexes (ESCs) can form, even though collisions will still be happening.

Each enzyme has its own specific optimum temperature at which it catalyses the reaction at the fastest rate. In practice, most enzymes operate in organisms at temperatures a little lower than their optimal, sacrificing some efficiency for stability.

Study tip

Proteins and enzymes are only *denatured* at temperatures *above* their optimum temperature. At very low temperatures they are said to be inactive.

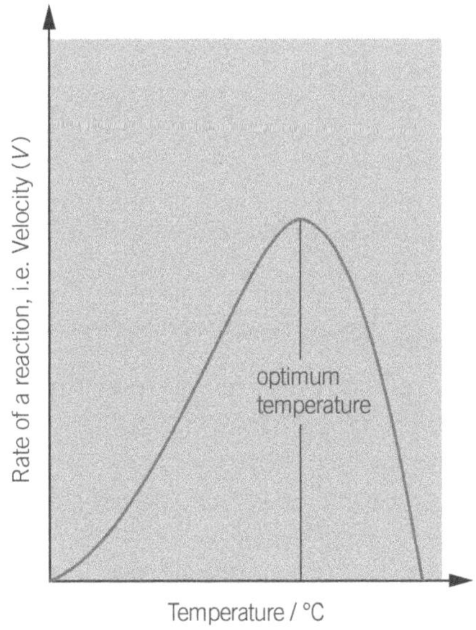

◀ **Figure 1** *A graph to show the effect of temperature on the rate of an enzyme-controlled reaction*

Substrate concentration

The rate of reaction for any enzyme can be affected by the concentration of the substrate present in the environment. If the concentration of enzyme remains constant, as the concentration of substrate is increased the rate of reaction will increase. This is because as the substrate concentration is increased the number of substrate molecules present in a given volume will be higher. As a consequence, there are more molecules randomly moving in the given area. This increases the chance of a collision occurring between the substrate molecule and the active site and therefore increases the chance of a successful collision occurring. The higher the number of successful collisions occurring in a given period of time, the faster the rate of reaction will be.

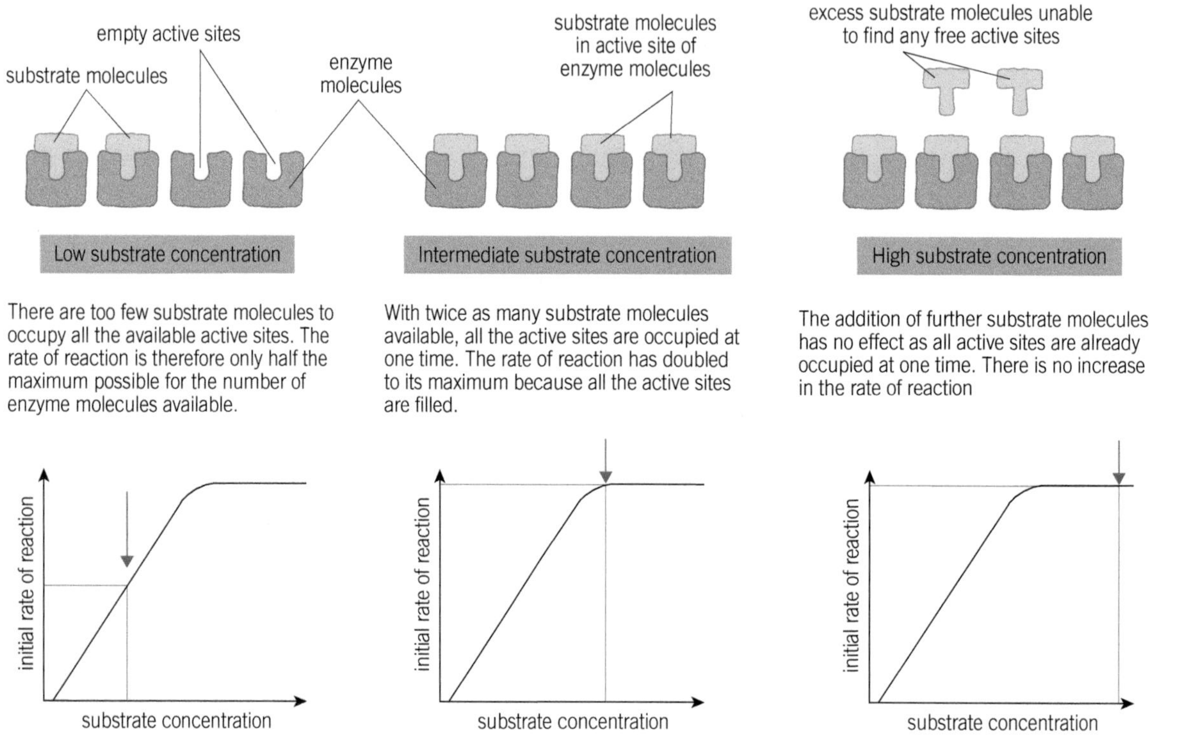
empty active sites
substrate molecules
enzyme molecules
substrate molecules in active site of enzyme molecules
excess substrate molecules unable to find any free active sites
Low substrate concentration
Intermediate substrate concentration
High substrate concentration
There are too few substrate molecules to occupy all the available active sites. The rate of reaction is therefore only half the maximum possible for the number of enzyme molecules available.
With twice as many substrate molecules available, all the active sites are occupied at one time. The rate of reaction has doubled to its maximum because all the active sites are filled.
The addition of further substrate molecules has no effect as all active sites are already occupied at one time. There is no increase in the rate of reaction
initial rate of reaction
substrate concentration
initial rate of reaction
substrate concentration
initial rate of reaction
substrate concentration

▲ **Figure 2** *A graph to show the effect of substrate concentration on the rate of an enzyme-controlled reaction*

It is important to remember that as the concentration of substrate increases, the total concentration of product produced will also increase proportionally.

After the substrate has bound to an active site, that particular active site will be unavailable for other substrate molecules. As the concentration of substrates increases, more and more of the active sites will be occupied at any one moment in time. This means that the increase in the rate at which the enzyme catalyses the reaction gets smaller and smaller as the substrate concentration increases until eventually there is no further increase. When all of the active sites are occupied excess substrate molecules have to wait for an active site to become vacant. At very high concentrations of substrate the rate of reaction will stop increasing and reach a plateau. In this instance the concentration of the enzyme is acting as the **limiting factor**. The enzymes are working at their maximum rate – this is called the V_{max}, where V indicates velocity and max refers to maximum.

Enzyme concentration

The rate of reaction will increase if the concentration of enzyme is increased, provided there is an excess of substrate molecules. This continues until the point where the number of enzyme molecules is greater than that of the number of substrate molecules. At this point, the substrate concentration is the limiting factor and the addition of any more enzyme will not increase the rate of reaction any further. Each enzyme can be reused indefinitely (providing it has not been denatured).

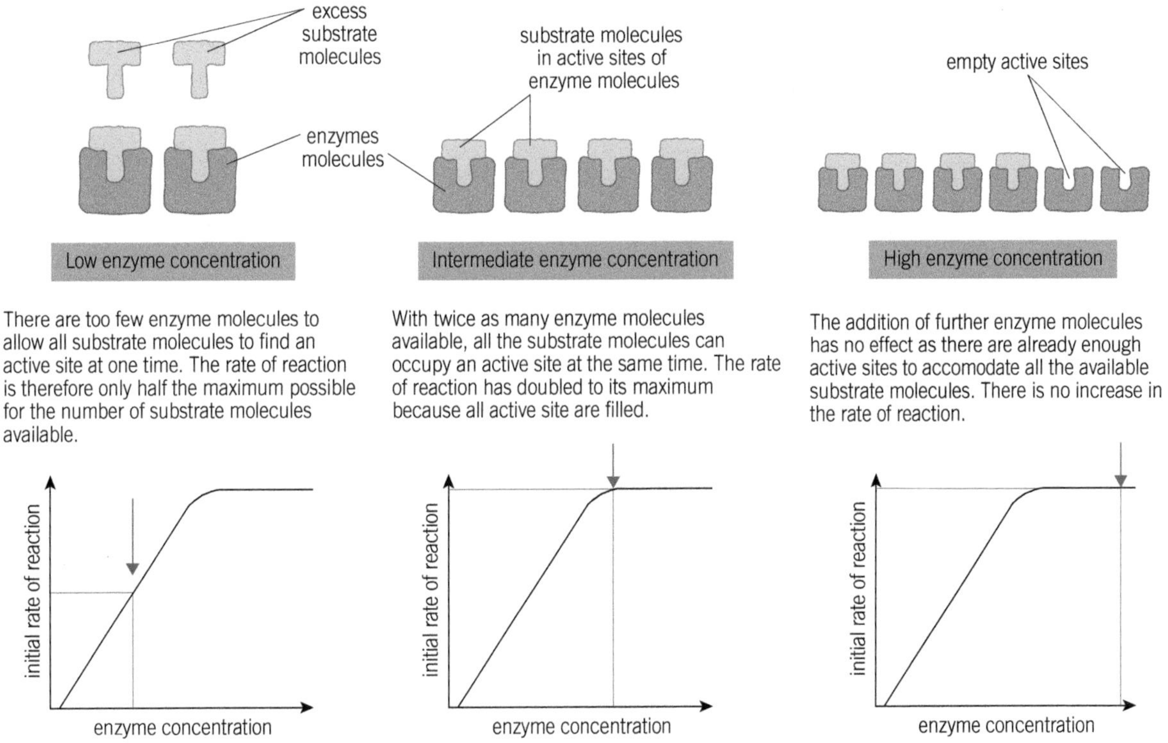
excess substrate molecules
enzymes molecules
substrate molecules in active sites of enzyme molecules
empty active sites
Low enzyme concentration
Intermediate enzyme concentration
High enzyme concentration
There are too few enzyme molecules to allow all substrate molecules to find an active site at one time. The rate of reaction is therefore only half the maximum possible for the number of substrate molecules available.
With twice as many enzyme molecules available, all the substrate molecules can occupy an active site at the same time. The rate of reaction has doubled to its maximum because all active site are filled.
The addition of further enzyme molecules has no effect as there are already enough active sites to accomodate all the available substrate molecules. There is no increase in the rate of reaction.
initial rate of reaction
enzyme concentration
initial rate of reaction
enzyme concentration
initial rate of reaction
enzyme concentration

▲ **Figure 3** *A graph to show the effect of enzyme concentration on the rate of an enzyme-controlled reaction*

pH

pH is a measure of the number of protons (H^+) in a solution. The more protons present, the more acidic a solution will be. The less protons present, the more basic it will be. A strongly acidic solution can have one hundred million million (100 000 000 000 000) times more hydrogen ions than a strongly basic solution. Scientists use **logarithmic scales** to make it easier to read and deal with such large numbers. The logarithmic scale is also called the pH scale. On the pH scale, a pH of 7 is neutral, a pH less than 7 is acidic and a pH greater than 7 is basic.

Each one-unit change in the pH scale corresponds to a tenfold change in hydrogen ion concentration. To be more precise, pH is the negative logarithm of the hydrogen ion concentration:

$$pH = \log \frac{1}{[H^+]} = -\log\ [H^+]$$

The square brackets around the H^+ automatically mean 'concentration' to a chemist. Each whole pH value below seven is ten times more

Synoptic link

You first met the concept of pH in Topic 2.5, Detecting and measuring proteins.

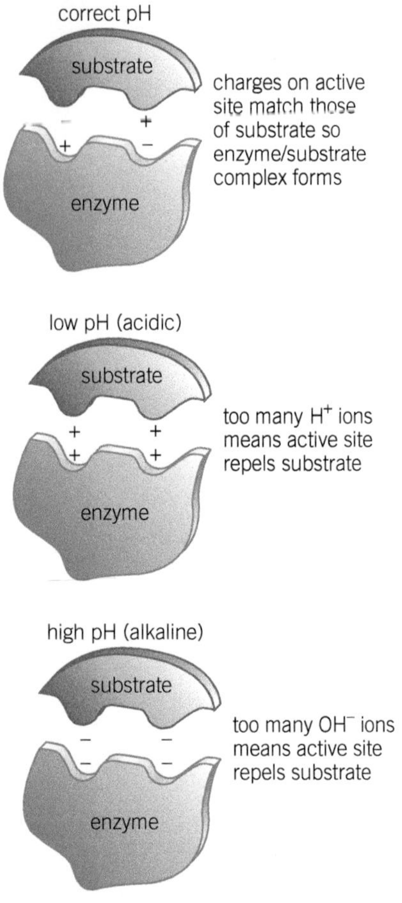

▲ **Figure 4** *Changes in pH can prevent the formation of enzyme-substrate complexes*

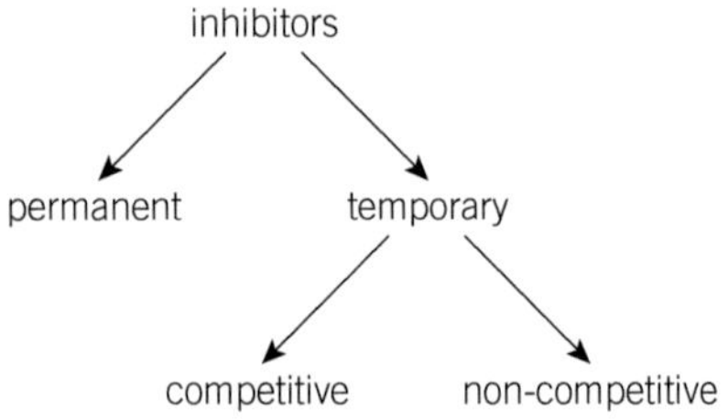

▲ **Figure 5** *Types of inhibitors*

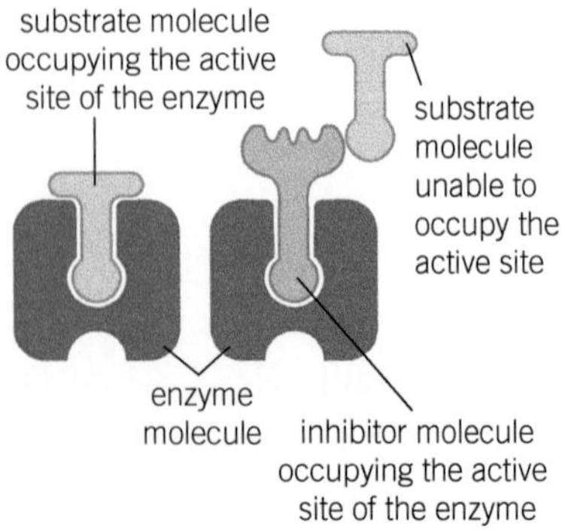

▲ **Figure 6** *Competitive inhibition*

acidic than the next higher value. For example, pH 4 is ten times more acidic than pH 5 and 100 (ten multiplied by ten) times more acidic than pH 6. The same holds true for pH values above 7, each of which is ten times more alkaline (another way to say basic) than the next lower whole value.

Each specific enzyme has an optimum pH at which it works at its maximum rate. Small changes in pH can affect the rate of reaction without denaturing the enzyme. Significant changes in pH can affect enzymes in two ways:

- Different R groups of amino acids carry different charges. These charges can be altered if the pH of the environment changes. This can affect the amino acids in the active site and may prevent the substrate(s) from binding to it. No ESCs will form and so the rate of reaction will decrease.
- A change in pH can cause the hydrogen and ionic bonds holding the secondary and tertiary structure of the enzyme in place to be broken. Beyond a certain pH the damage is irreversible and the enzyme is described as denatured.

Inhibitors

Inhibitors are molecules that interfere with the functioning of the active site of an enzyme either directly or indirectly. Inhibitor molecules can bind permanently or temporarily. Those that only make temporary attachments to the active site can be divided into two groups:

- **competitive inhibitors** – bind to the active site
- **non-competitive inhibitors** – bind to a binding site at a position away from the active site.

Competitive inhibitors

Competitive inhibitors have a chemical structure that is a similar shape to the shape of the substrate. This similarity enables them to occupy the active site and consequently block it, preventing the substrate from binding. The rate of the reaction is determined by the relative concentrations of the substrate and the inhibitor. As the attachment to the active site is not permanent, the inhibitor molecule will eventually be released. The active site will then be available for another molecule to attach to it – this could be the substrate or another inhibitor molecule.

An example of a competitive inhibitor is ethanol, which inhibits the enzyme alcohol dehydrogenase. Methanol, CH_3OH, is a poison, not because of what it does to the body itself, but because the enzyme alcohol dehydrogenase oxidizes it to formaldehyde, CH_2O, which is a potent poison. Treatment of methanol poisoning involves giving the patient ethanol, CH_3CH_2OH. The ethanol competes with methanol for the active site so as ethanol is added, less methanol can attach to alcohol dehydrogenase's active sites resulting in a slower rate of formaldehyde production.

Non-competitive inhibitors

Non-competitive inhibitors attach to a binding site separate from the active site of the enzyme. This site is called an allosteric site. This causes an allosteric effect and changes the shape of the enzyme, which distorts the shape of the active site. Consequently, the substrate molecules can no longer bind to the active site to form ESCs and the rate of reaction is reduced. An increase in the substrate concentration will not increase the rate of reaction as the substrate and inhibitor molecules are not competing for the active site.

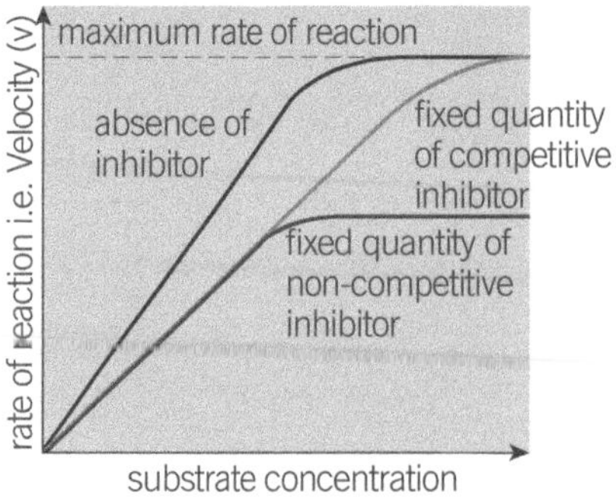

▲ **Figure 8** *Comparison of competitive and non-competitive inhibition*

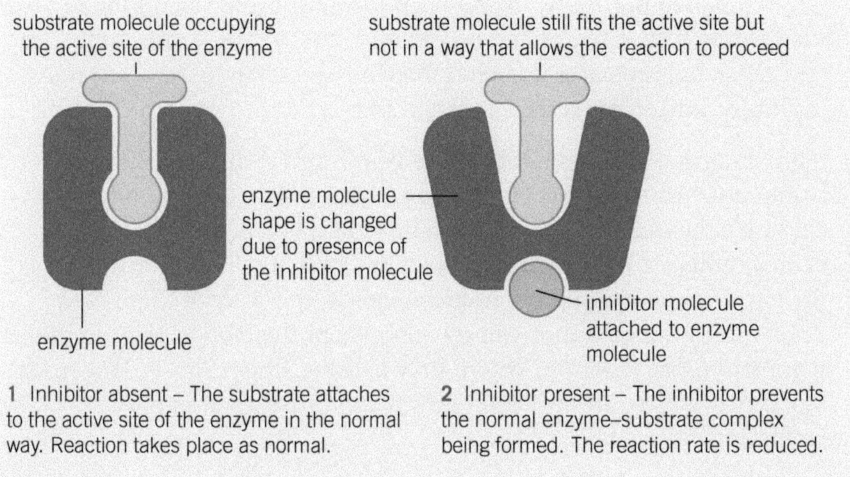

▲ **Figure 7** *Non-competitive inhibition*

The amino acid alanine non-competitively inhibits the enzyme pyruvate kinase. Alanine is one product of a series of enzyme-catalysed reactions, the first step of which is catalysed by pyruvate kinase.

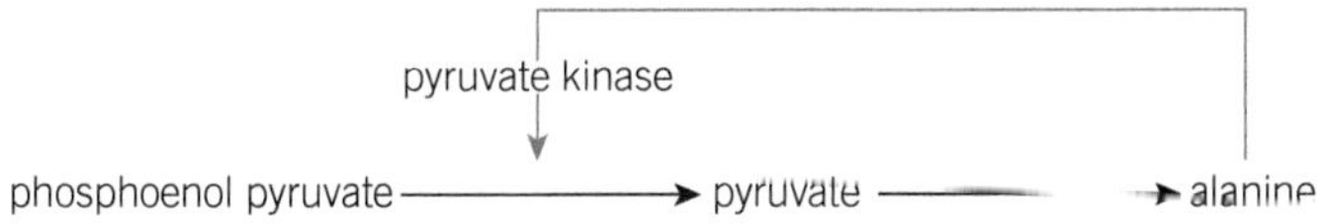

The advantage of the product of a series of enzyme-controlled reactions inhibiting one of the enzymes earlier in the chain is that it prevents the unnecessary build-up of product molecules. Once the cell has enough alanine it uses alanine to stop the series of reactions that would produce more unnecessary alanine. This is also termed 'end-product inhibition'.

Cofactors

Cofactors work by altering the shape of the enzyme so that it can bind more effectively with its substrate. Some cofactors help the enzyme transfer a particular group of atoms from one molecule to another. Chloride ions act as a cofactor for the enzyme salivary amylase. The chloride ions bind with the salivary amylase and slightly change its shape, an example of allostery. This makes it easier for the starch molecule to fit into the active site. In the absence of the cofactor the rate of reaction is reduced.

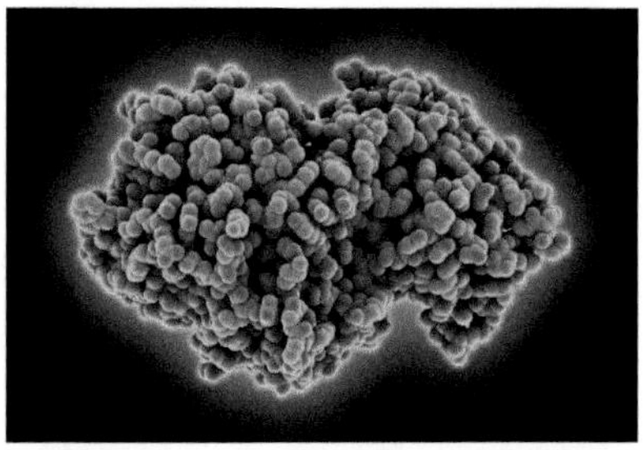

▲ **Figure 9** *A molecular model of salivary amylase – which has a calcium ion and chloride ion as cofactors in its structure*

Summary questions

1 Sketch a graph to show the effect of pH on pepsin, salivary amylase, and lipase. *(3 marks)*

2 A student decided that she would investigate the effect of temperature on the rate of respiration in yeast cells. Explain why it was important that she decided to use a solution containing glucose and a buffer in her investigation. *(3 marks)*

3 Psychrophiles are organisms that thrive permanently at near-zero temperatures by synthesising cold-active enzymes to sustain their cell cycles.
   a Suggest why these enzymes need to be produced. *(3 marks)*
   b Suggest how the structure of these enzymes may differ from enzymes found in humans. *(3 marks)*

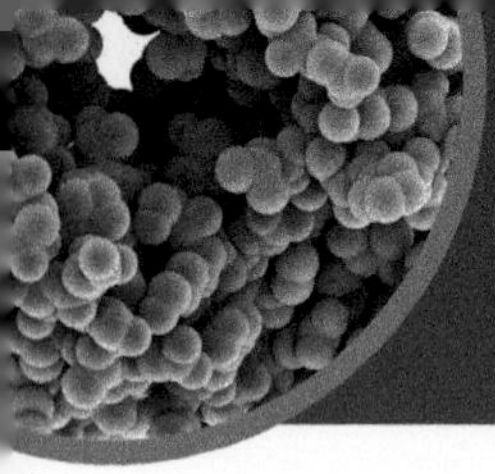

3.6 Blood clotting and enzymes

Specification reference: 2.1.3

Learning outcomes

Demonstrate knowledge, understanding, and application of:

- → how blood clots form
- → the role of thrombin in blood clotting.

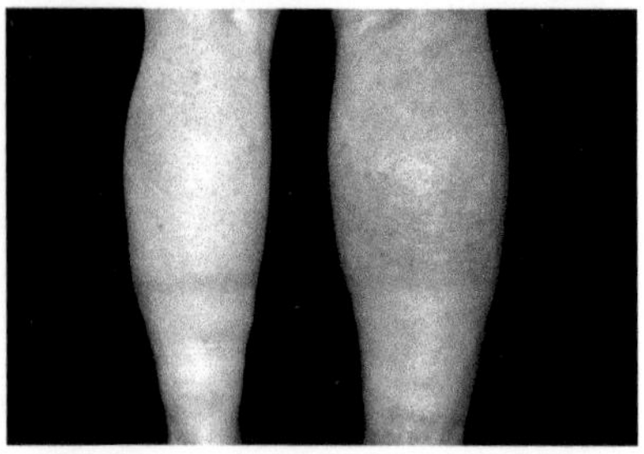

▲ **Figure 1** *Red and inflamed leg (on the right) in a 61 year old male patient due to deep vein thrombosis (DVT)*

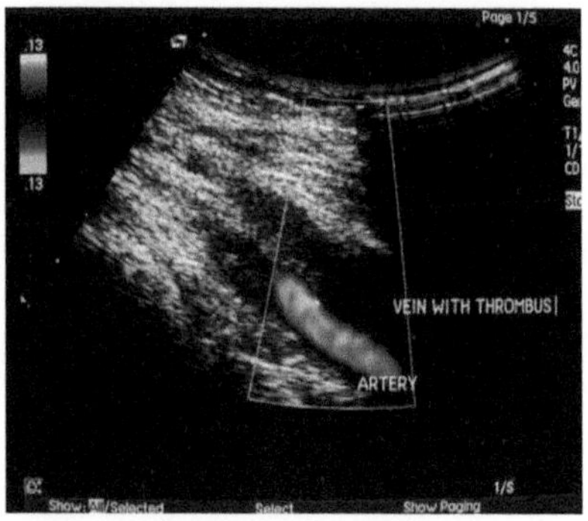

▲ **Figure 2** *A coloured Doppler ultrasound scan showing deep vein thrombosis (DVT) in a patient's leg. The thrombus (clot) blocking the vein is the dark area at centre right. Blood flow in an adjacent artery (orange) is slowed due to the clot*

Synoptic link

It will be valuable to read about the different components of the blood in Topic 1.4, Cells of the blood, before continuing.

Blood clotting

Blood clotting (also known as coagulation), is an essential process that prevents excessive bleeding when a blood vessel is damaged. A clot is formed from platelets and proteins in the blood plasma to seal the wound. The clot stops the bleeding and once the injury has healed the clot usually dissolves naturally. Sometimes clots can fail to dissolve naturally, or they can form on the inside of blood vessels when there has been no obvious injury. These situations can be very dangerous and it is vital that they are diagnosed accurately, promptly, and then treated appropriately.

If an abnormal blood clot forms in a vein it can result in less blood being returned to the heart. Blood can accumulate behind the clot causing pain and swelling. One example of this type of clot is deep-vein thrombosis (DVT). These clots often form in a major vein within the leg, arm, pelvis, and other large veins in the body. If these clots detach from the vein they can travel through the bloodstream to the heart and on to the lungs where they become stuck. This is dangerous as if it prevents sufficient blood flow it can cause a pulmonary embolism.

The clotting process

- Damage to the blood vessels exposes collagen fibres in the connective tissue.
- This activates platelets, which stick together and onto surfaces such as damaged blood vessel walls and fibrin fibres. The platelets clump together into a 'plug', forming an initial barrier to prevent further blood loss. Calcium ions are required for this stage of the clotting process.
- Leucocytes collect at the site of damage and the tissue just below the endothelium releases an enzyme called **thromboplastin**.
- Platelets break down and also release thromboplastin.
- Thromboplastin then catalyses the conversion of prothrombin (an inactive plasma protein) into **thrombin** (an active enzyme). This process also requires calcium ions.
- Thrombin, in the presence of calcium ions, hydrolyses fibrinogen (a large, soluble plasma protein) into fibrin (small, insoluble protein fibres).
- The fibrin fibres become tangled and form a mesh over the wound, which traps red blood cells forming a blood clot.
- The resulting clot is initially a gel, but when it is exposed to the air it dries to form a hard scab and prevents further blood loss and the entry of pathogens.

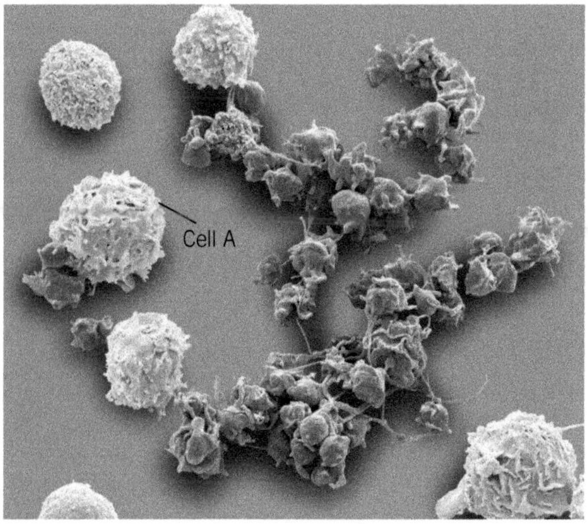

▲ **Figure 3** *A coloured SEM of white blood cells and aggregating platelets (red) (magnification ×3000)*

Study tip

Thromboplastin and thrombin are both enzymes so their rate of activity will be affected by pH, temperature, substrate concentration, and concentration of cofactors.

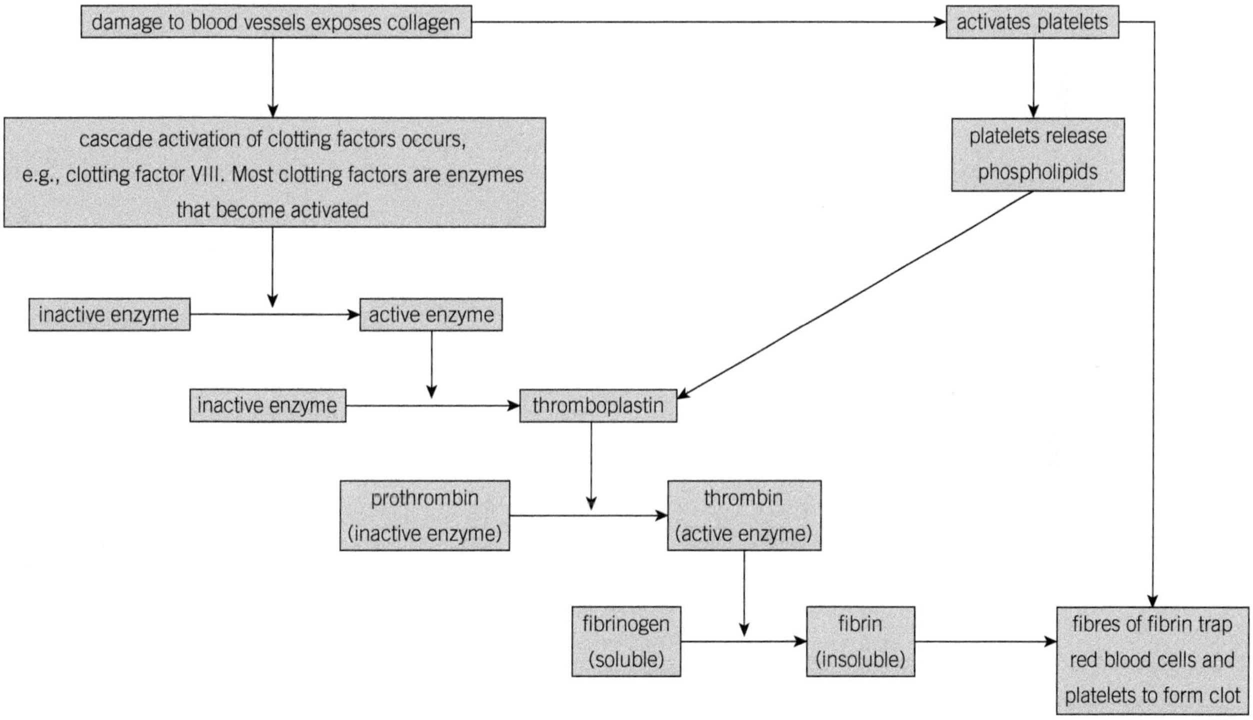

▲ **Figure 4** *Summary of the key steps in the blood clotting cascade*

Tranexamic acid

Tranexamic acid (TXA) has been commonly used in cardiac and orthopaedic surgery where it is used to reduce blood loss. It is also used to treat women who suffer from heavy uterine bleeding.

TXA is defined as an anti-fibrinolytic drug or a clot stabliliser. It prevents the breakdown of fibrin, a protein that is important in the coagulation of blood. It also inhibits the activation of plasminogen to plasmin, an enzyme that breaks down fibrin.

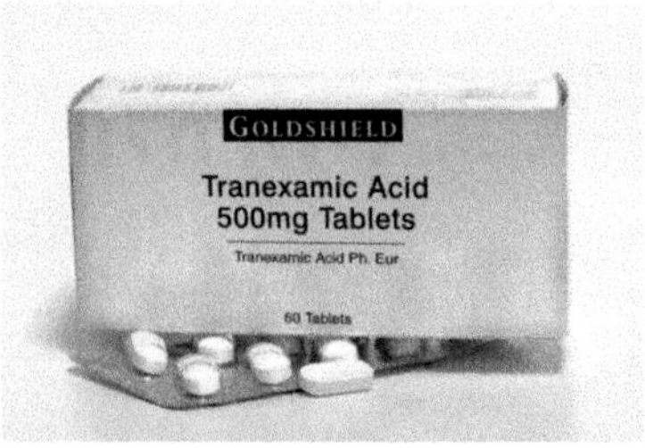

▲ **Figure 5** *Tranexamic acid tablets. This drug is used to treat haemophilia, a genetic disorder that prevents the proper clotting of blood, and menorrhagia, abnormally heavy menstruation*

The Military Application of Tranexamic Acid in Trauma Emergency Resuscitation Study (MATTERs) involved TXA administration in an out-of-hospital setting, on British military medevac helicopters in Afghanistan. Patients were included in this study if haemorrhagic shock was suspected and they were transfused at least one unit of blood. The data collected by MATTERs revealed a statistically significant survival benefit when receiving TXA – 17.4% in comparison to 23.9% mortality without TXA. However, MATTERs did also find that there were statistically significant increases in clotting complications of pulmonary embolisms and deep venous thromboses, yet the absolute numbers were quite small.

1 What is meant by the term anti-fibrinolytic?
2 Suggest how TXA may act on plasmin.
3 Suggest why a 5% solution of TXA is commonly used in dental treatments.

Summary questions

1 Calculate the diameter of Cell A shown in Figure 3 to the nearest µm. Show your working. *(2 marks)*

2 Copy and complete the table to summarise the roles of the two key enzymes involved in the process of blood clotting *(2 marks)*

Enzyme	Cofactors required	Substrate	Product

3 Thrombin belongs to a group of serine proteases, which also includes the enzyme trypsin. Trypsin breaks down proteins in the digestive system of many different organisms. Discuss the similarities and differences in the action of thrombin and trypsin. *(4 marks)*

3.7 Preventing blood loss

Specification reference: 2.1.3

Shock

Shock due to severe blood and fluid loss is referred to as hypovolemic shock. It is an emergency condition in which the heart is unable to pump enough blood around the body. This type of shock can cause many organs to stop working. Losing one fifth or more of the total amount of blood can cause hypovolemic shock.

Excessive blood loss can be due to bleeding from deep cuts or internal bleeding, for example from a ruptured spleen as a result of a car accident. In other situations, the amount of circulating blood may drop as a result of body fluid loss following burns, diarrhoea, excessive perspiration, or vomiting.

Learning outcomes

Demonstrate knowledge, understanding, and application of:

- → the risks of shock
- → first-aid procedures to prevent excessive blood loss.

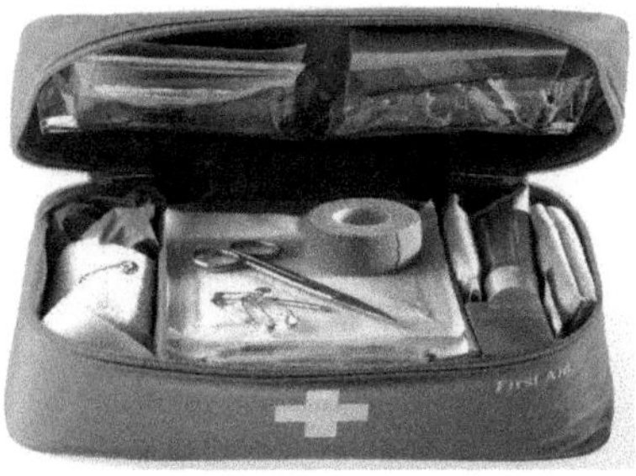

▲ **Figure 1** *A first-aid kit*

Preventing excessive blood loss

When a blood vessel is damaged in a normal situation, the blood clotting cascade is started and a blood clot is formed. However, if the wound is too large, or if an artery is cut, a blood clot is not usually able to stop the flow of blood and medical assistance is required.

First-aid procedure

If a person is bleeding profusely the following procedure should be followed:

1. Stay calm and carry out a primary survey.
   a. Danger – are you and the casualty free from danger? If not, make the situation safe first then assess the casualty.
   b. Response – is the casualty responsive? Can they hear or speak to you?
   c. Airway – open the casualty's airway by placing one hand on the casualty's forehead and gently tilting their head back, then lifting their chin using two fingers only.
   d. Breathing – look, listen, and feel for breathing for no more than 10 seconds – look to see if the chest is rising and falling, listen for breathing, and feel for breath against your cheek:
      - If the casualty is breathing normally, treat any life threatening injuries and place the casualty in the recovery position.
      - If the casualty's breathing becomes noisy, stop treatment and turn the casualty into the recovery position before continuing.
      - If the casualty is not breathing normally or you have any doubt whether breathing is normal begin CPR.
2. If possible obtain help from a doctor or trained first aider.
3. If available, put on sterile gloves.
4. Remove or cut clothing to expose the wound.

5 Check the site of the wound for any object embedded in it (e.g., glass).
- If there is no object in the wound, place a clean piece of cloth over the wound and press down firmly with the fingers or the palm of the hand. This will reduce the rate of blood loss.
- If there is an object in the wound, press at the sides of the wound, pushing the edges of the wound together. Do NOT remove the object.

6 If the wound is in a limb and it is possible to do so, raise the limb above the level of the heart.

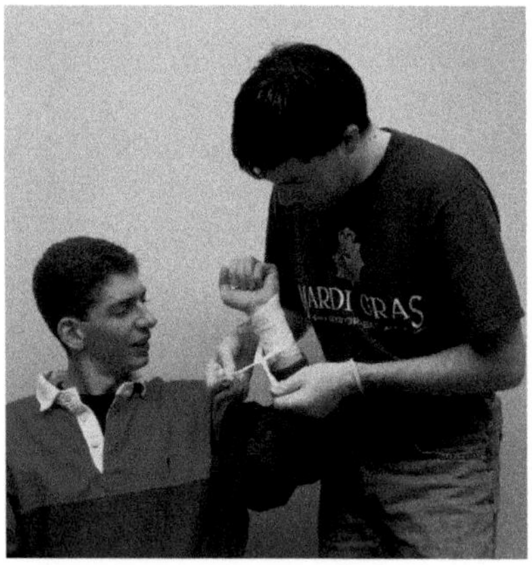

◀**Figure 2** *A trainee first aider ties a bandage around the wrist of a volunteer acting as a wounded casualty. The first aider is wearing disposable rubber gloves. A sterile surgical dressing (now covered) has been placed directly over the wound and bandaged firmly into place. This helps to stop the bleeding and allows the blood to clot. The first aider also keeps the casualty's arm raised above the level of his heart. This slows blood flow to the injured area*

7 If the casualty feels sick or faint (a natural reaction to shock) help the casualty to sit or lie down.

8 Use a bandage to maintain pressure on the wound. If blood soaks through the first dressing then apply a second dressing over the top of the first (do not remove the first dressing).

Why should you not remove objects present in wounds?

Summary questions

1 Suggest why it is important for the first aider to wear sterile, disposable gloves if they are available. *(3 marks)*

2 Suggest why it is important to raise the bleeding limb above the position of the heart, if possible. *(2 marks)*

3 Suggest why it is important not to remove the first dressing before applying the second dressing if blood has soaked through the first dressing. *(3 marks)*

3.8 Enzymes in medicine

Specification reference: 2.1.3

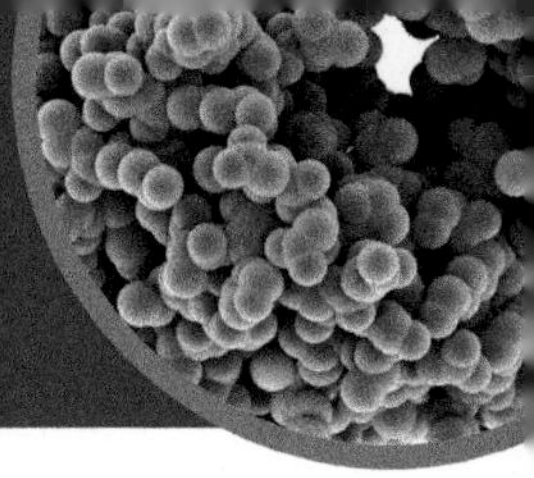

Uses of enzymes in medicine

Enzymes and enzyme inhibitors are used in many areas of medicine including diagnostics and treatment. A natural, reversible, competitive inhibitor found in the blood is antithrombin, a glycoprotein of 432 amino acids made in the liver. Antithrombin combines with thrombin and partially blocks its active site. This prevents the thrombin from forming fibrin. The role of antithrombin is to mop up any thrombin left over once the blood clot has formed and combine with it for approximately 20 minutes. After this time the blood clotting process is normally complete and no further thrombin will be being made.

Learning outcomes

Demonstrate knowledge, understanding, and application of:

→ the use of enzymes and enzyme inhibitors in medical diagnosis and treatment.

Diagnostic enzymes

Blood amylase

The pancreas and the salivary glands both produce an enzyme called amylase which hydrolyses carbohydrates. Amylase is released into the blood if the pancreas is inflamed or diseased. Testing for amylase in the blood is therefore used to help diagnose pancreatic disorders.

As well as producing amylase the pancreas also produces digestive enzymes and hormones such as insulin and glucagon. These digestive enzymes usually only become active when they reach the more alkaline conditions of the small intestine (where they are needed for food digestion). Sometimes these enzymes can become active in the pancreas (acute pancreatitis). When this happens the enzymes start digesting the tissue of the pancreas itself which causes swelling, haemorrhaging (bleeding), and damage to the pancreas and its blood vessels. Acute pancreatitis can be indicated by increased concentrations of blood amylase, urine amylase, and serum blood lipase.

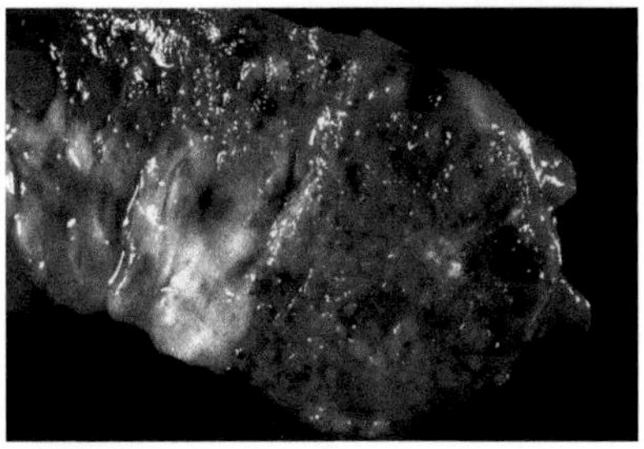

▲ **Figure 1** *A pancreas with acute pancreatitis (inflammation). Acute pancreatitis causes severe pain with shock. Its cause is not always known, but can be associated with gallstones or alcoholism. Treatments include surgical drainage and drugs to aid the pancreatic functions*

Lactic Acid Dehydrogenase (LDH)

Most tissues in the body have the enzyme lactic acid dehydrogenase (LDH). More LDH is released into the blood when tissues are damaged by disease or injury. It is therefore possible to use the presence LDH in the blood to screen for tissue damage. The level of damage can vary and can be categorised as:

- acute (e.g., due to traumatic injury)
- chronic (e.g., due to a long-term condition such as liver disease).

Conditions that can cause increased LDH in the blood include:

- liver disease
- heart attack
- cancer
- muscle trauma

- bone fractures
- certain types of anaemia
- infections, such as meningitis, encephalitis, and HIV.

Levels of LDH can also help determine how well chemotherapy is working during treatment for lymphoma and monitor progressive conditions such as muscular dystrophy.

LDH levels are measured from a sample of blood taken from a vein. An LDH test is useful in diagnosing tissue damage, but other tests are usually necessary to pinpoint the exact location of the damage. One such test is called the LDH isoenzymes test. LDH isoenzymes are five different forms of the LDH enzyme that are found in specific concentrations in different organs and tissues.

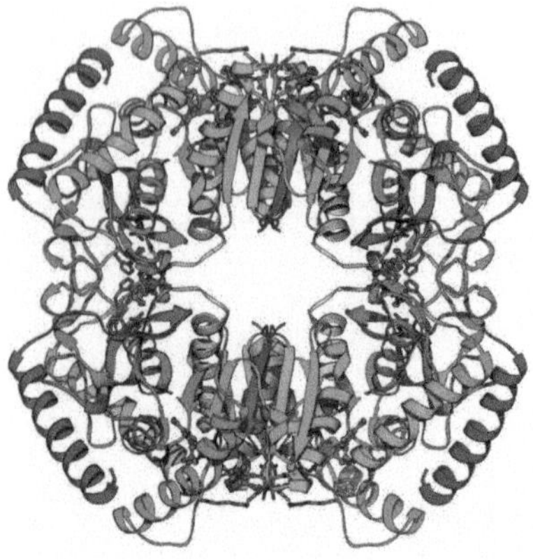

▲ **Figure 2** *A molecular model of lactate dehydrogenase enzyme*

- LDH-1 is found primarily in heart muscle and red blood cells.
- LDH-2 is concentrated in white blood cells.
- LDH-3 is highest in the lung.
- LDH-4 is highest in the kidney, placenta, and pancreas.
- LDH-5 is highest in the liver and skeletal muscle.

By measuring the blood levels of these different isoenzymes, doctors can determine the type, location, and severity of the cellular damage.

Enzyme inhibitor treatments

Aspirin

Aspirin is a medication that has two main actions in the body:

- an anti-prostaglandin (to treat inflammation, fever and provide pain relief)
- an anti-platelet agent.

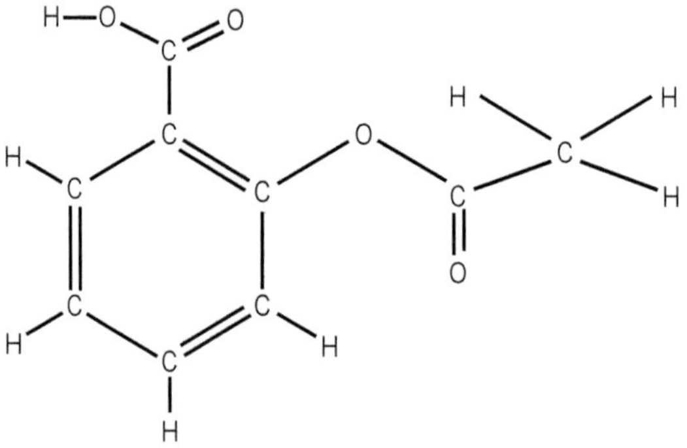

▲ **Figure 3** *Chemical structure of aspirin*

Both of these actions are the result of the effect of aspirin inhibiting the enzyme cyclo-oxygenase (COX). COX catalyses the formation of prostaglandins. Prostaglandins are a group of inflammatory chemicals that are synthesised from arachidonate, a fatty acid. The inhibitory effect of aspirin is achieved by the addition of an acetyl group (—CH_3CHO) on to an amino acid very close to the active site of COX. This stops the arachidonate from binding to the active site, subsequently preventing the formation of prostaglandins when the body is injured.

COX is also involved in stopping bleeding. COX activates thromboxane A2, a chemical which causes platelets to aggregate (stick together) to form a 'plug' over the wound. The reduction in the aggregation of platelets explains why aspirin is often inaccurately referred to as a 'blood thinner' or 'anti-platelet agent' and why bleeding is a side effect of aspirin.

Whilst significant bleeding is not common in people who take aspirin, it can occur. The 'blood thinning' effect of aspirin often means it is prescribed to prevent heart attacks and strokes.

▲ **Figure 4** *A white willow (Salix alba) branch with catkins (flowers) and leaves. The bark and leaves of this plant were originally the source of the active ingredient in aspirin, salicylic acid*

Warfarin

Warfarin is sometimes referred to as a 'blood thinner', but this is inaccurate as it does not affect the viscosity of blood. Instead warfarin inhibits the vitamin K-dependent synthesis of clotting factors II, VII, IX, and X, as well as the regulatory factors.

Warfarin inhibits the enzyme epoxide reductase, which causes a decrease in the the amount of available vitamin K and vitamin K hydroquinone in the tissues. This then inhibits the carboxylation activity of the glutamyl carboxylase and prevents the formation of prothrombin in the liver cells.

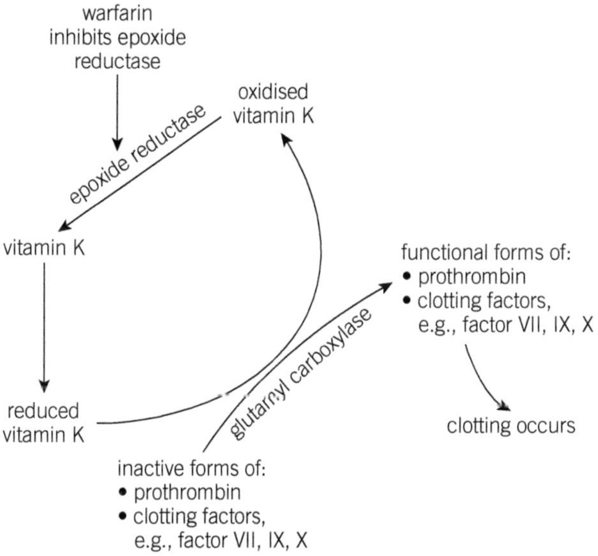

▲ **Figure 5** *An increase in the amount of warfarin leads to less carboxylated vitamin K. Consequently less factor II, VII, IX, and X is carboxylated and inhibition of clotting occurs*

Warfarin is therefore used to prevent blood clots from forming or increasing in size in the blood and blood vessels. It is prescribed for people with certain types of irregular heartbeat, people with prosthetic (replacement or mechanical) heart valves and people who have suffered a heart attack. Warfarin is also used to treat or prevent venous thrombosis (swelling and blood clot in a vein) and pulmonary embolism (a blood clot in the lung). One of the side effects of warfarin is that it can cause rapid and unnoticed bruising after minimal trauma.

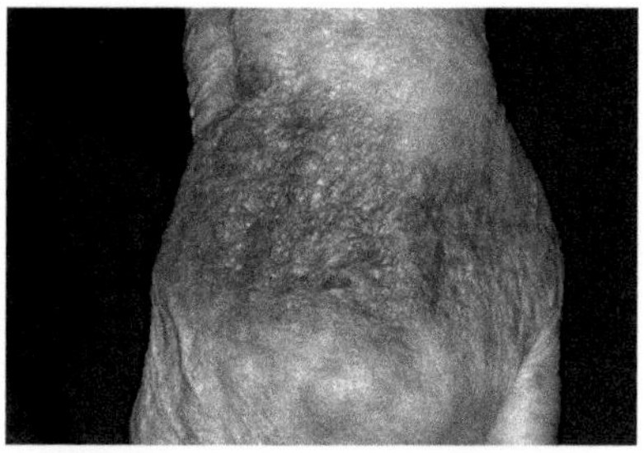

▲ **Figure 6** *The wrist of an elderly female patient, showing spontaneous bruising (bleeding) of the skin from the anti-clotting drug warfarin*

Streptokinase

Streptokinase is used in thrombolytic therapy. Along with tissue plasminogen activator (t-PA) it is a form of clot dissolving medication. Streptokinase is an enzyme secreted by many species of *Streptococcus* bacteria as a toxin. It is a fibrinolytic drug that hydrolyses plasminogen to plasmin (by cleaving a peptide bond), which breaks down fibrin and dissolves clots. Streptokinase is given intravenously (directly into a vein) as soon after a heart attack as possible to dissolve any clots in the arteries of the heart wall.

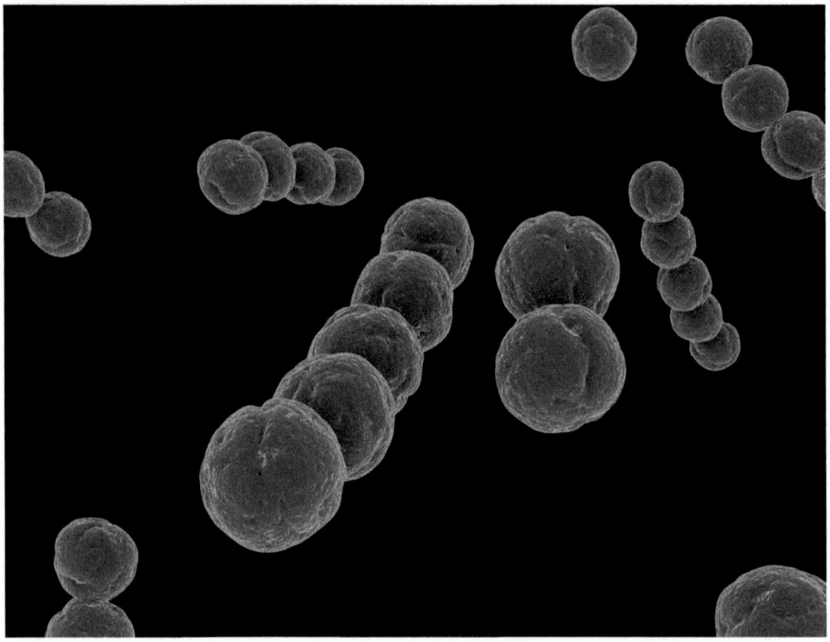

▲ **Figure 7** *Streptococci bacteria, approximately × 20 000 magnification*

However, as it is a bacterial enzyme it is possible for a patient to build up immunity to streptokinase so the drug is only administered again four days after the initial dose. This second dose may not be as effective and it may cause an allergic reaction. Consequently, it is normally only used for a patient's first heart attack. Today there are genetically engineered drugs that can be used instead, but they are usually more expensive, so streptokinase is still widely used.

Synoptic link

You find out how pathogens such as bacteria cause disease in Topic 11.1, Communicable diseases.

You will find out more about immunity and allergic reactions in Topic 12.2, Antibodies and immunity.

Summary questions

1 Veins and arteries vary in size from one patient to another and from one side of the body to the other. Drawing blood from some people may be more difficult than from others. Suggest two risks that may be associated with having a blood test. *(2 marks)*

2 Evaluate the use of enzymes in medicine. *(4 marks)*

3 Suggest how the isoenzyme LDH-5 may:

  a be the same *(1 mark)*

  b differ to the other forms of LDH *(2 marks)*

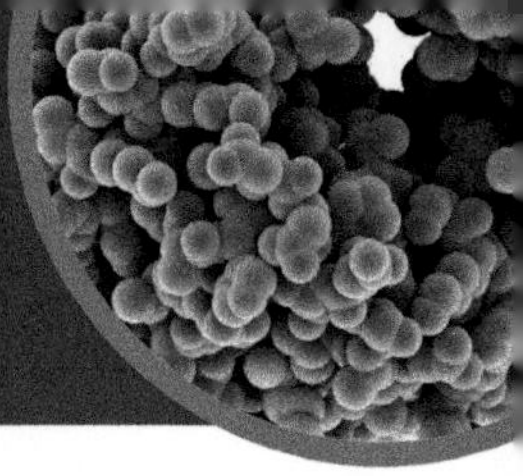

3.9 Blood donation

Specification reference: 2.1.3

Giving blood

25% of us will need a blood transfusion at some point in our lives, but currently only 4% of adults give blood. Each blood donation can help up to three people. Donated blood is used in long-term treatments, not just in emergency situations:

- 30% is used for general surgery
- 11% is used to treat gastro-intestinal bleeding
- 30% is used to treat anaemia
- 6% is used for treating maternity patients
- 23% is used for other uses.

Learning outcomes

Demonstrate knowledge, understanding, and application of:

→ how blood is donated

→ how blood products are stored and used

→ how and why blood is screened.

Blood donations

Although all blood is made of the same basic elements, there are differences in the blood between individuals. In fact, there are eight different common blood types. These are determined by the presence or absence of certain antigens (substances that cause an immune response if they are foreign to the body).

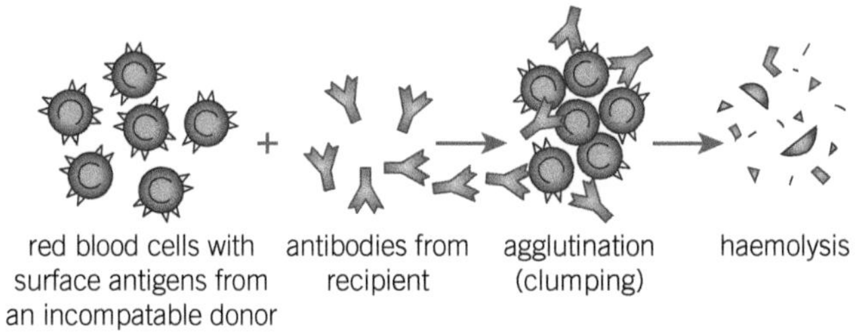

▲ **Figure 1** *The process of agglutination and haemolysis*

The four major **blood groups** are determined by the presence or absence of two antigens, antigen A and antigen B, on the surface of the person's erythrocytes:

- Group A – only the A antigen on the erythrocytes (and also B antibodies in the plasma)
- Group B – only the B antigen on the erythrocytes (and also A antibodies in the plasma)
- Group AB – both A and B antigens on the erythrocytes (but neither A nor B antibodies in the plasma)
- Group O – no A antigens or B antigens on the erythrocytes (but both A and B antibodies in the plasma).

Safe blood transfusions depend on careful blood typing and cross-matching. If a non-compatible blood type was transfused, its antigens could trigger an immune response. Blood typing involves mixing different samples of blood with antibodies. The results from a blood group test can be seen in Figure 2 – blood types are shown in rows and antibody serums are shown in columns.

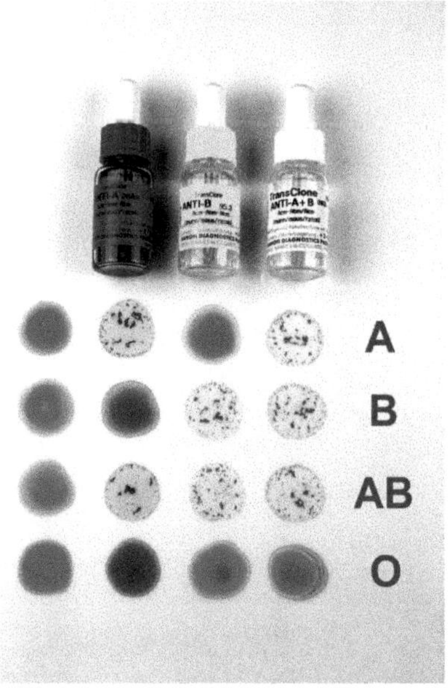

▲ **Figure 2** *Blood group test – the first column shows the blood's appearance before the tests*

As well as the A and B antigens, there is a third antigen found on the surface of the erythrocytes which is called the Rhesus (Rh) factor. Individuals are either Rh-positive (the antigen is present) or Rh-negative (the antigen is absent). In general, Rh-negative blood is given to Rh-negative patients and Rh-positive blood or Rh-negative blood may be given to Rh-positive patients. The most common blood type in the UK is O Rh-positive (O positive).

Who can give blood?

A person can donate blood, providing they are in good health, if they are between 17 and 66 years old (if it is their first donation) and weigh at least 50 kg. However, if they are female, younger than 20 years old, weigh less than 65 kg, and are under 168 cm in height, it will be necessary to estimate their blood volume before they can donate.

Male donors can give blood every 12 weeks and female donors can give blood every 16 weeks.

There are a variety of reasons why some people are not allowed to give blood:

- if donating blood could potentially harm the person donating, for example, if it could result in anaemia
- if the donated blood could potentially harm the patient receiving it, for example, if the potential donor has taken antibiotics in the last seven days or has had an infection in the last two weeks.

You may also be unable to give blood if you:

- have had an infection in the last two weeks
- have any of the following symptoms – chesty cough, sore throat and/or cold sore
- have had jaundice or hepatitis in the last year
- have had semi-permanent make up, a tattoo, or any skin piercing in the last four months
- have had acupuncture in the last four months that was not carried out within the NHS or by a qualified Healthcare Professional
- have received blood anywhere in the world since 1st January 1980
- have ever injected drugs
- are male and have had sex in the last 12 months with another man
- are female and have had sex in the last 12 months with a man who has had sex with another man.

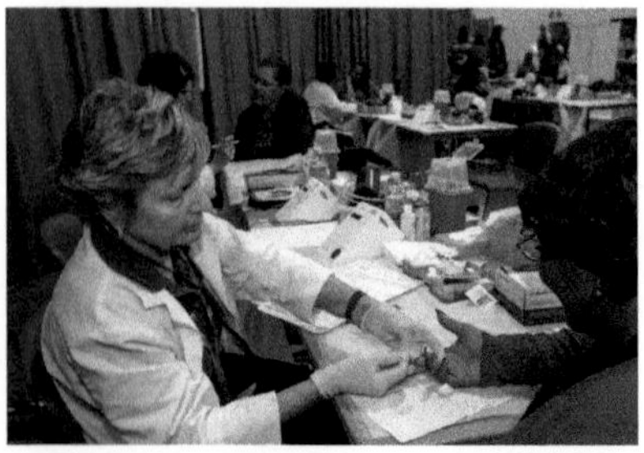

▲ **Figure 3** *A donor carer taking a sample of blood by pricking the skin on a woman's fingertips*

What happens during a blood donation session?

1 Welcome and preparation

Before donating blood, the donor should eat regular meals, drink plenty of non-alcoholic fluids, and avoid vigorous exercise. On arrival, the donor will be asked to drink a pint of water and read a welcome leaflet, which explains the importance of blood safety.

2 Health screening

A donor carer will then give the donor a private health screening to confirm the donor's identity and ask a number of confidential questions to ensure that it is safe for the donation to proceed.

A quick pin-prick test will also be carried out to ensure the donor has sufficient haemoglobin in their blood before donating.

3 The donation

Whilst on a specialised chair or bed, a **sphygmomanometer** will be used to maintain a small amount of pressure during the donation. A donor carer will locate a suitable vein on the donor's arm and clean the area with an antiseptic sponge before inserting the needle. During donation the donor is advised to carry out muscle tension exercises to maintain blood pressure.

The blood being collected is constantly weighed and measured. Collection stops automatically when a full donation of 470 ml has been collected; this usually takes between 5 and 10 minutes. A small sample of blood is also kept for testing and screening.

4 After the donation

After the donation, the needle is removed and a sterile pressure roll and dressing are applied to the arm. The donor is offered a selection of snacks and drinks. Donors are instructed to stay at the donor centre for at least 15 minutes and to have at least two drinks.

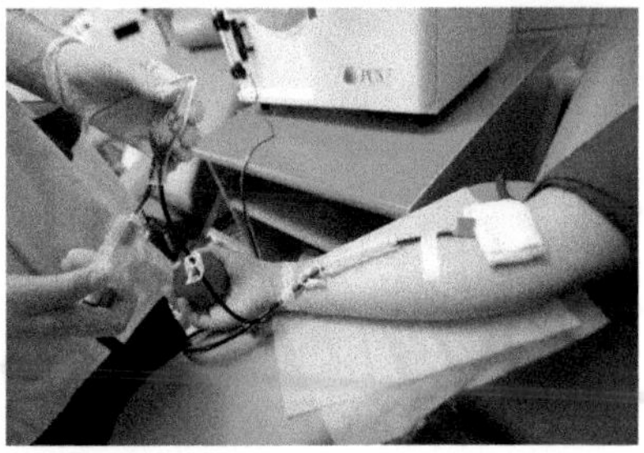

▲ **Figure 4** *A mobile donation unit*

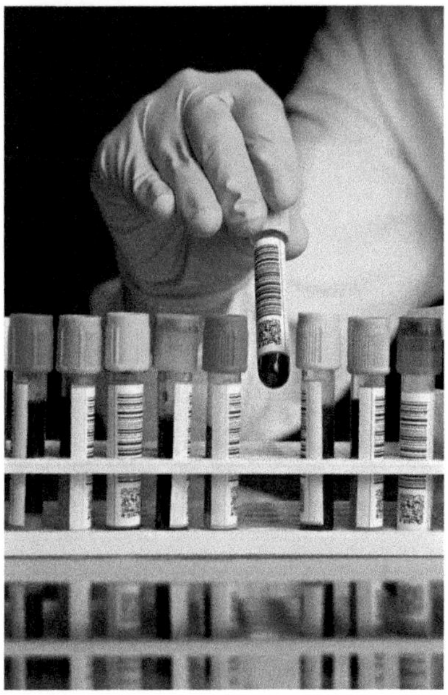

▲ **Figure 5** *Blood samples will be screened for microorganisms to prevent infections from being transmitted to the patients who receive the blood*

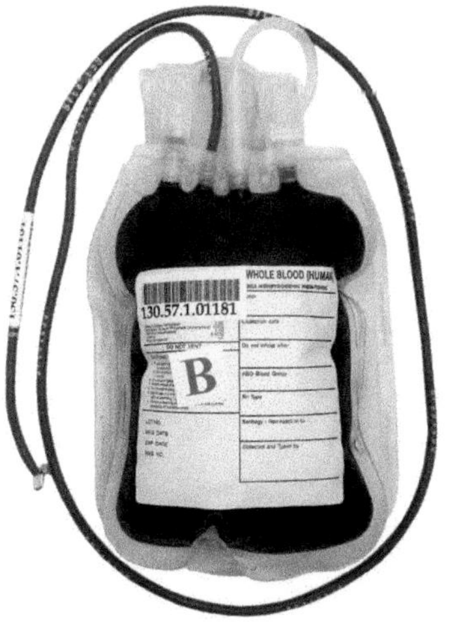

▲ **Figure 6** *Blood donation bag. The bag contains about half a litre of blood – an average adult has between 4.5 and 6 litres of blood*

How is blood stored?

It is essential that the blood is stored in the correct conditions to ensure that it can be used effectively.

The blood cold chain is a system for storing and transporting blood and blood products, within the correct temperature range and conditions, from the point of collection from blood donors to the point of transfusion to the patient.

Most of the blood collected from donations is transported to blood transfusion centres where it is processed into different forms. These different forms have specific uses.

▼ **Table 1** *How stored blood is used in medicine*

Type of blood storage product	Use	Shelf life
whole blood	This consists of plasma, white blood cells (leucocytes), and red blood cells (erythrocytes). This is rarely used today except in cases of severe blood loss.	up to 24 hours
leuco-depleted blood	This is whole blood with the leucocytes removed. It is used for patients who receive regular transfusions.	up to 42 days

▲ **Figure 7** *Technician at a blood transfusion centre placing bags of donated blood plasma onto a shelf in a refrigerator, where they will be stored for up to one year*

▼ **Table 1** *Continued*

Type of blood storage product	Use	Shelf life
packed red cells	This is also called red cell concentrate. The erythrocytes are separated from the rest of the blood. When it is needed, the concentrate is diluted with a solution containing sodium chloride, adenine, glucose, and mannitol (SAGM). No white blood cells or platelets are present in the transfusion. This is often used to treat patients with anaemia, or after surgery or childbirth.	up to 35 days
platelets	This is used to treat patients who have bone marrow failure, or following transplant, chemotherapy, or leukaemia treatments.	up to 7 days
clotting factors	Some people have genetic disorders that result in low or no clotting factors being produced. This means that their blood is slow to clot (e.g., Factor V Leiden thrombophilia is an inherited disorder of blood clotting).	variable
plasma	Plasma is blood where all cells and platelets have been removed. Fresh frozen plasma is made by centrifuging whole blood within two hours of it being collected. It can be used in cardiac surgery to reverse any anti-coagulant treatment, and when a woman has lost blood during childbirth. It can also be used to replace clotting factors after major transfusions or to treat liver disease.	1 year

How is blood screened?

The small sample of blood that is taken at the same time as the main donation is screened for a variety of infections:

- Syphilis – A sexually transmitted infection caused by the bacterium *Treponema palladium*. Donated blood is tested for the presence of specific antibodies to the bacterium. A positive test can indicate a current or previous infection. In either case the blood cannot be used.
- Hepatitis virus – Infection by either hepatitis B or C viruses can cause inflammation and damage to the liver. Two tests are carried out on the donated blood. One tests for presence of the hepatitis surface antigen (outer surface of the virus). The other tests for the presence of viral nucleic acid (indicating presence of the virus itself).

- Human immunodeficiency virus (HIV) – Donated blood is tested for the presence of a protein on the outer surface of the virus and to the antibody of the virus. If any of these tests are positive further tests are carried out to confirm the result. In contrast to many other infections, the antibodies produced do not protect against the virus.
- Additional tests – Depending on the individual circumstances of the donor further tests may be carried out, particularly in relation to skin piercing and travel.

In all cases a blood sample is added to antigens for these diseases. If antibodies are present they will attach to the antigens and give a positive result.

Synoptic link

You will learn more about infections, antigens and antibodies in Topics 12.1, The immune system and 12.2, Antibodies and immunity.

Summary questions

1 Explain why plasma and serum can be frozen but whole blood is stored at 4 °C. *(3 marks)*

2 The table below shows the number of people in 2011 acquiring different conditions as a result of blood transfusions. Draw an appropriate graph to present the data in the table. *(5 marks)*

Acquired condition	Number of people
Allergic reaction	180
Anaphylaxis	33
Fluid overload	71
Acute haemolytic transfusion reaction	10
Bacterial blood poisoning	0
Viral contaminated blood	0
Variant Creutzfeldt-Jakob disease (vCJD)	4

3 Suggest why:

a It is important that patients who have aplastic anaemia (a disorder where the bone marrow does not produce enough blood cells) who receive regular transfusions should be given leuco-depleted blood. *(3 marks)*

b It is common to use red cell concentrate to treat people with anaemia rather than whole blood. *(2 marks)*

Practice questions

The diagram shows some of the bonds which may be found in proteins.

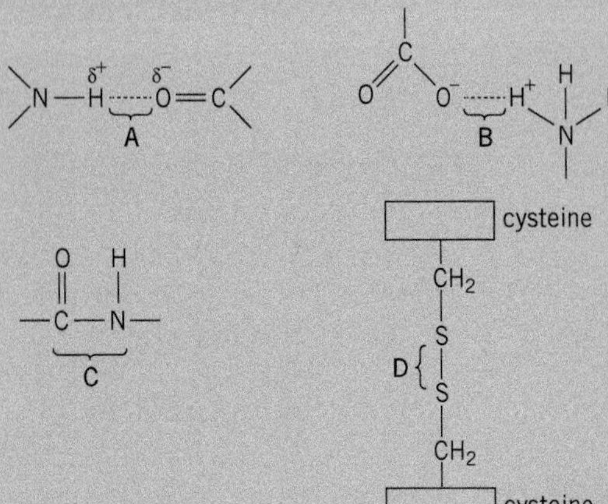

1 Which of the following statements about the bonds is/are correct?

Statement 1: bonds A,B,C are found in the tertiary structure of a protein.

Statement 2: bonds C and D are not broken by high temperatures.

Statement 3: bond A is found in alpha helix and beta pleated sheets.

A 1, 2 and 3

B Only 1 and 2

C Only 2 and 3

D Only 1 *(1 mark)*

2 Which of the bonds in the diagram (A, B, C, or D) is formed by a condensation reaction? *(1 mark)*

3 Paper chromatography can be used to identify amino acids. The Rf values of 3 amino acids were found to be as follows:

Amino Acid	Rf value
X	0.93
Y	0.85
Z	0.58

The amino acid methionine was run in the same solvent. The solvent front was measured at 25 cm Methionine then formed a 'spot' at 14.5 cm when a chemical was sprayed on the chromatogram.

a (i) Name the chemical sprayed on the chromatogram to visualise the amino acid. *(1 mark)*

(ii) Identify which amino acid, X Y or Z corresponds to methionine. Show your working. *(2 marks)*

b The diagram shows the R group of the amino acid phenylalanine.

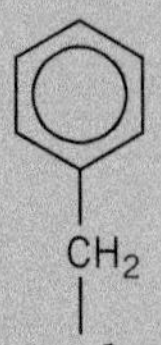

Draw a diagram of the complete amino acid. *(3 marks)*

4 When the enzyme catalase is added to hydrogen peroxide, the following reaction occurs:

$$H_2O_{2\,(l)} \xrightarrow{\text{(catalase)}} 2\,H_2O_{\,(l)} + O_{2\,(g)}$$

In an investigation into the effect of temperature on the rate of this reaction, a student set up the apparatus as shown in the diagram, using liquidised celery as a source of catalase.

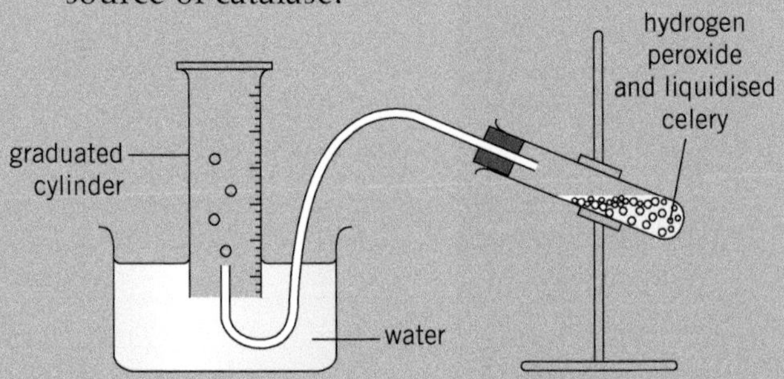

a (i) State the other variable that needs to be measured in order to calculate the rate of reaction *(1 mark)*

(ii) Identify one potential problem with using samples of liquidised celery as a source of catalase in this investigation **and** suggest a way to minimise this problem. *(2 marks)*

b Another student carried out a similar procedure and presented their results as a graph.

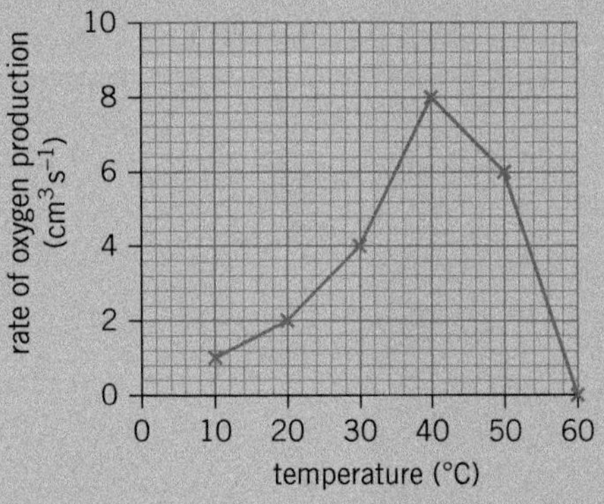

(i) Q_{10} is a measure of the increase in the rate of reaction for a 10 °C rise in temperature. It is calculated using the following formula:

$$Q_{10} = \frac{\text{rate at (t + 10 °C)}}{\text{rate at t °C}}$$

Where t + 10 °C is the rate at the higher temperature and t is the rate at the lower temperature.

Using the information in the graph, calculate Q_{10} between 15 °C and 25 °C. Show your working. *(1 mark)*

(ii) In the conclusion to this experiment, the student wrote the following:

As the heat increases, the reaction went faster until it got to its highest. After this the rate of reaction fell. This happened because the enzyme was killed and the hydrogen peroxide could not fit into the enzymes key site.

Suggest a more appropriate word to replace each of the underlined words. *(4 marks)*

5 If skin is damaged due to injury, a blood clot forms. The diagram shows how a blood clot forms if tissue is damaged.

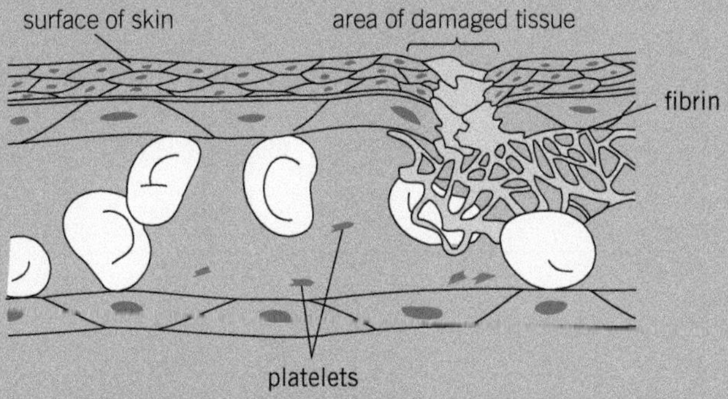

a Outline how each of the following is involved in the blood clotting process:

Damage to tissue, platelets, fibrin. *(3 marks)*

The blood clotting process is controlled by a series of reactions involving enzymes. The reaction that produces fibrin is catalysed by the enzyme thrombin.

b Explain why thrombin does not catalyse any other reaction in the blood clotting process. *(2 marks)*

c Some anti-clotting substances such as heparin, work by inactivating thrombin in the clotting process. The inactivation of thrombin stops the formation of fibrin and so blood does not clot.

The figure shows thrombin with its substrate and the molecule heparin.

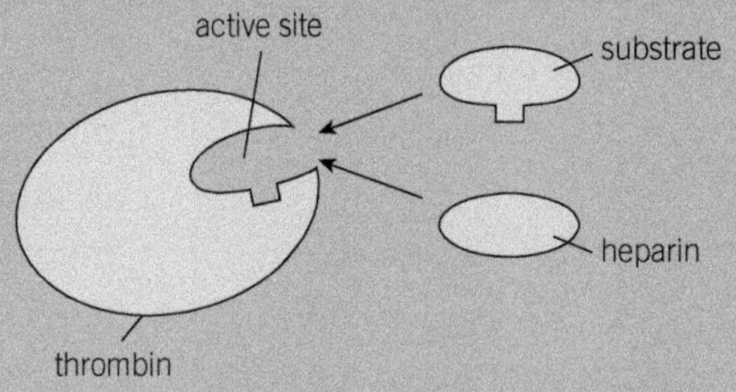

Using the information in the figure, suggest how heparin may inactivate thrombin. *(2 marks)*

d Patients who are taking heparin are sometimes advised not to take drugs such as aspirin. Suggest what justification could be given for this advice. *(2 marks)*

6 All blood donated in the United Kingdom is labelled according to its blood type.

Only 4% of adults in the UK are blood donors.

- In 2013 the population of the UK is 64.1 million
- 43% of people are blood group O.

a Calculate how many blood donors are likely to belong to blood group O. Give your answer to 3 significant figures. *(1 mark)*

b The table shows some results from blood typing carried out on four different blood samples. A tick indicates that agglutination has occurred. A cross indicates that no agglutination has occurred.

Blood sample	Anti B antiserum	Anti A antiserum	Anti D antiserum
1	✓	✓	✗
2	✗	✗	✓
3	✗	✗	✗
4	✓	✓	✓

(i) Which sample(s) correspond to blood group O? *(1 mark)*

(ii) Which sample(s) correspond to Rhesus negative blood? *(1 mark)*

4 NUCLEIC ACIDS

4.1 Nucleotides and nucleic acids

Specification references: 2.1.4

Learning outcomes

Demonstrate knowledge, understanding, and application of:

- → the structure of nucleotides
- → how to join nucleotides to form polynucleotides
- → the variety and importance of specialised nucleotides.

Study tip

Take care with purines and pyrimidines. It is a common error to confuse the two.

The discovery of nucleotides

Phoebus Levene (1869–1940), a Lithuanian-American biochemist, discovered the order of the components of nucleic acids, that is, phosphate–sugar–base. He coined the term **nucleotide**. He also discovered the sugar in RNA was ribose and the sugar in DNA was deoxyribose. Levene identified the way the nucleic acids RNA and DNA are put together.

Structure of nucleotides

Nucleotides (monomers) can be joined together to form polynucleotides (polymers). Individual nucleotides have a common structure composed of three distinct parts:

- Pentose sugar – two types occur, ribose, $C_5H_{10}O_5$, and deoxyribose, $C_5H_{10}O_4$.
- Phosphate group, PO_4^{2-} – found in all nucleotides. This group is slightly acidic and is the negatively charged part of the nucleic acid.

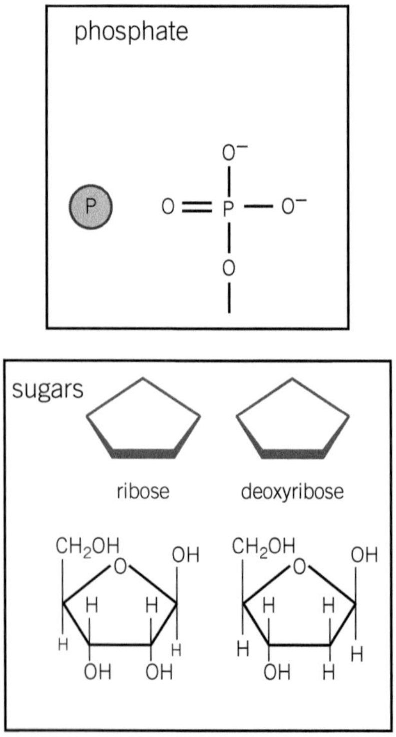

▲ **Figure 1** *Components of nucleotides*

Adenine (a purine)	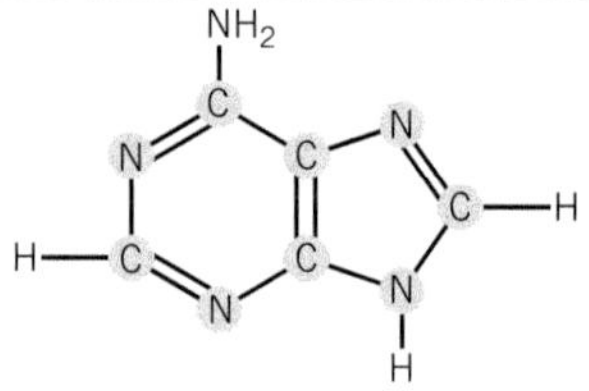	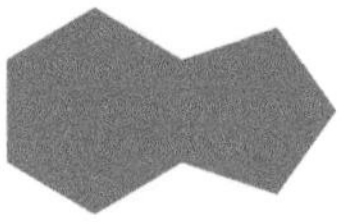 Adenine
Guanine (a purine)	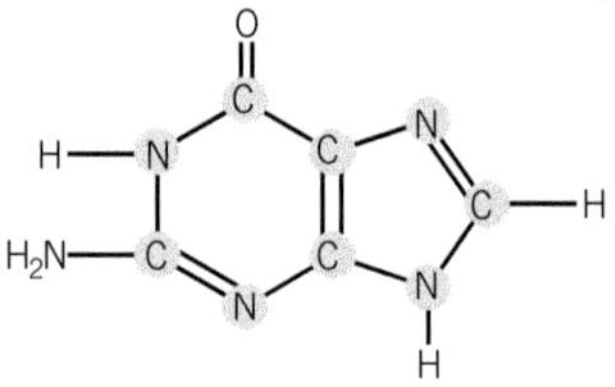	 Guanine
Cytosine (a pyrimidine)	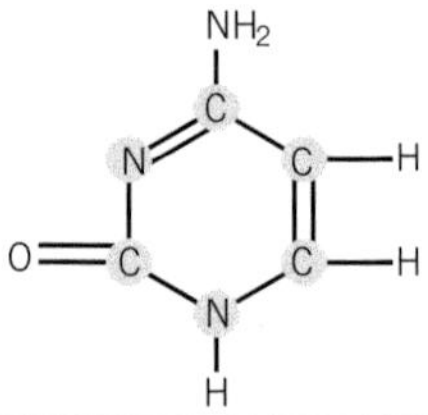	 Cytosine
Thymine (a pyrimidine)	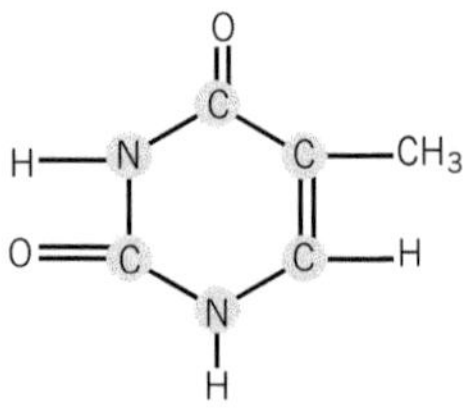	 Thymine

▲ **Figure 2** *Types of nitrogen containing bases in the nucleic acids DNA and RNA*

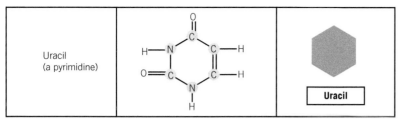

Uracil (a pyrimidine)	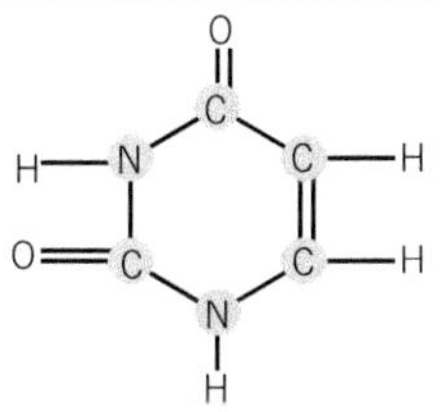	 Uracil

▲ **Figure 2** *continued*

- Nitrogen containing organic base – there are five different bases, which can be divided into two groups:
  - **purines** – these are double ring structures – two different purines are found in nucleic acids – guanine and adenine
  - **pyrimidines** – these are single ring structures – three different forms are found in nucleic acids – cytosine, thymine, and uracil.

Joining nucleotides

The bonds holding the phosphate group to the sugar and the base to the sugar in all nucleotide molecules are products of condensation reactions. Water is eliminated when they form. In both cases, the oxygen to form the water comes from the —OH groups of the pentose sugar molecule.

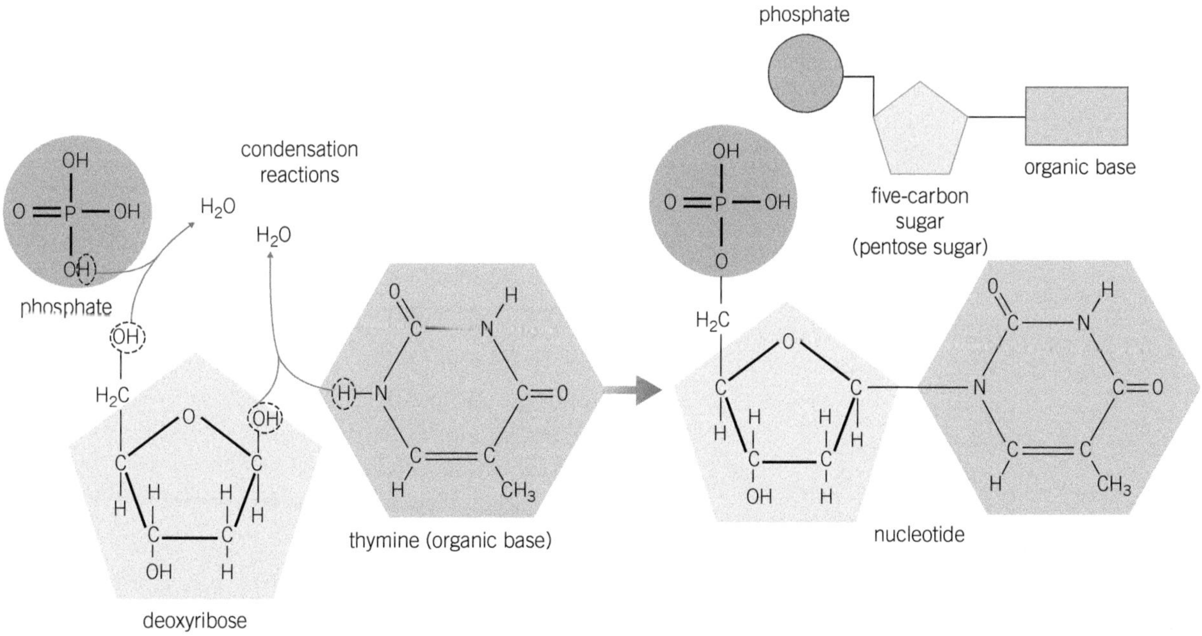

▲ **Figure 3** *Formation of a DNA nucleotide using the nitrogenous base thymine*

Two individual nucleotides can be joined by a condensation reaction to form a dinucleotide. If this process is repeated many times a polynucleotide is formed (for example, RNA, DNA).

Summary questions

1 How many condensation reactions are needed to form a dinucleotide from individual pentose sugars, nitogenous bases, and phosphate groups? Explain your answer. *(4 marks)*

2 Compare the structure of cytosine and thymine. *(4 marks)*

3 Explain what is meant by the term 'universal energy carrier'. *(3 marks)*

The role of nucleotides

There are many biologically important molecules that contain nucleotides.

▼ **Table 1** *Molecules that contain nucleotides*

Molecule	Abbreviation	Function
deoxyribonucleic acid	DNA	contains the genetic information of cells
ribonucleic acid	RNA	several forms exist – all of which play an important role in protein synthesis
adenosine monophosphate	AMP	coenzymes, important in releasing energy for metabolic activities, muscular contraction, cell signalling, and so on
adenosine diphosphate	ADP	
adenosine triphosphate	ATP	
nicotinamide adenine dinucleotide	NAD	coenzymes – electron (hydrogen) carrier, important in the Krebs cycle and electron transport chain
flavine adenine dinucleotide	FAD	coenzymes – electron (hydrogen) carrier, important in the Krebs cycle and electron transport chain
coenzyme A	CoA	an important enzyme in respiration – combines with pyruvate to form acetyl CoA to transfer acetyl groups to the Krebs cycle

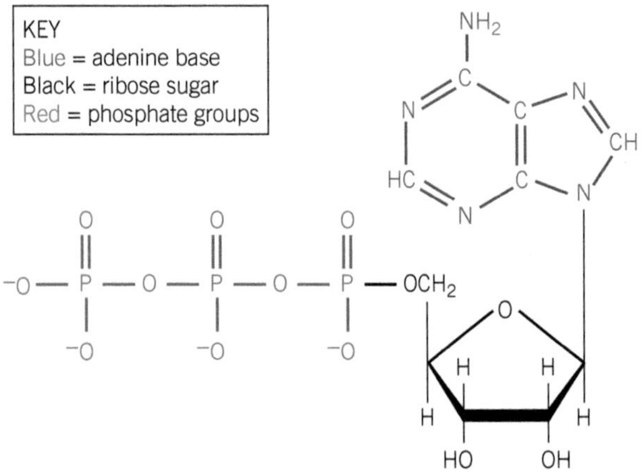

▲ **Figure 4** *The components of ATP*

Adenosine triphosphate (ATP)

Adenosine triphosphate (ATP) is a phosphorylated nucleotide. It is a short term energy store found in all cells. It is easily transported within cells and is referred to as the universal energy carrier.

ATP can be hydrolysed to release energy and form adenosine diphosphate (ADP). Further hydrolysis of ADP forms adenosine monophosphate (AMP). AMP and ADP can be reconverted to ATP by the addition of phosphate groups by phosphorylation.

Synoptic link

Hydrolysis and condensation reactions are covered in detail in Topic 2.3, Carbohydrates in cells and biofluids.

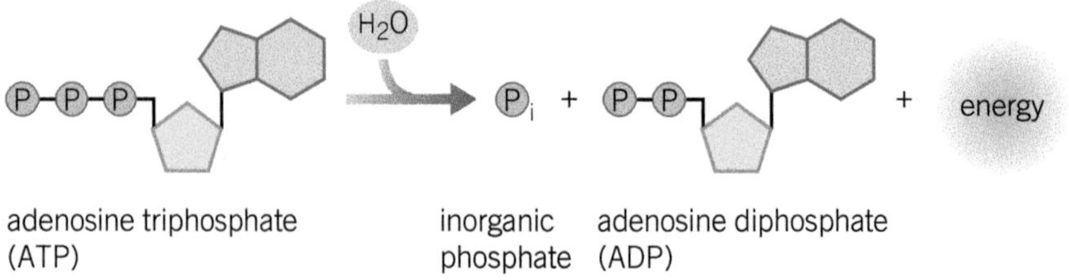

▲ **Figure 5** *The hydrolysis of ATP*

4.2 Polynucleotides – DNA

Specification references: 2.1.4

The discovery of DNA

Rosalind Franklin was a British biophysicist and X-ray crystallographer at King's College London. She made vital contributions to the understanding of the molecular structures of DNA (deoxyribonucleic acid), RNA, viruses, coal, and graphite. Franklin studied X-ray diffraction images of DNA, which led to Francis Crick and James Watson determining the structure of the DNA double helix. Unfortunately, Franklin's images of X-ray diffraction confirming the helical structure of DNA were shown to Watson without her approval or knowledge. This image provided valuable insight into the DNA structure, but Franklin's scientific contributions to the discovery of the double helix are often overlooked.

The structure of DNA

DNA stands for **deoxyribonucleic acid**. It is a double-stranded polymer of nucleotides made from the pentose sugar deoxyribose and the organic bases adenine, guanine, cytosine, and thymine. Each nucleotide chain is very long and can be made from many millions of nucleotides.

Chargaff's data

Edwin Chargaff was an Austrian biochemist who analysed DNA samples from a range of species. Chargaff's work led to the discovery of two crucial rules which were important in determining the structure of DNA:

1 In natural DNA, the number of guanine units equals the number of cytosine units and the number of adenine units equals the number of thymine units. In human DNA, for example, the four bases are present in these percentages: A = 31.0% and T = 31.5%, G = 19.1% and C = 18.4%.

2 The composition of DNA varies from one species to another, in particular in the relative amounts of A, G, T, and C bases.

▼ **Table 1** *The composition of DNA in several species*

Source of DNA	Group	Percentage of base within DNA (%)			
		Adenine	Guanine	Cytosine	Thymine
human	mammal	31.0	19.1	18.4	31.5
salmon	fish	29.7	20.8	20.4	29.1
sea urchin	invertebrate	32.8	17.7	17.4	32.1
wheat	plant	27.3	22.7	22.8	27.1
yeast	fungus	31.3	18.7	17.1	32.9
Mycobacterium tuberculosis	bacterium	15.1	34.9	35.4	14.6

Learning outcomes

Demonstrate knowledge, understanding, and application of:

→ the structure of DNA

→ the properties of DNA.

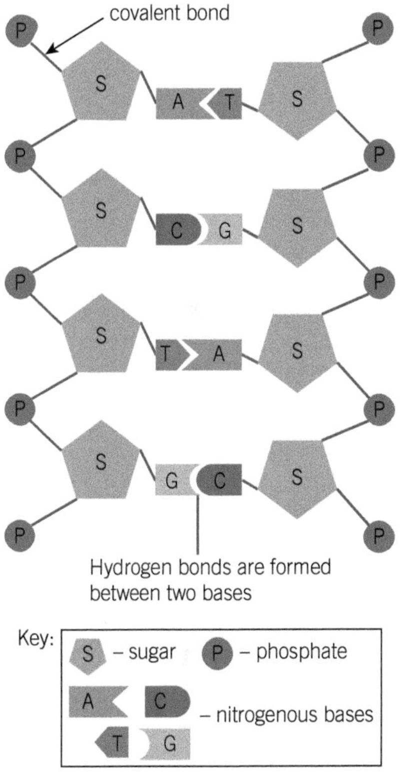

▲ **Figure 1** *A simplified diagram of DNA showing complementary base pairing*

Synoptic link

You first encountered hydrogen bonding in Topic 2.1, The properties of water.

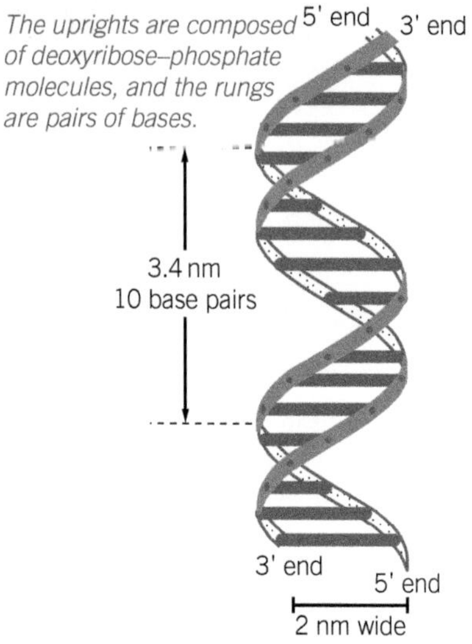

▲ **Figure 2** *The DNA double helix*

> **Study tip**
>
> Remember complementary base pairing using:
>
> A pairs with T: Alan Titchmarsh (famous horticulturist)
>
> G pairs with C: George Clooney (famous actor).

Properties of DNA

Stable molecule

Each polynucleotide strand is a polymer of nucleotides held together by *covalent bonds* called **phosphodiester bonds**. These bonds are strong and give the sugar-phosphate backbone strength. In eukaryotes the DNA is further stabilised by being coiled around proteins known as histones to form the chromosomes.

Antiparallel stands

The two polynucleotide strands are parallel to each other but they run in opposite directions hence they are described as **antiparallel**. One strand runs in the 5' to 3' direction and the other runs in the 3' to 5' direction.

Double helix

The two polynucleotide strands are wound together in a double helix.

Complementary base pairing

The two strands are held together by hydrogen bonding between the nitrogen containing bases. Adenine is always paired with thymine and guanine with cytosine. This is called **complementary base pairing** as A and T complement each other by forming *two* hydrogen bonds between them and G and C complement each other by forming *three* hydrogen bonds between them. Complementary base pairing was the conclusion drawn from Chargaff's evidence.

Worked example: Calculating base ratio

Calculating the ratios of bases led to the complementary base pairing theory. Base ratios can also be used to classify prokaryotes. Using Table 1, calculate the base ratio $\frac{A+G}{T+C}$ for humans.

Step 1: A + G = 31.0 + 19.1 = 50.1

Step 2: T + C = 31.5 + 18.4 = 49.9

Step 3: Ratio = $\frac{50.1}{49.9}$ = 1.004

Step 4: Ratio corresponds to 1 : 1

Calculate the base ratio $\frac{A+T}{G+C}$ for Mycobacterium.

Step 1: A + T = 15.1 + 14.6 = 29.7

Step 2: C + G = 35.4 + 34.9 = 70.3

Step 3: Ratio = $\frac{29.7}{70.3}$ = 0.422 : 1

Step 4: 0.4 : 1

Summary questions

1 Compare the base composition of *Mycobacterium tuberculosis* with the base composition of the eukaryotes in Table 1. *(2 marks)*

2 If the DNA of a new species if found to contain 32.6% of its base pairs as cytosine, calculate what percentage of the DNA will be adenine. Show your working. *(3 marks)*

3 Suggest a reason why the base pairings of cytosine to adenine and guanine to thymine are not found. *(2 marks)*

4 Bluetongue is caused by the pathogenic virus, Bluetongue virus (BTV).
The virus particle consists of ten strands of double-stranded RNA surrounded by two protein shells.
State what the base pairing would be in the section of double-stranded RNA with one strand having the sequence as shown: (Topic 4.5, Polynucleotides – RNA, may help with this question)
AUG GUC AAU CGG CGU AUC *(1 mark)*

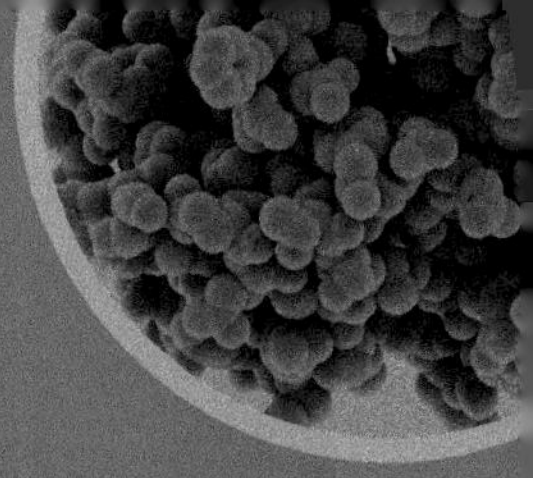

4.3 Semi-conservative DNA replication

Specification references: 2.1.4

Why must DNA be replicated?

The cells that make up any organism are always produced from pre-existing cells by the process of cell division. Before the cell can divide it is essential that the nucleus divides. This can occur if the total amount of DNA within the nucleus has been doubled, that is, the DNA must be copied exactly and very precisely. This is to ensure that all subsequent daughter cells contain the correct genetic information to produce the required proteins. For example, in the human small intestine the cells are in a state of continual replacement with the entire lining being replaced once a week on average.

Semi-conservative DNA replication

DNA replicates by a process called **semi-conservative DNA replication**, which takes place in the S phase of the cell cycle. Before eukaryotic cells can divide, there are four key requirements for DNA replication:

- four different types of DNA nucleotides – activated nucleotides with thymine, adenine, guanine, and cytosine nitrogen-containing bases
- the original DNA molecule to act as two template strands
- specific enzymes, including DNA polymerase and DNA ligase, to catalyse the reaction
- ATP to provide energy for the process.

The process of DNA replication is shown in the diagram below.

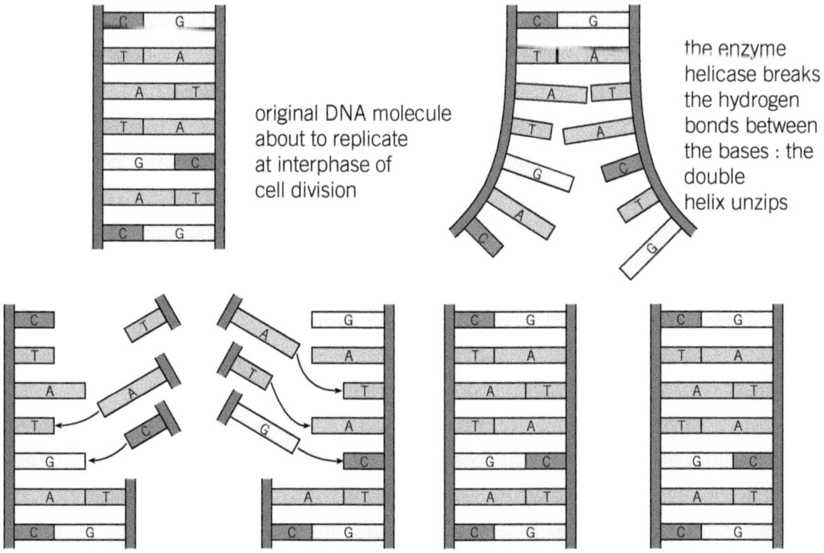
original DNA molecule about to replicate at interphase of cell division
the enzyme helicase breaks the hydrogen bonds between the bases : the double helix unzips
free nucleotides attach to each template DNA where bases are compatible
the nucleotides join up producing two identical DNA molecules

▲ **Figure 1** *The key stages in semi-conservative DNA replication*

Learning outcomes

Demonstrate knowledge, understanding, and application of:

→ how DNA replicates

→ the synthesis of a new DNA polynucleotide strand

→ the principle of semi-conservative DNA replication.

Synoptic link

You will find out more about the cell cycle in Topic 8.1, The cell cycle and mitosis.

Study tip

Remember semi-conservative replication can be broken down into:

semi = half (half of the genetic material is kept in the new DNA molecule)

conservative = the genetic information is kept the same (conserved), that is, the sequence of bases is the same

replication = the original DNA is copied, that is, there is twice as much DNA and each is the same as the original (a replica).

Summary questions

1. The bases on one template strand of a DNA molecule are ATTGGCCGTATGCGT. State the sequence of bases on the opposite strand of DNA. *(1 mark)*
2. With the aid of an annotated block diagram – show how two DNA nucleotides can be joined together *(5 marks)*
3. Figure 2 shows three principles that were first thought to be possible methods for DNA replication. These are called conservative replication, dispersive replication and semi-conservative replication. Describe which principle applies to A, B, and C. *(6 marks)*

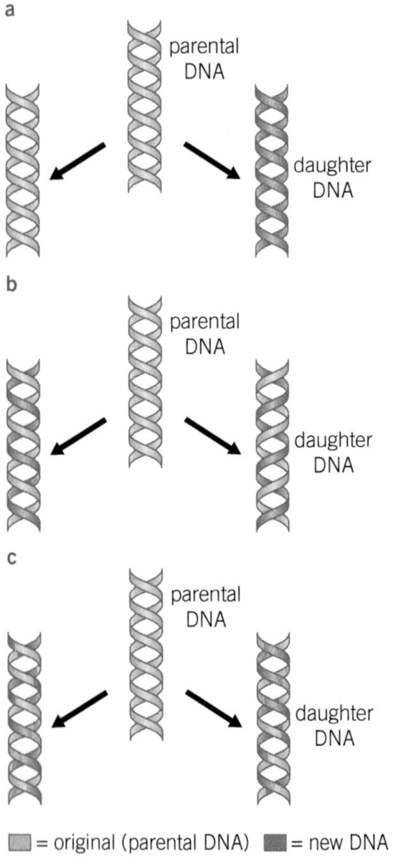

▲ **Figure 2** *Three possibilities for DNA replication*

1. The histones are removed from DNA
2. The DNA molecule is unwound by the enzyme DNA helicase which also breaks the hydrogen bonds between the base pairs (the DNA 'unzips')
3. The two DNA polynucleotide strands are exposed and act as template strands.
4. A DNA polymerase enzyme binds to each of the template strands
5. The DNA nucleotides are activated by the addition of two phosphate groups from ATP
6. The activated DNA nucleotides diffuse through the nucleoplasm and align next to their complementary bases on the exposed template strands
7. Hydrogen bonds join two complementary bases together
8. DNA polymerase joins each new nucleotide to its neighbour by catalysing a condensation reaction to form a 'phosphodiester bond'
9. Synthesis is slightly different on the two strands as they are antiparallel. DNA ligase is required to join up short nucleotide sequences (Okazaki fragments) on the 'lagging' strand while a continuous sequence of nucleotides is produced on the leading strand
10. The double stranded DNA molecules is rewound and wrapped around histones to form the chromosome copies.

The original DNA has now been replicated. Each new DNA molecule retains one of the DNA polynucleotide strands from the parent DNA molecule and also contains one newly synthesised DNA polynucleotide strand. The two new molecules of DNA both have exactly the same order of nitrogen-containing bases, which is vital for protein synthesis. There are specialised enzymes that 'proofread' the DNA to check that the DNA has been copied accurately. If a mistake is found the enzyme corrects it.

Accuracy of semi-conservative replication

Cellular error-checking mechanisms ensure near perfect reliability for DNA replication. The accuracy of DNA replication is essential in ensuring accurate protein production. If an enzyme is not made correctly this can have significant consequences for the cell or organism. In general, DNA polymerases are highly accurate, with an error rate of less than one mistake for every 10 million nucleotides added. In addition, some DNA polymerases also able to proofread as they go and remove nucleotides from the end of a growing strand to correct mismatched bases.

After replication has finished, DNA mismatch repair mechanisms monitor the DNA for errors, and can distinguish between mismatches in the newly synthesised DNA strand from the original strand sequence. These checking mechanisms mean that there is less than one mistake for every one billion nucleotides added. This is equal to a person copying 100 large (1000 page) dictionaries word for word, including punctuation, and only making one error in the whole process! The extent of proofreading in DNA replication determines the **mutation rate**, which is different in different species.

Mutations

A **mutation** is a change of the nucleotide sequence of the **genome** of an organism. Mutations may or may not cause noticeable changes in the **phenotype** (observable characteristics) of an organism. Mutations contribute to both normal and abnormal biological processes, for example, cancer, evolution, and immune system development. They occur randomly and spontaneously, but the risk of a mutation increases with exposure to chemical **mutagens** or radiation. Environmental factors influence the rate of mutations but do not usually determine the direction of the mutation. For example, exposure to harmful chemicals such as benzene may increase the mutation rate, but will not cause more mutations that make the organism resistant to those chemicals. Hence, mutations are random – whether a particular mutation happens or not is unrelated to how useful that mutation would be.

Other specialised enzymes involved in DNA replication

DNA polymerase only works in the 5'–3' direction of a new polynucleotide strand. As a DNA molecule is made from two DNA polynucleotide strands, both of these must be precisely copied. DNA polymerase synthesises new DNA on the 5'–3' strand, which is called the leading strand. This occurs in a continuous process in the direction of the replication fork.

The other strand of the DNA molecule is antiparallel and this runs in the 3'–5' direction. The polynucleotide strand that is formed on this template strand is called the lagging strand. This is produced by small sections of DNA nucleotides (fragments) being formed simultaneously in the opposite direction. The fragments are then joined later by the enzyme DNA ligase.

DNA polymerase also plays an important role in proofreading the newly synthesised DNA molecule. When an incorrect base pair is recognized, DNA polymerase moves backwards by one base pair of DNA. The 3'–5' *exonuclease* activity of the enzyme allows the incorrect base pair to be removed. The polymerase can then re-insert the correct base and replication can continue forwards.

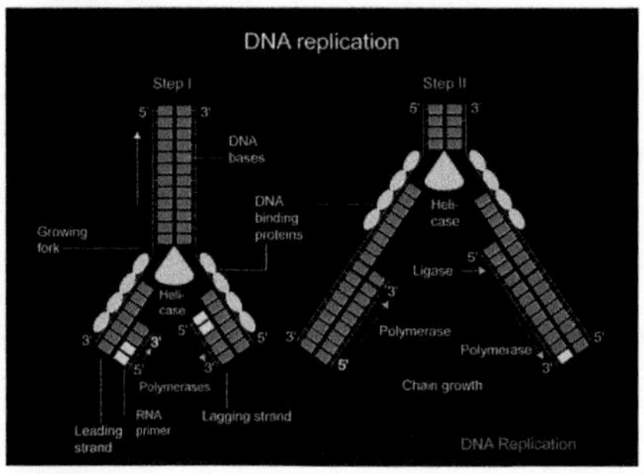

▲ **Figure 3** *DNA replication fork*

DNA helicase is an enzyme that consists of six globular polypeptide chains arranged in a sphere. The subunits arrange themselves so that one strand of the DNA molecule passes through the centre of the sphere and the other strand remains outside of the sphere. Energy from ATP is used to move the helicase along the DNA molecule and break the hydrogen bonds between the complementary bases. This unzips the DNA molecule. In order to unzip the DNA it must be unwound, so the helicase enzyme does this at the same time as it moves along the DNA molecule. The enzyme DNA topoisomerase stabilises the DNA molecule ahead of the helices enzyme.

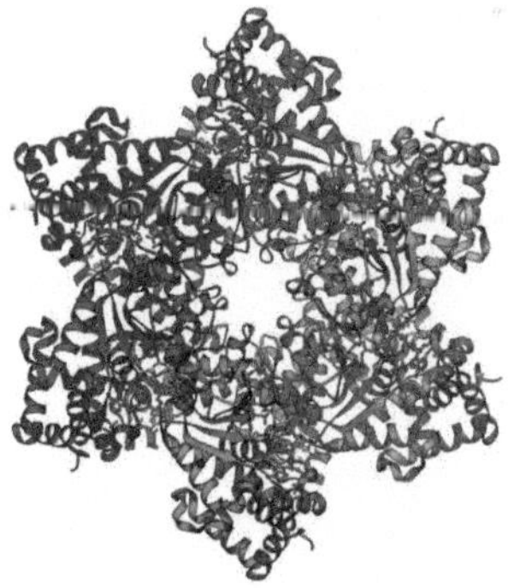

▲ **Figure 4** *DNA helicase structure*

1 Suggest why the DNA polymerase only works in one direction.

2 a Suggest what is meant by the term *'exonuclease'*.

  b Suggest how an exonuclease may work.

3 State and explain what level of protein structure DNA helicase has.

4.4 The genetic code

Specification references: 2.1.4

Learning outcomes

Demonstrate knowledge, understanding, and application of:

- → the genetic code
- → the features of the genetic code:
  - triplet
  - degenerate
  - non-overlapping
  - universal.

Study tip

Only the sequence on one of the DNA strands is used in the making of a protein. This strand is referred to as the template or antisense strand. The other strand is referred to as the sense strand.

The genetic code

You now know how DNA is able to copy itself prior to cell division. DNA also acts as a store of genetic information. It is found in the nucleus, super-coiled into structures called **chromosomes**. Distributed across the chromosomes are the thousands of different regions, called **genes**, which code for the production of a polypeptide chain. The length of DNA making up a gene carries the instructions to make one specific protein. This information is referred to as the genetic code.

Triplet code

There are four bases that are used to code for 20 naturally occurring amino acids. If one base coded for one amino acid only four amino acids would have a specific code. If two bases were used it would be possible to generate 16 different combinations; that is, unique codes for 16 different amino acids:

AA, AT, AC, AG, TA, TT, TC, TG, CA, CT, CC, CG, GA, GT, GC, and GG.

However, by using combinations of three bases there are 64 possible combinations, which is more than enough to code for the 20 different amino acids. These DNA triplets are called **codons**. The codon for methioine acts as a start codon.

▼ **Table 1** *The triplet code showing DNA codons from the sense strand*

First base		Second base				Third base
		T	C	A	G	
T		Phe	Ser	Tyr	Cys	T
		Phe	Ser	Tyr	Cys	C
		Leu	Ser	Stop	Stop	A
		Leu	Ser	Stop	Trp	G
C		Leu	Pro	His	Arg	T
		Leu	Pro	His	Arg	C
		Leu	Pro	Gln	Arg	A
		Leu	Pro	Gln	Arg	G
A		Ile	Thr	Asn	Ser	T
		Ile	Thr	Asn	Ser	C
		Ile	Thr	Lys	Arg	A
		Start	Thr	Lys	Arg	G
G		Val	Ala	Asp	Gly	T
		Val	Ala	Asp	Gly	C
		Val	Ala	Glu	Gly	A
		Val	Ala	Glu	Gly	G

Degenerate

The genetic code is described as degenerate as there is more than one codon (triplet of DNA bases) for most amino acids. For example, leucine has six codons, lysine has two and serine has four. There are three codons that are not amino acid codes – TAA, TAG, and TGA. These are stop or nonsense codons, which are important in protein synthesis (Topic 4.6, Protein synthesis – transcription).

Non-overlapping

The sequence of bases is read so that each base is only part of one codon, that is, each triplet is read separately. For example, the sequence ATGGCTCCAAAT is read as ATG-GCT-CCA-AAT and not read as ATG-TGG-GGC-CTC, and so on.

Universal

The genetic code is the same in (almost) all organisms – plants, animals, and bacteria. There are some differences in protoctists and the DNA of mitochondria and chloroplasts.

Humulin production

The production of human insulin, Humulin, shows how the genetic code is universal. Due to this feature genes can be transferred from one species to another and the instructions can still be 'understood'.

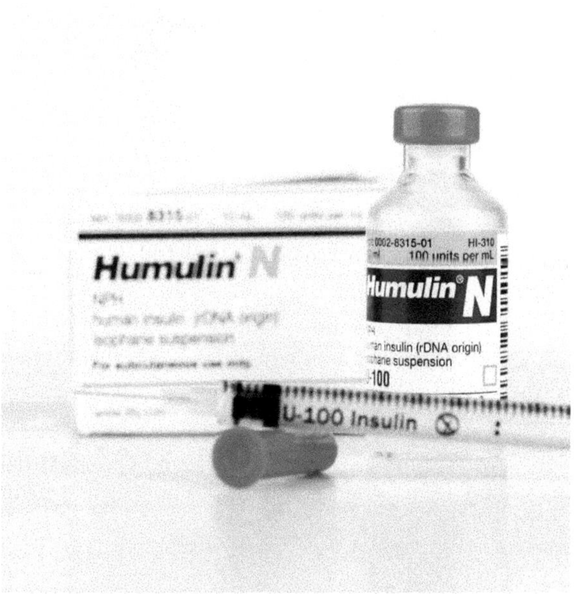

▲ **Figure 2** *Humulin – genetically engineered human insulin, which was first produced in 1982 using* Escherichia coli

The β-cells of the pancreas, which normally produce the hormone insulin, are destroyed in some people with diabetes. In the past, diabetes was

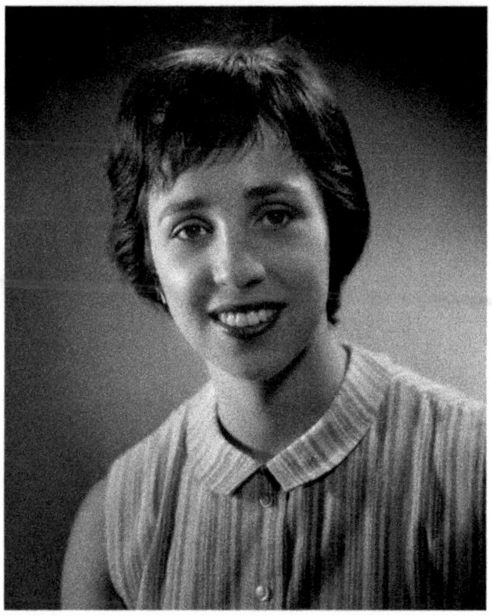

▲ **Figure 1** *Maxine Singer is an American molecular biologist. From 1956, she worked at the Laboratory of Biochemistry at the National Institutes of Health, where her synthesis work contributed to Marshall Nirenberg's experiments that uncovered the nature of the genetic code*

Study tip

Take care when you describe the genetic code as being degenerate. Some students confuse the amino acid and the codon – saying that some codons have more than one amino acid is incorrect.

Synoptic link

You will look at the role of the genetic code in protein synthesis in Topic 4.7, Protein synthesis – translation.

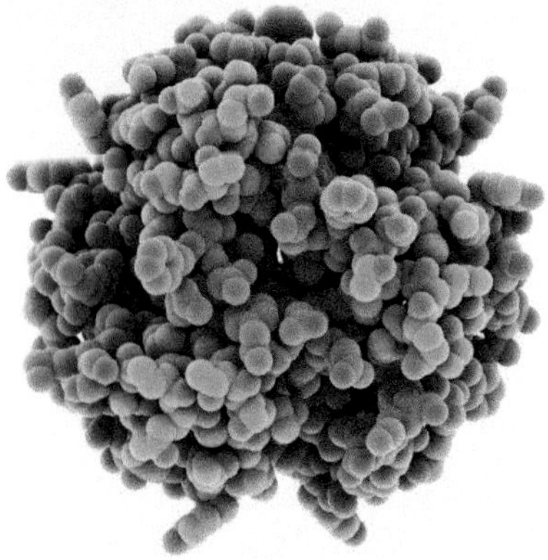

▲ **Figure 3** *Insulin molecule*

treated by injecting insulin that had been extracted from the pancreases of pigs and cattle. Insulin from pigs (porcine) has one amino acid different in its sequence and insulin from cattle (bovine) has three differences. Despite the differences in the amino acid chain the final protein hormone binds successfully to the human insulin receptor. This then causes the blood glucose levels to fall, as with normal human insulin.

However, over time some diabetics developed an allergy to the porcine and bovine insulin. This led to research into alternative methods of producing insulin. Arthur Riggs worked with Genentech to express the first artificial gene in bacteria. His work was critical to the modern biotechnology industry because it enabled the large-scale manufacturing of protein drugs, including insulin. The group succeeded in producing insulin in 1978, and in 1979, Riggs received the Juvenile Diabetes Foundation Research Award for this work.

The gene for insulin is inserted into a ring of DNA found in *E. coli*. The instructions in the gene can then be read by the enzymes in the bacterium and the protein insulin is made.

1 Using the information in Figure 3 and your previous knowledge, describe the structure of insulin.
2 Suggest why part of the molecule is shown in pink and part is in blue.
3 Suggest how the pink and blue parts of the molecule are held together.
4 Suggest why the group at Genentech used bacteria as the organism to produce insulin.
5 Give three advantages of using Humulin to treat diabetics.

Summary questions

1 Using the information in Table 1 identify which two amino acids only have one DNA codon. *(1 mark)*

2 a Use Table 1 to determine the amino acid sequence for the following DNA sense strand sequence:
ATG GTC CCC TCA CGT GGC GAC. *(1 mark)*
b Predict what would happen if the 11th base in the sequence was replaced with the base guanine. *(2 marks)*

3 Evaluate the advantages and disadvantages if DNA was an overlapping code instead of a non-overlapping one. *(2 marks)*

4.5 Polynucleotides – RNA

Specification references: 2.1.4

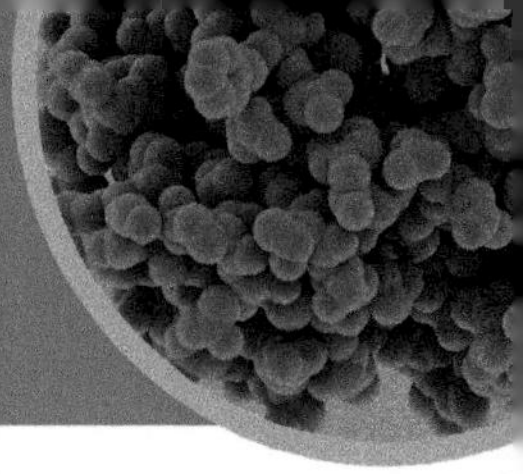

Ribonucleic acid

Ribonucleic acid (RNA) is a polymer of mononucleotides such as DNA. However, the nucleotides used to make RNA are different from those used to make DNA. Each RNA nucleotide is made from three components:

- pentose sugar – ribose
- phosphate group
- one of four different nitrogen-containing bases – adenine, uracil, guanine, and cytosine.

Learning outcomes

Demonstrate knowledge, understanding, and application of:

→ the general ribonucleic acid (RNA) structure

→ different types of RNA

→ the function of specific types of RNA.

Types of RNA

RNA is a single-stranded molecule found in three different forms.

Messenger RNA (mRNA)

Messenger RNA (mRNA) carries the genetic information from the nucleus to the site of protein synthesis – the ribosomes in the cytoplasm. One molecule of mRNA carries the instructions to make one specific polypeptide chain. The length of mRNA varies according to the number of amino acids in the polypeptide chain. mRNA is usually about 2000 nucleotides long, arranged in a single helix.

Different cells produce different proteins, including hormones and enzymes, which are specific to their function. For example, β-cells of the pancreas synthesise insulin and so these cells produce large quantities of mRNA that codes for insulin. Palisade mesophyll cells synthesise large quantities of RuBisCo, an important enzyme used in photosynthesis, so these cells produce large quantities of mRNA that codes for RuBisCo.

mRNA can leave the nucleus via the nuclear pores and enter the cytoplasm where it associates with ribosomes. The mRNA acts as a template for protein synthesis. Most mRNA is easily broken down and so it only exists when the cell is actively producing proteins.

Transfer RNA (tRNA)

Transfer RNA (tRNA) is much smaller than mRNA, usually only around 80 nucleotides long. It is a single-stranded molecule arranged into a clover-leaf shape.

There are several types of tRNA molecule, each of which carries a specific amino acid. One side of the tRNA molecule is longer than the other and this is where the amino acid is attached. The amino acid is attached to the tRNA by a specific enzyme. At the base of the tRNA molecule is a sequence of three RNA nucleotide bases which form the **anticodon**. Each tRNA molecule can only bind to one specific amino acid and therefore each anticodon is specific to each amino acid.

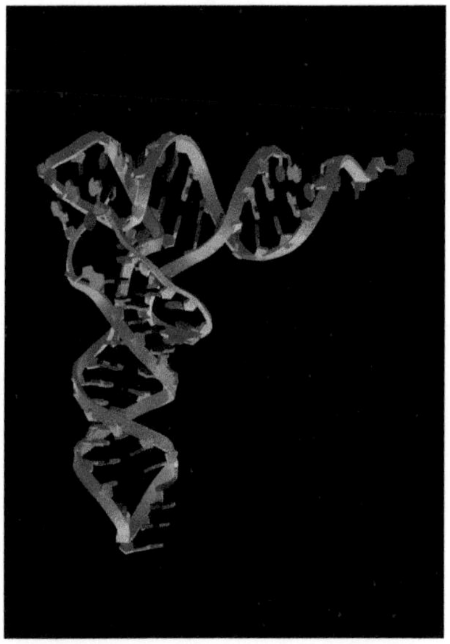

▲ **Figure 1** *Computer artwork of a tRNA molecule*

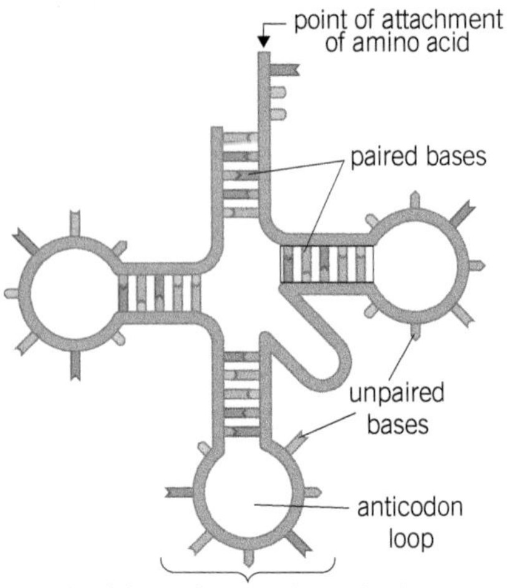

▲ **Figure 2** *The clover-leaf structure of a tRNA molecule*

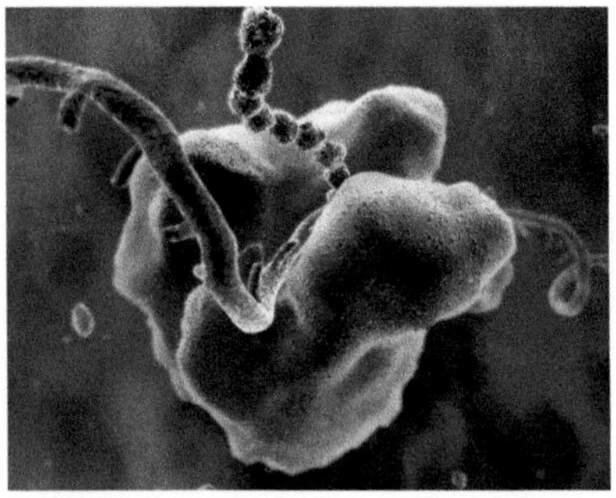

▲ **Figure 3** *Computer artwork. Messenger ribonucleic acid (mRNA, purple) passes between the two ribosome subunits and provides the instructions for the assembly of a protein (polypeptide) chain (yellow) from amino acids*

During protein synthesis the anticodon of a tRNA molecule binds to a codon on the mRNA by complementary base pairing.

Ribosomal RNA (rRNA)

Ribosomal RNA (rRNA) and proteins are assembled to form organelles called ribosomes. Each ribosome has two subunits:

- a large sub-unit that joins amino acids to form a polypeptide chain
- a small sub-unit that reads the mRNA.

Each subunit is composed of one or more rRNA molecules and a variety of proteins. The ribosome holds the mRNA so that six bases are exposed at one time. A group of three bases on the mRNA is called a codon, so two codons are exposed on the ribosome.

Summary questions

1 The diagram below shows part of a mRNA molecule.

A U G C U C A G G C C U

a State how many codons are shown in this section of mRNA. *(1 mark)*
b Explain what is specified by the sequence of codons in a mRNA molecule. *(1 mark)*
c List the anticodons that would be required to read this section of mRNA. *(1 mark)*

2 a Describe the similarities in the formation and structure of RNA and DNA. *(5 marks)*
b Copy and complete the table to compare DNA and RNA: *(3 marks)*

Feature	DNA	mRNA	tRNA
Number of polynucleotide chains			
Relative size			
Overall shape			
Sugar			
Organic nitrogen-containing bases			
Location			
Relative quantity in the cell			
Relative stability of the molecule			

3 Explain why there will be no tRNA with the anticodon ACU. *(2 marks)*

4.6 Protein synthesis – transcription

Specification references: 2.1.4

Protein synthesis

All cells must produce proteins in order for them to function. For example, specialised cells found in the liver are called hepatocytes. Hepatocytes contain several enzymes which are found exclusively in the liver. For example, alanine transaminase (ALT) catalyses the transfer of an amino group from L-alanine to α-ketoglutarate. This is an important step in the conversion of alanine to glucose in gluconeogenesis, which you will cover later in the course.

Protein synthesis can be divided into two stages.

The process of transcription

The function of most genes is to determine the sequence of amino acids in a polypeptide chain that then folds to form an active protein. The first stage of protein synthesis is called **transcription**.

Transcription is the synthesis of a mRNA molecule using the reference strand of the DNA molecule as a template. It can be summarised as follows:

1. DNA helicase unwinds a specific section of the DNA molecule (called a cistron) by breaking the hydrogen bonds between the two DNA strands.
2. RNA polymerase binds to a specific binding site on the DNA at the start of the gene.
3. RNA polymerase moves along one of the two strands of the DNA, which is known as the template strand.
4. RNA nucleotides are activated by the addition of two phosphate groups from ATP.
5. Activated RNA nucleotides diffuse through the nucleus until they align next to complementary bases on the template (anti-sense) strand of the DNA.
6. RNA polymerase forms covalent bonds between the adjacent RNA nucleotides called phosphodiester bonds.
7. As the RNA polymerase adds RNA nucleotides one at a time to form the mRNA, the DNA strands rejoin behind the enzyme. Consequently only about 12 base pairs on the DNA are exposed at any one point in time.
8. When the RNA polymerase reaches a specific sequence of bases on the DNA that form a 'stop' codon, the enzyme detaches and the formation of the mRNA is complete.

Learning outcomes

Demonstrate knowledge, understanding, and application of:

→ protein synthesis

→ the process of transcription.

Synoptic link

Proteins carry out many different functions in cells and organisms. Look back at Chapter 3, Protein and enzymes, to remind yourself.

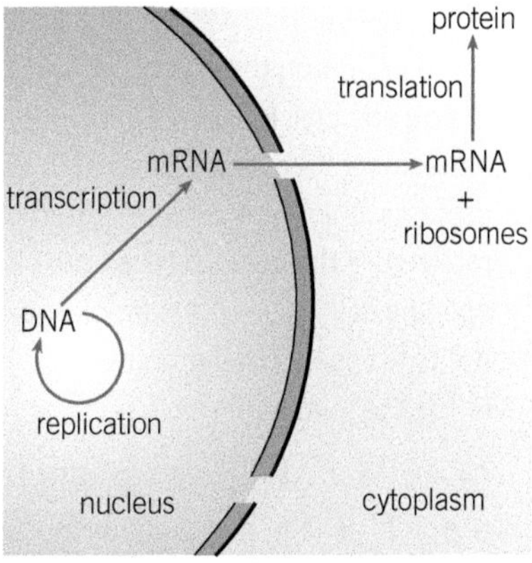

▲ **Figure 1** *An overview of protein synthesis*

Study tip

The sense strand of DNA has the same base sequence as the RNA (except T is replaced with U in mRNA).

The anti-sense (template) strand acts as the template for transcription and has a complementary base sequence to both the mRNA and the DNA sense strand.

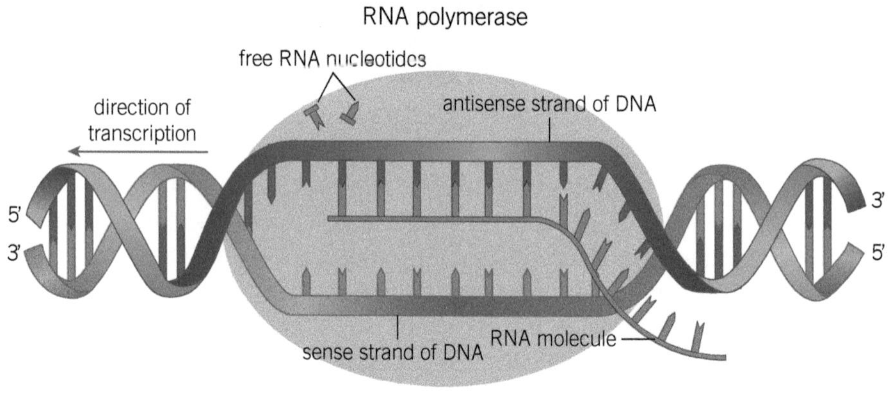

▲ **Figure 2** *The role of RNA polymerase in transcription*

Switching on genes – promoters

There are many thousands of genes in eukaryotic cells. In any one cell, however, only a small number of these genes need to be expressed, that is, used to make proteins. There are complex control mechanisms that make sure the correct genes are switched on for each specific cell type – for example, genes to produce ALT are expressed in hepatocytes but not expressed in cells lining the stomach.

Genes have a stretch of DNA called a promoter. This region is usually about 40 nucleotide bases long and identifies where the process of transcription should start. There is also another shorter length of DNA called the operator that controls when the gene is transcribed. The length of DNA that includes the promoter, the operator, and the actual gene is called an operon.

Although the promoter is usually found next to the protein coding part of the DNA, there may be other sections of DNA that also help control when the gene is expressed that may be located thousands of nucleotides away from it.

1 Promoters are activated by small proteins called transcription factors (TFs). There are many different TFs coded for in eukaryotic cells. Suggest what all these TF proteins have in common.

Summary questions

1 Describe how mRNA leaves the nucleus and why this cannot occur by diffusion. *(2 marks)*

2 Explain why DNA is double stranded if the code itself is only on one strand. *(3 marks)*

3 Suggest why translation occurs faster in prokaryotes than eukaryotes. *(2 marks)*

4.7 Protein synthesis – translation

Specification references: 2.1.4

Translation

Translation is the second stage of protein synthesis. It converts the coded information of mRNA into the correct sequence of amino acids in a polypeptide chain. Translation takes place in the cytoplasm using ribosomes.

Learning outcomes

Demonstrate knowledge, understanding, and application of:

→ how a polypeptide is chain produced

→ the process of translation.

Amino acid activation

Each amino acid is carried by its own unique tRNA molecule (Topic 4.5, Polynucleotides – RNA). The tRNA molecule attaches itself to its specific amino acid with the aid of an enzyme and ATP. This is called amino acid activation.

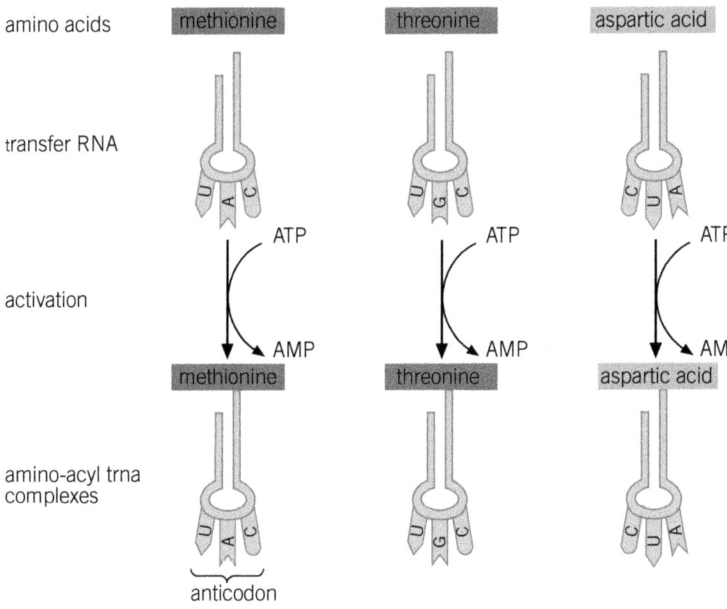

▲ **Figure 1** *Amino acid activation*

The process of translation

1. The mRNA moves into the cytoplasm through the nuclear pore.
2. The mRNA *attaches* to the groove between the two subunits of the ribosome.
3. The ribosome binds at the 5′ end of mRNA (translation occurs in the 5′ to 3′ direction).
4. The first amino acid-tRNA complex with the correct complementary **anticodon** binds to the first mRNA codon by hydrogen bonding between complementary bases.
5. Another amino acid-tRNA complex binds to the vacant mRNA codon (i.e., two tRNA molecules are bound to the ribosome at any one point in time).
6. A peptide bond is formed between the two adjacent amino acids. The energy to form the peptide bond is provided by breaking of the bond between the amino acid and the amino acid-tRNA complex.

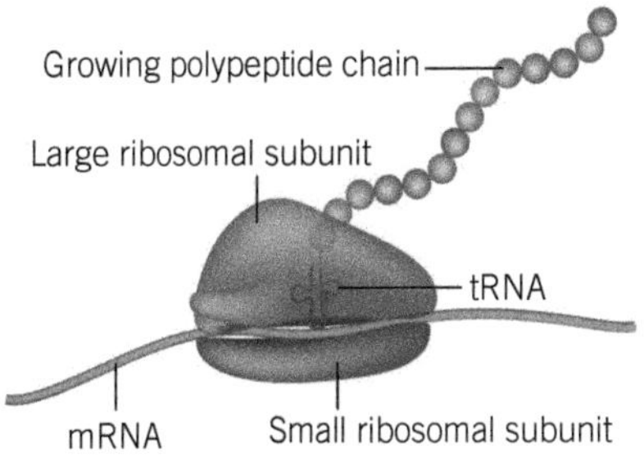

▲ **Figure 2** *A ribosome attached to mRNA*

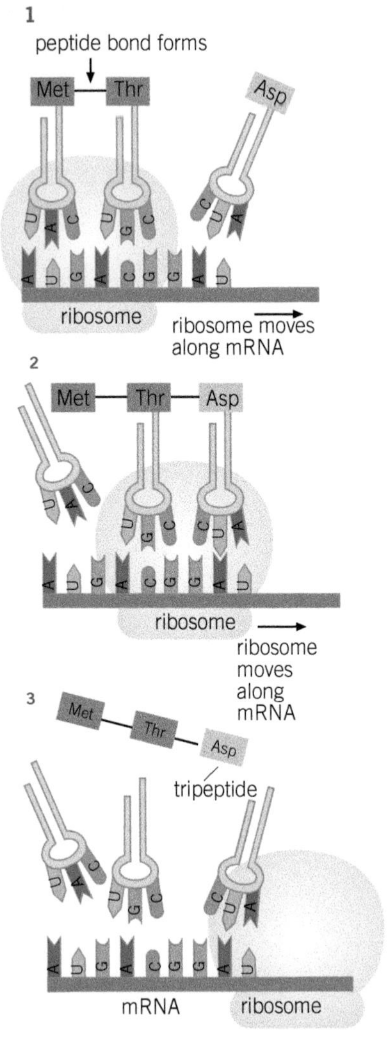

▲ **Figure 4** *The process of translation*

> **Synoptic link**
>
> You may need to revisit the structure of proteins. This is covered in Topic 3.3, Protein structure – haemoglobin.

7 The ribosome moves along one mRNA codon to next exposed mRNA codon.

8 The empty tRNA molecule is released and returns to cytoplasm to become activated again.

9 A new amino acid-tRNA complex binds to the vacant mRNA codon as in step 5. This continues at a rate which can be as high as 40 amino acids per second in some prokaryotes.

10 Steps 5 to 9 continue until the ribosome reaches a stop codon (UAA, UAC, UGA). There are no tRNA molecules with complementary anticodons for these codons.

11 This causes the ribosome to fall off the mRNA molecule.

12 The polypeptide chain is released and the initial amino acid methionine is removed.

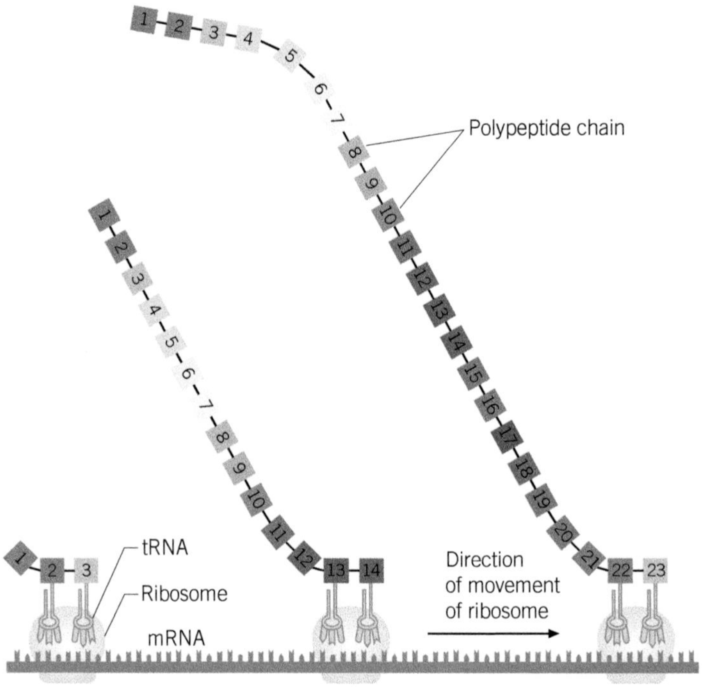

▲ **Figure 3** *The formation of a polypeptide chain*

Modification of the polypeptide chain

After the polypeptide chain has been released from the ribosome the chain is usually modified before it becomes a functional enzyme:

- The polypeptide chain is coiled or folded to form the secondary structure (α-helix or β-pleated sheet).
- The secondary structure is coiled further to form the tertiary structure (globular or fibrous protein). The first amino acid, methionine, is removed (remember all genes start with the initiation codon AUG, which codes for methionine).
- The polypeptide chain is coiled or folded to form the secondary structure.
- The secondary structure is folded further to form the tertiary structure.
- Different polypeptide chains and any necessary prosthetic groups are linked to form the quaternary structure in the Golgi apparatus.

Worked example: Calculating the length of a DNA molecule

Figure 5 shows a coloured transmission electron micrograph of deoxyribonucleic acid (DNA, blue) transcription coupled with translation in the bacterium *Escherichia coli*. During transcription, complementary messenger ribonucleic acid (mRNA) strands (red) are synthesised and immediately translated by ribosomes (green). The enzyme RNA polymerase recognises a promoter on the DNA strand and moves along the strand building the mRNA. mRNA is the intermediary between DNA and its protein product.

The magnification of this electron micrograph is ×150 000. Using the line X–Y, calculate the actual length of the DNA molecule (shown in blue).

Step 1 Measure the line X–Y in mm = 70 mm

Step 2 Convert line to µm = 70 × 1000 = 70 000 µm

Step 3 Divide the length of the line by the magnification factor = $\frac{70000}{150000}$

The length of the DNA molecule is approximately 0.460 µm.

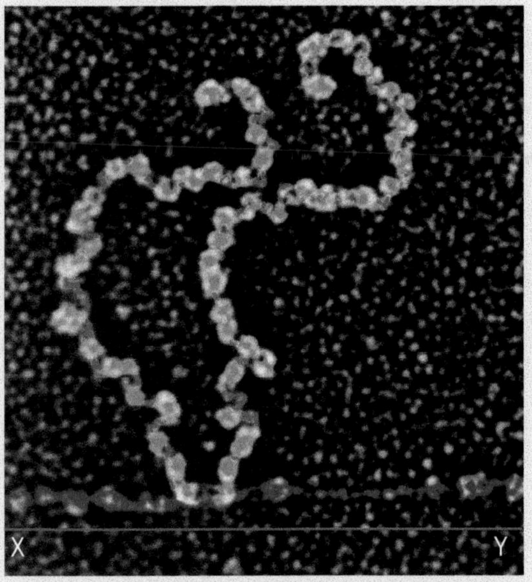

▲ **Figure 5** *Coloured transmission electron micrograph of DNA transcription and translation, × 150 000 magnification*

Summary questions

1 A mRNA molecule has 63 codons, but the protein produced from it only has 61 amino acids.

a Suggest why this might be the case. *(2 marks)*

b Calculate how long it would take to translate this mRNA strand (see step 9 on previous page). Show your working. *(1 mark)*

2 The enzyme methionine amino peptidase removes the initial amino acid methionine from polypeptides following translation. Suggest why mutations in the gene for this enzyme can be lethal in some cells. *(3 marks)*

3 Copy and complete the table below to outline the processes of semi-conservative replication, transcription, and translation. *(10 marks)*

Feature	Semi-conservative replication	Transcription	Translation
Free nucleotides are activated			
How much DNA is involved			
Number of new molecules per cycle			
Hydrogen bonds are broken between complementary base pairs			
Cytosine pairs with guanine			
Adenine pairs with thymine			
Phosphodiester bonds are made			
Peptide bonds are made			
Location of process			
Product			

Practice questions

1 Which of the following monomers is used in the synthesis of the polymer RNA?

A ATP **B** ADP **C** AMP **D** Adenine (*1 mark*)

2 The diagram shows a tRNA molecule.

Paired bases hold the molecule in a clover leaf shape

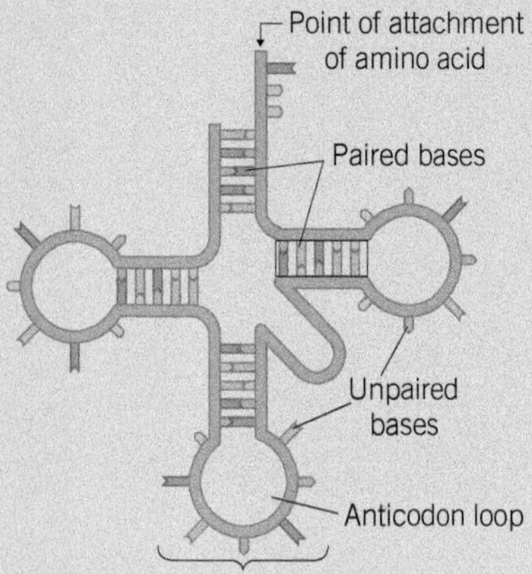

Which of the following statements is/are true?

Statement 1: the paired bases form hydrogen bonds with each other.

Statement 2: the paired bases will have a purine paired with a pyrimidine.

Statement 3: the paired bases will be adenine with thymine and cytosine with guanine.

A 1, 2 and 3 **B** Only 1 and 2

C Only 2 and 3 **D** Only 1 (*1 mark*)

3 The following are some statements about DNA:

- DNA is found in the nucleus of a cell.
- During interphase DNA replicates.
- DNA is involved in the transcription stage of protein synthesis.

The statements **A** to **H** (in the table), refer to events that may take place during:

- DNA replication **only**
- transcription **only**
- **both** DNA replication **and** transcription
- **neither** DNA replication **nor** transcription.

Copy and complete the table by marking the appropriate boxes with a tick (✓) if the event takes place or a cross (✗) if it does not take place (*8 marks*)

	Statement	DNA replication	Transcription
A	Nucleotides line up along an exposed DNA strand.		
B	The whole of the double helix 'unzips'.		
C	Uracil pairs with adenine.		
D	A tRNA triplet pairs with an exposed codon.		
E	Both DNA polynucleotide chains act as templates.		
F	Adjacent nucleotides bond, forming a sugar-phosphate backbone.		
G	The original DNA molecule is unchanged after the process.		
H	Adenine pairs with thymine.		

4 DNA can be isolated in the school laboratory using a wide variety of plant materials.

a One recommended source of DNA for isolation is ripe strawberries.

- ripe strawberries contain cellulase enzymes
- commercially grown strawberries are octoploid.

Evaluate the use of strawberries as source of DNA for extraction in schools. (*3 marks*)

b Extraction of DNA in the school laboratory requires the use of an extraction buffer. A typical extraction buffer can be made up as follows:

Washing-up liquid, 10 cm^3
Sodium chloride, 3 g
Distilled water, 100 cm^3

(i) Explain the role of detergents in the extraction of DNA from within cells. (*2 marks*)

(ii) Technicians frequently make up solutions in a concentrated form. The solution is then diluted to obtain the correct concentrations.

A technician provided 1 litre (dm^3) of concentrated extraction buffer. The buffer required a 1 in 5 dilution to obtain the correct concentrations of washing up liquid and sodium chloride.

Copy and complete the table to show the composition of the concentrated extraction buffer. *(2 marks)*

Volume of water (dm^3)	Volume of washing up liquid (cm^3)	Mass of sodium chloride (g)
1		

c DNA is hydrophilic and soluble in water. The role of sodium chloride in the extraction buffer is to make the DNA molecules less hydrophilic. This is due to positive sodium ions binding to negative charges on the DNA molecule.

Suggest which part of a DNA molecule is negatively charged. *(1 mark)*

d The final stage in DNA extraction involves the addition of ice cold ethanol poured gently down the side of the tube. DNA precipitates at the interface between the ethanol and the extraction buffer.

Suggest why DNA is less soluble in ethanol than in the extraction buffer. *(1 mark)*

5 RNA polymerase and DNA polymerase are both enzymes. *Describe and explain the difference between the functions of these two enzymes. *(6 marks)*

6 During research into the mechanism of DNA replication, bacteria were grown for many generations in a medium containing only the 'heavy' isotope of nitrogen ^{15}N. This resulted in all the DNA molecules containing only ^{15}N. This is illustrated in the graph.

a

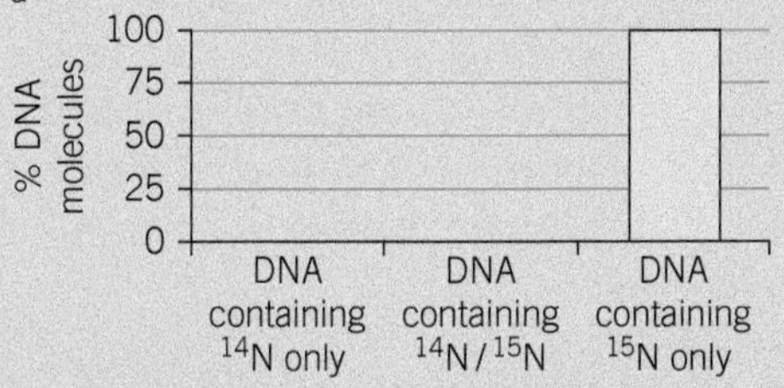

These bacteria were then grown in a medium containing only 'light' nitrogen ^{14}N. After the time taken for the DNA to replicate once, the DNA was analysed. The results are shown in the graph.

b

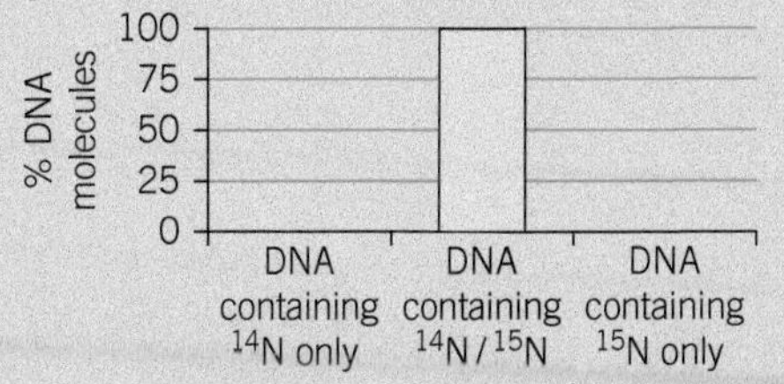

a Explain how these data support the semi-conservative hypothesis of DNA replication. *(3 marks)*

The bacteria continued to grow in the 'light' nitrogen ^{14}N until the DNA had replicated once more. The DNA molecules were analysed. The results are shown in the graph.

c

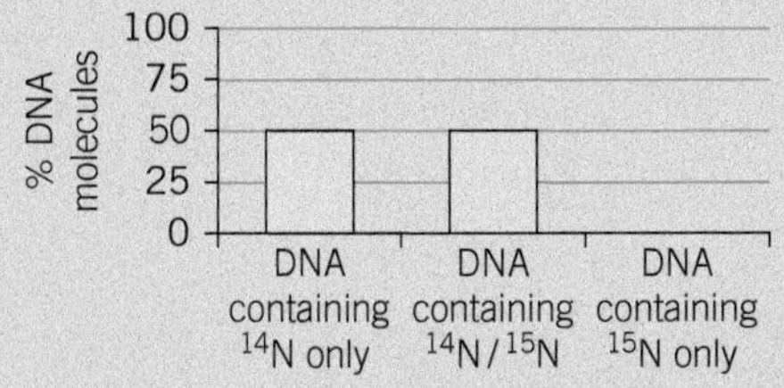

The diagrams here show simple representations of DNA molecules indicating the nitrogen content of each.

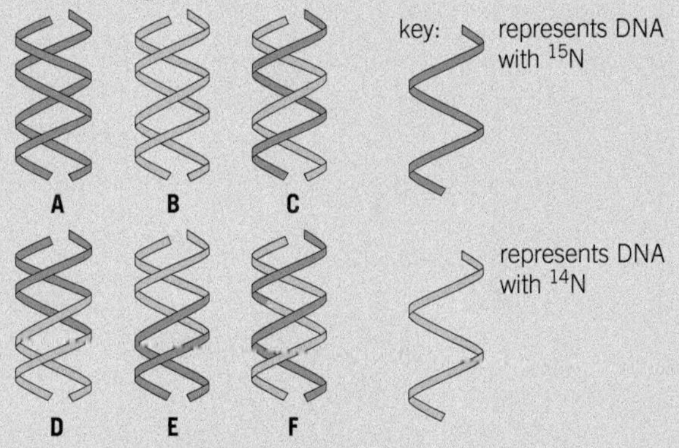

b With reference to the diagram, select the letter or letters which best represent the bacterial DNA in graphs a, b, and c. *(3 marks)*

The bacteria continued to grow in 'light' nitrogen ^{14}N medium until the DNA had replicated once more. The DNA molecules were analysed.

c Copy and complete a bar chart to indicate the expected results of the composition of these DNA molecules (using the previous graphs as a template). *(3 marks)*

5 THE HEART AND MONITORING HEART FUNCTION

5.1 The heart

Specification references: 2.2.1

Learning outcomes

Demonstrate knowledge, understanding, and application of:

- → the need for a mass transport system
- → the internal structure of the heart
- → the external structure of the heart
- → cardiac muscle as a unique muscle tissue
- → blood flow through the heart.

Synoptic link

Mass transport, where everything moves in one direction is discussed in greater detail in Topic 6.1, The transport system in mammals.

The need for a mass transport system

Small organisms gain all the nutrients and oxygen they need to survive by simple diffusion across the cell surface membrane. Simple diffusion supplies enough oxygen when the organism has a large surface area to volume ratio or when the metabolic rate is low. However, larger multicellular organisms cannot survive relying only on diffusion because their surface area to volume ratio is much smaller and they may have a high basal metabolic rate. Animals move around more to gain food, increasing their basic energy needs. In larger multicellular organisms, structures increase the available exchange surface and a mass flow system is needed to move substances from the exchange surface to other parts of the organism.

Plants require substances obtained from the environment to manufacture their own food – however, they do not require a mass transport system for oxygen and carbon dioxide.

Mass flow helps to move substances around the body quickly and will shorten the diffusion distance. Where mass flow systems need a high efficiency, for example in mammals, a pumping mechanism (the heart) is required to speed up the delivery of oxygen and nutrients to cells.

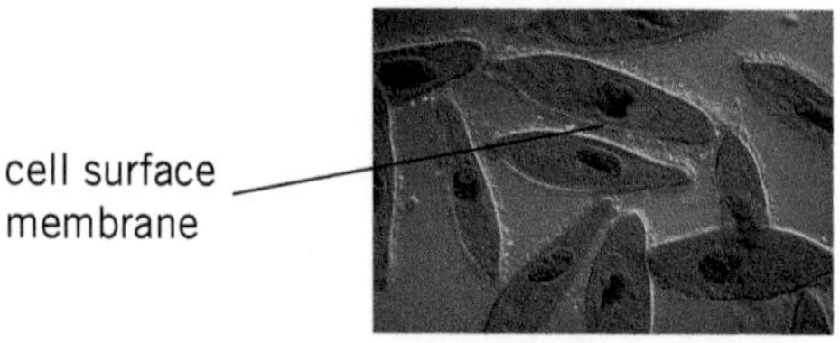

▲ **Figure 1** *The cell membrane controls what chemicals enter and leave. It allows the entry of oxygen for respiration. Excretion happens simply by waste products diffusing out through the membrane, ×100 magnification*

Structure of the mammalian heart

The heart pumps blood around the body in a series of coordinated contractions. There are actually two pumps side-by-side separated by a muscular wall which runs down the middle of the heart and effectively divides the system into two.

The left pump receives blood from the lungs into the left **atrium** and moves it through the left **ventricle**, which pumps it out to the body via the **aorta**. The right pump receives blood from the body into the right atrium and moves it through the right ventricle, which pumps it to the lungs via the **pulmonary artery**.

The external structure of the heart

The heart is a highly muscular pump. In humans, it is normally about the size of your clenched fist. You cannot easily see the heart chambers externally, although the two atria may be visible as flattened structures on top of the ventricles. These are flat when the heart is empty of blood because they have thin walls and so collapse. The two ventricles appear as a single structure from the outside and have extremely thick muscular walls, which feel solid when pressed. The major blood vessels may be seen curving over the top of the heart. The coronary artery runs diagonally across the surface of the heart with its many branches. These supply the heart muscle with oxygen and nutrients and are the vessels that may easily become blocked and cause angina pectoris, a heart attack, or cardiac arrest.

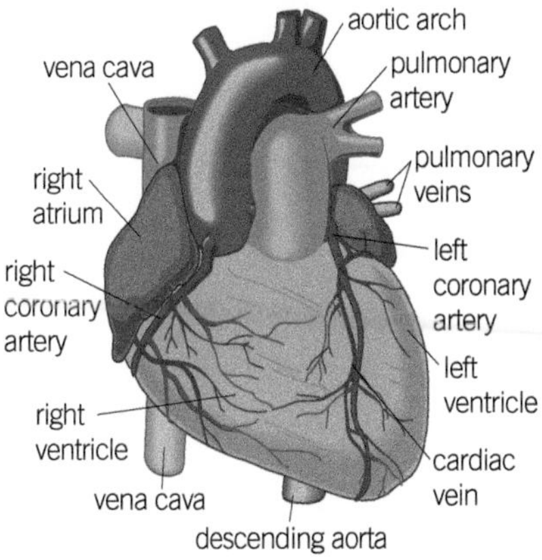

▲ **Figure 2** *The external structure of the heart*

Heart disease

Heart disease is one of the biggest killers in the western world. Coronary heart disease (CHD) accounts for half of all known heart disease deaths. It is caused by blockage of the coronary arteries, which run across the surface of the heart, as you have learnt already. These blood vessels supply the heart muscle with oxygen and nutrients such as glucose. When a fatty plaque, called an atheroma, builds up on the inside walls of the coronary arteries the blood flow is restricted and may eventually become blocked. At this point a heart attack (myocardial infarction) may occur.

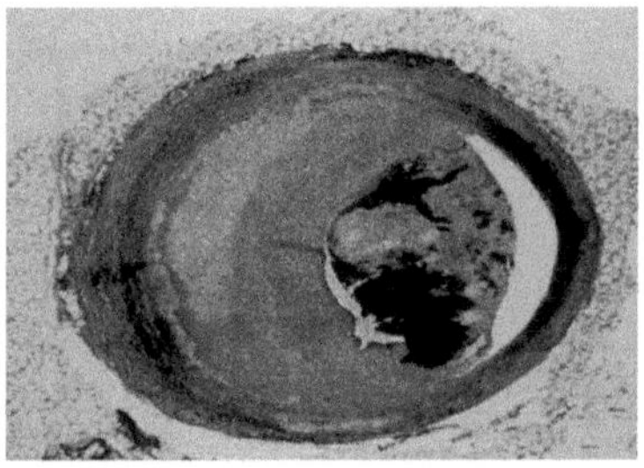

▲ **Figure 3** *Human coronary artery with a fatty atheroma partially blocking the lumen, approximately ×12 magnification*

Atheroma is caused by a build-up of white blood cells that have taken up low density lipoproteins (LDLs) and become deposited into the inner lining of the arteries. Deposits of cholesterol, dead fibrous tissue, and muscle cells from the artery wall also accumulate and narrow the lumen, restricting the blood flow.

1 What effect will the build up of atheroma, in the coronary artery, have on the muscle tissue of the heart?

The internal structure of the heart

The four chambers of the heart are divided into the left and right side by the middle wall or septum.

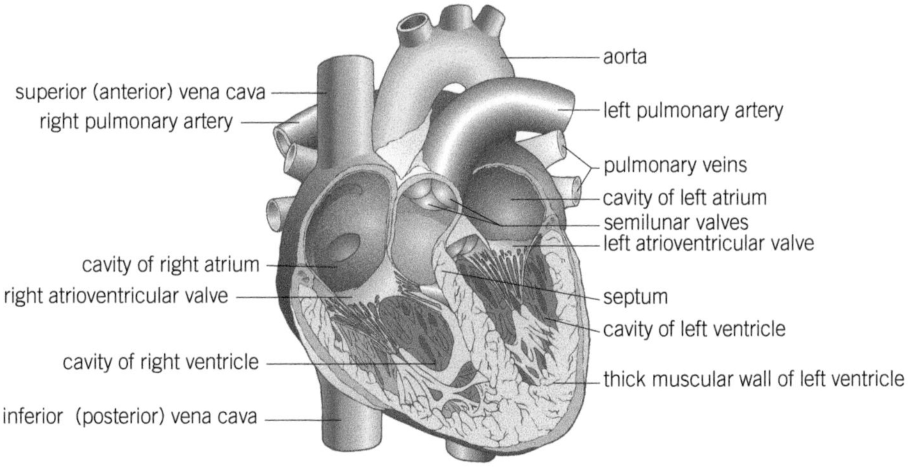

▲ **Figure 4** *The internal structure of the heart*

- The two atria are at the top of the heart with the two ventricles below each atrium.
- Between the atria and ventricles are the **atrioventricular valves**. The one on the left side is called the bicuspid valve, while that on the right is called the tricuspid valve. They are both under considerable pressure when the muscular wall of the ventricle contracts and so they are held in place by the **chordae tendinae** (the heart strings), which prevent them being inverted. These are attached to the muscle wall of the ventricle.
- The wall of the left ventricle is very much thicker than the right and both ventricle walls are very much thicker than those of the atria.
- The two **pulmonary veins** open into the left atrium bringing oxygenated blood from the left and right lungs, whilst the two **vena cava** open into the right atrium bringing deoxygenated blood from the head and body.
- Leaving the left ventricle is the major artery, the **aorta**, which carries blood to all parts of the body, whilst the **pulmonary arteries** leave the right ventricle and carry blood only as far as the left and right lungs. This explains why the right ventricle wall is so much thinner than the left. It does not need to pump blood as far nor under such high pressure.
- At the opening of each major artery as it leaves the heart are two more valves, called the **semilunar valves**.
- The function of the atrioventricular and semilunar valves is to prevent backflow, so that the blood continues to flow in one direction regardless of the pressure changes.

Study tip

Cardiac muscle cells are specialised for their function in three ways:

- unlike skeletal muscles they do not fatigue, but contract regularly and repeatedly throughout the lifetime of the individual
- they are also myogenic – they have their own intrinsic beat and will contract to this rhythm even when isolated
- finally they are syncytial – interconnected with their neighbouring cardiac muscles cells allowing the rhythm of contraction to spread from one to another and coordinate synchronous contraction of the tissue.

Cardiac muscle

The heart wall consists of a special type of muscle called **cardiac muscle**. It will contract without the need of any nerve stimulation and will continue to contract at its own beat even when isolated. This type of muscle is said to be **myogenic**.

Study tip

Remember that the two sides of the heart contract together and pump the *same* volume (the stroke volume) of blood out through the aorta and the pulmonary artery in one beat.

Blood flow

Blood flows through the heart in a set sequence – it enters the two atria, from the pulmonary veins and vena cava, the atria contract and force blood into the two ventricles. The ventricles then contract and blood is pumped out into the main arteries (the aorta and the pulmonary arteries).

The atria always contract together and, after a short delay, the ventricles always contract together.

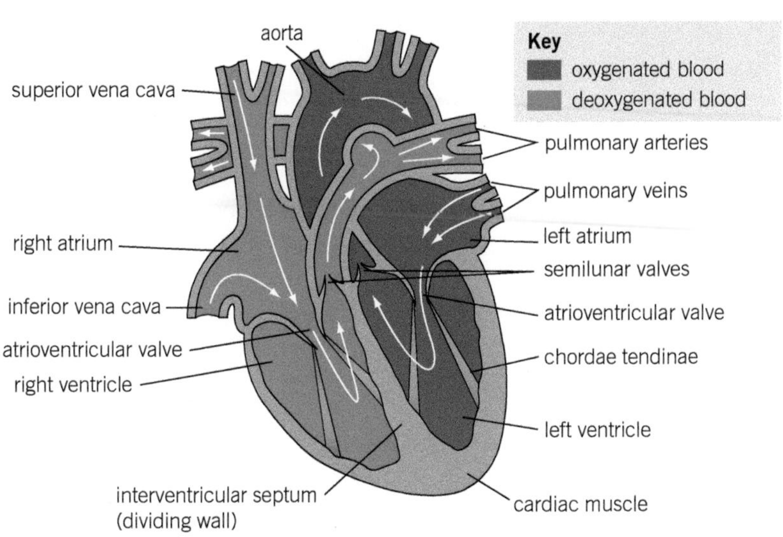

▲ **Figure 5** *The flow of blood through the heart*

Summary questions

1 Explain why a transport system is not required for respiratory gases in large multicellular organisms, such as trees. *(2 marks)*

2 Explain how heart muscle tissue is specialised for its function. *(3 marks)*

3 Both skeletal and cardiac muscle cells have a 'refractory period' following contraction. During this time, no further contraction is possible. The refractory period in cardiac muscle is much longer than in skeletal muscle. Why is a long refractory period an essential feature of cardiac muscle? *(2 marks)*

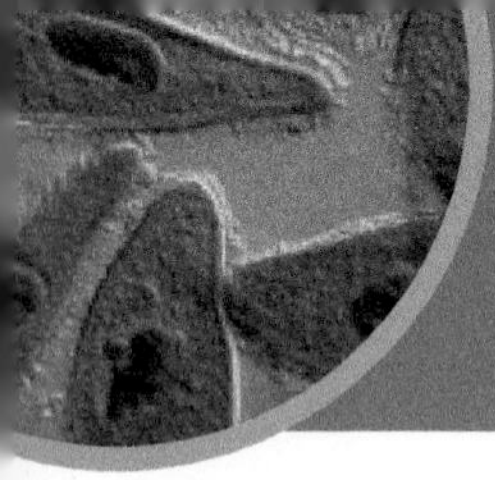

5.2 The cardiac cycle

Specification references: 2.2.1

Learning outcomes

Demonstrate knowledge, understanding, and application of:

→ the cardiac cycle and the pressure changes involved

→ the action of the valves

→ initiating the heart action

→ how the heart action is controlled.

The cardiac cycle

The cardiac cycle is a complete sequence of events beginning with atrial systole (atrial walls contract) followed by ventricular systole (ventricle walls contract) and followed finally by the resting heart or **diastole**. It includes the action of the heart muscle at each stage and is accompanied by movement of the heart valves and movement of the blood through the heart. There are about 72 of these cycles every minute.

Study tip

Don't confuse the cardiac cycle with the circulatory system, which is the term that describes the passage of blood through the heart and around the body.

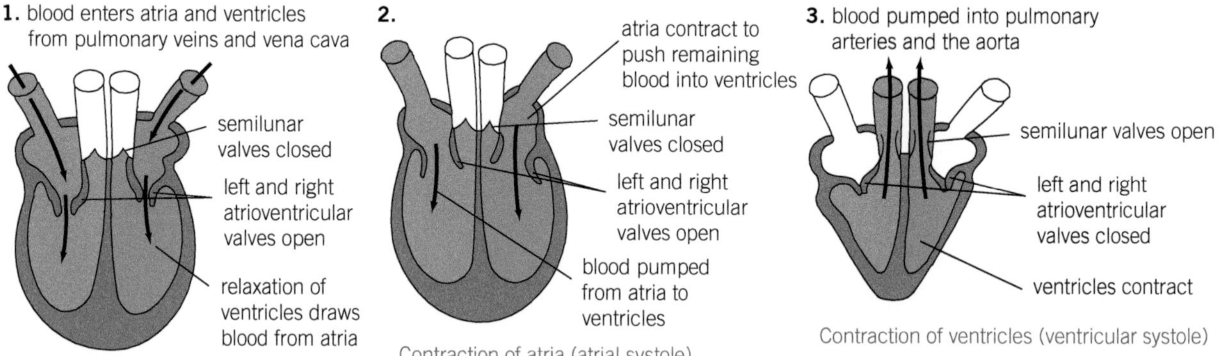

1. blood enters atria and ventricles from pulmonary veins and vena cava
semilunar valves closed
left and right atrioventricular valves open
relaxation of ventricles draws blood from atria
2.
atria contract to push remaining blood into ventricles
semilunar valves closed
left and right atrioventricular valves open
blood pumped from atria to ventricles
3. blood pumped into pulmonary arteries and the aorta
semilunar valves open
left and right atrioventricular valves closed
ventricles contract

Relaxation of heart (diastole)
Atria are relaxed and fill with blood. Ventricles are also relaxed.

Contraction of atria (atrial systole)
Atria contract, pushing blood into the ventricles. Ventricles remain relaxed.

Contraction of ventricles (ventricular systole)
Atria relax. Ventricles contract, pushing blood away from heart through pulmonary arteries and the aorta.

▲ **Figure 1** *The cardiac cycle*

▼ **Table 1** *The stages in the cardiac cycle*

Stage of the cycle	Muscle state	Blood movement	State of valves	Pressure changes
Atrial systole	Atrial walls contract	Passive flow of blood into the ventricles changes by atrial contraction forcing blood into ventricles. Atria emptied of blood.	AV valves pushed fully open by the blood flowing through.	Pressure rises in atria as they contract with a slight rise in ventricles as blood enters them.
Ventricular systole	Ventricle walls contract	Blood is forced into the main arteries. (Right ventricle into pulmonary artery and left ventricle into aorta.)	AV valves pushed shut as pressure in the ventricles rises above pressure in the atria; backflow prevented. Heart tendons hold valves in place to prevent the valves inverting.	Pressure in atria falls. Pressure in ventricles rises steeply as muscle contracts. Rise continues for 0.1 s then falls as blood is emptied from the ventricles. Pressure in arteries rises as blood enters from the ventricles and then falls.

Stage of the cycle	Muscle state	Blood movement	State of valves	Pressure changes
			Semilunar valves pushed fully open as blood enters the arteries.	
Diastole	Muscular walls relaxed	The atria fill with blood, opening the atrioventricular valves. Blood slowly and passively enters the ventricles.	Semilunar valves close since pressure in artery is higher than in the ventricle. Atrioventricular valves open.	Pressure in atria rises as they fill but then falls as AV valve opens. Pressure in ventricles falls but rises slightly again as they become full. Pressure in arteries falls but remains higher than the ventricles.

Pressure changes

During the cardiac cycle, the pressure changes in the atria, ventricles, and arteries can be mapped onto a graph. Usually this graph only shows the pressure changes in the left side of the heart since the changes are greatest on this side.

The graph in Figure 2 shows three lines that correspond to the left atrium, the left ventricle, and the aorta.

Study tip

To help you understand the graph, look at the points where the lines cross over each other. You should be able to relate these to the opening or closing of valves.

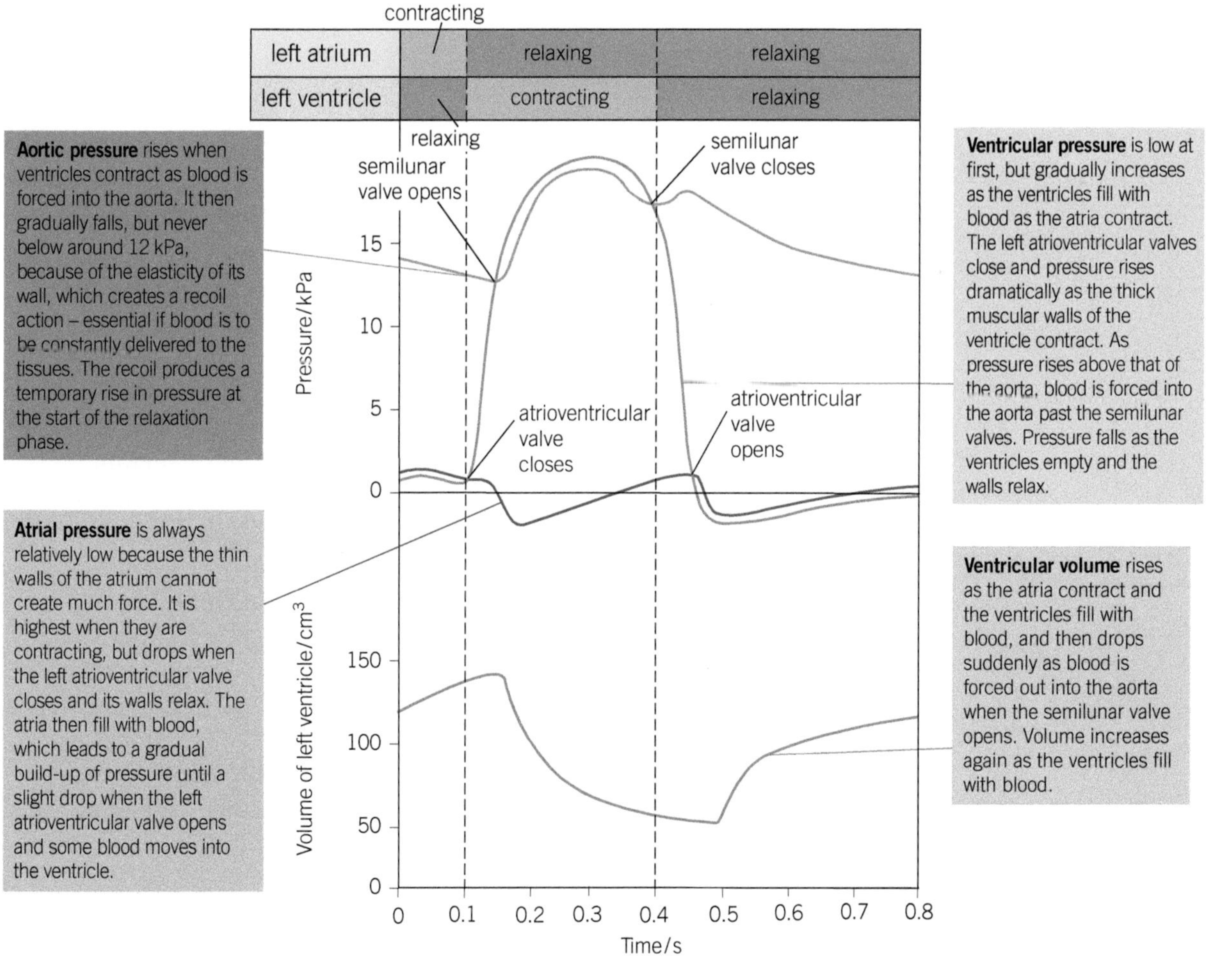

contracting
left atrium
relaxing
relaxing
left ventricle
contracting
relaxing
relaxing
semilunar valve opens
semilunar valve closes
Aortic pressure rises when ventricles contract as blood is forced into the aorta. It then gradually falls, but never below around 12 kPa, because of the elasticity of its wall, which creates a recoil action – essential if blood is to be constantly delivered to the tissues. The recoil produces a temporary rise in pressure at the start of the relaxation phase.
Ventricular pressure is low at first, but gradually increases as the ventricles fill with blood as the atria contract. The left atrioventricular valves close and pressure rises dramatically as the thick muscular walls of the ventricle contract. As pressure rises above that of the aorta, blood is forced into the aorta past the semilunar valves. Pressure falls as the ventricles empty and the walls relax.
Pressure/kPa
15
10
5
0
atrioventricular valve closes
atrioventricular valve opens
Atrial pressure is always relatively low because the thin walls of the atrium cannot create much force. It is highest when they are contracting, but drops when the left atrioventricular valve closes and its walls relax. The atria then fill with blood, which leads to a gradual build-up of pressure until a slight drop when the left atrioventricular valve opens and some blood moves into the ventricle.
Ventricular volume rises as the atria contract and the ventricles fill with blood, and then drops suddenly as blood is forced out into the aorta when the semilunar valve opens. Volume increases again as the ventricles fill with blood.
Volume of left ventricle/cm³
150
100
50
0
0 0.1 0.2 0.3 0.4 0.5 0.6 0.7 0.8
Time/s

▲ **Figure 2** *Graph showing the pressure changes in the left side of the heart during the cardiac cycle*

Study tip

Make sure you can explain the movement of the valves in terms of pressure differences between atria and ventricles and between ventricles and arteries.

Synoptic link

It may help to look at other aspects of biology where pressure changes affect the direction of flow and the outcome, for example Topic 7.1 on gas exchange.

When the pressure in the chamber rises, the volume of blood in that chamber falls if the valve is open and so blood will flow down a **pressure gradient**, from an area of high pressure to an area of lower pressure.

If you listen to the heart using a stethoscope, two sounds can be heard – a 'lub' sound, which is followed a fraction of a second later by a 'dub' sound. The 'lub' sound of the heart beat occurs as the AV valves snap shut and as the semilunar valves close the 'dub' sound is created. These two sounds give the normal heart sound of 'lub-dub'.

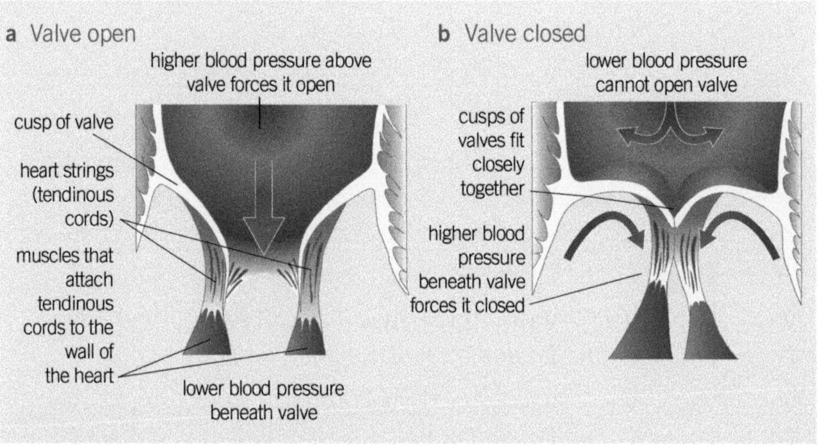

▲ **Figure 3** *How the valves close during the cardiac cycle*

Controlling the heart action

It is important to remember that the heart muscle is myogenic and so contracts spontaneously. Within the muscular wall of the right atrium is the **sinoatrial node** (SA node), perhaps better known as the pacemaker. The SA node functions as the initiator and controller of the heart muscle contraction and, from here, an electrical impulse spreads rapidly through the walls of both atria. This causes the walls of the two atria to contract triggering atrial systole. A second node, called the **atrioventricular node** (AV node) lies in the septum between the two atria, close to where the valves are situated. The layer of fibrous tissue that forms the valves prevents the wave of contraction from passing on immediately and a short delay occurs allowing the atria to empty the blood into the ventricles. The AV node picks up the electrical signal and, after a short delay, transmits it on to the ventricles. From the AV node there are conductile fibres called the **bundle of His**, which carry the electrical impulse across the fibrous ring and on to the **Purkinje tissue**. The Purkinje tissue carries the impulse to the apex of the ventricles and then, through bundle branches, up through the walls of the ventricles initiating the contraction of the ventricles from the apex upwards. This contraction is called ventricular systole.

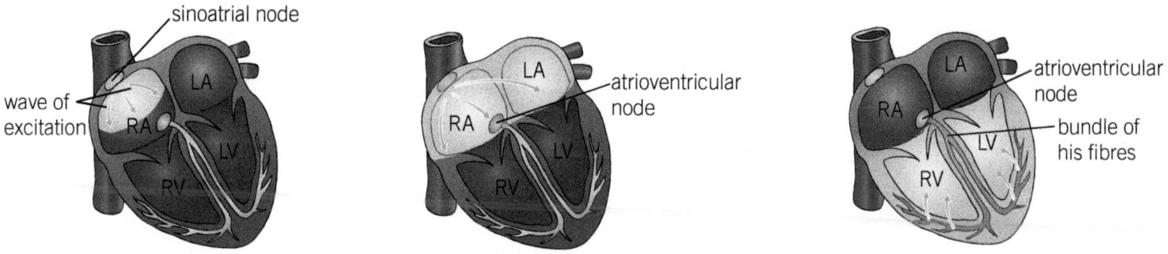

a wave of electrical activity spreads out from the sinoatrial node

b wave spreads across both atria causing them to contract and reaches the atrioventricular node

c atrioventricular node conveys wave of electrical activity between the ventricles along the bundle of His and releases it at the apex, causing the ventricles to contract

▲ **Figure 4** *Control of the cardiac cycle*

Although heart muscle is myogenic, there are two nerves that can regulate the heart rate. These are the accelerator nerve and the vagus nerve. Neurotransmitter chemicals and hormones such as adrenaline will also regulate the heart rate.

Atrial kick

For various reasons, the atria can start to contract 'out of time' with the ventricles. The contractions are rapid and unsynchronised and this is called atrial fibrillation. 'Atrial kick' occurs at the end of atrial diastole and ventricular diastole when the open tricuspid valve allows blood to fill both right atrium and right ventricle simultaneously. The ventricle receives about 85% of its blood volume during this phase, whilst atrial systole only contributes an additional 15% of the blood volume to the ventricle – the 'atrial kick'. This topping up of the ventricles by the atria becomes more significant in older patients and the contribution is closer to 35% of the blood volume. This is one reason why elderly patients are affected more by atrial fibrillation.

1 Assuming the ventricle receives 70 cm^3 of blood, what volume of blood entering the ventricles is delivered by the atrial kick in a young person and an elderly person?

Summary questions

1 List all the structures a single blood cell will flow through as it completes one circuit of a mammalian body. *(3 marks)*

2 Describe what happens in the aorta once the blood enters during ventricular systole. What special features of the artery wall allow this to happen? *(3 marks)*

3 Explain the role of the valves in maintaining blood flow through the heart. *(3 marks)*

5.3 Monitoring the heart

Specification references: 2.2.1

Learning outcomes

Demonstrate knowledge, understanding, and application of:

→ the effect of heart rate on cardiac output

→ the effects of exercise

→ measuring pulse rate and using secondary data

→ factors affecting heart rate.

Why monitoring is important

Heart disease (which you learnt about in Topic 5.1, The heart) and vascular disease are the biggest causes of death in the UK and other Western countries. Some health issues known as 'risk factors' affect the chances of such diseases occurring. Monitoring the heart is one way of determining the effect and implication of the risk factors on an individual. This topic looks at different ways of monitoring the heart.

The effect of heart rate on cardiac output

Stroke volume is the volume of blood pumped out of the left ventricle during each cardiac cycle, which is approximately 60–80 cm^3. **Cardiac output** is the volume of blood pumped out of the left ventricle each minute. **Heart rate** is the number of beats per minute (bpm), which for an average person at rest is about 72 bpm.

Cardiac output can be calculated using the following equation:

Cardiac output = stroke volume × heart rate

You can measure your heart rate by taking your **pulse rate**. The pulse is the number of 'pulses' felt in the artery each minute. These pulses are caused by the expansion of the artery wall, as the blood is pumped into the main arteries with each ventricular systole, followed by the elastic recoil of the artery wall as the blood pressure drops when the ventricle is in diastole (at rest). It is usually measured at the wrist where the radial artery lies close to the bone and can easily be felt. It can also be felt at the neck, by pressing two fingers onto the carotid artery, which carries blood to the brain. The carotid artery is one of the branches from the aorta, which divide off soon after the aorta leaves the left ventricle.

Study tip

In an explanation of how pulse rate is normally measured you should state clearly which artery is being used and specify that two fingers are used rather than a thumb, since a 'pulse' occurs in the thumb.

During strenuous exercise, the body muscles contract much more and so the volume of blood transporting oxygen to the muscles needs to increase dramatically. As a result of both an increase in heart rate and an increase in the stroke volume, cardiac output is increased. Normally the stroke volume will be around 70 cm^3 per beat, but this may increase to up to 150 cm^3 per beat. The heart rate may also increase from a normal 72 bpm to 140 bpm or more. The heart muscle also contracts more strongly and contraction of the skeletal muscles compresses the veins, and so blood is returned to the heart more quickly.

Study tip

When answering a question on cardiac output or heart rate or pulmonary ventilation remember to include the correct units ($cm^3\,min^{-1}$).

Worked example: Calculating cardiac output

At rest, the cardiac output will be approximately 4320 $cm^3\,min^{-1}$. During strenuous exercise this can increase to a much higher output. If the stroke volume is increased to 150 cm^3 and the heart rate to 140 bpm, calculate the cardiac output.

- Cardiac output = stroke volume × heart rate = 150 × 140 = 21 000 $cm^3\,min^{-1}$

Starling's Law and other factors affecting cardiac output

Echocardiograms can be used to measure the ejection fraction of the ventricles. At the end of ventricular systole, the left ventricle holds approximately 100 cm^3 of blood. Between 50% and 80% of the blood is normally ejected, giving a stroke volume of 50 to 80 cm^3. The English physiologist, Ernest Starling, demonstrated that as cardiac muscle is stretched it contracts more forcefully. During exercise, blood is returned to the heart more quickly as contracting skeletal muscles compress veins. This in turn increases the volume entering the ventricles and increases the end diastolic pressure – stretching the muscular walls of the ventricles. In addition the cardiac muscle contracts more forcefully, causing it to eject a greater volume of blood. This is known as Starling's Law.

As heart rate increases there is a corresponding increase in cardiac output up to a point. At a high heart rate of above 150 bpm, cardiac output actually declines. This is because at a high heart rate there is less time for the heart to fill so the stroke volume falls, leading to a decline in cardiac output. If a graph of heart rate against cardiac output is drawn it will form almost a bell shaped curve.

1 Explain why cardiac output declines at very high heart rates.

Factors affecting heart rate

There are many things that affect heart rate. Some aspects cannot be controlled:

- Age – children have a faster heart rate than adults and as people get older the heart rate gets slower.
- Inheritance – some people inherit the tendency to have a lower or higher heart rate. Some inherited diseases may also affect the heart rate – for example, cystic fibrosis, which increases both heart rate and pulmonary blood pressure.

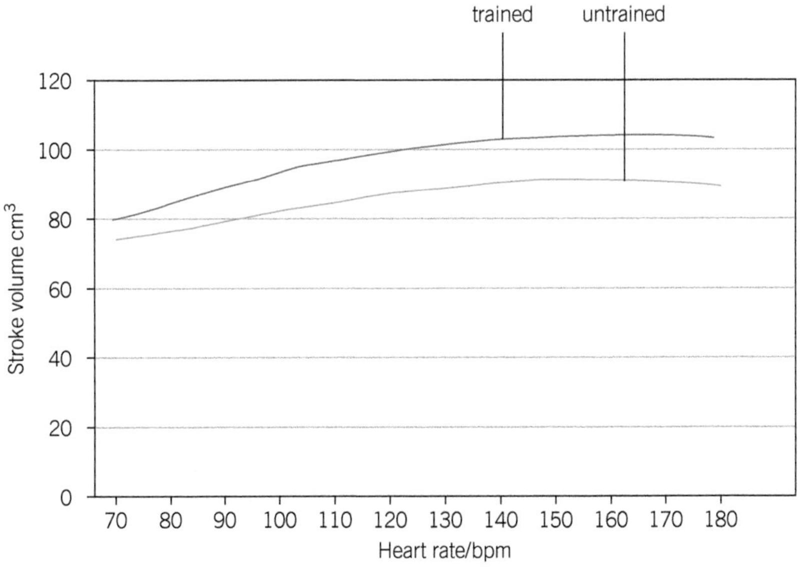

Figure 1 *This graph shows that a trained athlete can increase the stroke volume of the heart and so increase cardiac output for the same heart rate*

Testing factors affecting the heart rate

You can investigate the factors affecting heart rate, such as exercise, but you need to plan your investigation carefully.

- You will need to know your resting pulse rate by recording your pulse using the same location, such as the radial artery on the inside of your wrist. You should consider repeating the count to ensure the result is a reliable measure and allow a mean calculation.
- When changing a factor and investigating the impact the change has on the pulse rate, it is essential to control as many other variables as possible.
- You will need to think about how many other people you will test and whether they will all be the same age, sex, and have similar lifestyles. Any of these variables will affect the results significantly when you are testing the effect of another variable, such as different levels of exercise.
- Another way of improving the quality of the data you may use is to use *secondary data*, which is data collected by someone else. This could involve research data or data from a census. The advantage of secondary data is that it can provide a source of larger, high quality data that will give a much better basis on which to make an informed conclusion.

In a study on the effect of exercise on heart rate in a group of subjects, measurements were taken at rest and then taken again after a set period of exercise at a given intensity.

What statistical test could you use to analyse this data? Explain your choice.

Some aspects affecting heart rate we can control:

- Activity level – an increased level of activity increases heart rate for the period of activity and a recovery period afterwards, but a sustained long-term activity regime usually reduces the heart rate as the heart muscle becomes stronger and the heart more efficient.
- Smoking – chemicals such as nicotine cause the heart rate to increase due to the stimulation of release of noradrenaline neurotransmitter which affects the SA node.
- Diet – the effect of diet on heart rate is an active area of research. It is believed that eating a healthy diet rich in fruits, vegetables, and starchy foods helps to improve the health of the heart. It is generally believed that saturated fats increase the risk of strokes and heart attacks, although some evidence has conflicted with this.

Data for the effect of different variables on heart rate

In the data in Table 1, it is evident that with increased age there is a reduction in heart rate.

▼ **Table 1** *The effect of age on normal resting heart rate*

Age of individual	Resting heart rate/beats per minute
Baby aged 0–12 months	120–160
Children aged 1–10 years	65–140
Teenagers aged 10–18 years	65–100
18 years–adult	60–100

In the data in Table 2, the more active lifestyle in the trained athlete causes a reduction in the heart rate due to increased efficiency of the heart muscle as the muscle becomes more developed (cardiac hypertrophy) and so contracts more strongly. The increased efficiency improves stroke volume so that the heart rate can reduce whilst still allowing the same volume of blood to be pumped out.

▼ **Table 2** *The effect of lifestyle on resting heart rate*

Level of activity	Resting heart rate/beats per minute
Sedentary lifestyle	75 – 120
Moderately active	65 – 100
Highly trained athlete	40 – 60

Summary questions

1 State two variables that would need to be controlled in order to investigate the effect of increasing activity levels and heart rate. *(2 marks)*

2 Figure 2 shows the effect of exercise on the heart rate of two boys. They started to cycle at 2 minutes and continued for 5 minutes, before both stopping at 7 minutes.

   a Compare the effect of exercise on the heart rate of the two boys between 2 and 6 minutes. *(2 marks)*

   b Calculate the percentage increase in heart rate of the two boys for the same time period. *(2 marks)*

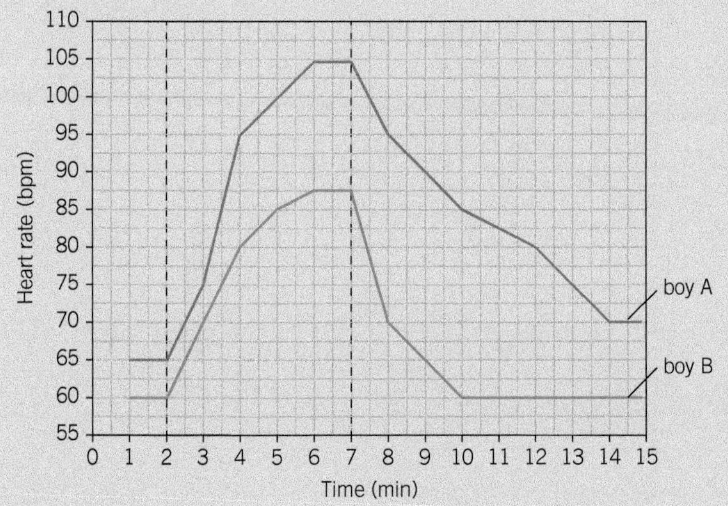

▲ **Figure 2**

3 Ventricular stroke volume measurements can be used to assess cardiac function. One method uses ultrasound to measure the volume of blood in the ventricles at **two** points in the cardiac cycle. The stroke volume is calculated by subtracting one volume from the other.

a Suggest at what points in the cardiac cycle the two volumes are measured *(1 mark)*

b Using your answer to (a) suggest the equation that could be used for calculating the stroke volume. *(1 mark)*

5.4 Using and interpreting an electrocardiogram

Specification references: 2.2.1

Interpreting an electrocardiogram (ECG)

Every television program about hospitals and medical emergencies will, at some point, show someone on a heart monitor. You have probably looked at the machine and wondered what it showed, apart from whether the patient was still alive! The machine is a **cardiac monitor** and the process is electrocardiography. Electrocardiography shows the electrical changes in the heart during the cardiac cycle displayed as an ECG or **electrocardiogram**.

A typical ECG trace has an initial gentle rise called the P phase which corresponds to atrial systole followed by a QRS wave which corresponds to ventricular systole. The final wave, called the T wave, is linked to the ventricular diastole. A normal ECG trace shows these distinct peaks and troughs occurring in a regular pattern in regular time intervals.

Changes in the waves and the size of the peaks and troughs in the ECG, allows a cardiologist (heart specialist) to detect heart conditions.

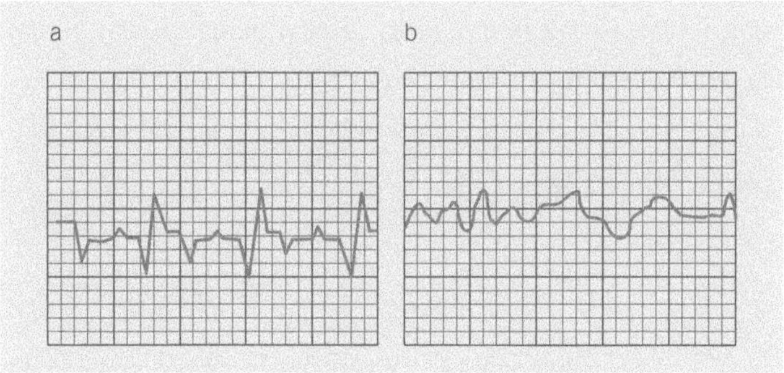

▲ **Figure 2** *An ECG showing a heart attack (a) and one showing fibrillation (b)*

A myocardial infarction, also known as a heart attack, occurs when the blood supply to part of the heart muscle is partly or totally blocked such that the muscle cells in that region are deprived of oxygen and stop contracting. During a myocardial infarction the troughs and peaks of the ECG become less distinct and are not as regular. Cardiologists will also look for evidence that the S-T portion of the ECG wave is higher than normal (ST elevation) to confirm a myocardial infarction. Atrial fibrillation can also be diagnosed from an ECG trace. In atrial fibrillation, the walls of the atria contract to an abnormal rhythm, preventing efficient ventricular filling. In Figure 2, ventricular fibrillation is occurring in (b) while (a) shows an ECG taken during a heart attack. Ventricular fibrillation is a cause of cardiac arrest and sudden cardiac death as the ventricles are unable to carry out their function.

Learning outcomes

Demonstrate knowledge, understanding, and application of:

- → using an electrocardiogram (ECG)
- → interpreting an ECG
- → treating a heart-attack patient.

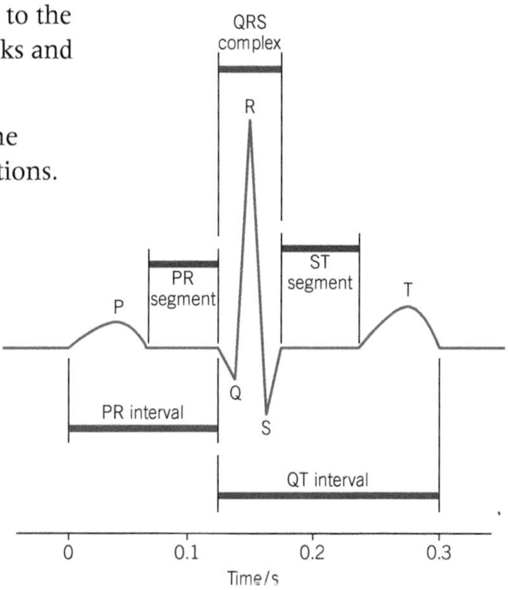

▲ **Figure 1** *Linking the cardiac cycle to the typical ECG trace*

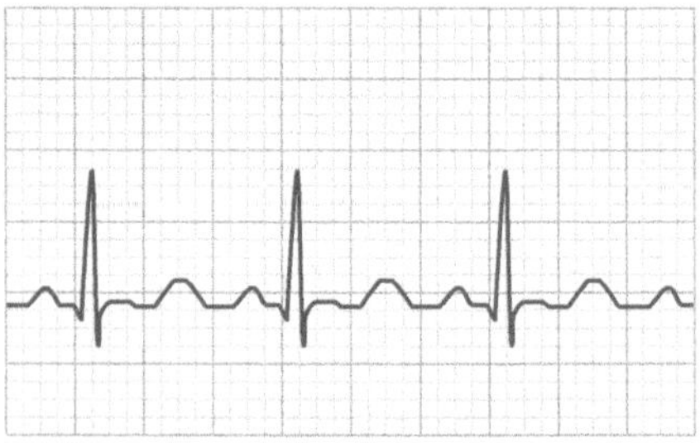

▲ Figure 3 *An ECG showing tachycardia*

Tachycardia

Tachycardia is a high resting heart rate (above 100 bpm). The ECG shows little gap between the T wave of one cycle and the P wave of the next.

Bradycardia

Bradycardia is a very slow resting heart rate between 40 and 60 bpm. The gap between each peak in the trace is very long, causing a long gap between the T wave and the next P wave.

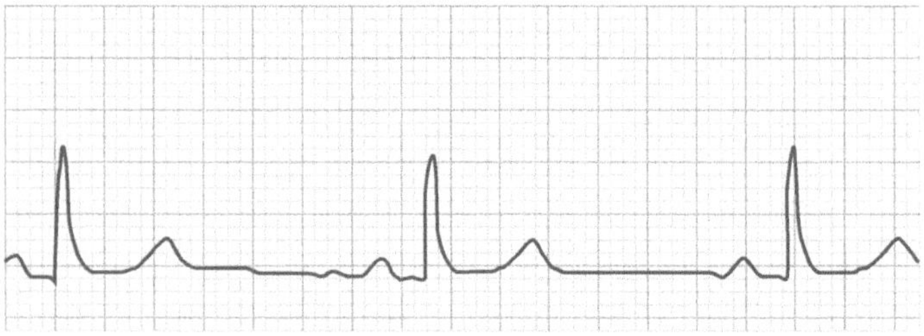

▲ Figure 4 *An ECG showing bradycardia*

Study tip

Use the graphs to understand the differences in the ECG traces. Relate this to the type of appropriate treatments that may be used.

How to recognise a possible heart attack

The following are signs of a possible heart attack:

- The person may experience heavy crushing pressure on the chest, possibly with pain spreading to the jaw, neck, and down one or both arms.
- They may experience breathlessness and discomfort in the abdomen with profuse sweating, but a cold ashen-looking skin and blueness at the lips.
- They will have a rapid, weak pulse which may be irregular.
- There may be nausea and/or vomiting and possibly an unexpected collapse.

Step by step of what to do to assist someone having a heart attack

Remember that this is no substitute for proper first aid training but it may help to save a life.

- Call 999 for emergency help and tell ambulance control you suspect a heart attack.
- Sit the person in the 'W' position – semi-recumbent (sitting up at about 75° to the ground) with knees bent.
- If it is available, there is no allergy issue and they are over 16 years of age, give them a 300 mg aspirin tablet to chew slowly.
- If the patient has any medication for angina, such as tablets or a spray, assist them to take it.
- Constantly monitor and record breathing and pulse rate until help arrives.

If the person becomes unconscious they might be in cardiac arrest. A cardiac arrest is when the heart is no longer performing any useful function as a pump to circulate blood around the body. This is usually because the heart has entered a state known as ventricular fibrillation in which rapid uncoordinated contraction of the muscles of the ventricle walls result in no effective movement of blood.

Treatment of cardiac arrest

If you witness a cardiac arrest there may be only a few minutes to help the patient. To perform CPR (cardiopulmonary resuscitation) effectively, training and frequent practice are essential. However aid provided by a member of the public until the emergency services arrive may save someone's life. Cardiac arrest can occur following a heart attack.

Remember to immediately ring for help, don't leave the patient and constantly reassure them.

- Check for breathing by tilting the head backwards.
- Place the heel of one hand in the centre of the victim's chest (the lower half of the sternum, or breastbone) and the heel of your other hand on top of the first, interlocking your fingers together.
- Keep your elbows straight and press down vertically, bring your bodyweight over your hands to make it easier.
- Press down firmly and quickly to about 4 to 5 cm downwards, relax and then repeat the compression. After each compression, release all the pressure on the chest without losing contact between the hands and the sternum.
- Aim for a rate of about 100 compressions per minute. You can help your timing and counting by saying out loud 'one and two and three and four...'
- Do this 30 times, then give mouth-to-mouth resuscitation, also known as Expired Air Resuscitation (EAR) twice, and continue this 30 : 2 procedure until help arrives.

Using a defibrillator

During a cardiac arrest, the supply of blood to a region of cardiac muscle is disrupted, depriving that area of oxygen. The cardiac muscle in the atrial or ventricular walls can start to contract in a rapid, disorganised way. This results in insufficient filling of the ventricles due to atrial fibrillation or absence of ventricular diastole allowing the ventricles to fill. As a result, blood is not pumped out of the heart and the patient will stop breathing.

A first aider can use a **defibrillator** by applying it to the chest using two pads connected to the defibrillator machine. The pads are placed in a diagonal line across the chest with the heart position approximately in the middle.

A reading on the machine will show if the heart is fibrillating and if it is, an electrical shock or discharge is applied to momentarily stop the chaotic electrical activity of fibrillation and allow the heart to recover its own coordinated rhythm led by the SA node.

Summary questions

1. Explain why there is a delay between the P wave and the QRS wave. *(2 marks)*
2. On Figure 2, the time interval between P waves was measured at 550 milliseconds. Calculate the resting heart rate of this patient and give your answer to the nearest whole number. *(2 marks)*
3. The time period between the QRS and T waves increases with age. Suggest what effect this has on the maximum heart rate. *(2 marks)*

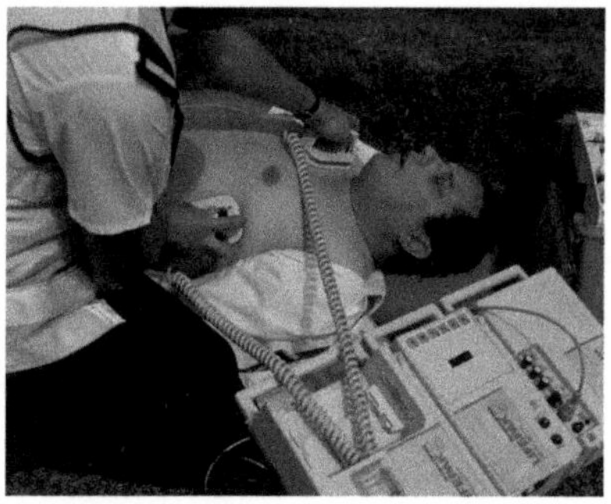

▲ **Figure 5** *Applying a defibrillator to the chest of a cardiac arrest patient*

Practice questions

1 Shown here is a diagram of a mammalian heart.

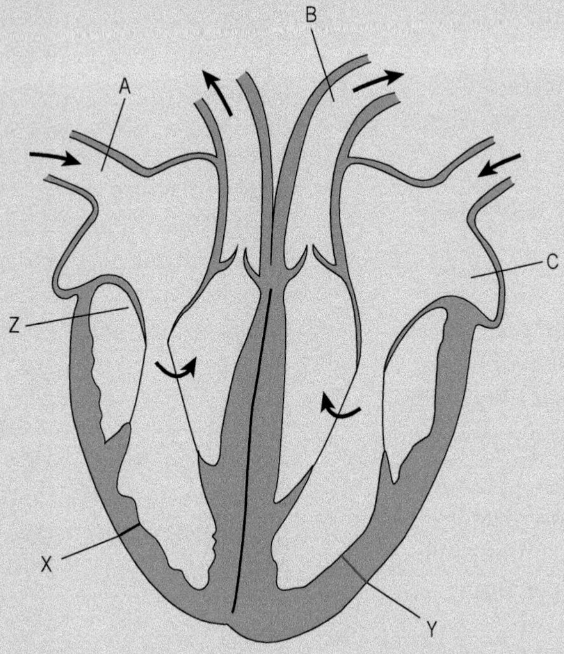

magnification: x 0.75

Which of the following is the actual width of the cardiac muscle at line Y given to 2 significant figures ?

A 0.67 mm

B 67 μm

C 6.7 μm

D 6.7 mm *(1 mark)*

2 Identify structure Z on the diagram.

A right semilunar valve

B right atrioventricular valve

C left semilunar valve

D left atrioventricular valve. *(1 mark)*

3 Which of the following statements about the diagram is/are correct:

Statement 1: vessel B returns blood to the systemic circulation.

Statement 2: chamber C receives blood from the systemic circulation.

Statement 3: vessel A receives blood from the pulmonary circulation.

A 1, 2 and 3

B Only 1 and 2

C Only 2 and 3

D Only 1 *(1 mark)*

4 The diagram shows an internal view of the heart viewed from above.

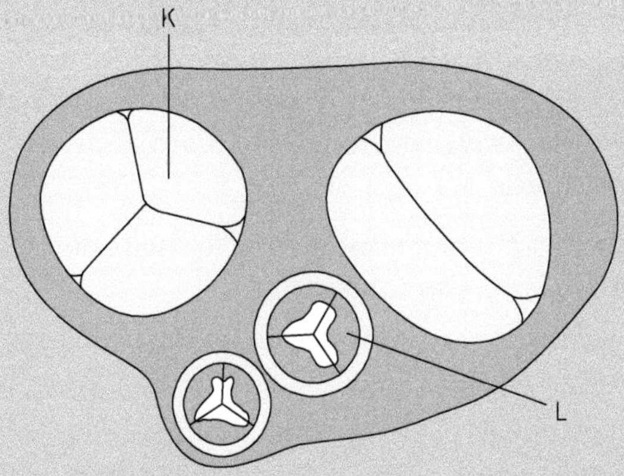

*Explain when and how valves K and L open and close during the cardiac cycle. *(6 marks)*

5 The graph shows the electrical changes during the cardiac cycle measured in a person at rest.

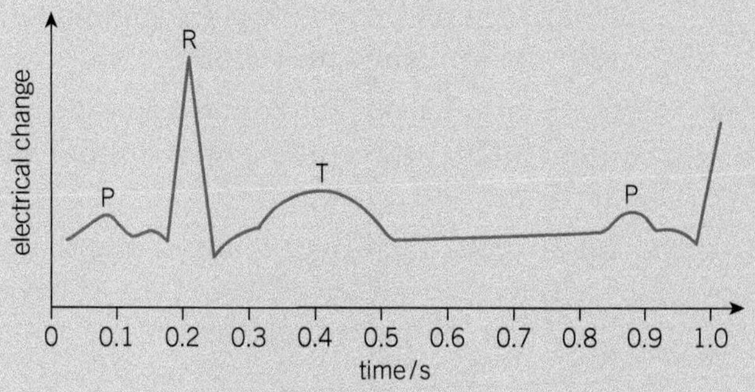

a (i) State what type of trace is shown in the graph. *(1 mark)*

(ii) Using the graph, calculate the heart rate of the person being measured. *(2 marks)*

b During a period of exercise, the cardiac output of the same person was estimated to be 10 $dm^3 min^{-1}$. Assuming that the stroke volume was 75 cm^3, calculate the person's heart rate during the exercise period. Give your answer to 3 significant figures. *(2 marks)*

c Investigators have found that, during exercise, both heart rate and stroke volume increase. The increase in stroke volume is due to the cardiac muscle in the ventricle walls contracting more strongly because it is stretched. This is a property of cardiac muscle.

(i) Suggest what is responsible for stretching the cardiac muscle and at what stage in the cardiac cycle this occurs. *(2 marks)*

(ii) Give **one** further property and one structural feature of cardiac muscle *(2 marks)*

6 The diagram shows the conduction pathways in the human heart.

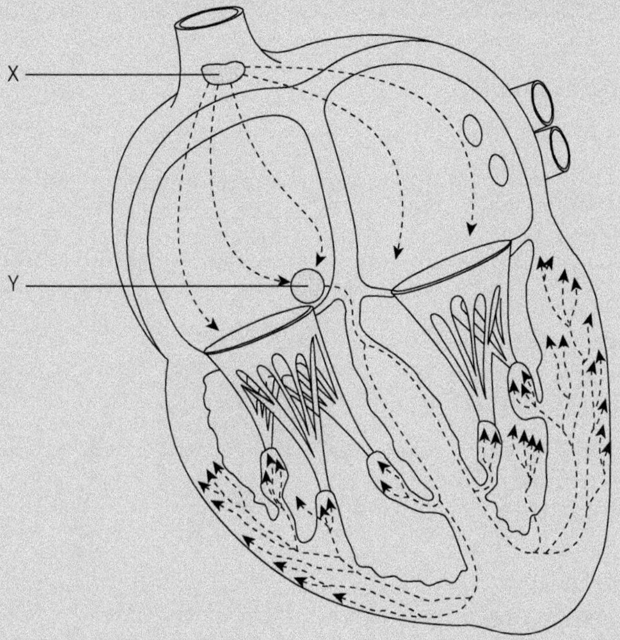

a Identify X and Y. *(2 marks)*

b Explain how heart action is coordinated to ensure that:

(i) atria contract before ventricles *(2 marks)*

(ii) ventricles eject the maximum volume of blood during each beat. *(2 marks)*

c Problems with the conduction pathway can lead to a condition known as fibrillation.

- Ventricular fibrillation can be a feature of a cardiac arrest.
- In ventricular fibrillation, the ventricles contract rapidly and are not synchronised with the atria
- Where ventricular fibrillation occurs, a person usually collapses and a pulse cannot be detected.

(i) Suggest why no pulse can be detected when ventricular fibrillation occurs. *(2 marks)*

(ii) Outline how a defibrillator is used to treat a suspected cardiac arrest *(3 marks)*

7 The table shows a breakdown of the estimated cost of CHD in the UK in 1999.

	Cost of CHD in UK (£ million per year)
Direct healthcare	1730
Informal care	2416
Productivity loss	2909
Total	7055

a Using the information in the table, calculate what the percentage of the total cost of CHD is due to productivity loss.

Show your working. Give your answer to the nearest whole number. *(2 marks)*

b If left untreated, CHD may lead to a heart attack.

Describe the first aid treatment you would give to a person suspected of having a heart attack who is still conscious *(3 marks)*

6 TRANSPORT SYSTEMS IN MAMMALS AND PLANTS

6.1 The transport system in mammals

Specification references: 2.2.2

Learning outcomes

Demonstrate knowledge, understanding, and application of:

- → the closed double circulatory system
- → blood pressure in the systemic and pulmonary systems
- → the structure and function of blood vessels.

Study tip

In a mass transport system, everything within the system moves in one direction but some means of assisting the movement is required. When discussing mass transport, remember to include the mechanism.

Synoptic link

It may be useful to look back at Topic 5.1, The heart for further information on mass transport.

The circulatory system

All organisms need to exchange materials with their environment to allow the uptake of oxygen and essential nutrients. As mentioned in Topic 5.1, once organisms reach a certain size, diffusion is no longer sufficient to provide all body cells with the oxygen and nutrients that they need. As a result, a specialist exchange surface is needed to exchange substances such as oxygen and nutrients with the medium outside. A mass transport system then becomes vital to carry these substances from the exchange site to the rest of the body cells. This is the circulatory system in mammals, which transports oxygen and nutrients to all the cells.

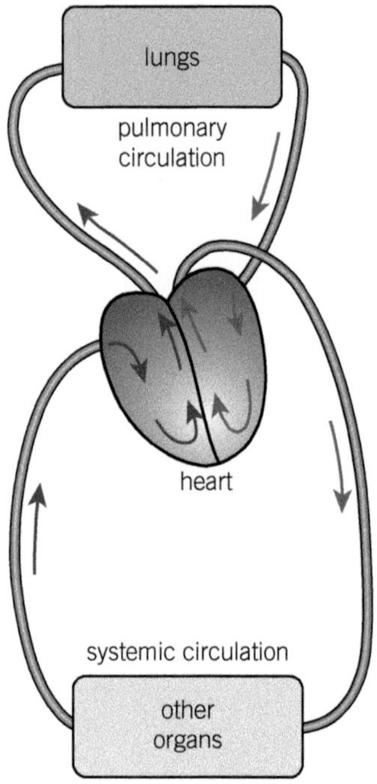

▲ **Figure 1** *A simple double circulatory system*

A closed and double circulatory system

There are two important aspects to the circulatory system in mammals that make it very efficient:

- It is a closed system, which means the blood always moves within blood vessels and can be maintained at pressure.
- It is a **double circulation**, which means there are two circuits – one is the circuit between the lungs and the heart (the pulmonary circuit) and the other is between the body and the heart (the systemic circuit).

In a double circulation deoxygenated blood returns from the body cells to the right side of the heart and from there on to the lungs, where gas exchange occurs. From the lungs oxygenated blood returns to the left side of the heart and leaves the left ventricle of the heart to the body tissues and cells. The two systems are separate from each other.

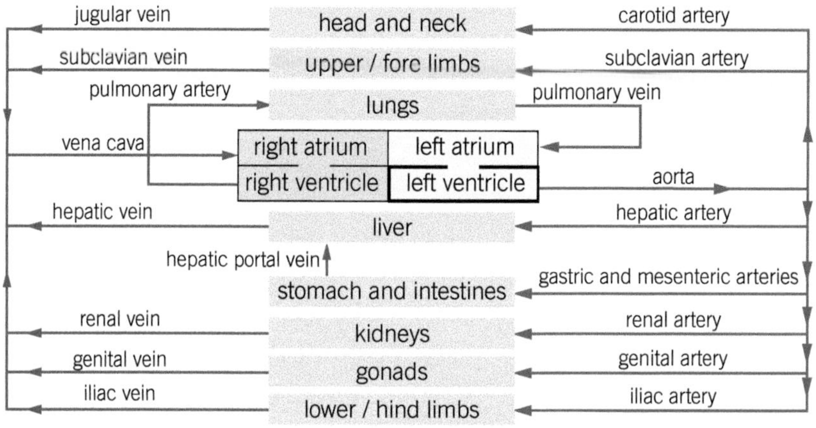

▲ Figure 2 *The human circulatory system is a closed and double circulation*

Advantages of a double system

- Blood pressure can be maintained around the body rather than falling due to resistance to flow as it moves through the lungs and other tissues.
- Oxygenated blood does not mix with deoxygenated blood making oxygen uptake from the exchange site more efficient.
- Delivery of oxygen and nutrients to the tissues is increased in efficiency.
- The blood pressure in the systemic and pulmonary circulations can be maintained at different pressures.

Advantages of a closed system

- Blood pressure can be maintained all through the body.
- Pressures can differ in the pulmonary and systemic systems.
- The blood supply to different organs can be varied depending on their functions and their needs at any one time. For example, the gut needs greater blood volumes during digestion of food and the muscles require greater volumes during exercise.
- Lower volumes of transport fluid, the blood, are needed to keep the system moving than would be needed if the system was an open one that allowed the blood to bathe all the tissues and organs.

Structure and function of arteries and veins

The blood is carried in a number of different types of blood vessels. Each type has a number of different features. The main vessels are the **arteries**, **veins**, and **capillaries**. Table 1 explains the differences in structure, function, and flow rate between the vessels.

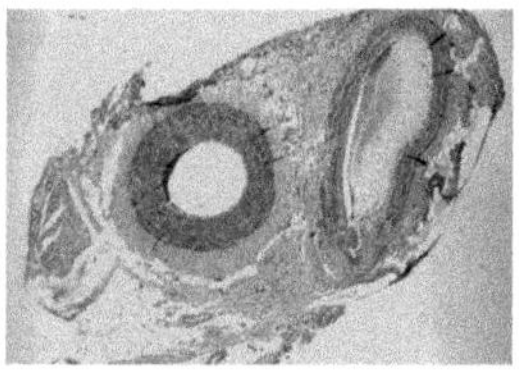

▲ Figure 3 *Cross sections through a human artery and vein, × 20 magnification*

▼ **Table 1** *The structure, function, and flow rate of the different blood vessels*

Blood vessels	Structure	Function	Blood flow
Arteries	Diameter is larger than 10 µm and the lumen is narrow. Overall the wall is thick, with large amounts of smooth muscle and elastic fibres. Inner endothelial lining is smooth to reduce friction. No valves along the length.	Arteries keep the pressure high as the blood is distributed throughout the circuit. The artery wall stretches during ventricular systole and recoils during ventricular diastole.	Take blood away from the heart under high pressure. The high pressure pulses as the pressure fluctuates between 120 mm Hg and 80 mm Hg.
Veins	The diameter varies but is very much larger than 10 µm. The wall is thin with a small amount of muscle and elastic tissue but veins have a wide lumen. Semilunar valves found all along the length.	Blood is carried under low pressure and the thin wall makes it easier to compress the vessel when the skeletal muscles contract around them. This aids blood flow through the veins.	Return blood to the heart. Flow is slow because the pressure is low. No pulse. The valves prevent backflow and keep the flow moving in one direction, which is necessary with the low pressure.
Capillaries	Capillaries are only 7–10 µm in diameter. The wall is a single layer of squamous epithelial cells with no elastic or muscle tissue.	They connect arterioles and venules, forming a vast network within the tissues. The size of the network provides a large overall cross-sectional area for exchange of materials between the blood and surrounding tissues.	Allow the flow between the arterioles and venules. Rate of flow is slow as the pressure is low.
Arterioles	Thin wall mainly of muscle and some elastic fibre.	Carry blood between arteries and capillaries. Regulate the flow and distribution.	Smooth muscle contracts to constrict the lumen or relaxes to dilate the lumen. This allows blood pressure and distribution to be regulated.
Venules	Very thin wall of muscle and elastic fibres.	Carry blood from capillaries back to the veins.	The wall has only a thin layer of muscle which does not provide any assistance in blood flow. Valves along the length of the venule prevent backflow.

Summary questions

1 State the advantages that a closed double circulatory system gives over a single open circulatory system. *(3 marks)*

2 Outline the role of the elastic fibres and smooth muscle in artery walls. *(2 marks)*

3 Explain how skeletal muscles and semilunar valves assist in the movement of blood through veins. *(2 marks)*

6.2 Blood pressure and tissue fluid

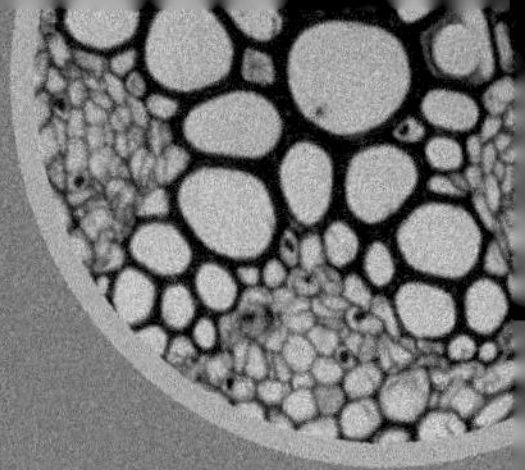

Specification references: 2.2.2

Blood pressure

When the left ventricle wall contracts (systole), blood is squeezed out into the aorta. This causes the **blood pressure** in the arteries to rise to its highest level. When the ventricle wall relaxes (diastole), the blood pressure falls in the arteries. The artery walls contain high levels of elastic fibres, which affect the blood pressure by gradually smoothing the fluctuations out by the time blood reaches the arterioles. The fibres stretch during systole and recoil in diastole, propelling the blood forward. It is this stretch and recoil that is felt as a pulse. Systolic pressure, therefore, reflects the force generated by the left ventricle contracting and diastolic pressure reflects the elasticity of the artery walls.

The contraction and relaxation of skeletal muscles acts to compress and then decompress the veins. As the veins are compressed pressure rises and the presence of valves ensures that blood flow is in one direction – towards the heart. As the veins decompress pressure falls and more blood is able to flow from the direction of the capillary beds through the valves and into the veins.

Measuring blood pressure

Blood pressure is measured using an instrument called a **sphygmomanometer**. This uses a measure of pressure called kilopascals (kPa). An older measure of pressure is millimetres of mercury (mm Hg), which gives more familiar blood pressure readings such as $\frac{120}{80}$ mm Hg.

Hypertension

As you saw in Topic 5.2, your blood pressure varies during the day according to the levels of activity or stress you experience. The average blood pressure in the population is $\frac{120}{80}$ mm Hg. High blood pressure (**hypertension**) is having a blood pressure persistently higher than this average.

There are several different stages of hypertension. The National Institute for Health and Care Excellence (NICE) defines them as follows:

- Stage 1 hypertension – clinic blood pressure of $\frac{140}{90}$ mm Hg or higher and subsequent ambulatory blood pressure monitoring (ABPM) or home blood pressure monitoring (HBPM) average of $\frac{135}{85}$ mm Hg or higher.
- Stage 2 hypertension – clinic blood pressure of $\frac{160}{100}$ mm Hg or higher and subsequent ABPM or HBPM average of $\frac{150}{95}$ mm Hg or higher.

Learning outcomes

Demonstrate knowledge, understanding, and application of:

→ blood pressure

→ how blood pressure is measured and the importance of measuring it

→ the formation of tissue fluid.

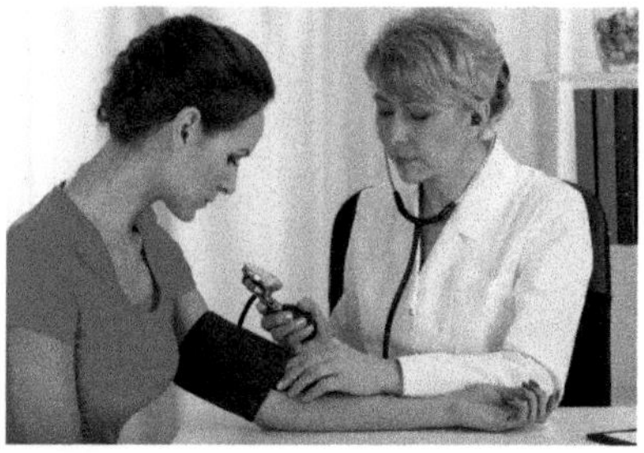

▲ **Figure 1** *A doctor taking the blood pressure of a patient*

Study tip

The unit mm Hg is a measure of pressure and refers to the height of mercury, in millimetres, in the column of a sphygmomanometer. The SI unit for pressure is kPa (kilopascals). You should know both of these units.

Synoptic link

You learnt about blood pressure in Topic 5.2, The cardiac cycle.

How to measure blood pressure

Make sure the person has been sitting down for five to ten minutes before you start as any activity will raise blood pressure. Blood pressure is normally taken from the left arm using an inflating cuff attached to the instrument. The cuff should be attached securely, but not too tightly, and then pumped up until it exerts sufficient pressure to stop the blood flow in the brachial artery. Now use the stethoscope to listen for the sounds of blood flow and slowly release the pressure from the cuff. The sounds that are heard through the stethoscope are called the Korotkoff sounds and will commence as soon as the pressure is equal to that in the artery at systolic pressure. When the sounds disappear, the pressure is equal to that in the artery at diastole.

The readings are given as two values, e.g., $\frac{110}{70}$ mm Hg. The top value is the **systolic reading**, and the bottom value the **diastolic reading**. They represent the pressure readings taken from the column of mercury in the sphygmomanometer – the traditional instrument that most doctors prefer to use because it is very accurate. Electronic blood pressure monitors can be attached to the finger or arm and used for continuous (or ambulatory) monitoring of the blood pressure. Data can be collected remotely and over longer periods of time.

Suggest how blood pressure measured with a sphygmomanometer compares to the blood pressure in the aorta. Explain your answer.

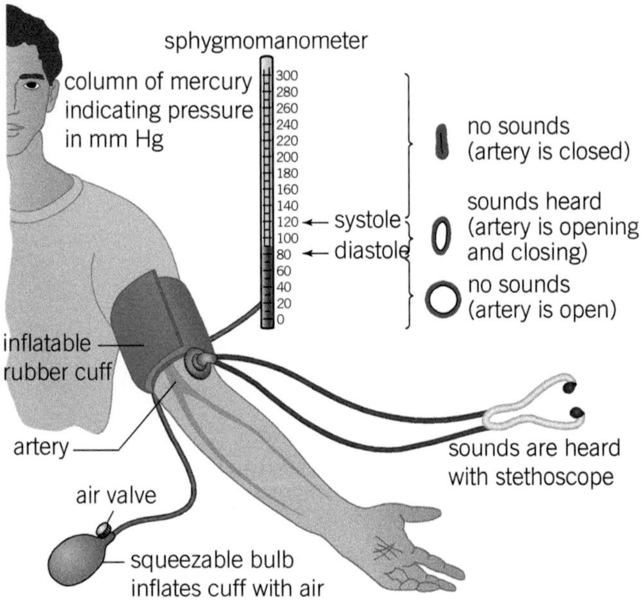

▲ **Figure 2** *Blood pressure reading and a table to explain the process*

- Severe hypertension or hypertension crisis – clinic systolic blood pressure of 180 mm Hg or higher, or a clinic diastolic blood pressure of 110 mm Hg or higher.

Hypertension is one of the most important, preventable causes of premature morbidity and mortality in the UK. It can result in:

- damage to the lining of the artery walls, resulting in an aneurysm (rupture of the artery wall) or a blood clot, leading to a heart attack or stroke
- damage to the heart valves
- kidney disease as the capillaries within the kidney are damaged by the constant high pressure.

Hypotension is persistently lower than the average blood pressure. Health problems that may result from this condition include:

- weakness and tiredness
- dizziness and fainting
- coma and possible death.

Risk factors that contribute to hypertension

There are a number of risk factors that increase the chance of having hypertension:

- smoking
- being obese or overweight
- a diet high in salt
- excess alcohol consumption
- high levels of stress
- age – the older the individual, the greater the risk. The fall in oestrogen levels after the menopause contributes to this risk in females
- gender – males have an increased risk when compared to females
- low levels of exercise or a completely sedentary life style.

1 Use this list to construct an ideal lifestyle that reduces the risk of hypertension.

Tissue fluid

The capillary network forms a close contact with all the cells of the tissues through which it runs. The high pressure (**hydrostatic pressure**) at the arteriole end of this capillary network forces water and dissolved soluble molecules out of the blood plasma into the surrounding area to form **tissue fluid**, which bathes the cells.

Synoptic link

You will find about more about the role of NICE in Topic 15.2, Medicinal drugs and clinical trials.

Study tip

Ambulatory blood pressure monitoring (ABPM) and home blood pressure monitoring (HBPM) are used to measure blood pressure outside of doctors' surgeries and clinics, since for some people the act of visiting a doctor can raise their blood pressure slightly.

In ABPM a portable blood pressure monitor is worn around the arm, which measures and records blood pressure periodically throughout the day and night.

In HBPM the patient is given a blood pressure monitor to use at home and instructed to measure their own blood pressure at regular intervals.

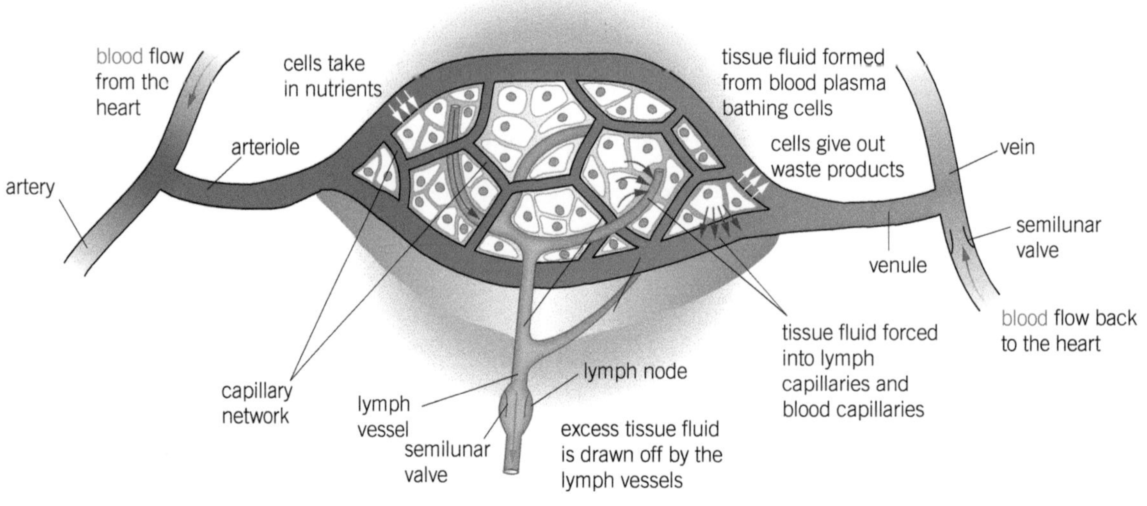

▲ **Figure 3** *The formation of tissue fluid*

Synoptic link

Look back at Topic 2.6, Osmosis in cells if you need to revise the concepts of osmosis and water potential.

Summary questions

1. State the difference between tissue fluid and blood plasma. *(2 marks)*
2. Explain what it means when a person is described as *hypertensive* by a doctor. *(2 marks)*
3. Drugs known as calcium ion channel blockers can be used to treat cardiovascular disorders. One side effect of some forms of the drug is oedema in the hands and ankles due to peripheral vasodilation. Suggest the location of the calcium channels targeted by these drugs. *(3 marks)*

Tissue fluid allows an exchange of materials – oxygen and nutrients such as glucose enter the cells and carbon dioxide and other waste materials, such as urea, leave the cells. This exchange can occur because the capillary wall is made of extremely thin, flat squamous epithelial cells which have numerous gaps between them. The presence of these gaps (fenestrations) makes the walls permeable to small molecules but not to larger protein molecules or red blood cells. The capillary walls are essentially leaky, allowing constant exchange of materials.

The exchange is facilitated by the slow movement of the blood through the many small capillaries and by the higher hydrostatic pressure at the arteriole end of the network. Larger protein molecules remain in the blood plasma and do not leave the capillary. These molecules cause an osmotic force called **oncotic pressure** (or colloid osmotic pressure). This force is negative and tends to 'pull' the water in. At the arteriole end of the capillary network, hydrostatic pressure is positive and greater than oncotic pressure, resulting in a net outward or 'pushing' force.

At the venule end of the network, the hydrostatic pressure has fallen due to the loss of the fluid. Oncotic pressure is now higher than hydrostatic pressure, resulting in a net inward force, drawing fluid back into the capillary. This fluid has lost much of the oxygen and nutrients it was carrying and now contains the waste materials that were lost into the tissue fluid from the cells.

Some tissue fluid will not return to the blood here but instead drains into the **lymphatic system**. The lymph vessels are a separate system of vessels that carry the excess tissue fluid, as lymph, back to the veins in the neck where it is returned to the blood.

6.3 Structure of vascular tissue in plants

Specification references: 2.2.4

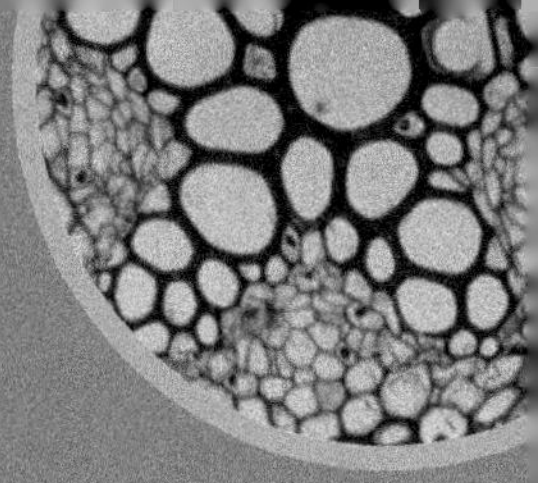

The need for two transport systems in plants

As described in Topic 6.1, the transport system in mammals, once organisms reach a certain size, diffusion is no longer sufficient to provide all their cells with the nutrients that they need. This means that in plants, like in mammals, a mass transport system is vital to carry substances from the exchange sites to the rest of the cells.

Plants require a source of water and a source of nutrients (minerals). Water movement depends on physical processes and always occurs in one direction in plants from roots to leaves. However, the direction of movement of **plant assimilates**, such as sugars, will vary. In one area of the plant, assimilates may be synthesised through photosynthesis and released into the transport system – this area is known as a **source**. Other areas take up assimilates from the system and are known as a **sink**. As a result, the movement of assimilates may be in either direction within the plant. The result is that two transport systems are required – one for water and minerals and one for assimilates. Collectively these transport systems are known as **vascular tissue**.

There is no need for a specialised transport mechanism for respiratory gases in plants because plants have a low metabolic rate, meaning that simple diffusion is efficient enough. Photosynthesising cells also produce oxygen and use carbon dioxide, giving other cells the oxygen they need and removing carbon dioxide *in situ*.

Learning outcomes

Demonstrate knowledge, understanding, and application of:

- → the need for two transport systems in plants
- → the organisation of vascular tissue in monocotyledons and dicotyledons.

The vascular tissue

In land-based plants, vascular tissue is found in all organs throughout the plant. It consists of **xylem** and **phloem** tissue organised into **vascular bundles**. In addition there are usually groups of packing cells (parenchyma) and some **cambium**, which contains meristematic cells that are capable of cell division. The vascular bundle is surrounded by a sheath of photosynthetic cells.

Flowering plants can be divided into two groups: **monocotyledons** (e.g., cereals and other grasses) and **dicotyledons** (e.g., potatoes and other broadleaved plants). These groups have different arrangements of vascular tissue in both the root and the stem. The classification of dicotyledons and monocotyledons is based on the number of primary leaves (cotyledons) that grow from their seeds. The seed of a dicotyledon grows two primary leaves whereas a monocotyledon only grows one.

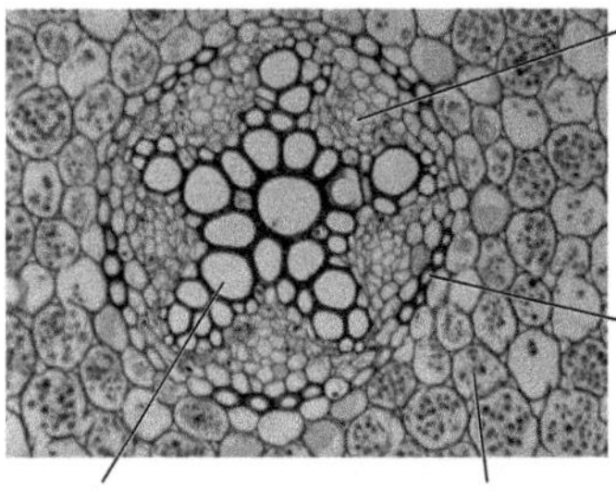

▲ **Figure 1** *Transverse section (T.S.) of a vascular bundle in a dicotyledonous root, approximately ×100 magnification*

Xylem and phloem in the root

In dicotyledonous roots, the vascular tissue is found in a central core. The xylem tissue generally forms the arms of an 'X' and the phloem is

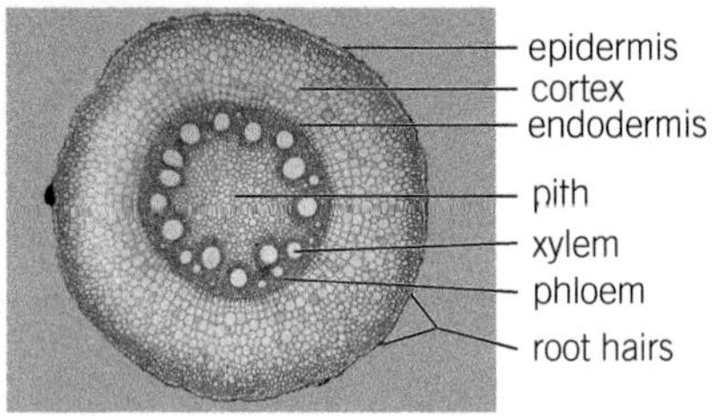

▲ **Figure 2** *T.S. of a monocotyledonous root, approximately ×5 magnification*

found as patches between the arms. Around the whole vascular tissue is a layer of cells called the endodermis, which has an important role in the passage of water in the root.

The arrangement of vascular tissue in monocotyledonous roots is different. In monocotyledonous roots the vascular tissue is found towards the centre of the root, arranged as a circle of xylem and phloem tissue as shown in Figure 2.

Xylem and phloem in the stem

The arrangement of vascular tissue in a stem is quite different from the root. A sunflower stem (*Helianthus*) is a typical example of a dicotyledonous stem. The epidermis is the outermost layer and under that is a complete layer called the cortex. This is a relatively thin layer of smaller **parenchyma** cells packed quite tightly together. Below this the vascular bundles are organised into a complete circle around a large central region or pith. The xylem is on the inside of each bundle, with the phloem on the outside. Sandwiched between the two tissue layers is the cambium.

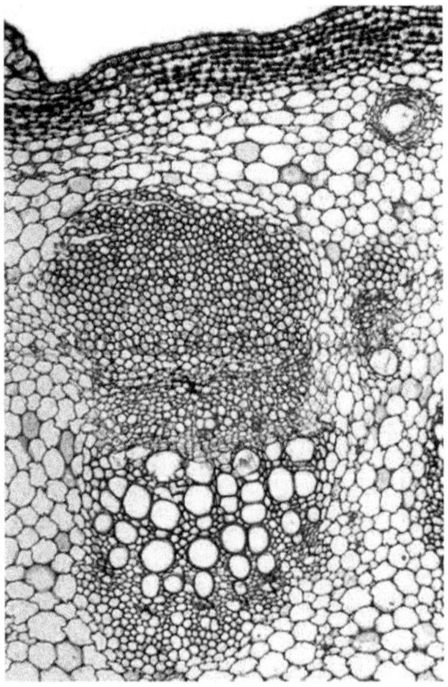

▲ **Figure 3** *Light micrograph T.S. of a dicotyledonous stem , ×18 magnification*

In a monocotyledonous stem, the arrangement differs slightly with the vascular bundles rarely forming a typical ring. Instead they are scattered more evenly across the whole stem. The arrangement of tissue within each bundle still tends to be the same with the xylem tissue towards the inner part of the stem and the phloem tissue to the outside, with the cambium in between.

Xylem and phloem in the leaves

In a dicotyledonous leaf, the xylem and phloem form the veins of the leaf. The veins arise from the central vein or midrib and form a branching network, which becomes increasingly smaller the further away from the midrib. The exact arrangement of veins varies between species and may be used as part of the classification of plants.

In a monocotyledonous leaf there are rarely any central midribs. Instead the veins usually run parallel along the length of the leaves, which tend to be long and narrow.

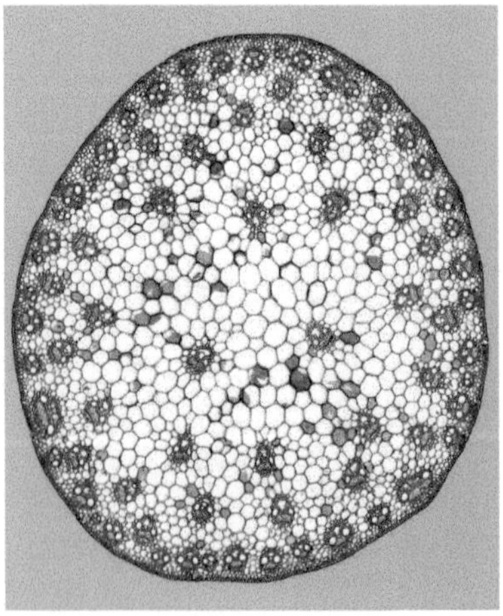

▲ **Figure 4** *Light micrograph T.S. of a monocotyledonous stem ×22 magnification*

Xylem structure

The xylem tissue is made up of a number of different cell types. The main type is the **xylem vessel element**, the structure of which is well adapted to the function of transporting water. The cells are long and thin and have thickened walls impregnated with a substance called lignin. Lignin is impermeable to water and so the cell contents of the vessel elements quickly die when the xylem is first formed. The cells are arranged one above the other in columns, the walls between them having been broken down. This leaves a long, thin column, called a xylem vessel, which offers little resistance to the passage of water.

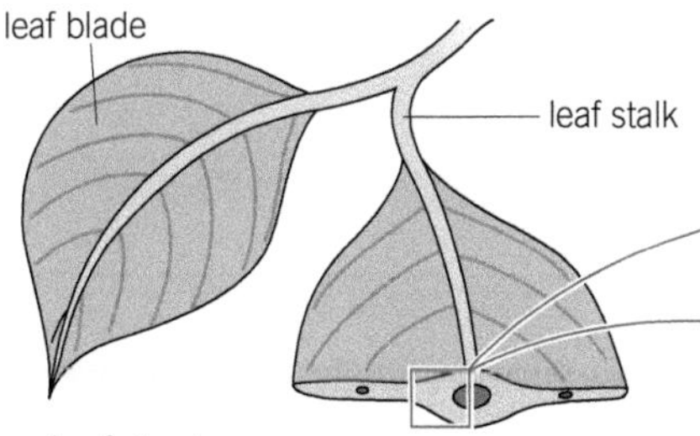

a *Leaf structure*

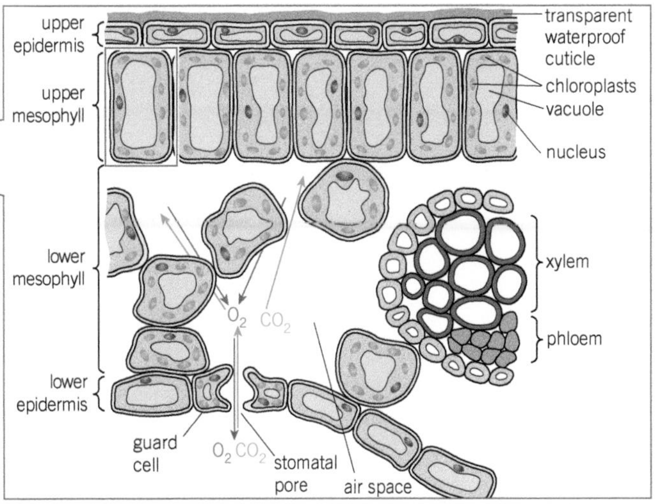

b *Vertical section through a dicotyledonous leaf*

▲ Figure 5 *T.S. of a dicotyledonous leaf*

The lignin typically forms specific arrangements in the vessel wall with some plants having ring shapes (annular), some spiral, and others forming a patchwork (reticulum). The lignin keeps the xylem vessels open and prevents them collapsing inwards even during drought conditions. It also provides support and strengthens the walls.

The xylem vessel wall has pores where there is no lignin. These pores are called bordered pits and they allow water to move horizontally between one xylem vessel and another. This ensures that water movement can continue even when there is a blockage in one vessel.

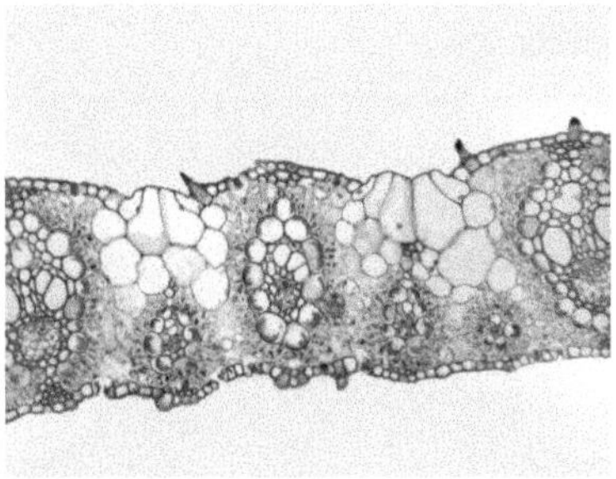

▲ Figure 6 *T.S. of a monocotyledonous leaf, ×59 magnification*

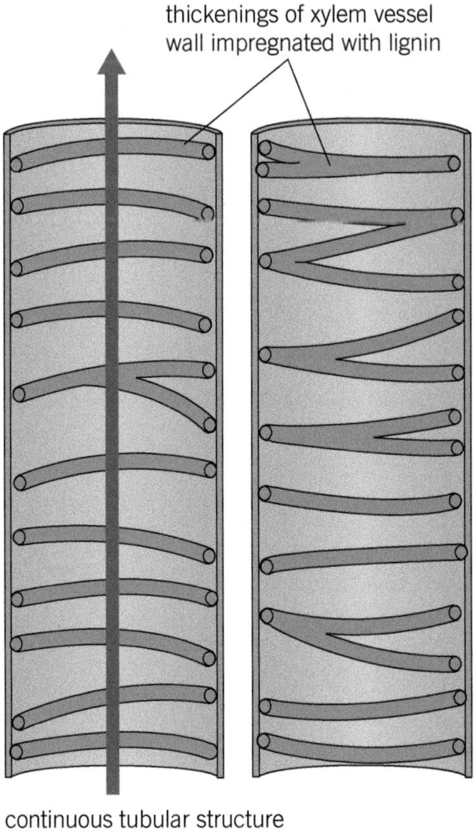

◀ Figure 7 *Longitudinal section (L.S.) of xylem tissue*

Study tip

Do not confuse xylem and phloem with the cells making up these tissues. This is important when labelling drawings of plant sections viewed through microscopes. Through the low-power lens only tissue layers will be visible, and therefore only tissues can be labelled. Through the high-power lens cells should be visible and cells can therefore be labelled.

Phloem structure

Phloem tissue is made up of two types of cells – **sieve tubes** and **companion cells** – and is adapted to transport sugars (assimilates) around the plant.

Sieve tubes are long thin cells laid end to end. Their end walls are perforated by many small pores to form a sieve plate, which allows sugar sap to flow from one cell to the next. The cells contain very little cytoplasm and have no nucleus or tonoplast. These features mean the sugar sap can flow through the sieve tubes unimpeded.

Alongside the sieve tubes are the companion cells. Companion cells are small thin cells, approximately the same length as the sieve tubes but much thinner. Companion cells contain a large nucleus, dense cytoplasm and numerous mitochondria, which produce the ATP needed for all the metabolic processes in both the sieve tubes and companion cells. Between the companion cells and the sieve tubes there are many small gaps in the wall called plasmodesmata, which allow communication and exchange of materials between the sieve tube element and the companion cell.

Synoptic link

You learnt how to section and stain plant material in Topic 1.3, Differential staining and blood smears.

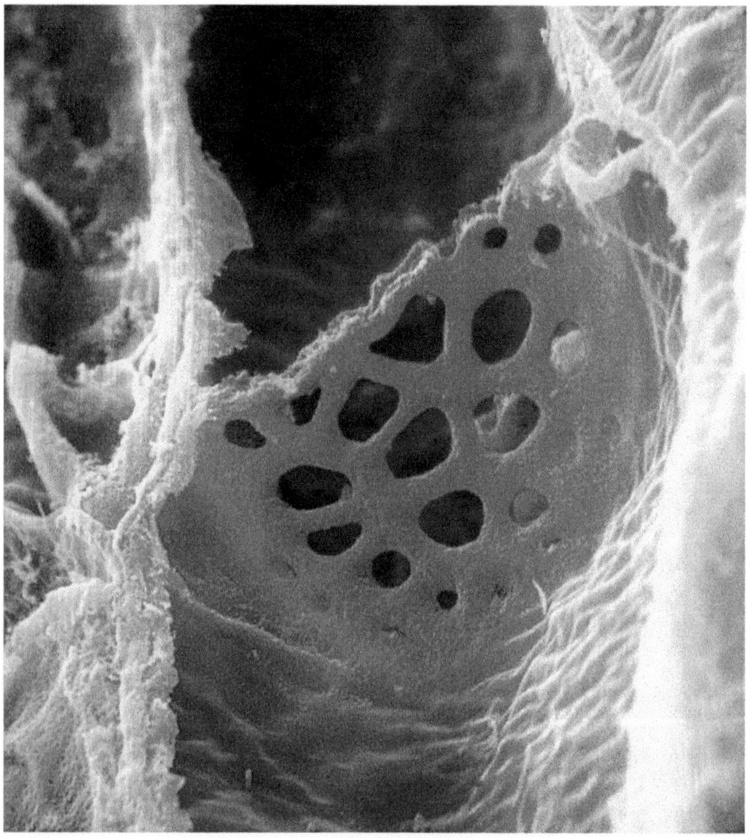

▲ Figure 8 *Scanning electron micrograph T.S. of phloem tissue , ×4 800 magnification*

Summary questions

1 Explain how the location of xylem and phloem tissue differs between the roots in cereals and in the roots of broad-leaved crops like potatoes. (*2 marks*)

2 Describe the features of the xylem vessel that make them especially well adapted to transporting water. (*3 marks*)

3 Sieve tubes contain a very thin layer of cytoplasm and have no nucleus or tonoplast, which means they are unable to carry out the usual processes of cell metabolism. Explain how these cells are still able to survive. (*2 marks*)

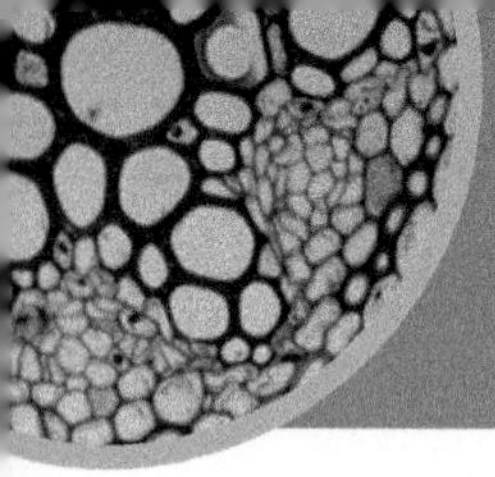

6.4 Transport of water in plants

Specification references: 2.2.4

Learning outcomes

Demonstrate knowledge, understanding, and application of:

- → the transport of water through plants
- → the process of transpiration
- → the physical processes involved in the transpiration stream.

Transport of water

Plant roots are the site where water and nutrients are taken into the plant from the soil. In order to transport the water absorbed by the roots, xylem vessels run from the root, up the stem and all the way to the leaves. In a large tree there may be up to 250 dm^3 of water carried from the roots to the leaves every hour.

The flow of water through the plant involves several different stages but is the inevitable consequence of having open pores (stomata) on the leaves and stem. The result is water vapour loss, but the benefits include the cooling effect of water flowing, the transport of mineral ions, and the transport of water needed for photosynthesis.

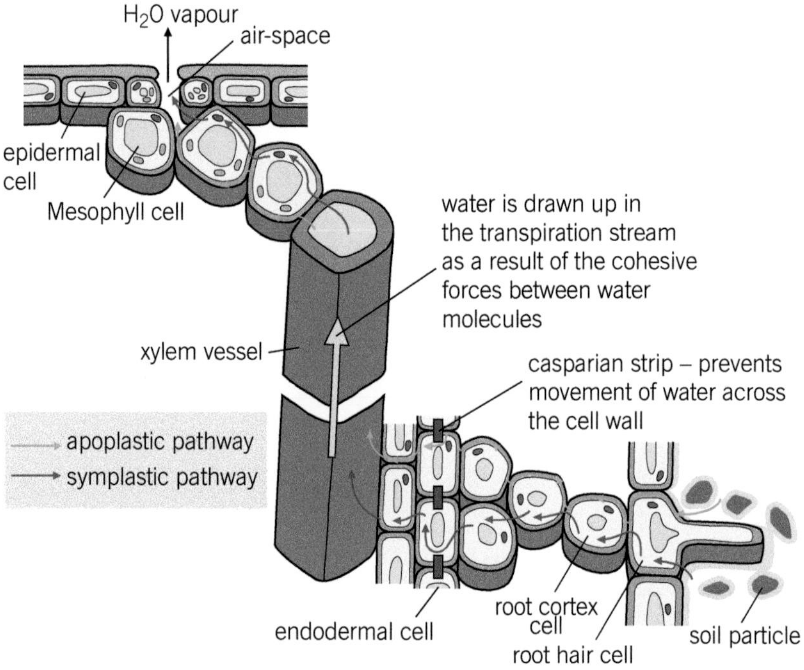

▲ **Figure 1** *Summary of water transport through a plant*

Synoptic link

You first encountered water potential and osmosis in Topic 2.6, Osmosis in cells.

Study tip

Transpiration is the loss of water by evaporation from the aerial parts of the plant. Don't confuse this with the transpiration stream which is the movement of water through the plant.

Transpiration

Transpiration is the loss of water vapour from the aerial parts of a plant. Most water vapour is lost by diffusion through open stomata (singular: stoma). Since stomata are generally open when the plant is photosynthesising, to allow gas exchange, transpiration is inescapable.

The water evaporates from the surfaces of the mesophyll cells in the air spaces within the leaf, lowering the water potential in the mesophyll cells. This causes water to move across the leaf from the xylem, through osmosis down a water potential gradient, to replace the water lost.

The transpiration stream

Water moves up the stem to replace the water leaving the xylem in the leaf. The force of cohesion between water molecules is strong enough to allow a continuous column of water to move up in the xylem vessels.

The pulling force created is called the 'transpiration pull' and it causes a negative pressure or tension within the xylem. This mechanism for water movement in the xylem is called the cohesion-tension theory. It is a passive process with no energy required to move the water molecules other than solar energy needed to evaporate the water molecules into the air spaces of the leaf. If the water column is broken, the water can move sideways through the bordered pits to another xylem vessel. There are also adhesion forces involved in the xylem vessels, which cause the water molecules to cling to the sides of the vessels and prevent the water column dropping back.

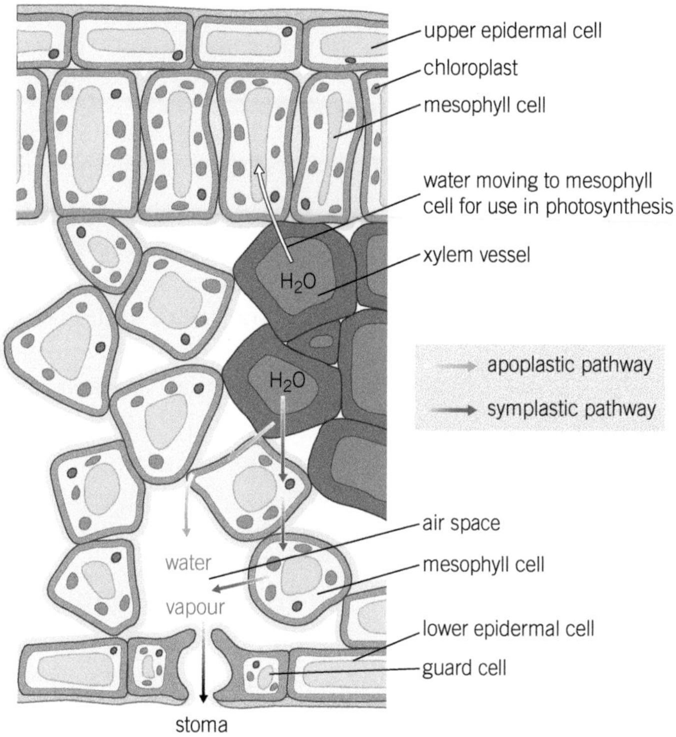

▲ Figure 2 *Movement of water from the xylem to the atmosphere, via the stoma*

Synoptic link

You first looked at the cohesive forces between water molecules in Topic 2.1, The properties of water.

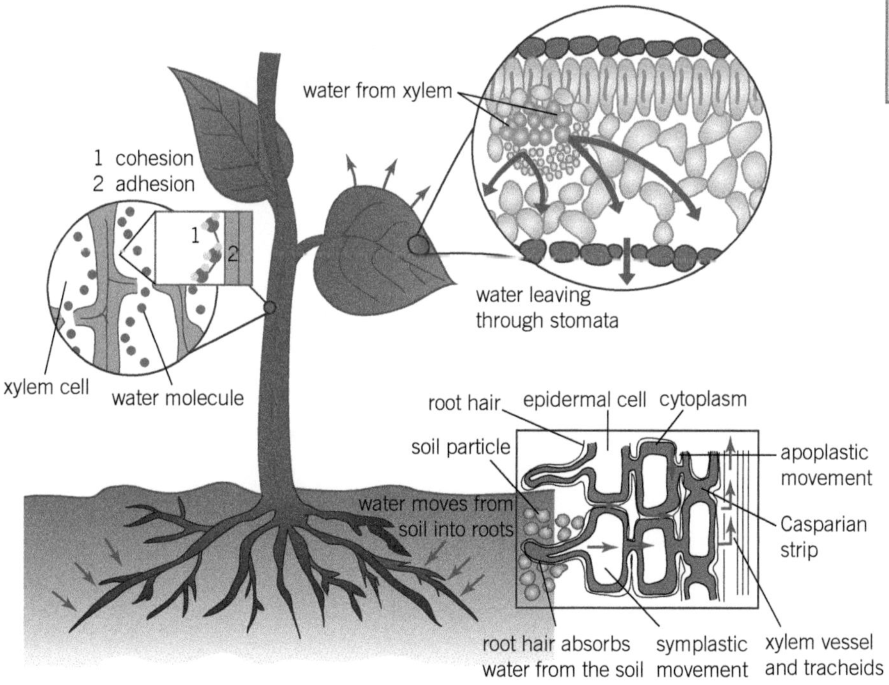

▲ Figure 3 *The transpiration stream*

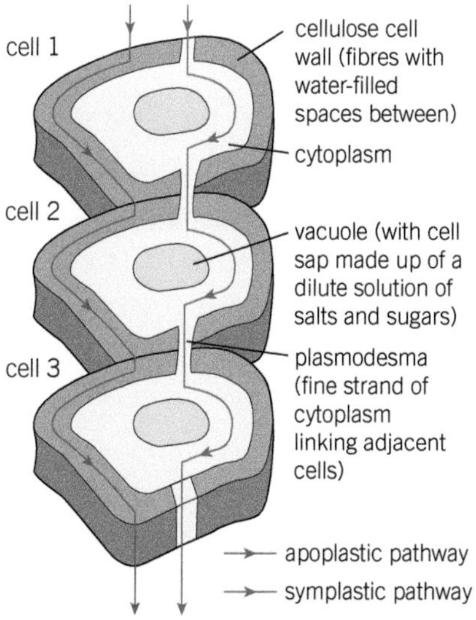

▲ Figure 4 *The apoplastic and symplastic pathways across the root cortex*

Synoptic link

You first encountered active transport across the cell membrane in Topic 1.11, Active movement across cell membranes.

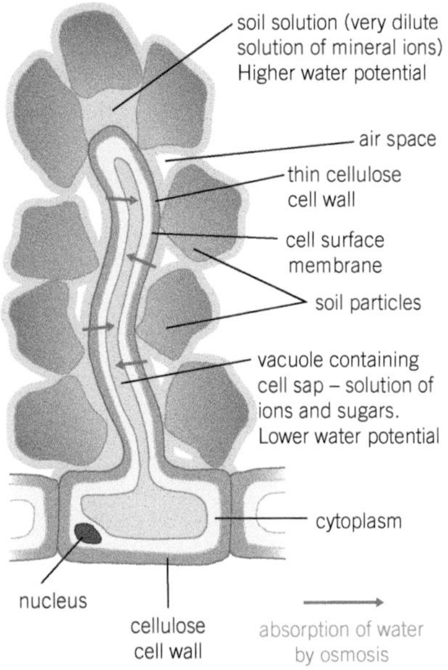

▲ **Figure 5** *Absorption of water by a root hair cell*

Summary questions

1 Describe the apoplastic pathway in plants. *(1 mark)*

2 Describe and explain the role of the Casparian strip in the plant root. *(3 marks)*

3 Cyanide is a poison which prevents respiration in cells. Suggest why root pressure is absent in roots treated with cyanide. *(2 marks)*

Movement of water between cells

Water moves down a water potential gradient between plant cells through three possible routes:

- The apoplastic pathway – water moves into the space outside the plasma membrane and through the cell walls and extracellular spaces.
- The symplastic pathway – water moves through the plasma membrane into the cytoplasm and can move from one cell to another through the plasmodesmata. A lower water potential in one cell caused by loss of water causes water to enter by osmosis from the neighbouring cells. This in turn lowers their water potential, so water moves from cell to cell through the continuous link of the cytoplasm down the water potential gradient.
- The vacuolar pathway – where water enters the cytoplasm and the vacuoles of the cells and moves down the water potential gradient from cell to cell in a similar way to the symplastic route.

Movement of water into the root

Plant roots are covered in tiny root hairs, which help to increase the surface area and are surrounded by soil particles and soil water. Water enters the root hairs by osmosis as shown in Figure 5. Water will then move across cells in the root cortex by both the apoplastic, symplastic and vacuolar pathways.

A layer of cells called the endodermis surrounds the vascular tissue in the root. The walls of these cells are impregnated with a layer of waterproof material called suberin, forming a structure called the Casparian strip (see Figure 6). The Casparian strip blocks the apoplastic pathway, forcing water into the cells and through the symplastic pathway to cross the endodermis.

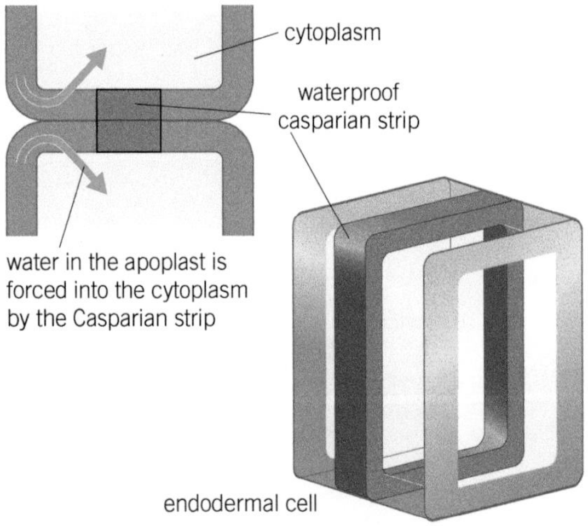

◀ **Figure 6** *Water movement across the endodermis, showing the Casparian strip*

Endodermal cells actively transport mineral salts into the xylem and can therefore regulate the water potential and control the flow of water through the symplastic pathway. The action of the endodermal cells is also responsible for a force called root pressure which tends to push water up the xylem. The plasma membrane of endodermal cells contains numerous transport proteins to move molecules in and out by active transport.

6.5 Factors affecting transpiration

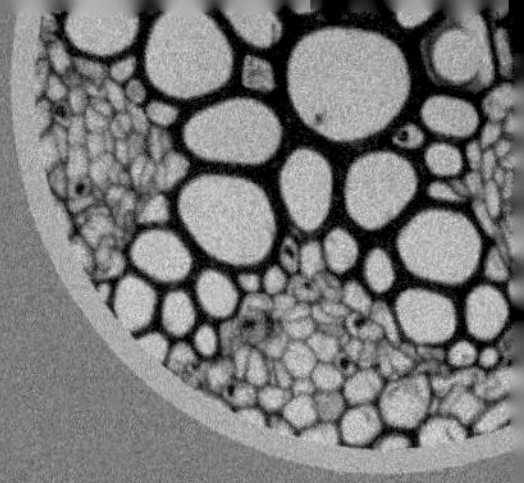

Specification references: 2.2.4

Factors affecting transpiration

The rate of transpiration can be affected by factors within the plant itself and by environmental factors, such as changes in the external conditions.

Factors within the plant

The size of the leaves and the number of leaves present will affect the surface area over which water vapour may be lost, and therefore the rate of transpiration. The size and number of stomata on each leaf surface are also important factors as this will influence how much water vapour can be lost in a given time.

Some plants have adaptations to minimise the effect of the water potential gradient between the inside of the leaf and the atmosphere. These include anatomical features, such as hairs on the leaf, sunken stomata, or rolled leaves (e.g., in marram grass). These features all allow a build-up of water vapour around the stomatal pore, reducing the water potential gradient between the inside of the leaf and the environment. Table 1 lists some of the main factors within the plant itself that affect the rate of transpiration.

▼ **Table 1** *Summary of factors within the plant that can affect transpiration rate*

Factor	How factor affects transpiration	Increase in transpiration caused by	Decrease in transpiration caused by
Number of leaves on plant	Leaf number affects surface area for loss of water vapour	Increased number of leaves	Fewer leaves
Number and size of stomata	The water loss depends on the number and size of stomata since this affects the diffusion of water from the air spaces in the mesophyll	A larger number and size of stomata means more diffusion of water from the air space, causing an increase in evaporation from the surface of the mesophyll cells	Fewer or smaller stomata reduces diffusion of water from the air space as does stomatal closure.
Presence and thickness of a cuticle	The waxy cuticle reduces water loss and a thicker cuticle will increase this effect	A thin cuticle or the absence of a cuticle	The presence of a waxy cuticle or a thicker cuticle

Learning outcomes

Demonstrate knowledge, understanding, and application of:

→ factors affecting transpiration

→ practical techniques for measuring transpiration rate.

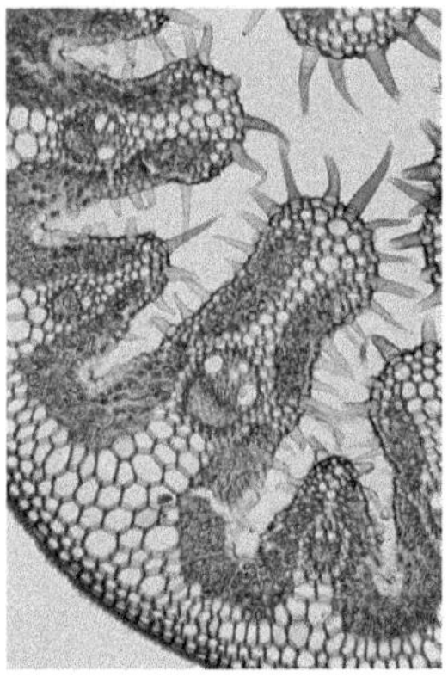

▲ **Figure 1** *Light micrograph of a marram grass leaf, ×50 magnification*

Synoptic link

Marram grass is a xerophyte (a plant specialised to survive extremely dry conditions). You will find out more about xerophytic adaptations in Topic 10.4, Adaptations in plants.

Environmental factors

There are many environmental factors that affect the rate of transpiration. Some of these are listed in Table 2. All of them produce their effect by altering the water potential gradient between the inside of the leaf and its immediate environment.

▼ **Table 2** *Summary of environmental factors affecting transpiration rate*

Factor	How factor affects transpiration	Increase in transpiration caused by	Decrease in transpiration caused by
Light	Stomata open in the light and close in the dark	Higher light intensity	Lower light intensity
Temperature	Alters the kinetic energy of the water molecules and the relative humidity of the air	Higher temperatures	Lower temperatures
Humidity	Affects the water potential gradient between the airspaces in the leaf and the atmosphere	Lower humidity	Higher humidity
Air movement	Changes the water potential gradient by altering the rate at which moist air is removed from around the leaf (Figure 2)	More air movement	Less air movement

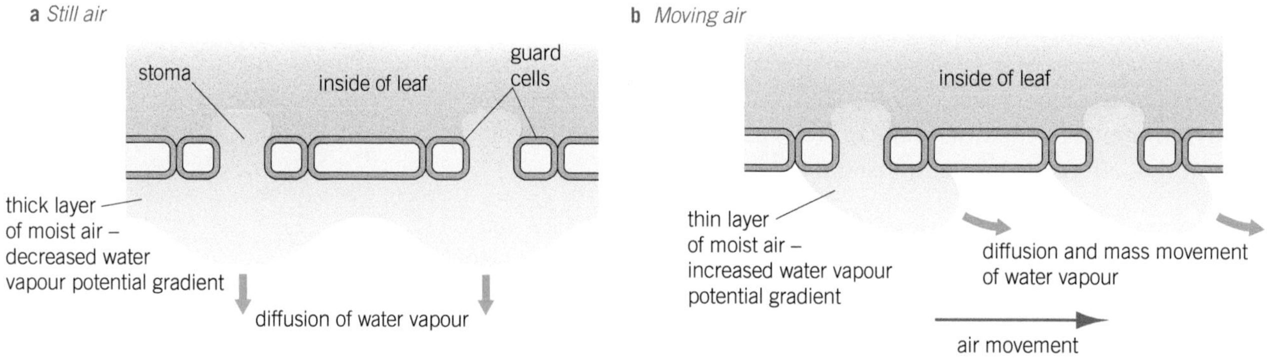

▲ **Figure 2** *Effect of air movement on the rate of transpiration*

Using a potometer to measure transpiration

A potometer is a simple apparatus used to measure the rate of water uptake from a cut shoot, however it can also be used to estimate the rate of water loss from a leafy shoot (i.e., the transpiration rate). This is based on the fact that approximately 99% of the water taken up by a leafy shoot is lost in transpiration.

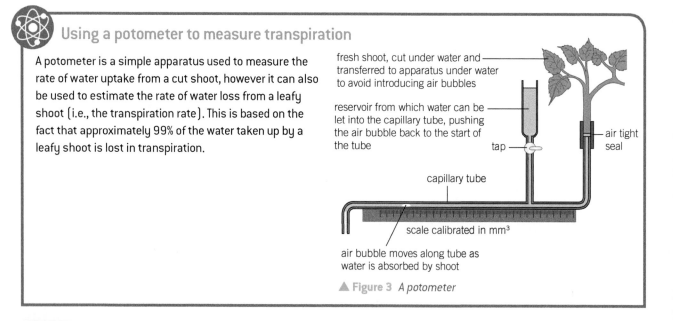

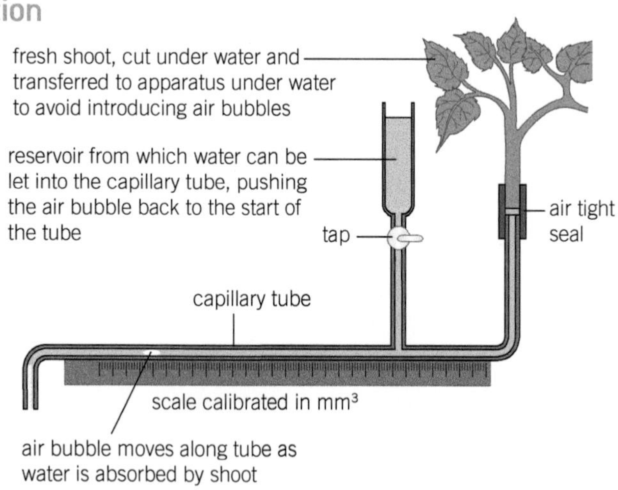

▲ **Figure 3** *A potometer*

When setting up a potometer there are several important things to note:

1. Make sure there are no air bubbles in the apparatus and cut the shoot with a slanting cut under water to avoid air bubbles.
2. Ensure the cut end of the shoot is a perfect fit for the rubber tubing connecting the shoot and the capillary tube.
3. The movement of the meniscus at the end of the column of water represents the water taken up by the leafy shoot as it loses water to the environment. This movement can be measured to represent the rate of loss of water by transpiration.
4. Allow the apparatus to stabilise for approximately 10 minutes when you change the conditions in which the potometer and leafy shoot are being used.

Manipulating variables:

5. Place the potometer and leafy shoot in a sunny position. After stabilising, begin measuring the position of the meniscus over a set time period.
6. Repeat the same method but this time leaving the potometer in darkened conditions.
7. Repeat but now using a fan to create windy conditions.

It may be preferable to set up several different potometers so that different conditions could be measured at a similar time. However, in this case it is vital that the type and size of the shoot, the number of leaves, and the type of potometer are all kept constant to avoid the influence of other variables.

The distance moved by the meniscus in the potometer estimates the water lost indirectly. How could water loss be estimated in this apparatus?

Summary questions

1. Explain why transpiration is sometimes referred to 'as a necessary evil'. *(2 marks)*

2. Describe how each of the following conditions would affect the transpiration rate, and explain why.
   a. increased temperature
   b. a reduction in wind
   c. reduced light intensity. *(3 marks)*

3. The volume of water taken up by a plant in a potometer is calculated using this formula:

   Volume = length of bubble × πr^2 (where r = radius of the capillary tube)

   In an experiment, the mean distance moved by the air bubble in the potometer was 14.0 mm in one minute. The capillary tube had a radius of 0.5 mm. Calculate the rate of water uptake in $mm^3\ hr^{-1}$. *(3 marks)*

6.6 Translocation

Specification references: 2.2.4

Learning outcomes

Demonstrate knowledge, understanding, and application of:

- → the transport of assimilates in phloem
- → the mechanism of translocation.

Movement of sugars

The phloem transports assimilates through all parts of the plant. This process is called **translocation** and the assimilates are organic compounds such as sugars and amino acids. Translocation of these compounds occurs from any part of the plant that has a surplus of them (a **source**) to other parts of the plant that need a supply (a **sink**).

However it is worth noting that sometimes a source can become a sink and vice versa. As a result, it is important that the transport of organic compounds can occur in both directions and that there are no one-way valves in the phloem to prevent this two-way flow. Sucrose is a good transport molecule as it is soluble and cannot be directly metabolised, so it can be transported without being used up in respiration on the way. Table 1 lists the parts of a plant that can be a source or a sink.

▼ **Table 1** *The parts of the plant that can be a source or a sink*

Sources	Sinks
Photosynthetic tissues: • mature green leaves • green stems. Storage organs that are unloading their stores: • storage tissues in germinating seeds • tap roots or tubers at the start of the growth season.	Roots that are growing or absorbing mineral ions using energy from cell respiration. Parts of the plant that are growing or developing food stores: • developing fruits • developing seeds • growing leaves • developing tap roots or tubers.

The flow relies on pressure gradients between a source and a sink, which in the case of translocation are created by active, energy-requiring processes.

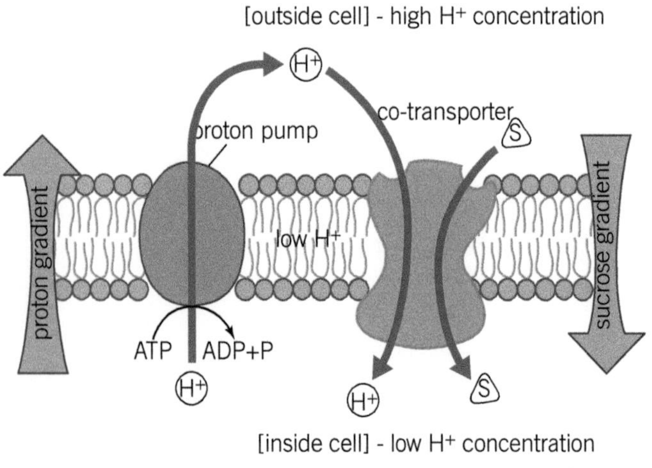

▲ **Figure 2** *Movement of sucrose across a sieve tube membrane*

Phloem loading

Sucrose is actively loaded into the phloem at the source. This process is called phloem loading. The companion cells alongside the phloem sieve tubes use ATP to actively transport protons (hydrogen ions) out of the companion cell cytoplasm into the surrounding tissue. This sets up a diffusion gradient which allows hydrogen ions to flow back down the concentration gradient into the companion cells through a co-transporter protein carrier, which also carries sucrose.

As the concentration of sucrose builds up, sucrose will flow through the plasmodesmata into the phloem sieve tube. This results in a lower water potential within the phloem sieve tube, which in turn draws water in by osmosis, increasing the hydrostatic pressure at the source.

At the sink the sucrose will be used by the cells and tissues. This lowers the sucrose concentration and so sucrose is drawn out of the phloem sieve tubes by diffusion down the concentration gradient. Sucrose may be used in respiration, once it has been hydrolysed into simple sugars, or it may be converted into complex carbohydrates such as starch for storage. In some cases sucrose may be actively transported out of the phloem sieve tubes when needed. As the sucrose is removed, the water potential is raised and water flows out of the phloem by osmosis. The loss of water reduces the hydrostatic pressure in the phloem at the sink.

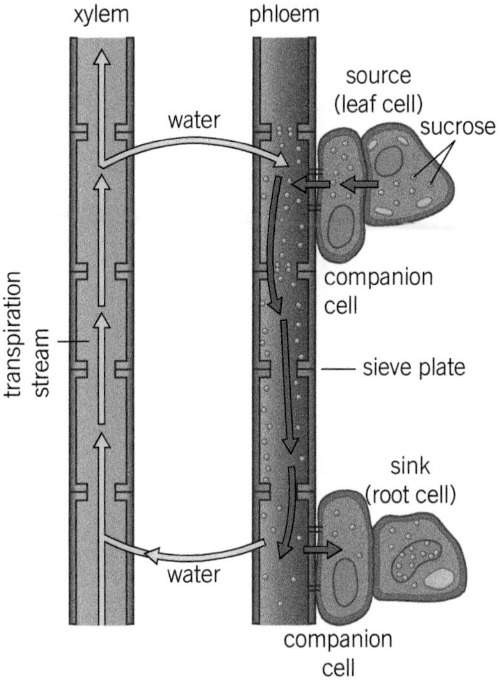

▲ **Figure 3** *Translocation and transpiration compared*

Summary questions

1 The phloem sieve tube elements have no cell organelles. What suggestion is there that the sieve tube is a living cell and how can it survive? *(2 marks)*

2 What evidence is there that translocation is the mechanism for transport of sugars? *(4 marks)*

3 Choose an appropriate presentation format to display the data in Table 2, including the standard error values.

▼ Table 2

Plant part	Mean carbohydrate content ($\mu g\,g^{-1}$ fresh mass ± standard error of mean)			
	sucrose	glucose	fructose	starch
Leaf blade	1,312 ±212	210 ±88	494 ±653	62 ±25
Vascular bundle in the leaf stalk, consisting of xylem and phloem	5,757 ±1,190	479 ±280	1,303 ±879	<18
Tissue surrounding the vascular bundle in the leaf stalk	417 ±96	624 ±714	1,236 ±1,015	<18
Buds, roots, and tubers (underground storage organs)	2,260 ±926	120 ±41	370 ±242	152 ±242

(3 marks)

4 Suggest reasons for the distribution of carbohydrates in the different parts of the plant shown in Table 2. *(6 marks)*

Practice questions

Shown here is a photomicrograph of two blood vessels. X is the pulmonary vein and Y is the pulmonary artery.

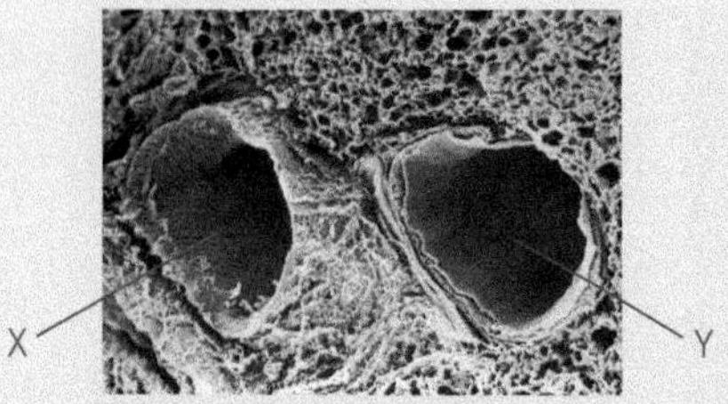

1 Which of the following statements correctly describes the image?

A A TEM of a transverse section through the blood vessels.

B A TEM of a longitudinal section through the bloods vessels.

C An SEM of a transverse section through the blood vessels.

D An SEM of a longitudinal section through the blood vessels. (*1 mark*)

2 Which of the following statements about X is correct?

A It carries deoxygenated blood to the lungs.

B It carries deoxygenated blood to the left atrium.

C It carries oxygenated blood to the lungs.

D It carries oxygenated blood to the left atrium. (*1 mark*)

3 Which of the following statements about Y is **not** correct?

A The systolic pressure is $\frac{25}{10}$ mm Hg.

B The semilunar valve prevents backflow from Y into the right ventricle.

C Y carries deoxygenated blood to the left atrium.

D The flow of blood in Y is pulsatile. (*1 mark*)

4 The diagram shows some of the hydrostatic pressure (HP) values found in capillaries which contribute to the exchange of materials between the capillary and the surrounding tissues.

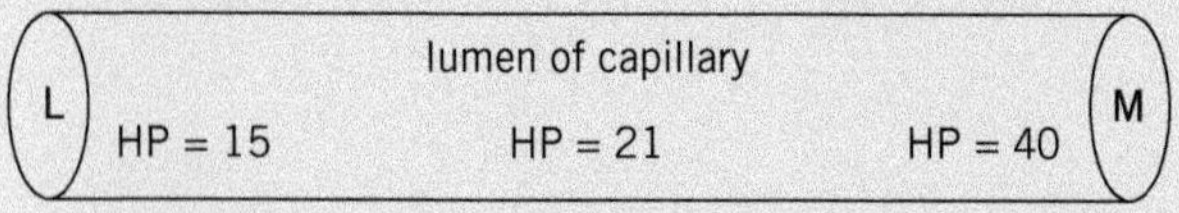

Which of the following statements about this diagram is/are correct:

Statement 1: the vessels supplying blood to M would be arterioles.

Statement 2: the COP in the capillary would be negative.

Statement 3: there would be a net loss of fluid as blood moved from L to M.

A 1, 2 and 3

B Only 1 and 2

C Only 2 and 3

D Only 1 (*1 mark*)

5 The photo shown here is an electron micrograph of xylem tissue in the stem of a plant.

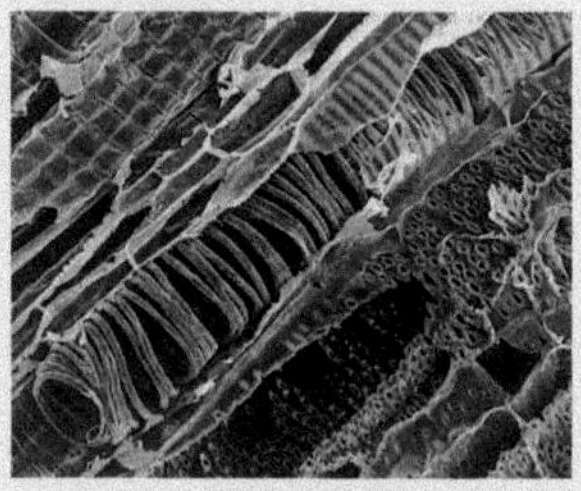

a State one function of xylem tissue (*1 mark*)

b The spiral band in the xylem vessel shown in the image contains a substance called lignin.

State the function of this spiral band of lignin and explain why it is important that the xylem vessel becomes lignified in this way. (*3 marks*)

c Explain the function of the pits seen in the image. (*2 marks*)

6 The diagram shows an aphid feeding from a plant stem. The aphid feeds by inserting its tube-like mouthparts into the tissue that transports sugar solution. Some details of this transport tissue are shown in the vertical section.

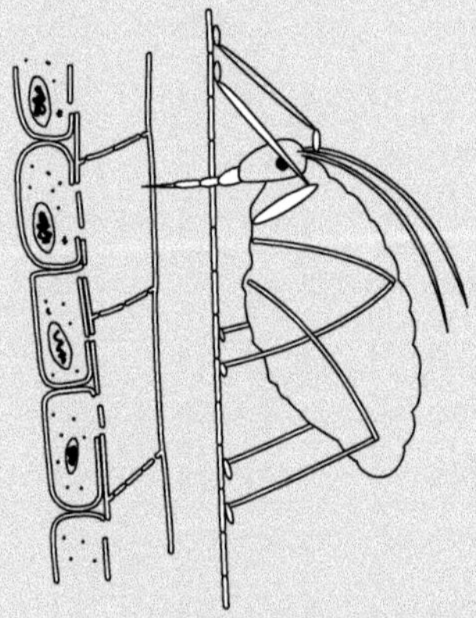

a (i) Name the sugar most commonly transported through the stem of a plant **and** the tissue that transports the sugar. (*1 mark*)

(ii) Sugar molecules are actively loaded into the transport tissue. Describe how active loading takes place. *(3 marks)*

b A classic experiment investigated the effect of temperature on the rate of sugar transport in a potted plant.

Aphid mouthparts were used to take samples of sugar solution from the transport tissue in the stem. The sugary solution dripped from the mouthparts. The number of drips per minute was counted. The procedure was repeated at different temperatures. The table shows the results obtained.

Temperature (°C)	Number of drips per minute
5	3
10	6
20	14
30	26
40	19
50	0

Suggest brief explanations for these results. *(3 marks)*

7 The diagrams show the internal structure of the heart and its associated circulatory system in a simplified form. The left diagram represents the system for a mammal and the right diagram that for a frog (an amphibian).

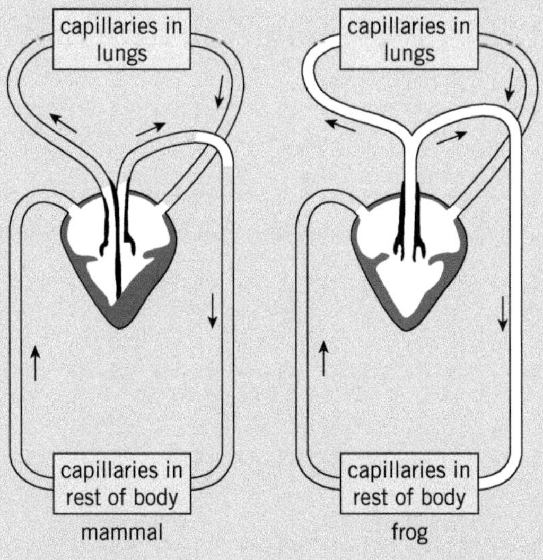

Both systems are described as closed systems. The mammalian system is also described as a complete double circulation but the frog as a partial double circulation.

a State what is meant by a closed system. *(1 mark)*

b Use the information in the diagrams suggest why the mammalian system is called a complete double circulation whilst that of the frog is called a partial double circulation. *(3 marks)*

c Suggest why the system shown for the frog may be less effective at supplying the body tissues with oxygen. *(2 marks)*

8 Accident and emergency units in hospitals regularly monitor the blood pressure of a patient when admitted following an injury.

Blood needs to flow through the circulatory system at a certain pressure to ensure that there is efficient exchange of oxygen and nutrients between the blood and body cells. The diagram shows blood pressure measurements and how they should be interpreted.

a Name the instrument used to measure blood pressure. *(1 mark)*

b Explain why a person's blood pressure should be measured when they are resting. *(1 mark)*

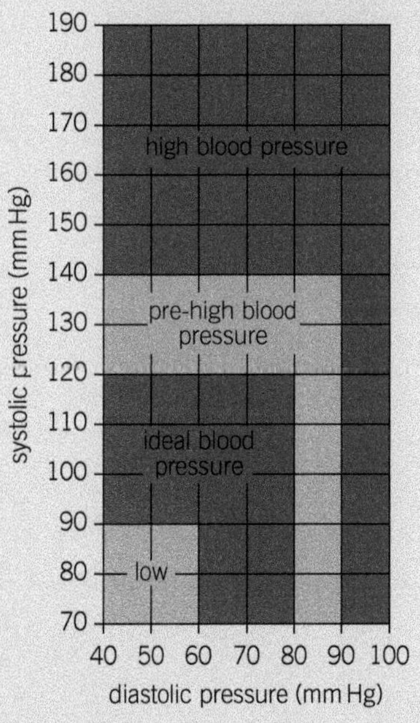

c Describe what is meant by systolic blood pressure. *(2 marks)*

d Using the diagram give a blood pressure measurement (in mmHg) for a person who may:

(i) have hypertension *(1 mark)*

(ii) have suffered severe blood loss *(1 mark)*

7 MAMMALIAN GAS EXCHANGE

7.1 The mammalian gas exchange system

Specification references: 2.2.3

Learning outcomes

Demonstrate knowledge, understanding, and application of:

→ the relationship between cells, tissues, and organs in the mammalian gas exchange system

→ the process of gas exchange in alveoli.

▲ **Figure 1** *Ciliated cells lining the airways in the trachea, approximately ×3570 magnification*

The mammalian gas exchange system

A **tissue** is a group of similar cells, of one or more different types, which are specialised to carry out a function. Tissues are grouped into **organs**. The tissues collectively enable the organ to carry out its function(s). The lungs are one example of many organs within the mammalian body, each organ having a specific function.

The gas exchange system includes organs such as lungs, bronchi, and bronchioles. Ciliated epithelial tissue lines the bronchi and bronchioles. Its function is to keep the airways free of debris. The ciliated cells remove mucus which would, if left, reduce ventilation.

The mammalian gas exchange system is a complex structure that combines the need to increase the surface area for gas exchange with a mechanism to ventilate this surface. It also has an extremely close relationship with the circulatory system, which carries the exchanged gases to and from the lungs. The system is made up of a series of airways that are lined with specialised cells making up a number of different types of tissues.

The primary function of the gas exchange system is diffusion of the respiratory gases into and out of the blood stream. The rate of diffusion possible is determined by, amongst other factors, the ratio

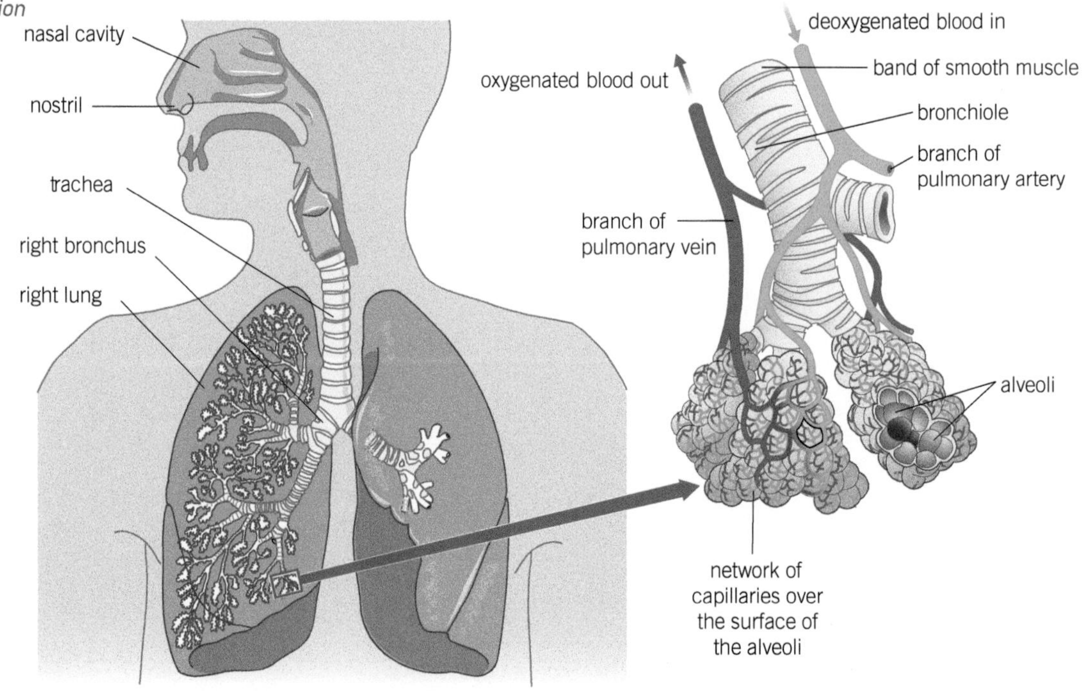

▲ **Figure 2** *The structure of the gas exchange system in humans*

between the surface area and the volume of the organism, with a high ratio giving a higher rate of diffusion. Large multicellular organisms have a small ratio with a long diffusion distance. However, the large number of alveoli in lungs is an example of how an organ is specialised to provide the additional surface area (cells in contact with the environment) needed to allow efficient gas exchange.

The trachea

The **trachea** is kept permanently open with 'C' shaped rings of cartilage around its walls. Under the cartilage is a layer of smooth muscle, elastic fibres, glandular tissue, connective tissue, and blood vessels. The trachea is lined with a layer of ciliated epithelial cells and goblet cells (ciliated epithelial tissue), which produce mucus. The ciliated cells have many small hairs called cilia which beat in a rhythm, moving the mucus and any trapped dirt and bacteria back up the trachea to the throat where it is swallowed and destroyed by the acid in the stomach.

Inside the lungs

During normal inhalation, air moves in and down the trachea. At the bottom the trachea branches into two **bronchi**. From these the air then moves into many branches of **bronchioles** and on into the blind ends called the **alveoli**. The clusters of alveoli are surrounded by capillaries increasing the surface area for gas exchange.

The walls of the bronchioles contain smooth muscle and elastic fibres. The smooth muscle can contract, narrowing the lumen of the airway, while the elastic fibres allow recoil back to normal size when the muscle relaxes.

The alveoli form the gas exchange surface. They have a wall consisting of a single layer of **squamous epithelial** cells and elastic fibres for expansion and recoil. In addition the wall also contains some collagen and stretch receptors which provide sensory input to assist in controlling the breathing cycle.

There is a liquid lining the alveoli, which contains **surfactant**, a detergent-like phospholipid, produced by some of the cells in the alveolar wall (septal cells). Surfactant reduces the surface tension of the water and makes it easier to inflate the lungs and stops the surfaces sticking together. It also acts as an antibacterial substance. The liquid allows efficient gas exchange across the surface by allowing oxygen to dissolve and then diffuse into the blood, but it does not increase the rate of diffusion as such.

Table 1 shows how the gas exchange system meets the requirements for efficient gas exchange.

Study tip

Remember that cells in tissues do not need to be exactly the same type, for example, in the airways the lining tissue consists of ciliated cells and mucus producing cells, which together make up the ciliated epithelial tissue.

Synoptic link

Revisit Topic 6.1, The transport system in mammals if you need to revise the role of the circulatory system. The principles of diffusion and factors affecting diffusion rates are covered in Topic 5.1, The heart.

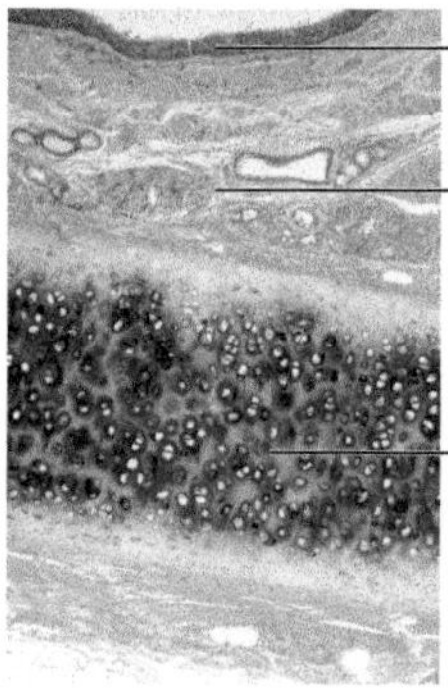

▲ **Figure 3** *Light micrograph section through the wall of the trachea, ×20 magnification*

Synoptic link

When an asthmatic is exposed to allergens in air, their airways can become constricted. You will cover this in more detail in Topic 15.1, Pollutants and lung disease.

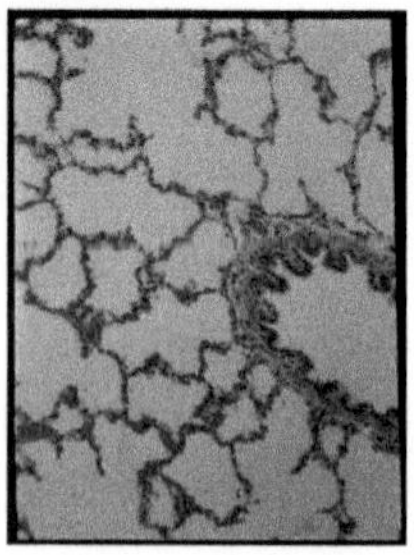

▲ **Figure 4** *Photomicrograph of lung tissue showing a bronchiole and alveoli, approximately ×40 magnification*

Study tip

Remember, liquid water has a high surface tension due to the hydrogen bonding between water molecules.

▼ **Table 1** *Adaptations of the mammalian gas exchange system*

Feature of a good gas exchange site	Adaptation of the gas exchange system
A large surface area	Many alveoli. The elastic fibres allow them to expand during inhalation, further increasing the surface area. A large surface area enables many molecules to cross at the same time.
A short diffusion distance	The squamous epithelial layer is extremely thin (approximately 0.6 – 0.7 μm thick). The capillary walls consist of a single layer of cells so the total distance between the blood in the capillary and the air in the alveoli is extremely small so molecules can cross in less time.
A steep diffusion gradient	The pulmonary circulation rapidly moves oxygenated blood away and brings deoxygenated blood. Pulmonary ventilation replaces the carbon dioxide rich air in the alveoli with oxygen rich air maintaining the diffusion gradient.

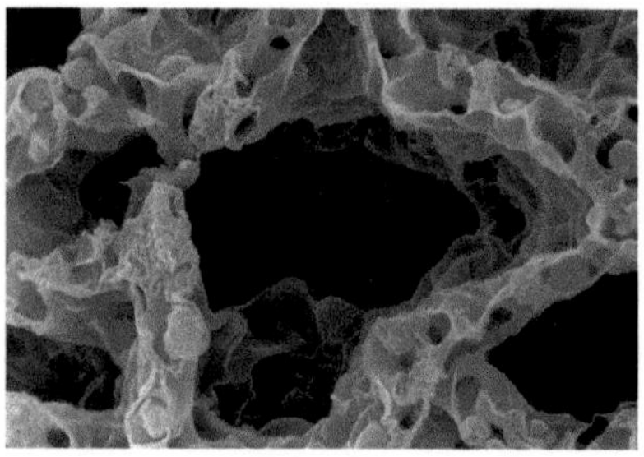

▲ **Figure 5** *An alveolus and a capillary showing how gas exchange is possible, approximately ×2 200 magnification*

Summary questions

1 Explain why the trachea and bronchi are supported by 'C shaped' rings of cartilage. *(2 marks)*

2 Bronchioles have a diameter of approximately 0.5 mm. Use this to calculate the magnification of the photomicrograph shown in Figure 4. *(2 marks)*

3 In a mixture of gases, each gas has a 'partial pressure' which reflects the relative concentration of gas in the mixture.

a Use the data in Table 2 to explain why blood in the pulmonary vein has the same partial pressure of oxygen and carbon dioxide as the air in the alveoli. *(1 mark)*

b Explain the differences in partial pressures of oxygen and carbon dioxide in the pulmonary artery compared to the alveolar air. *(2 marks)*

▼ **Table 2**

Gas	Alveolar air (kPa)	Blood in pulmonary artery (kPa)	Blood in pulmonary vein (kPa)
Oxygen	13.8	5.3	13.8
Carbon dioxide	5.3	6.1	5.3

7.2 Pulmonary ventilation

Specification references: 2.2.3

Ventilation in the lungs

Lung ventilation is vital to maintain the diffusion gradient. Muscle contraction causes the thorax to move and so changes its volume and the volume of the lungs within it. This in turn causes the pressure changes that allow ventilation of the lungs.

During inspiration:

- The ribs move up and out because the **external intercostal** muscles between the ribs contract.
- The muscular **diaphragm** contracts and flattens.
- As a result the lungs increase in volume so the air pressure within drops below the pressure outside the lungs, drawing air in.

During forced expiration:

- The ribs move down and in, caused by contraction of the **internal intercostal** muscles.
- The muscular diaphragm relaxes and reverts to its domed shape.
- This reduces the lung volume so pressure of air rises above the pressure outside the lungs, forcing air out.

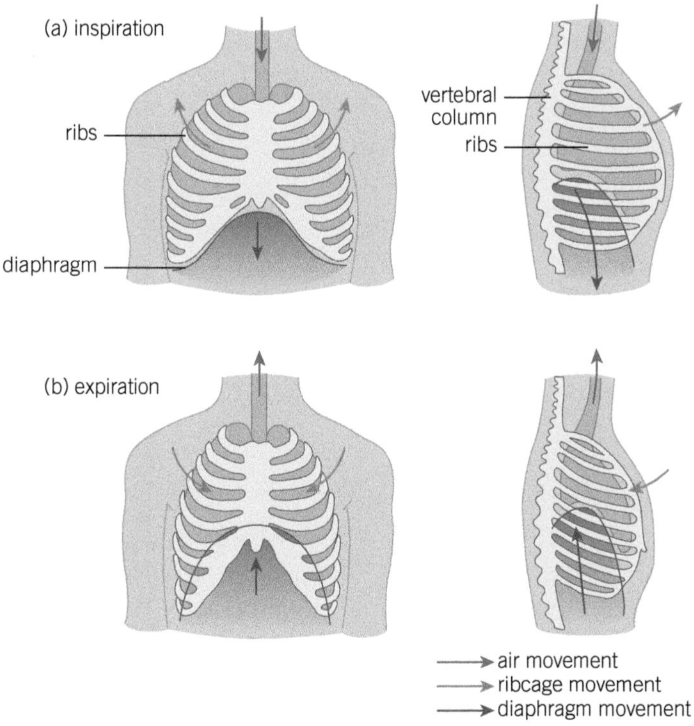

◀**Figure 1** *Ventilation in the lungs. The arrows show the direction of air movement during inspiration and expiration*

Lung capacity

The capacity of the lung depends on your body size, your level of activity and your general health. On average, lung capacity is about six litres. However during normal breathing much of this capacity is not used.

Learning outcomes

Demonstrate knowledge, understanding, and application of:

→ pulmonary ventilation and the parameters

→ using air resuscitation in cases of respiratory arrest.

Tidal volume is the volume of air that moves in and out of the lungs during a normal breath. It is normally only about 0.5 dm^3. The deepest possible breath in followed by the deepest possible exhalation is called the **vital capacity** and is normally about 3–5 dm^3. After the really deep exhalation there is still some air left in the lungs, about 1.5 dm^3, which is necessary to prevent the lungs collapsing and the internal walls sticking together. This air left in the lungs is called the **residual volume**. The number of breaths taken per minute is the **breathing rate** or ventilation rate.

Pulmonary ventilation

Pulmonary ventilation is the movement of fresh air into the lung and stale air out of the lung and is usually measured over one minute. It can be calculated using the following equation:

Pulmonary ventilation = tidal volume × breathing rate

Measuring pulmonary ventilation

There are two key measurements to assess pulmonary ventilation.

Forced expiratory volume in one second (FEV_1) is the volume of air that can be breathed out in the first second of forced exhalation. FEV_1 values are measured and compared to the predicted value for someone of the same gender, age and height. The formula used to predict FEV_1 is:

$FEV_1 = (4.3 \times \text{height} - 0.029 \times \text{age}) - 2.49$

Where height is measured in metres and age is in years.

Peak expiratory flow rate (PEFR) is the maximum rate of forcing air out through the mouth.

Checking FEV_1 and PEFR enables a doctor to decide if the airways are restricted or reduced.

1 Calculate the predicted FEV_1 for a male aged 18 who is 1.5 m tall.

Normal values of PEFR

These are the values that should be found in a normal healthy individual and are used by medical practitioners as part of the diagnosis of poor pulmonary health such as seen in an asthmatic or a sufferer of COPD (chronic obstructive pulmonary disease):

- In an adult male of 25, the volume is between 570 and 640 $dm^3 min^{-1}$, depending on body size.
- In an adult female of 25, the volume is between 420 and 460 $dm^3 min^{-1}$, depending on body size.

This contrasts with a reduced PEFR of 400 $dm^3 min^{-1}$ or less in males and 200 $dm^3 min^{-1}$ or less in females during an episode of reaction to allergens.

Synoptic link

In Topic 15.1, Pollutants and lung disease you will consider the effects of pollutants on the respiratory system and learn more about the effects of the risk factors.

The PEFR will increase with age from a young child to a maximum PEFR at the age of around 30–35 years before falling from that peak to a lower level at the age of 85.

Treating respiratory arrest

Respiratory arrest occurs when someone stops breathing. This may be caused by a number of things:

- an obstruction in the airway blocking the trachea or bronchi
- a drug overdose that has resulted in the nervous system and the breathing system being depressed sufficiently to stop all together.
- an asthma attack, severe pneumonia, severe shock, or a heart attack.

How to perform expired air resuscitation

- Firstly call for help and if available wear gloves and a mask to protect yourself and the patient.
- Carefully roll the person onto their back to avoid injury.
- Remove any obstruction visible in the mouth using a sweeping motion of your finger
- Hold the head back by gently pressing onto the forehead and lifting the chin, to move the tongue from the back of the throat and open the airway.
- Pinch the nostrils closed with a thumb and forefinger so the air does not escape and then make a seal over the mouth with your mouth.
- Blow gently into the mouth and watch that the chest rises. If it does not tilt the head back again and try again. Or check again for obstructions to the airways.
- If the chest does rise up, blow gently until the chest rises, pause, wait for it to fall, and then blow again.
- After two breaths, check the pulse. If there is a pulse continue blowing into the mouth.
- If there is no pulse you must perform cardiopulmonary resuscitation (CPR) to try to keep the blood circulating. Prompt action must be taken to help the person.

Suggest how this procedure would differ if carried out on a small child.

Synoptic link

It may be useful to look back at Topic 5.4, Using and interpreting an electrocardiogram for details of the steps involved in CPR.

Summary questions

1 Outline the difference between tidal volume and vital capacity. (*2 marks*)

2 Calculate the pulmonary ventilation in a patient with a tidal volume of 500 cm^3 and a breathing rate of 9 breaths per minute. (*2 marks*)

3 If vital capacity is 3.5 dm^3 and tidal volume is 0.5 dm^3, calculate the percentage of the total lung capacity normally used at rest and the percentage that is used during deepest possible breathing. (*3 marks*)

Practice questions

1 Which of the following structures in the gas exchange system contain cartilage?

A alveoli

B bronchi

C bronchioles

D alveolar duct *(1 mark)*

2 Which of the following is a correct description of the role of goblet cells in the secretion of the glycoprotein mucin?

A they secrete mucin by endocytosis into the bronchi.

B they secrete mucin by active transport into the bronchi.

C they secrete mucin by exocytosis into the alveoli.

D they secrete mucin by exocytosis into the trachea. *(1 mark)*

3 The organs in the mammalian gas exchange system have a number of tissues and cells in common.

Which of the following statements is/are true?

Statement 1: smooth muscle is found in the trachea, bronchi and some bronchioles.

Statement 2: ciliated epithelial tissue is found in all bronchioles.

Statement 3: secretory cells are found in the alveolar walls.

A 1, 2 and 3

B Only 1 and 2

C Only 2 and 3

D Only 1 *(1 mark)*

4 The approximate diameter of an alveolus is 250 μm.

a Calculate potential gas exchange surface of a person who is estimated to have 3×10^9 alveoli. Give you answer in m^2 to 3 significant figures. Show your working. *(2 marks)*

b Suggest why the **actual** area available for gas exchange may be different to the figure calculated in (a). *(2 marks)*

c Tidal volume is the volume of air that moves in and out of the lungs during one breath. In a healthy young adult, this volume is approximately 500 cm^3

Explain why only 350 cm^3 of air reaches the gas exchange surface. *(2 marks)*

d Tidal volume is one of the parameters affecting pulmonary ventilation.

State one further parameter needed to calculate pulmonary ventilation in $cm^3\,min^{-1}$ *(1 mark)*

5 Asthma affects many people in the UK.

- People with asthma may find it useful to monitor their peak expiratory flow rate (PEFR) using a peak flow meter.
- PEFR varies with the height and gender of a person, but also changes with age.

a Describe how a peak flow meter is used to measure PEFR. *(3 marks)*

The graph shows the normal range of PEFR values for men and women with a height of 175 cm.

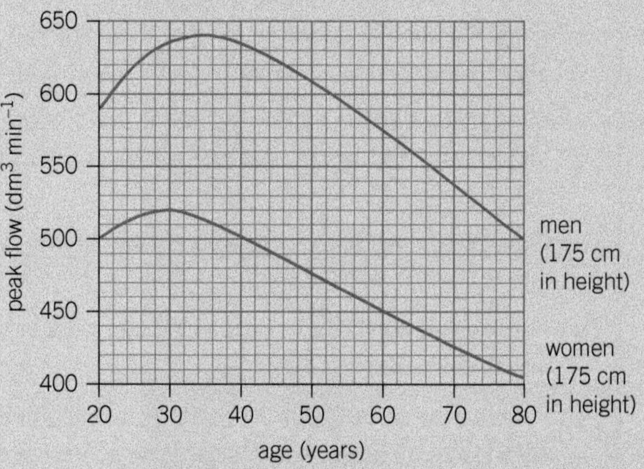

b (i) Using the graph, describe how the PEFR changes in men from the age of 20 years and suggest reasons for these changes *(4 marks)*

(ii) Using the graph, calculate the percentage decrease in PEFR for a woman between 40 and 60 years of age.

Show your working and give your answer to the nearest whole number. *(2 marks)*

c Severe asthma may cause respiratory arrest.

(i) Name two other causes of respiratory arrest *(2 marks)*

(ii) Describe how expired air resuscitation (EAR) is used to help an adult in respiratory arrest. *(3 marks)*

6 The vital capacity and the forced expiratory volume of a person with asthma were measured over a period of 23 days.

The forced expiratory volume is the volume of air that can be breathed out in one second. On day 4 of the investigation, the person breathed in an allergenic substance.

The results are shown in the graph.

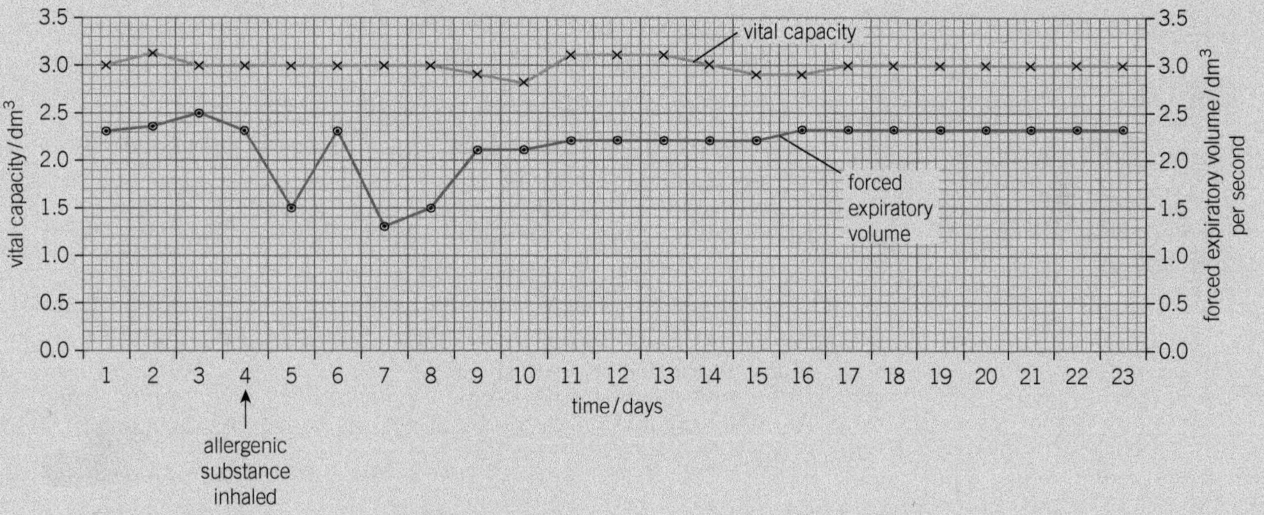

6 a (i) Calculate for day 1 the percentage of the vital capacity that was breathed out in one second.

Show your working and give your answer to the nearest whole number. *(2 marks)*

(ii) Using the data in the graph, describe the effect of the allergenic substance on the forced expiratory volume and the vital capacity. *(3 marks)*

7 The drawing here shows a transverse section of part of a bronchiole from a healthy lung.

a (i) Name tissues X and Y *(2 marks)*

(ii) Identify structure Z *(1 mark)*

b *Describe how the tissues in the gaseous exchange system contribute to the functioning of the lungs. *(6 marks)*

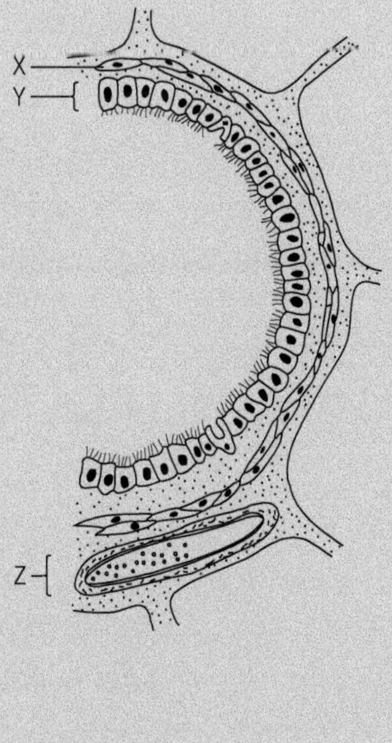

8 THE DEVELOPING CELL

8.1 The cell cycle and mitosis

Specification references: 3.1.1

Learning outcomes

Demonstrate knowledge, understanding, and application of:

→ the processes that occur during interphase

→ mitosis and cell division being only a small percentage of the cell cycle.

Synoptic link

You learnt about the content and structure of cells in Chapter 1, Microscopy – the light microscope.

You have studied the content and structure of cells in Chapter 1. The content of cells, however, is not fixed. Cells divide to enable growth and repair of tissues. Before dividing, cells must grow and produce new organelles and molecules. These processes of growth, synthesis, and division occur in a particular order, which is known as the **cell cycle**.

The stages of the cell cycle

The cell cycle involves several phases:

1 **Interphase**, in which cells grow and synthesise additional protein, organelles, and DNA in preparation for nuclear and cell division.
2 **Mitosis**, in which the nucleus of the cell divides.
3 **Cytokinesis** or cell division, which produces two genetically identical daughter cells.

The duration of one cycle varies between species and cell types. For example in humans a rapidly proliferating cell type would have a cell cycle lasting 24 hours, whereas a liver cell can take more than a year. Some specialised cells, such as neural cells in the brain, do not complete the cell cycle or divide.

You will now take a closer look at what is happening during each stage of the cell cycle.

Interphase

Interphase comprises approximately 90% of the cell cycle. It can be divided into three parts.

- G_1 – The first growth phase. The cell grows in size. Proteins are synthesised, enabling new organelles to be produced.
- S – The synthesis phase. DNA is replicated.
- G_2 – The second growth phase. Cell growth and protein synthesis continue. Organelles grow and divide. The cell builds up its energy stores.

Mitosis (nuclear division)

As Figure 1 shows, mitosis occupies only a small part of the cell cycle. It consists of several stages (prophase, metaphase, anaphase, and telophase), which you will examine in more detail in Topic 8.2, Mitosis.

Cell division

After nuclear division, the cell divides in a process called cytokinesis.

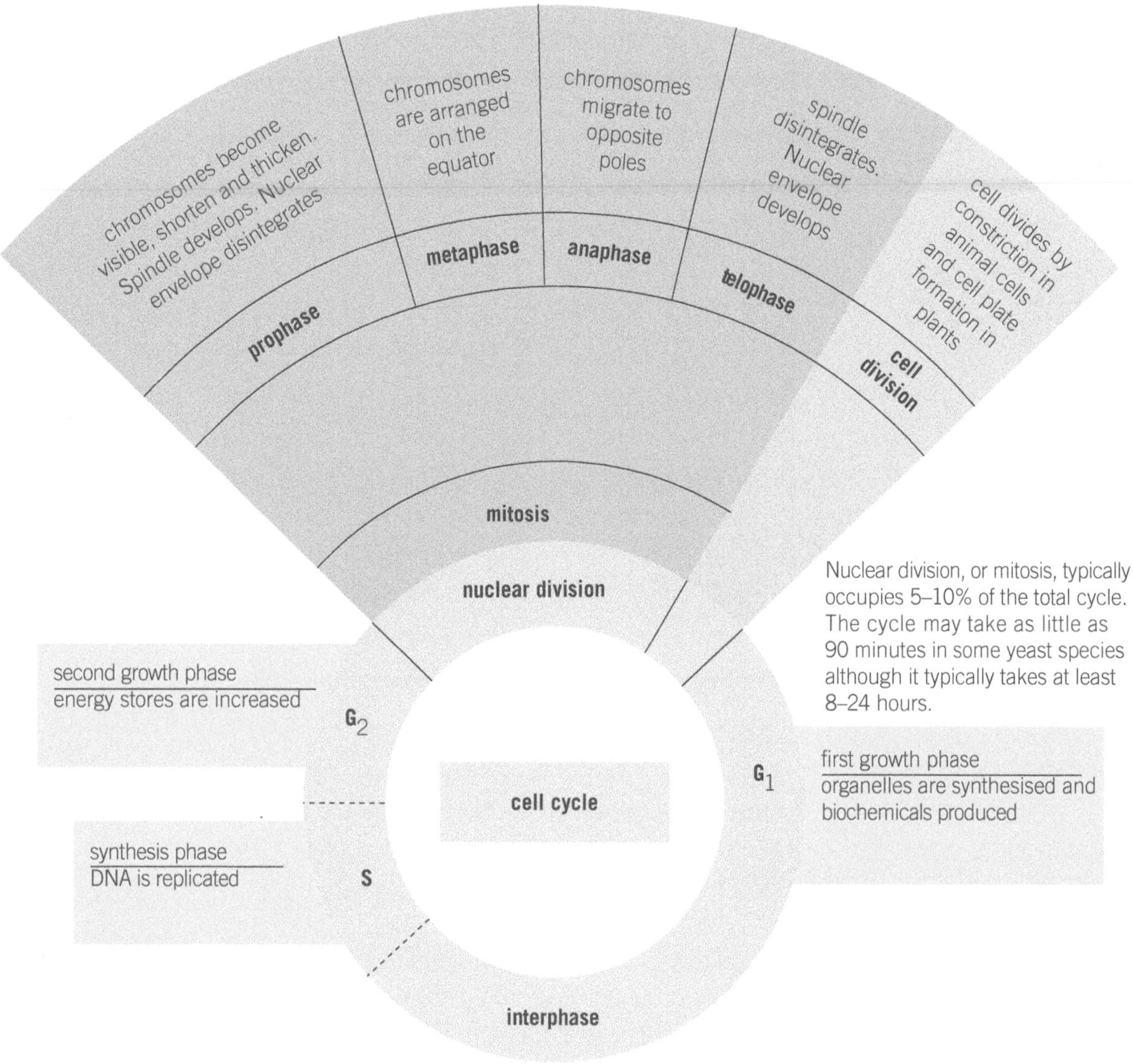
chromosomes become visible, shorten and thicken. Spindle develops. Nuclear envelope disintegrates
chromosomes are arranged on the equator
chromosomes migrate to opposite poles
spindle disintegrates. Nuclear envelope develops
cell divides by constriction in animal cells and cell plate formation in plants
prophase
metaphase
anaphase
telophase
cell division
mitosis
nuclear division
second growth phase
energy stores are increased
G2
synthesis phase
DNA is replicated
S
cell cycle
G1
first growth phase
organelles are synthesised and biochemicals produced
interphase

Nuclear division, or mitosis, typically occupies 5–10% of the total cycle. The cycle may take as little as 90 minutes in some yeast species although it typically takes at least 8–24 hours.

▲ **Figure 1** *The cell cycle*

How can we analyse the cell cycle?

Flow cytometry is a useful tool for counting and analysing cells. You looked at the theory behind this technology in Topic 1.5. One use of flow cytometry is in cell cycle analysis – determining the length of each phase of the cell cycle.

Prior to analysis, the DNA in the cells is stained with a fluorescent dye. The stained cells are passed through the flow cytometer and the fluorescence intensity is recorded. Cells in G_1 will have the least DNA and therefore the least fluorescence. As you learnt in this unit, DNA is replicated during the S phase, therefore cells at this stage of the cycle will contain more DNA and produce more fluorescence than cells in G_1. Cells in the G_2 phase that are undergoing mitosis will have twice the DNA levels and therefore double the fluorescence of G_1 cells.

Flow cytometry can therefore be used to determine the proportion of an organism's cells at each phase of the cell cycle. This enables the length of each phase to be calculated. This is one method for estimating the length of a phase:

$$\textit{length of phase} = \frac{T_c \times \ln(f_p + 1)}{\ln 2}$$

(Where T_c = cell cycle duration, f_p = the fraction of cells in a phase, ln = natural logarithm.)

For example, imagine a cell cycle that lasts 24 hours, in which 30% of cells are found to be in the S phase. You could calculate the length of the S phase as follows:

$$\textit{length of S phase} = \frac{24 \times \ln(0.3 + 1)}{\ln 2} = \frac{24 \times 0.26}{0.69} = 9.0\ \textit{hours}$$

1 Calculate the length of the G_1 phase in a cell cycle that lasts 18 hours and in which 45% of cells are in the G_1 phase.
2 Changes in the length of cell cycle phases can indicate health problems. Suggest what changes in the length of the cell cycle might indicate.

Synoptic link

You first encountered flow cytometry in Topic 1.5, Counting cells.

Synoptic link

You will learn more about cancer in Topic 14.3, Cancer.

Study tip

Interphase is sometimes referred to as the 'resting phase'. This is a misleading term. As you have learnt, although nuclei and cells are not dividing, a great deal of chemical activity takes place during interphase.

Cell cycle control

The order and timing of processes in the cell cycle are under tight control. Proteins called cyclins regulate the cell cycle. Cyclins activate enzymes known as cyclin-dependent kinases (CDKs), which control each step in the cycle. Several cyclins exist, each initiating a different phase of the cell cycle.

The cell cycle can be stopped when errors are detected. For example, if DNA is damaged, a protein called p21 binds to CDKs to halt the G_1 stage of interphase. This enables the cell to repair the DNA. Sometimes, as a result of mutations to particular genes, this repair process fails to function correctly and DNA is not successfully repaired. This can result in cancer.

Summary questions

1 List the similarities and differences between the first and second growth phases of interphase. *(3 marks)*

2 G_1 is the longest stage of the cell cycle. Assuming a cell cycle completes in 48 hours and 42% of cells are found in the G_1 phase, calculate how many hours are spent in G_1. Give your answer to two significant figures. *(1 mark)*

3 Explain the importance of cyclins and CDKs in providing checkpoints during the cell cycle. *(3 marks)*

8.2 Mitosis

Specification references: 3.1.1

The importance of mitosis

You learnt in the previous topic that cells spend most of their time growing and synthesising molecules, as well as performing their specialised functions. The division of a cell takes up a relatively small proportion of its cycle. Yet cell division is a crucial process. Here you will examine the steps involved in nuclear and cellular division.

Mitosis produces two nuclei that contain identical genetic material. This ensures that, following cell division, the two daughter cells are exact copies of the original cell. The replication of genetically identical cells is important for the following processes:

- Growth in multicellular organisms – Cells that have differentiated can produce identical copies and grow into a tissue in which all the cells have the same structure and function.
- Repair in tissues – Damaged cells need to be replaced by cells that perform the same function and have an identical structure.
- Replacement – Some cells, such as red blood cells, have a limited lifespan and must be replaced regularly with new cells.
- Asexual reproduction – Some organisms, such as plants and fungi, can reproduce using mitosis to give genetically identical offspring (although most can also reproduce sexually).

The stages of mitosis

Mitosis is preceded by interphase. DNA replicates during interphase, which means a cell at the beginning of mitosis has two copies of all its genetic material. This ensures that, when cell division occurs, the two daughter cells will both contain the same DNA.

Mitosis consists of four stages, each with their own distinctive features and processes.

Prophase

During interphase, DNA in the nucleus is uncoiled and not visible as **chromosomes**. In prophase, DNA condenses (forms coils and supercoils using proteins called histones) and chromosomes become visible. Each chromosome consists of two **chromatids**. These chromatids are identical and are a result of DNA replication during interphase. Chromatids are held together by a **centromere**.

Structures called centrioles move to opposite ends of the cell and begin to produce a system of microtubules known as the **spindle fibres**.

The nucleolus disappears, and the nuclear envelope breaks down towards the end of prophase, which means chromosomes are free to move within the cell's cytoplasm.

Learning outcomes

Demonstrate knowledge, understanding, and application of:

- → the changes in appearance of a cell during the stages of mitosis
- → the preparation of plant tissue to observe the stages of mitosis.

Study tip

You should be careful when using the terms 'strand' and 'molecule'. A DNA molecule contains two polynucleotide strands. Each chromosome consists of two chromatids. Each chromatid is a double-stranded DNA molecule, not an individual strand.

Metaphase

The spindle fibres produced by the centrioles attach to the centromere on each chromosome during metaphase. Chromosomes are pulled to the centre of the cell (often referred to as the equator).

Anaphase

The microtubules that make up the spindle fibres are pulled back towards the centrioles. This causes the centromeres to divide and the two chromatids on each chromosome are pulled to opposite poles of the cell.

Telophase

The changes that took place during prophase are reversed in telophase. The spindle fibres break down and disappear. Two new nuclear envelopes begin to form around the two identical sets of chromatids, which can now be referred to as 'daughter chromosomes'. The DNA starts to uncoil again and the chromosomes disappear.

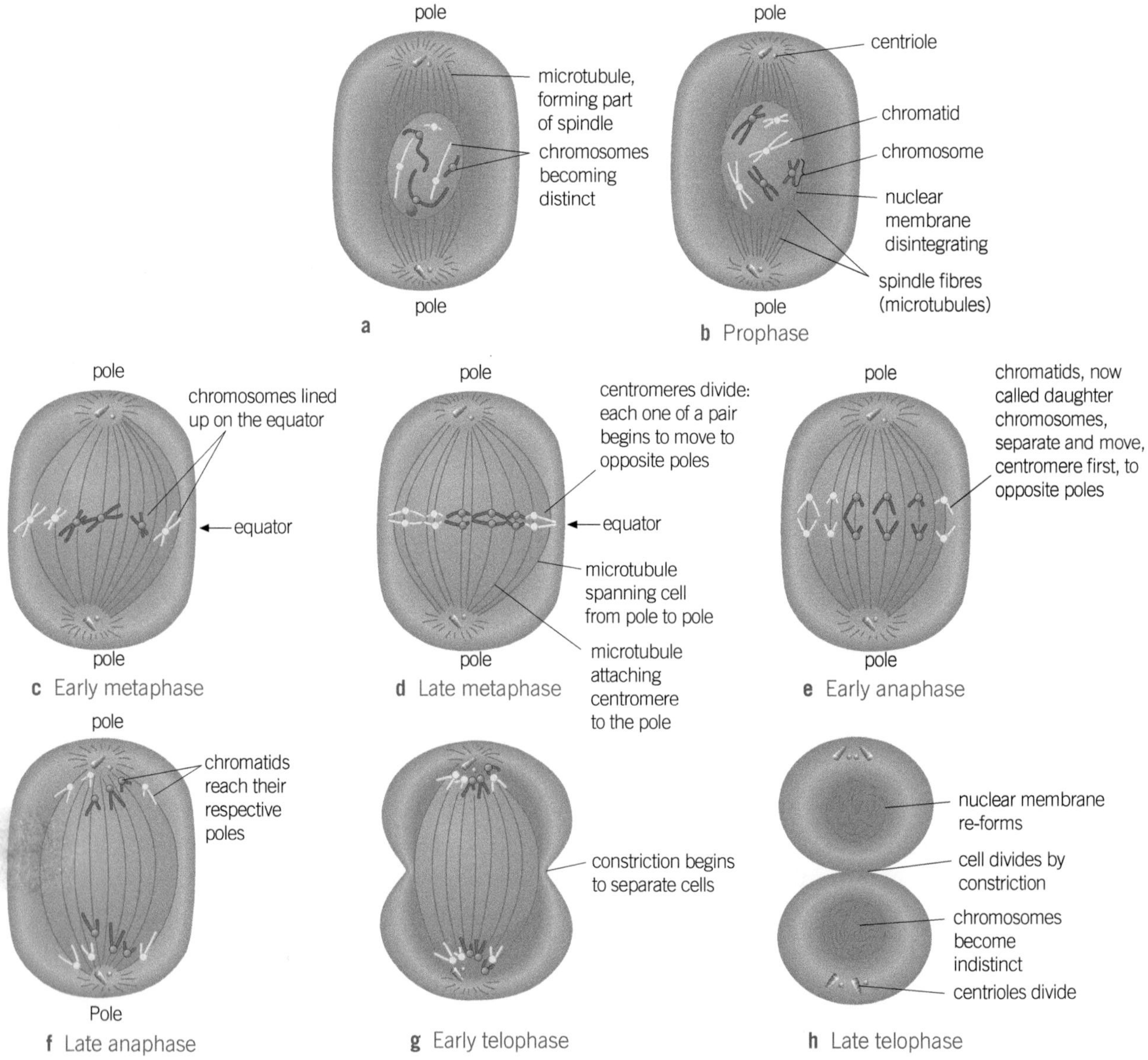

▲ **Figure 1** *Mitosis in an animal cell*

Cell division

After telophase, the cell divides. Two new cells are formed, each containing the same genetic material. Each cell receives approximately half of the organelles and cytoplasm from the original cell. The mechanism of cell division is called cytokinesis.

Comparing mitosis in plant and animal cells

There are some slight differences in mitosis between plant and animal cells and these are illustrated in Table 1.

▼ **Table 1** *Difference between mitosis in plant and animal cells*

Animal cell mitosis	Plant cell mitosis
Occurs in most tissues	Only occurs in meristematic tissue
Cell becomes rounded before division	No shape change
Centrioles play a role	No centrioles
Spindle disappears before cytokinesis	Some of the spindle remains during cytokinesis
Microfilaments involved	Microfilaments do not play a major role

Viewing mitosis under a microscope

Using a light microscope, it is possible to view cells that are undergoing mitosis. Tissue with a high rate of cell division, such as plant root tips (meristems), is usually selected. The tissue is treated with a stain that binds to DNA. This makes the chromosomes easier to see under the microscope. Figure 2 shows cells at different stages of mitosis, as seen under a microscope.

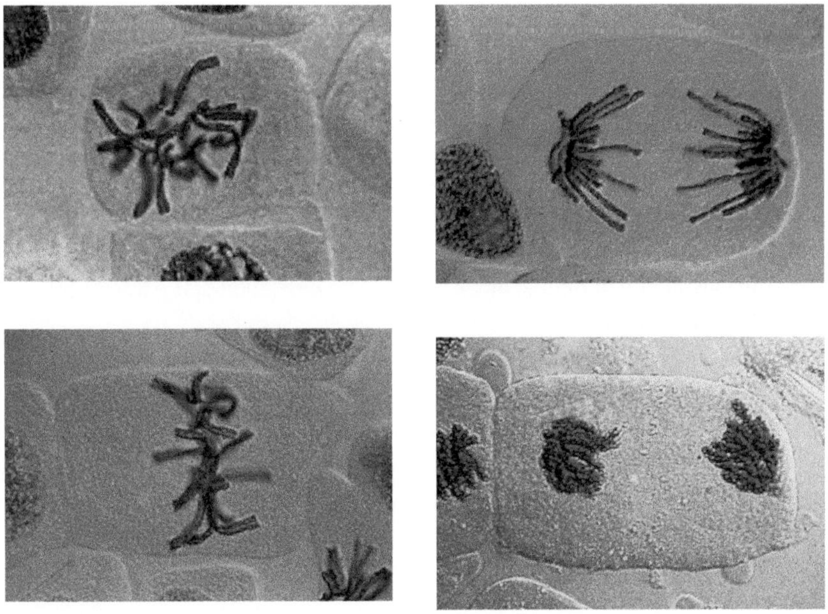

▲ **Figure 2** *The four stages of mitosis seen under a light microscope, approximately ×1000 magnification*

Synoptic link

You first learnt about microscopy in Topic 1.1, Microscopy – the light microscope.

Preparation of plant tissue to view under a microscope

One method for preparing plant tissue for microscopy involves the following steps:

1. Remove the end of a root tip (approximately 1 cm in diameter).
2. Place the root tip in ethanoic acid for 10 minutes.
3. Place the root tip in HCl (1 mol dm^{-3} concentration) in a 60 °C water bath for 6–7 minutes.
4. Rinse the tip in distilled water to remove excess acid.
5. Dry on filter paper.
6. Apply DNA stain to the plant tissue and leave for approximately 10 minutes.
7. Rinse and dry the root tip again.
8. Transfer the root tip to the centre of a clean microscope slide and add a drop of water.
9. Use a razor blade to cut off the unstained part of the root.
10. Cover the root tip with a coverslip and then carefully push down on the coverslip with the wooden end of a dissecting probe to squash and spread the plant tissue.

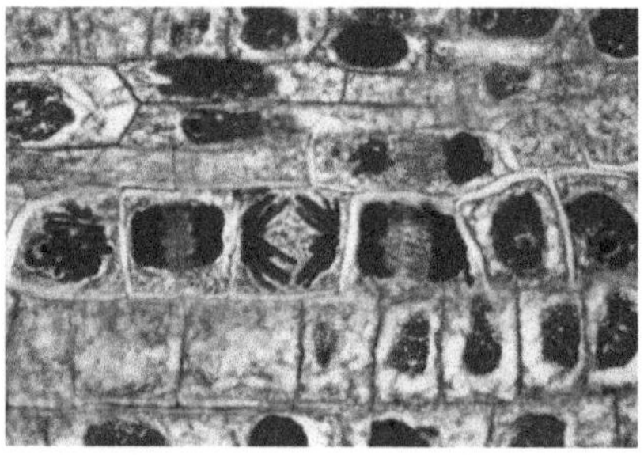

▲ **Figure 3** *Cells from the shoot tip of* Coleus *plant, approximately ×400 magnification*

Suggest the purpose of the acids in steps 2 and 3 of the procedure.

Summary questions

1. Outline three potential hazards that would be encountered when preparing plant tissue for microscopy. *(3 marks)*

2. The calculation of mitotic index provides an indication of the rate of cell division in a tissue.

$$\textit{mitotic index} = \frac{\textit{number of cells in mitosis}}{\textit{total number of cells}}$$

Figure 3 shows a tissue in which some cells are undergoing mitosis (shown by the dark red stain) and other cells that are not dividing. Calculate the mitotic index of this tissue. Include only those cells that have boundaries entirely visible in the photograph. Give your answer to three significant figures and show the steps in your calculation. *(1 mark)*

3. Suggest which disease might result in an increase in mitotic index. Explain your answer. *(2 marks)*

8.3 Apoptosis

Specification references: 3.1.1

Programmed cell death

We have looked at how cells replicate earlier in this chapter. However, the destruction of damaged or unwanted cells has the same importance as the production of new cells. More than 50 billion cells are destroyed each day in an average adult human. The process by which these cells are removed is known as **apoptosis**.

Cells can be destroyed either by apoptosis or necrosis. Necrosis is a damaging form of cell death caused by infection or trauma. It involves the rupture of cell surface membranes and the release of hydrolytic enzymes. Several pathogenic bacteria act by causing necrosis in tissues and this can be fatal.

In contrast, apoptosis is often referred to as programmed cell death. Unlike necrosis, it is a regulated, controlled process. Apoptosis is controlled by a range of cell signals, including hormones and proteins known as cytokines, which are released by cells of the immune system. Most cells divide by mitosis approximately 50 times before undergoing programmed cell death. The rate of cell death should balance the rate of cell production via mitosis in adult organisms.

Learning outcomes

Demonstrate knowledge, understanding, and application of:

→ the stages and features of apoptosis

→ the importance of apoptosis in growth and repair.

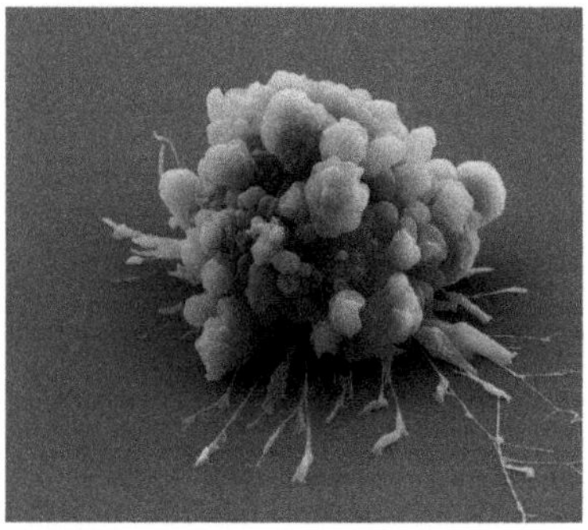

▲ **Figure 1** *A cancer cell undergoing apoptosis, ×4 000 magnification*

The importance of apoptosis

Cell death is as significant to the development and maintenance of an organism as cell proliferation. In particular, apoptosis is important during the growth and development of young organisms and for destroying damaged cells.

Destruction of damaged cells

Apoptosis can be initiated in cells that are infected with viruses or have damaged DNA. Cells that have damaged DNA increase the production of a protein called p53, which is an example of a tumour suppressor. Apoptosis is initiated by p53, thereby preventing cells with faulty DNA from replicating, which prevents the development of cancer.

Synoptic link

You will learn more about p53 and tumour suppressor genes in Topic 14.3, Cancer.

Development

Apoptosis is a destructive process, but it is essential for development, especially during early fetal growth. As an organism grows, more cells are produced than are required. These surplus cells can be pruned by apoptosis to shape the structures within a fetus. For example, apoptosis removes excess cells from fetal feet and hands to produce distinct toes and fingers. Other examples of apoptotic processes in development include:

- the formation of connections between neurones in the brain
- the destruction of potentially harmful immune cells during the development of the immune system.

Synoptic link

The stages in apoptosis can be followed using flow cytometry, which you learnt about in Topic 1.5, Counting cells. Alterations to the permeability of the cell surface membrane during apoptosis allow dyes to be taken up that are normally excluded. Flow cytometry enables the proportion of cells undergoing apoptosis to be estimated.

Study tip

Remember that in organs and tissues both apoptosis and mitosis will be happening. If a region of tissue or an organ remains the same size then the two processes are balanced. If the rate of apoptosis exceeds mitosis, then the organ will decrease in size. The decrease in size of an organ is called atrophy.

The events in apoptosis

Apoptosis, like cell division, can be controlled by signals from either inside or outside the cell. The process is highly regulated, and the sequence of events is the same each time:

1 The cell shrinks and the nucleus condenses (pyknosis).
2 Enzymes break down the cytoskeleton.
3 The cell surface membrane alters and forms bulges known as blebs.
4 The nucleus breaks down (karyorrhexis).
5 The cell is broken into fragments, which are held in vesicles called apoptotic bodies.
6 Phosphatidylserine, which is a phospholipid normally found on the inside of cell membranes, is present on the outside of the vesicles.
7 Phosphatidylserine binds to receptors on immune cells called macrophages.
8 Macrophages engulf the cell fragments.

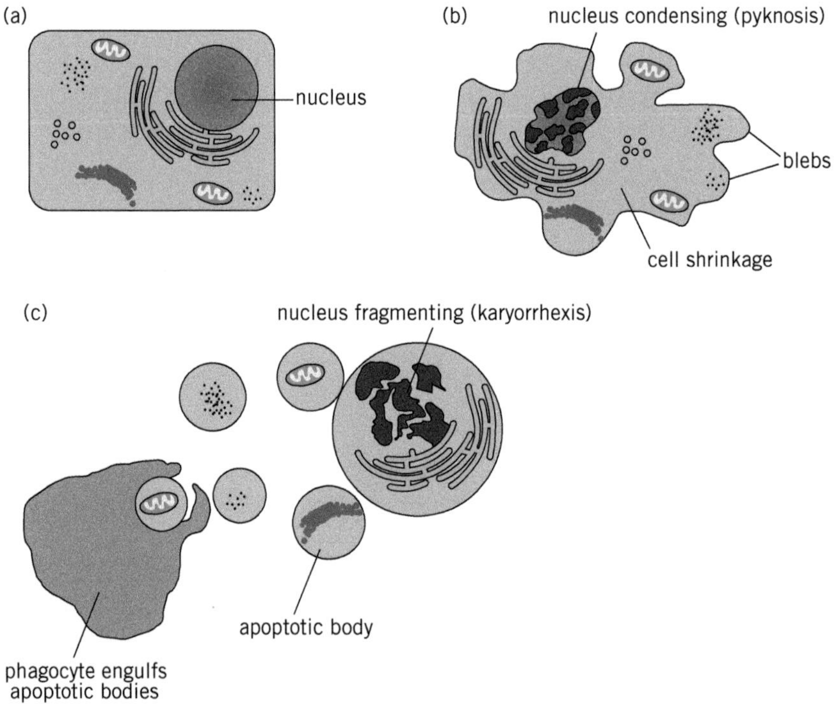

▲ **Figure 2** *The process of apoptosis*

Summary questions

1 Describe three roles of apoptosis in organisms. *(3 marks)*

2 Explain the differences between apoptosis and necrosis. *(4 marks)*

3 Outline the role of phosphatidylserine in apoptosis. *(2 marks)*

8.4 Stem cells

Specification references: 3.1.1

What are stem cells?

You examined the general structure and features of cells in Topic 1.6. Most cells in multicellular organisms are specialised cells. This means their structures are fine-tuned to enable them to perform specific roles. Even in adults, however, some cells remain unspecialised. In this topic you will learn about these unspecialised cells, which are known as **stem cells**.

A cell becomes specialised for a role when it undergoes **differentiation**. Although every cell in an organism contains the same DNA, some genes 'switch off' during differentiation. This produces the shape and structure of a particular cell. For example, palisade cells in plants have a high concentration of chloroplasts to enable photosynthesis, whereas red blood cells lack organelles, which enables them to contain a high concentration of haemoglobin.

The undifferentiated cells from which cells differentiate are called stem cells. An early embryo consists entirely of stem cells. However, groups of stem cells are present in different tissues in adults as well.

Learning outcomes

Demonstrate knowledge, understanding, and application of:

- the differentiation of stem cells into specialised cells
- the difference between totipotent, multipotent, and pluripotent stem cells
- current applications and uses of stem cells.

Synoptic link

You learnt about palisade cells in Topic 1.6, Cell ultrastructure and about red blood cells in Topic 1.4, Cells of the blood.

Types of stem cells

Totipotent

Totipotent stem cells are found in very early embryos. They are able to differentiate into any type of cell. Totipotent stem cells can divide to form a whole organism and extra-embryonic membranes, such as the chorion and amnion.

Pluripotent

Like totipotent cells, **pluripotent stem cells** can differentiate into any cell type that makes up the body. However, unlike totipotent cells, they cannot divide to form a whole organism. They are found in embryos that have grown to the 50–100 cell stage.

Multipotent

Multipotent stem cells can differentiate into a limited range of different cell types. For example, multipotent stem cells in adult bone marrow can differentiate into various types of blood cell, but they would not be able to differentiate into nerve cells or skin cells. Bone marrow stem cells are called haemocytoblasts. These stem cells are round and have relatively little cytoplasm in relation to the size of their nuclei.

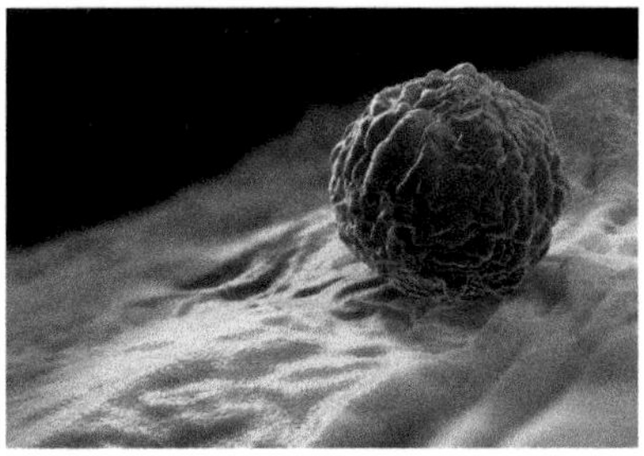

▲ **Figure 1** *Artwork showing a human embryonic stem cell*

Uses of stem cells

Multipotent adult stem cells

Bone marrow transplants can be used in the treatment of leukaemia and other blood and bone cancers. Bone marrow is a source of multipotent adult stem cells. A bone marrow transplant from a donor replaces the stem cells that are destroyed during cancer treatment.

Research has been conducted to explore the possibility of using neural stem cells from adults to treat conditions of the nervous system such as Parkinson's and Alzheimer's disease.

Pluripotent embryonic stem cells

Pluripotent stem cells can be removed from embryos grown *in vitro*. These cells are called embryonic stem cells (ESCs). In the future, ESCs could be used for several aspects of medical research and treatment:

- cell replacement therapies – ESCs could, for example, be differentiated into pancreatic beta cells to treat type I diabetes, dopamine-producing cells to treat Parkinson's disease or heart muscle cells to treat heart disease
- testing potential drugs *in vitro* (i.e., using tissue grown in a lab rather than a living organism)
- studying the development of diseases *in vitro*.

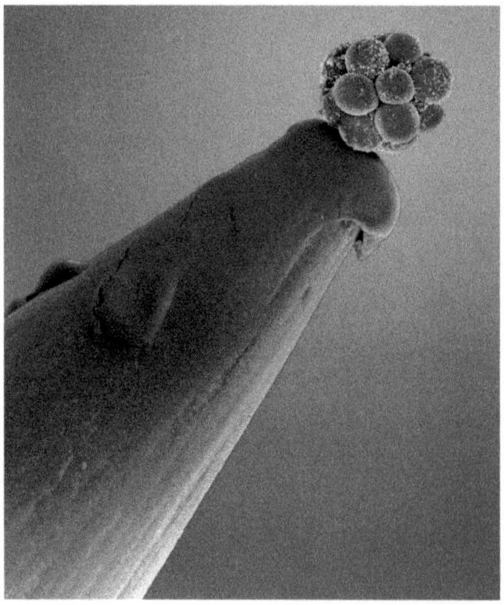

▲ **Figure 2** *A human embryo at the 16-cell stage, ×130 magnification*

Problems with potential stem cell therapies

ESCs are obtained from embryos that are younger than 14 days old. Some people view early embryos as a ball of undifferentiated cells that lack human qualities. Others think the use of cells from human embryos is wrong, regardless of the age of the embryo. They worry that using stem cells for therapies could lead to reproductive cloning of humans. Another concern is the potential for stem cell transplants to produce tumours.

A scientist called Robert Lanza announced in 2006 that it is possible to obtain embryonic stem cells without destroying an embryo. His technique might reduce the ethical concerns surrounding the use of ESCs. The ban on government funding of stem cell research in the USA was lifted in 2009 although there have been several unsuccessful appeals against this decision since that date.

Scientists have also developed techniques for reprogramming differentiated adult cells back into undifferentiated pluripotent stem cells. The scientist Shinya Yamanaka added proteins called transcription factors to human skin cells. These cells were reset and became stem cells. This technique does not require the use of embryos and therefore overcomes the ethical objections held by some people. The reprogrammed pluripotent cells are known as induced pluripotent stem cells (iPSCs). However, problems have been encountered with the use of iPSCs. There were worries about their potential to produce tumours known as teratomas. The success rate for converting adult cells into stem cells was also initially very low. If these issues can be overcome, the widespread therapeutic use of iPSCs might be possible.

Summary questions

1. Describe the difference between totipotent and pluripotent stem cells. *(2 marks)*
2. Explain why bone marrow transplants are given to leukaemia patients who have undergone chemotherapy. *(3 marks)*
3. Evaluate the future use of stem cell therapies, taking into account both the potential benefits and the risks. *(4 marks)*

Practice questions

1 Mesenchymal stem cells can differentiate into adipose, bone, cartilage, and connective tissue cells, but not other types of cell.

What type of stem cells are mesenchymal stem cells?

A unipotent

B multipotent

C pluripotent

D totipotent *(1 mark)*

2 Plant tissue can be prepared and viewed under a microscope to examine cells at different stages of mitosis.

Which statement(s) is/are (a) correct description(s) of the preparation of plant tissue to be viewed under the microscope?

1 A root tip is removed and soaked in sodium hydroxide for 10 minutes.

2 The cells' DNA is stained using a chemical such as orcein.

3 The root tip is placed on a microscope slide, covered with a drop of water and a cover slip is placed over the tissue.

A 1, 2 and 3

B Only 1 and 2

C Only 2 and 3

D Only 1 *(1 mark)*

3 Apoptosis involves a regulated sequence of events that results in cell death.

Which of the following statements is/are true?

1 Pyknosis is the shrinkage and condensation of the cell nucleus.

2 Blebbing is the breakdown of the cell nucleus.

3 Phosphatidylserine on the inside of vesicles binds to macrophage receptors.

A 1, 2 and 3

B Only 1 and 2

C Only 2 and 3

D Only 1 *(1 mark)*

4 One method for estimating the length of a cell cycle phase is:

$$\text{Length of phase} = \frac{T_c \times \ln(f_p + 1)}{\ln 2}$$

(where T_c = cell cycle duration, fp = the fraction of cells in a phase, ln = natural logarithm)

What is the length of the G_2 phase in a cell cycle that lasts 16 hours and in which 35% of cells are in the G_2 phase?

A 3 hours

B 6 hours

C 8 hours

D 7 hours *(1 mark)*

5 a The figure shows some drawings of a cell during different stages of mitosis. Place stages P, Q, R, S and T in the correct sequence.

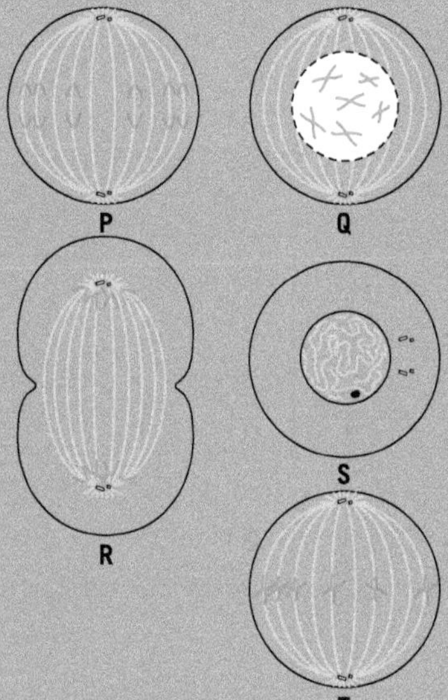

(5 marks)

b Mitosis is part of the cell cycle.

(i) Name one process that occurs during stages G_1 and G_2 of the cell cycle. *(1 mark)*

(ii) The genetic information is copied and checked during stage S of the cell cycle. Suggest what might happen if the genetic information is not checked. *(2 marks)*

[question 5, OCR F211, June 2009]

6 Describe the potential uses of stem cells in disease treatment and the problems that have been encountered when researching potential stem cell therapies. *(6 marks)*

7 In plants, dividing cells can be found in meristematic tissue.

a In an investigation, a student observed the cells in a stained section of meristematic tissue. The student counted how many cells could be seen in each stage of the cell cycle. The results are shown in the table.

Stage of the cell cycle	Percentage of cells in stage (%)
Interphase	82.00
Prophase	4.34
Metaphase	3.23
Anaphase	3.23
Telophase	7.20

(i) Explain why the meristematic tissue needed to be stained for this investigation. *(2 marks)*

(ii) Name the type of nuclear division that occurs in a plant meristem *(1 mark)*

b Using the results shown in the table, calculate the percentage of the cell cycle taken during nuclear division. Show your working. *(2 marks)*

[question 4, OCR F211, June 2011]

8 Women who drink excessively during pregnancy are at risk of giving birth to infants with malformations of the eye. Development of the retina in the eye depends upon the processes of both mitosis and apoptosis.

A research study used rats as an experimental model to find out about the effect of prenatal exposure to alcohol on mitosis and apoptosis in the developing retina.

a The graph shows the effect of prenatal exposure to alcohol on the number of cells undergoing mitosis and apoptosis in the development of the retina.

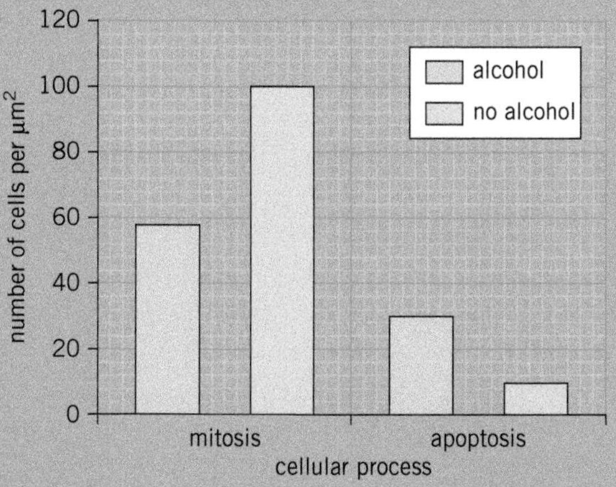

(3 marks)

(i) Using the information in the graphs, describe the effect of exposure to alcohol on mitosis and apoptosis in the cells of the developing retina. *(3 marks)*

(ii) Suggest one role of mitosis and one role of apoptosis in the growth and development of the retina. *(2 marks)*

(iii) With reference to the figure, suggest and explain how prenatal exposure to alcohol may affect the thickness of the retina in the eye. *(2 marks)*

b The table shows diagrams of the different stages of mitosis.

Using the information in the diagrams, copy and complete the table by adding a description of two changes that take place in the cell at each stage.

stage	description of **two changes** that take place
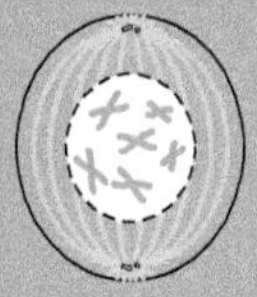	
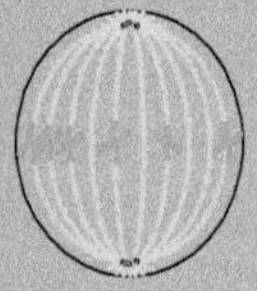	
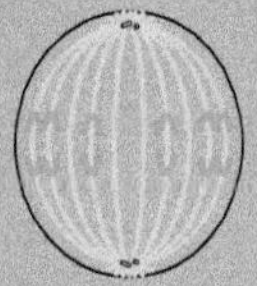	
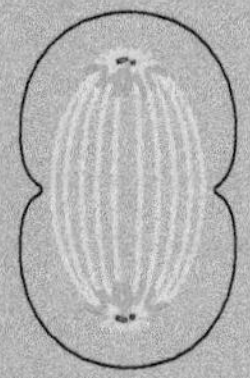	

(8 marks)

[question 4, OCR F222, June 2012]

9 MEIOSIS, GROWTH, AND DEVELOPMENT

9.1 Meiosis

Specification references: 3.1.2

Learning outcomes

Demonstrate knowledge, understanding, and application of:

- → the stages of meiosis in plants and animals
- → the processes in meiosis that produce genetic variation
- → the significance of meiosis in the production of haploid gametes in plants and animals.

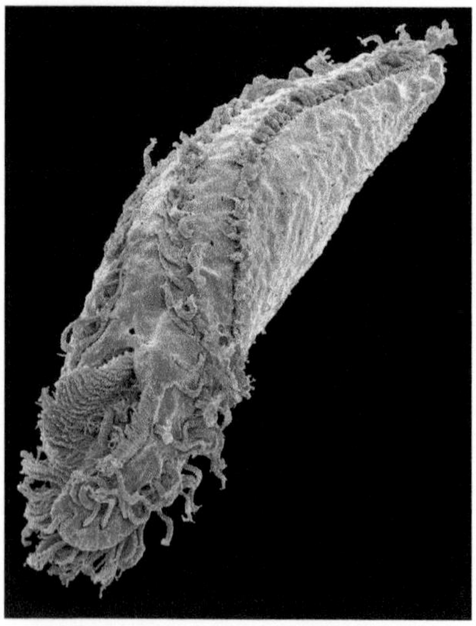

▲ **Figure 1** *An* Oxytricha *protozoan, which has a remarkably high diploid number of chromosomes, ×800 magnification*

The importance of meiosis

In Topic 8.2, you learnt that mitosis is nuclear division that produces two identical nuclei. It occurs during the growth and repair of body tissues, during which the replication of genetically identical cells is important. The formation of reproductive cells, however, creates genetic variation. The production of reproductive cells therefore involves a different type of nuclear division, **meiosis**.

Some organisms undergo asexual reproduction to produce offspring. For example, bacteria reproduce asexually using a process called binary fission, and some eukaryotes use mitosis. Offspring produced asexually are genetically identical to each other and to the parent. Only DNA mutations can introduce genetic variation in these organisms.

Sexual reproduction increases genetic variation – offspring are genetically different from each other. Each parent produces reproductive cells, known as **gametes**. Two gametes, one from each parent, fuse to form a zygote, which develops into an embryo. You will look at how genetic variation arises in meiosis later in this topic.

Plants can reproduce both asexually and sexually, with meiosis being responsible for the production of pollen in the anthers and ovules in the ovary. Plants grown from seeds will be genetically different to the parent plant, whereas plants that have reproduced asexually through bulbs or tubers will be genetically identical to the parent plant.

In most organisms, somatic cells ('body cells') are **diploid**. A zygote must be diploid because it will divide by mitosis to form the body cells of an organism. Diploid cells contain one set of chromosomes from the mother and one set from the father. In humans, 23 chromosomes come from each parent. Diploid human cells therefore have 46 chromosomes, comprising 23 pairs, with two copies of each chromosome. Gorillas have a diploid number of 48 – the ciliated protozoan *Oxytricha trifallax* has 15 600 chromosomes, whereas the diploid number of mosquitoes is only 6.

Meiosis produces gametes with only one copy of each chromosome (i.e., half the number found in diploid body cells). Gametes are called **haploid** cells. The haploid number in humans is 23, in gorillas it is 24, and in mosquitoes it is 3. Meiosis is important because it ensures that after fertilisation, when the two haploid gametes fuse, the resultant zygote is diploid. This prevents chromosome numbers from doubling every generation.

Correlations in biology

You might be wondering why a single-celled organism like *Oxytricha trifallax* has 15 600 chromosomes, whereas humans have a total of just 46 chromosomes. Clearly no relationship exists between the complexity of species and chromosome number. These two variables are said to be *uncorrelated*.

However, correlations are widespread in biology – so the identification and interpretation of these relationships is an important skill. Correlations can be negative (when one variable decreases, the other increases) or positive (when one variable increases, the other one also increases).

A correlation can exist because one variable affects the other (i.e., if one variable is increased, this causes a second variable to increase). However, this is not always the case. A useful phrase to remember is 'correlation does not imply causation.' Two variables can be correlated without influencing each other. Additional variables might affect both of the variables being studied. This needs to be taken into account when designing scientific studies.

Species complexity is not correlated with chromosome number, but there is a correlation between the complexity and the type of DNA possessed by a species. Genomic DNA comprises genes that code for polypeptides, but non-coding DNA is also present. Non-coding DNA has many functions, including the regulation of gene transcription. As Figure 2 shows, the higher the proportion of non-coding DNA, the greater the complexity of a species. These two variables are positively correlated. More than 98% of DNA in humans, for example, is non-coding.

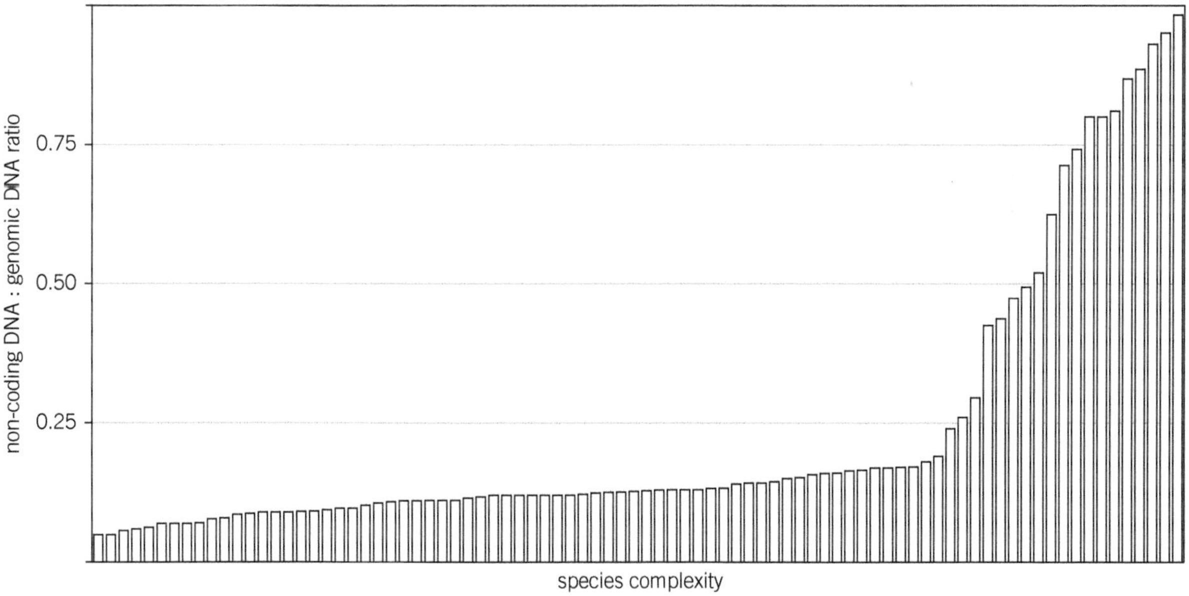

▲ **Figure 2** *The relationship between species complexity and the proportion of non-coding versus genomic DNA*

1. Suggest an explanation for the correlation between species complexity and the proportion of non-coding DNA.
2. What features of an experimental design enable causation to be tested?

The stages in meiosis

Meiosis, like mitosis, follows interphase. Unlike mitosis, which consists of a single nuclear division, meiosis is a two-stage division.

In the first division, meiosis I, **homologous chromosomes** are separated to produce two haploid cells. A homologous pair of chromosomes consists of one paternal chromosome and one maternal chromosome. A pair of homologous chromosomes each has the same genes, although the versions of the gene (i.e., alleles) might be different. In the second division, meiosis II, chromatids separate to produce four haploid gametes.

Prophase I

Homologous chromosomes pair up in prophase I. They interweave to form a structure called a bivalent. Sections of DNA can be swapped between the two chromosomes in a homologous pair. This is called crossing over, a process you will read more about later in this topic.

Other processes that occur in prophase I are the same as those occurring in the mitotic prophase – the spindle forms, the nucleolus disappears, and the nuclear envelope disintegrates.

Metaphase I

In metaphase I, the bivalents line up across the cell equator and spindle fibres attach to the chromosomes' centromeres.

Anaphase I

In anaphase I, the homologous chromosomes are separated. Spindle fibres pull one member of a pair to one pole of the cell and the other chromosome is pulled to the opposite pole.

Telophase I

Once the cell has reached telophase I, two haploid sets of chromosomes are present at opposite sides of the cell. In animal cells, two new nuclear envelopes will form around the two sets of chromosomes and the cell will divide by cytokinesis. The cell then enters prophase II after a brief interphase. Most plant cells proceed directly from anaphase I into meiosis II.

Prophase II

Prophase II resembles mitotic prophase, except the number of chromosomes will be haploid rather than diploid. For example, human cells will contain 46 chromosomes in mitotic prophase and 23 chromosomes in prophase II of meiosis. The nuclear envelope and nucleolus disappear, chromosomes condense, and the spindle reforms.

Metaphase II

In metaphase II, chromosomes line up on the cell equator and spindle fibres attach to the centromeres.

Anaphase II

During anaphase II, centromeres divide and sister chromatids are separated and pulled to opposite poles by the spindle fibres.

Telophase II

In telophase II, nuclear envelopes reform around the four haploid sets of daughter chromosomes.

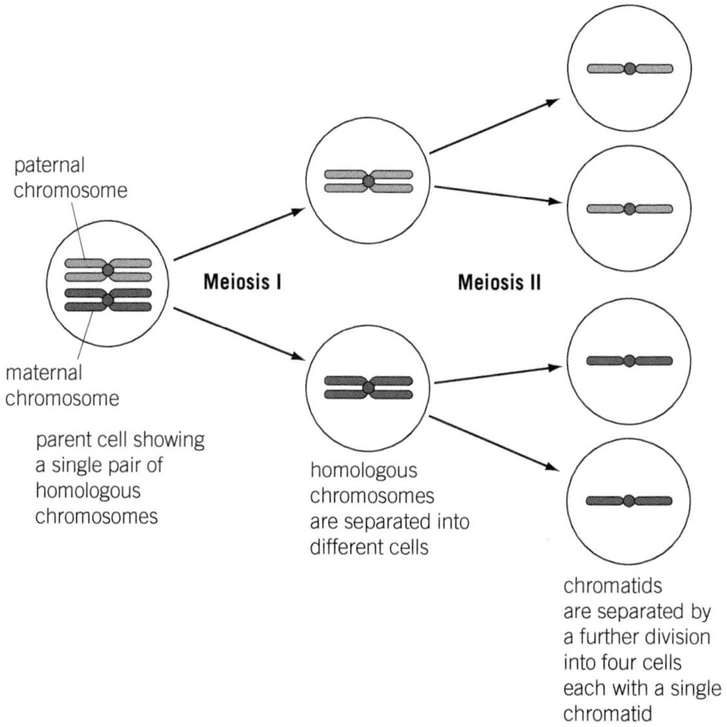
paternal chromosome
maternal chromosome
Meiosis I
Meiosis II
parent cell showing a single pair of homologous chromosomes
homologous chromosomes are separated into different cells
chromatids are separated by a further division into four cells each with a single chromatid

▲ **Figure 3** *A summary of meiosis*

How meiosis produces genetic variation

Two processes in meiosis contribute to genetic variation – **crossing over** and **independent assortment** of chromosomes.

Crossing over

When homologous chromosomes pair up during prophase I, chromatids twist around adjacent chromosomes. The points at which chromatids interweave are known as **chiasmata** (singular – chiasma). Genetic material can be exchanged at chiasmata in a process called crossing over. This is illustrated in Figure 4. The exchange of DNA creates new combinations of alleles.

Independent assortment

Homologous chromosomes are separated during anaphase I of meiosis. Homologous chromosomes are not genetically identical. The side of the cell to which each chromosome is pulled is random. Therefore many different haploid combinations of chromosomes are possible. This is known as independent assortment.

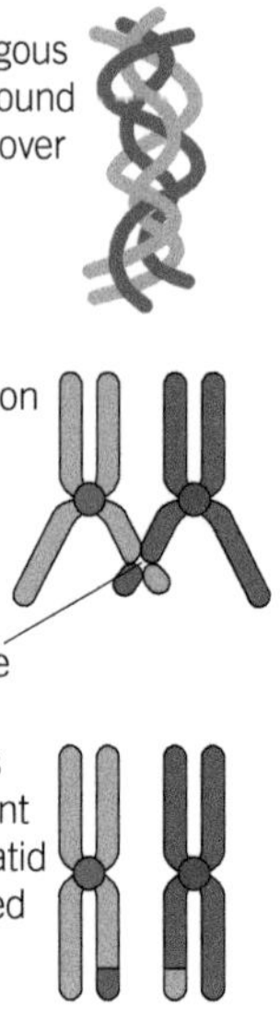
chromatids of homologous chromosomes twist around one another, crossing over many times
simplified representation of a single cross over
point of breakage
result of a single cross over showing equivalent portions of the chromatid having been exchanged

▲ **Figure 4** *Crossing over*

A similar process occurs during anaphase II of meiosis. Sister chromatids, which are no longer identical after crossing over, separate during anaphase II. Again, the pole to which each sister chromatid moves is random and many combinations of chromatids are possible in the final haploid gamete cells. This is known as independent assortment of chromatids.

After meiosis, further genetic variation is added through the process of fertilisation. The combination of two haploid sets of chromosomes in fertilisation is random.

The number of possible combinations of chromosomes in a gamete can be calculated using the term 2^n, with 'n' being the haploid number of the species.

Worked example: Calculating the number of chromosome combinations

In humans, the haploid number is 23. Calculate the number of possible chromosome combinations.

- Number of possible combinations = $2^n = 2^{23} = 8\,388\,608$

The number of possible chromosome combinations in a zygote, following fertilisation, is even more staggering. This is given by the term $(2^n)^2$ (again, 'n' represents the haploid number of the species).

Study tip

You might be asked to explain how meiosis produces genetic variation. Remember that fertilisation, although it contributes to variation, is not part of meiosis.

Summary questions

1. A cell from a giant panda, *Ailuropoda melanoleuca*, was found to have 21 chromosomes. Is this likely to be a haploid or a diploid cell? Explain your answer. *(2 marks)*
2. Tigers (*Panthera tigris*) have a diploid chromosome number of 38. Calculate the number of different chromosome combinations that are possible in a tiger gamete. *(1 mark)*
3. Describe and explain the ways in which genetic variation can occur during meiosis. *(4 marks)*

9.2 Pregnancy and fetal development

Specification references: 3.1.2

Pregnancy

You learnt about the process of meiosis in the previous topic. Meiosis produces gametes, and two gametes fuse during sexual reproduction to produce a zygote. This zygote is able to divide to form an embryo and then a **fetus** inside the mother's womb. In other words, pregnancy ensues. Here we look at the ways in which the development of a fetus can be monitored during pregnancy and the care available to pregnant women.

Learning outcomes

Demonstrate knowledge, understanding, and application of:

→ the programme of antenatal care in the UK

→ the changes to diet recommended for pregnant women

→ the effects of smoking and alcohol on fetal development.

Pre-conceptual care

Doctors will give medical advice to a woman wishing to become pregnant. This is known as **pre-conceptual care** and includes the following advice:

- Both parents will be advised to give up smoking. As you will learn later in this topic, chemicals in cigarette smoke can damage a baby's development.
- A mother will be advised to limit her alcohol consumption. Alcohol can impede the development of a fetus. You will look at this in more detail later in the topic.
- Women should check whether they are immune to rubella (German measles) prior to pregnancy. If a pregnant woman were to become infected with the rubella virus, it could cross the placenta and infect the fetus. The rubella virus can cause deafness and brain damage to a baby. Women who lack immunity should be vaccinated more than three months before conception.

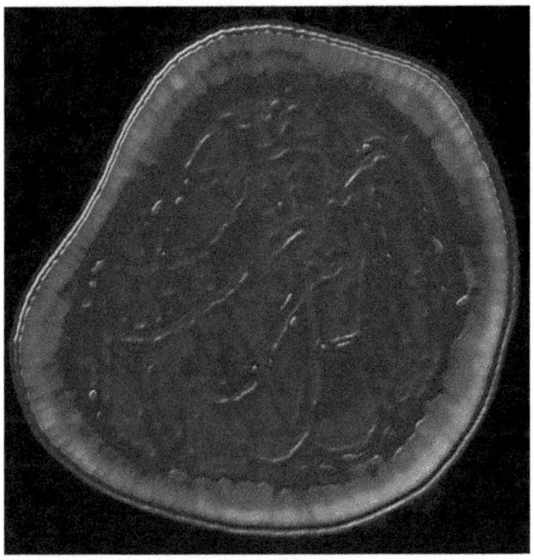

▲ **Figure 1** *A transmission electron micrograph of the rubella virus. Pregnant women should ensure they are vaccinated against this pathogen, ×324 000 magnification*

Antenatal care

The care received by pregnant women is known as **antenatal care** (also called post-conceptual or prenatal care). The first appointment is usually after 11 or 12 weeks of pregnancy. The care offered to women might include:

- Dietary advice, which you will look at in more detail on the next page.
- Advice on avoiding specific infections that could damage the unborn baby. For example, toxoplasmosis is a parasitic infection that can be acquired from raw meat or unpasteurised dairy products. Pregnant women should avoid these foods because toxoplasmosis is very dangerous in young children.
- Testing urine for glucose, which could indicate gestational diabetes, and for protein, which could indicate kidney disease or high blood pressure.
- Monitoring blood pressure to ensure that the mother is not developing a dangerous condition known as pre-eclampsia.

- Blood tests for the mother to check her blood group to determine her rhesus status and in case a transfusion is required during delivery. The blood is also tested for infections such as syphilis, rubella, and HIV.
- Tests for genetic defects in the fetus, which you will read about in Topic 9.3, Monitoring fetal development.
- Ultrasound scans to measure fetal growth. Again, you will examine this technique in greater detail in Topic 9.3.

Diet during pregnancy

A woman will gain approximately 10–15 kg during pregnancy. The maintenance of a balanced diet is important during this period. This ensures the baby is receiving the correct proportions of each nutrient group. Table 1 illustrates the dietary reference values (DRV) for pregnant and non-pregnant women.

▼ Table 1 *Dietary reference values for pregnant and non-pregnant women*

	Energy (kcal)	Protein (g)	Calcium (mg)	Iron (mg)	Vitamin A (μg)	Vitamin C (mg)	Folic acid (μg)
Adult female	1940	45	700	14.8	600	40	200
Pregnant female	2140	51	700	14.8	700	50	300

Roles of nutrients

Some nutrients, such as folic acid and protein, need to be consumed in greater amounts during pregnancy. Other nutrients, such as calcium and iron, are not needed in significantly larger quantities. The recommended intake for pregnant women depends on the specific role of each nutrient in the developing fetus.

Protein

The growth of the baby, uterus, and placenta all require the mother to consume additional protein. The amino acids from this protein are used to synthesise various cell structures, as well as haemoglobin, enzymes, and antibodies.

Calcium

Calcium is required for the development of strong teeth and bones and the functioning of the nervous system in the fetus. However, the mother does not need to drastically increase her intake of calcium while pregnant.

Iron

This is required for the synthesis of haemoglobin.

Vitamin A

Rhodopsin, which is a pigment in the rod cells of the eye, is produced from vitamin A. Other functions include roles in gene transcription and the immune system.

Vitamin C

Among vitamin C's functions is collagen formation. Collagen is a structural protein and an important component of many tissues, including skin, tendons, bones, and blood vessels.

Folic acid

Folic acid is sometimes referred to as vitamin B9. It is essential for DNA synthesis, cell division, and red blood cell production. A shortage of folic acid in the diet of a pregnant woman can hinder brain development in her baby and increases the risk of spine abnormalities such as spina bifida.

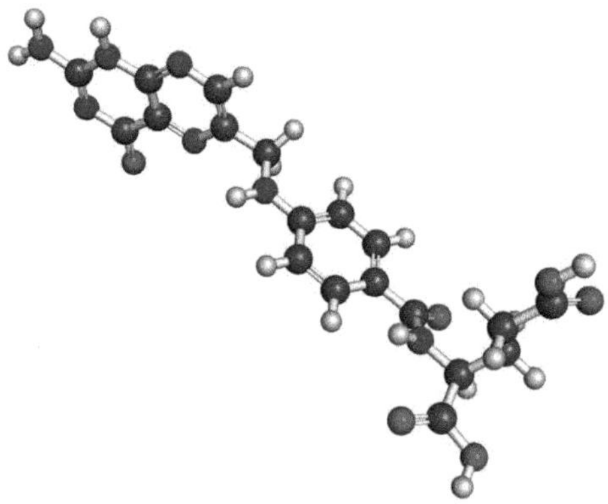

▲ **Figure 2** *A molecular model of folic acid, which is an important nutrient in the diet of a pregnant woman*

The effects of smoking and alcohol during pregnancy

Essential molecules (e.g., oxygen, amino acids, and glucose) pass across the placenta from mother to baby. However, harmful molecules, such as alcohol (ethanol) and chemicals from tobacco smoke, can also cross the placenta.

Alcohol

Advice from the department of health advises women to avoid alcohol consumption altogether. If they do opt to drink then they are advised to stick to fewer than two units per week. This is the equivalent of one pint of beer. A higher consumption increases the risk of a baby developing language and speech difficulties. The babies of women who drink more than six units per day have a high risk of developing fetal alcohol syndrome (see extension).

Chemicals from cigarettes

The smoke from cigarettes contains several chemicals that can impede the development of a fetus, notably carbon monoxide and nicotine. The combined effect of these chemicals is to increase the probability of premature birth. The baby's lungs will be poorly developed. The risk of still birth or death in infancy is also raised.

▲ **Figure 3** *Women are advised to minimise their consumption of alcohol and to stop smoking during pregnancy*

Carbon monoxide

Carbon monoxide binds to haemoglobin more readily than oxygen. This reduces the supply of oxygen from the mother to the fetus.

Nicotine

Blood vessel diameter is narrowed by nicotine. This further reduces the supply of oxygen across the placenta to the fetus. The fetus's heart will beat faster because nicotine stimulates the release of the hormone adrenaline, which raises heart rate.

What is fetal alcohol syndrome?

If a pregnant woman consumes an alcoholic drink, the ethanol absorbed into her blood will pass into her baby's circulatory system through the placenta.

Ethanol can damage the fetus in several ways. The combined effects of ethanol are called fetal alcohol syndrome (FAS). The principal effect is permanent damage to the central nervous system, resulting in psychological or behavioural problems after birth, and sometimes learning difficulties. Other signs of FAS include distinctive facial features, such as small eye openings and a thin upper lip, and growth deficiency.

A study of 400 000 women, all of whom consumed alcohol during pregnancy, suggested that having more than 15 drinks per week reduced the birth weight of the baby. Consuming 1–2 drinks per week does not appear to present a significant risk to the fetus. However, it is worth bearing in mind that women will differ in how they metabolise and process ethanol.

1. Suggest what additional information would be needed to assess the validity of the study discussed above.
2. Suggest how ethanol produces permanent psychological and behavioural problems in a baby.

Summary questions

1. Name two nutrients that would need to be consumed in greater amounts during pregnancy and explain why this increased consumption is necessary. *(4 marks)*
2. Describe the precautions a pregnant woman is advised to take when choosing what she consumes. *(4 marks)*
3. Explain how regular smoking by a pregnant woman can lower the birth weight of her baby. *(5 marks)*

9.3 Monitoring fetal development

Specification references: 3.1.2

Measuring fetal growth

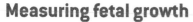

One aspect of antenatal care, which you considered in the previous topic, is the monitoring of fetal development. Doctors will monitor the growth of a fetus and can test for certain disorders.

The size of a fetus can be measured using an **ultrasound** scan. A small handheld transducer is moved backwards and forwards over the abdomen, which is usually covered in a lubricating jelly. The transducer emits sound waves into the mother's body. These waves are reflected back by the fetus and the placenta, producing an image that can be viewed on a monitor screen.

The image provides several important pieces of information:

- how many babies a mother is carrying
- the structure of organs (which can be seen after 18 weeks)
- the blood flow through the umbilical cord, which indicates whether the baby is receiving sufficient oxygen and nutrients
- the position of the placenta – a placenta that is too close to the cervix can cause excessive bleeding during labour, which means a Caesarean delivery would be recommended
- measurements of the fetus – two of the principal measurements are the crown-rump length (from the top of the fetus's head to its bottom) and the biparietal diameter (the width of the fetus's head at its widest point).

Learning outcomes

Demonstrate knowledge, understanding, and application of:

→ the analysis of secondary data from fetal growth charts

→ the use of ultrasound, amniocentesis and chorionic villus sampling

→ the use of karyotypes.

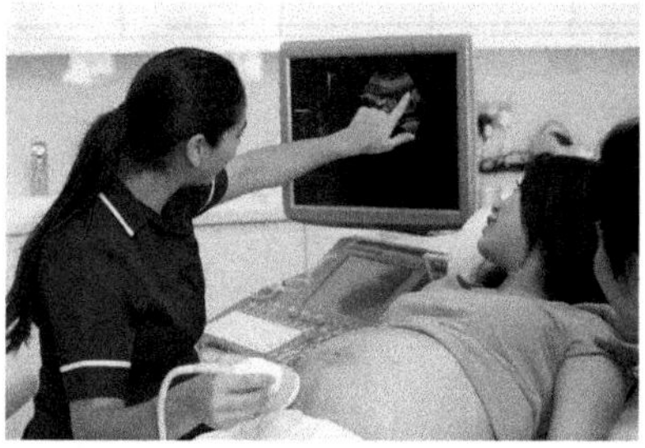

▲ **Figure 1** *An ultrasound scan being performed*

Analysing growth data

The measurements of a fetus can be compared to the expected values at a particular age. This enables doctors to monitor whether the fetus is growing normally. The growth chart below shows the expected head circumference measurements of babies during gestation.

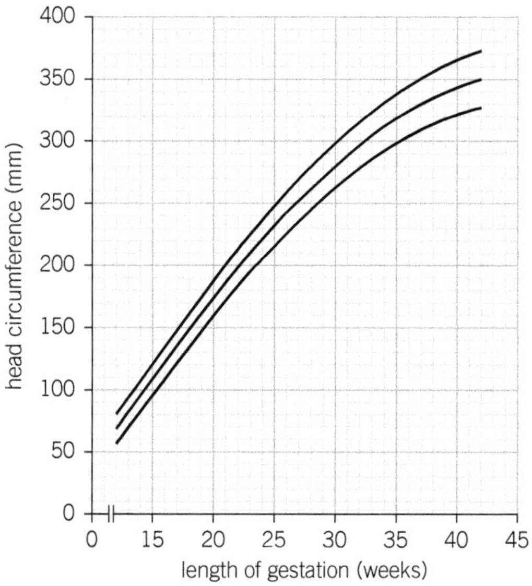

◀ **Figure 2** *Head circumference growth chart. The top line represents the 95th percentile, the middle line represents the 50th percentile and the lower line is the 5th percentile*

In Figure 2 three lines represent percentiles. 95% of fetuses will have measurcments below the 95th percentile – only 5% will have readings higher than this line. 50% of observations will be above the 50th percentile, which is also known as the median, and 50% will be below. Only 5% of fetuses will have values lower than the 5th percentile. Measurements below the 5th percentile indicate the fetus is growing slower than expected. Doctors are then in a position to assess what could be causing the slow growth rate.

Worked example: Analysing growth charts

Estimating growth rates from a graph is an important mathematical skill. For example, what is the growth rate at 20 weeks for a baby following the growth pattern of the 50th percentile?

Step 1 Draw a tangent (a straight line that follows the line of the graph at the point of analysis).

Step 2 Form a right-angled triangle, with the tangent as the hypotenuse.

Step 3 Read the change in the *x*-axis and *y*-axis variables based on the other two sides of the triangle. In this case, a triangle could be formed that shows the following changes – 18 to 22 weeks (*x*-axis) and 150 to 200 mm (*y*-axis). The size of the triangle does not matter. Try to make it as easy as possible to work out the changes in each variable, and ensure that the tangent follows the trajectory of the graph line as closely as possible.

Step 4 Divide the *y*-axis change by the *x*-axis change (i.e., the change in growth measurement divided by time). In this case, this will be $\frac{50\,\text{mm}}{4\text{ weeks}}$ = 12.5 mm week^{-1}.

Sampling cells from a fetus

Sometimes fetal DNA needs to be obtained for tests. Extracting tissue directly from a fetus would be dangerous and difficult. Instead, fetal cells can be obtained from the developing placenta, via a technique called **chorionic villus sampling**. Alternatively, cells can be obtained from the fluid in which a fetus grows, using a technique called **amniocentesis**. Both methods require ultrasound to locate the position of the fetus and the placenta. A needle is then inserted to remove the sample.

Chorionic villus sampling (CVS)

CVS can be carried out earlier in the pregnancy than amniocentesis. The procedure is usually conducted at 10–15 weeks. CVS has a higher risk of miscarriage (approximately 1%) than amniocentesis. It also carries a slight risk of producing fetal deformities.

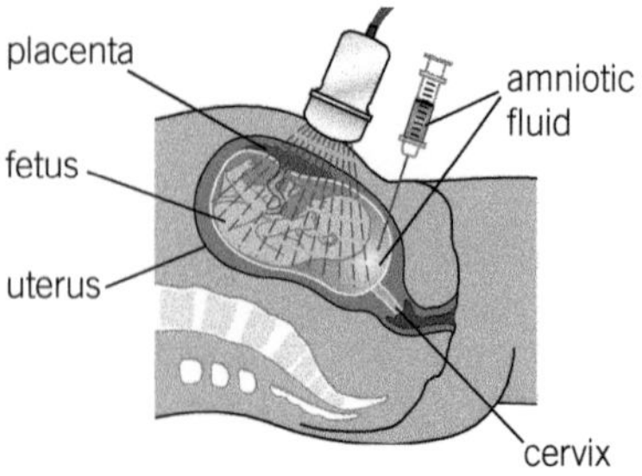

▲ **Figure 3** *Amniocentesis*

Amniocentesis

Amniocentesis is used between week 15 and 20 of gestation. Although the procedure carries a small risk of inducing miscarriage, the probability of this occurring is lower than for CVS. Fewer cells are sampled using amniocentesis compared to the quantities obtained with CVS.

Karyotyping

The fetus's chromosomes can be analysed after fetal cells have been extracted using either CVS or amniocentesis. A **karyotype** is the number and appearance of chromosomes in a cell. A karyogram is a picture of all the chromosomes in a cell.

The fetal cells are cultured in a laboratory and stimulated to divide by mitosis. A chemical called colchicine is added, which stops mitosis at metaphase. The chromosomes can then be photographed to produce a karyogram. A stain is added to improve their visibility. The chromosomes can be sorted into pairs, and doctors are able to determine sex and assess whether the fetus has any chromosomal mutations.

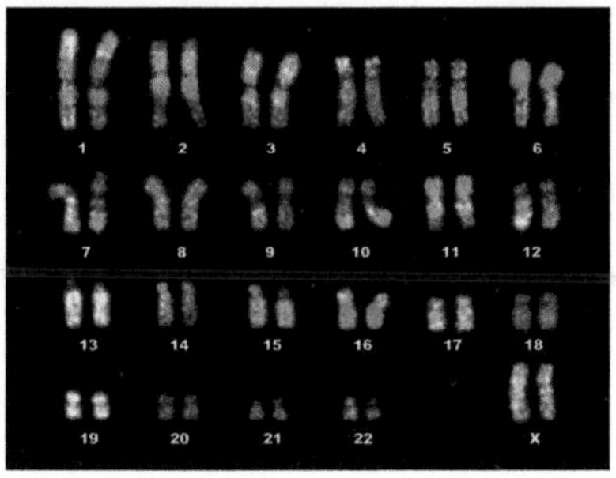

▲ **Figure 4** *A fluorescently stained karyogram of a female*

Types of chromosomal mutations

Down's syndrome

One chromosomal mutation is **Down's syndrome**. If a fetus has Down's syndrome, a karyogram would show three copies of chromosome 21 instead of two. Down's syndrome is characterised by distinctive facial features and some degree of learning disability and delayed development.

Turner's syndrome

Sex is determined by the sex chromosomes inherited by a fetus, with females being XX and males being XY.

Babies with **Turner's syndrome** have only one X chromosome and no Y chromosome. This would be written as XO. Symptoms include short stature and a lack of menstruation.

Klinefelter's syndrome

Two X chromosomes and one Y chromosome on a karyogram indicates that the fetus has **Klinefelter's syndrome**. Symptoms include tall stature and a feminised physique.

Study tip

You have now encountered the term 'syndrome' in several contexts. A syndrome is a collection of symptoms that characterise a condition.

Summary questions

1 Suggest what a karyogram showing 47 chromosomes could indicate about a fetus. *(1 mark)*

2 Outline two advantages and two disadvantages of sampling fetal cells using CVS rather than amniocentesis. *(4 marks)*

3 Using Figure 2, calculate the growth rate at 15 weeks of a baby following the growth pattern of the 5th percentile. *(2 marks)*

Practice questions

1 A karyogram is a picture of all the chromosomes in a cell. Which of the following statements about karyogram formation is/are true?

1 Fetal cells from chorionic villus sampling or amniocentesis can be used in karyotyping.

2 Colchicine is added to halt mitosis at prophase.

3 Each chromosome is stained and separated from its homologous pair.

A 1, 2 and 3

B Only 1 and 2

C Only 2 and 3

D Only 1 *(1 mark)*

2 Chemicals in cigarette smoke can harm the development of fetuses.

Which of the following statements about the chemicals in cigarettes is/are true?

1 Carbon monoxide binds to haemoglobin with greater affinity than oxygen.

2 Nicotine stimulates the production of adrenaline, which increases heart rate.

3 Nicotine narrows the lumen diameter of blood vessels.

A 1, 2 and 3

B Only 1 and 2

C Only 2 and 3

D Only 1 *(1 mark)*

3 The American bison (*Bison bison*) has a diploid chromosome number of 60.

How many different chromosome combinations are possible in a gamete of this species?

A 3.00×10^1

B 9.00×10^2

C 1.07×10^9

D 1.15×10^{18} *(1 mark)*

4 Crossing over during meiosis contributes to genetic variation.

Which is the correct description of crossing over? *(1 mark)*

A Chiasmata form when chromatids on non-homologous chromosomes exchange genetic material in prophase I.

B Chiasmata form when chromatids on homologous chromosomes exchange genetic material in prophase I.

C Chiasmata form when chromatids on non-homologous chromosomes exchange genetic material in metaphase I.

D Chiasmata form when chromatids on homologous chromosomes exchange genetic material in metaphase I.

5 The graph shows a fetal growth chart.

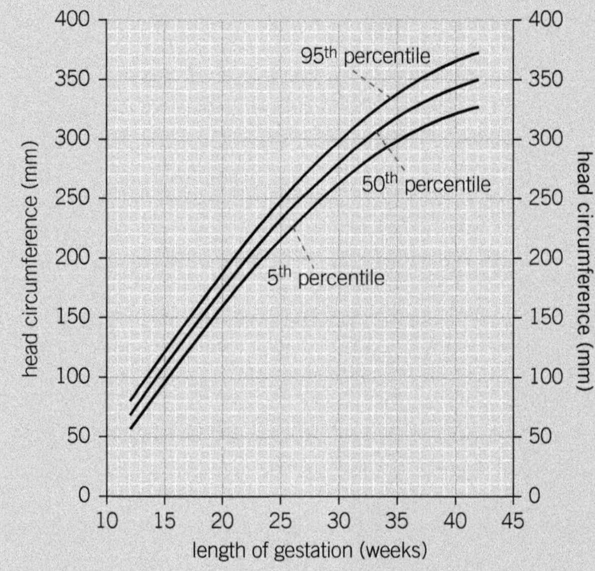

Calculate the growth rate for a baby following the growth pattern of the 50th percentile at:

A 15 weeks

B 40 weeks *(6 marks)*

6 The diagram shows the stages that occur during the second division of meiosis.

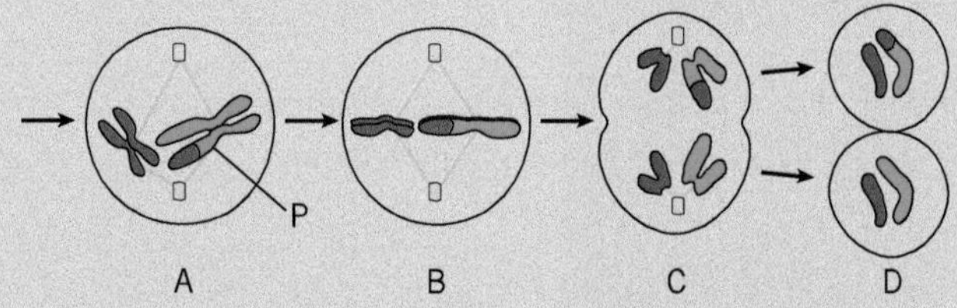

a Identify the stages labelled B and C. *(2 marks)*

b The chromosome labelled P in the diagram contains a section of genetic material from the chromosome that was homologous to P.

The two homologous chromosomes exchanged genetic material during the first division of meiosis.

(i) Identify the stage in the first division of meiosis in which exchange of genetic material occurs. *(1 mark)*

(ii) The figure here shows the appearance of chromosome P.

Draw a diagram to show the appearance of the chromosome homologous to P after the exchange of genetic material.

On your diagram, you should label the position of the centromere.

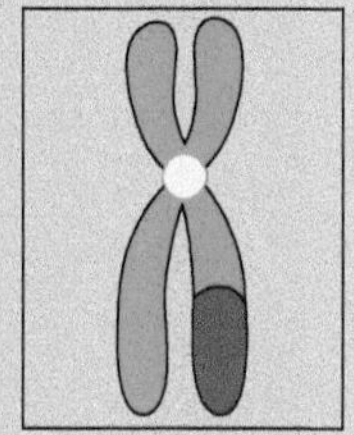

chromosome P *(3 marks)*

c Explain the importance of meiosis in the human life cycle. *(3 marks)*

[question 3, OCR F222 June 2009]

7 The table shows the Dietary Reference Values (DRV) for several nutrients for women between 19 and 50 years of age. It also shows how these values change during pregnancy.

	Protein (g day $^{-1}$)	Iron (mg day $^{-1}$)	Vitamin A (µg day $^{-1}$)	Vitamin C (µg day $^{-1}$)	Folic acid (µg day $^{-1}$)
Female aged 19–50 years	45	15	600	40	400
Pregnant female aged 19–50 years	51	15	610	40	600

a Using information shown in the table, caclulate the percentage increase in the recommended DRV for folic acid during pregnancy. Show your working. *(2 marks)*

b Explain why women are advised to increase their intake of the following nutrients during pregnancy:

(i) folic acid *(2 marks)*

(ii) protein *(2 marks)*

(iii) vitamin A *(2 marks)*

c Suggest why it is **not usually** necessary to increase the recommended DRV for iron during pregnancy. *(2 marks)*

d Women are advised to give up alcohol and stop smoking during pregancy.

Describe **and** explain the possible effects on the fetus of a mother drinking alcohol **and** smoking during pregnancy. *(6 marks)*

[question 3, OCR F222 June 2010]

8 Pregnant women are offered many different tests to monitor the health of their developing baby.

The photo shows an ultrasound scan of a fetus at 13 weeks of pregnancy.

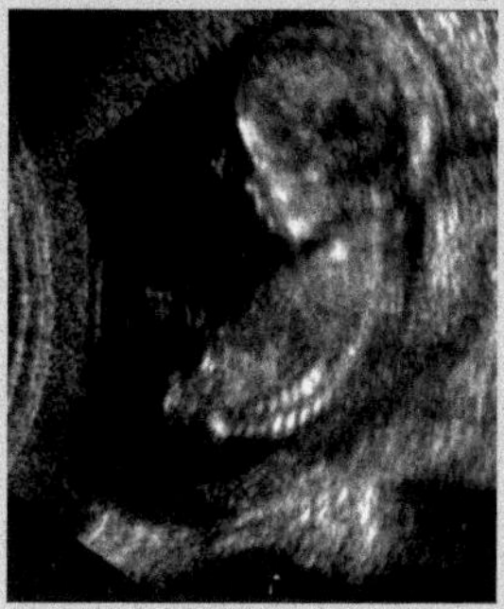

a Ultrasound scans are used to visualise the fetus during an amniocentesis test. Describe how an amniocentesis test is carried out. *(4 marks)*

b The fetal cells obtained by amniocentesis may be used to produce a karyotype.

(i) Give two reasons why a doctor might advise a woman to have an amniocentesis test. *(2 marks)*

(ii) State an alternative test that could be used instead of amniocentesis to obtain a karyotype and suggest **one** advantage of this test compared to amniocentesis. *(2 marks)*

c Both the ultrasound and karyotype results are used to detect disorders in the developing fetus.

(i) Name one disorder that could be detected by the **ultrasound** scans. *(1 mark)*

(ii) Name two disorders that could be detected by the **karyotype**. *(2 marks)*

(4 marks) [question 6, OCR Jan 2011 F222]

10 EVOLUTION AND CLASSIFICATION

10.1 Classification

Specification references: 3.1.3

Learning outcomes

Demonstrate knowledge, understanding, and application of:

- the biological and phylogenetic species concepts
- classification of organisms into the taxonomic hierarchy of domain, kingdom, phylum, class, order, family, genus, and species
- the evidence used to classify organisms
- the use of DNA barcoding in classification.

Scientists estimate that at least eight million **species** are alive on Earth today. An even greater number of extinct species are thought to have once existed. Naming and grouping this vast array of organisms makes it easier for us to study them. **Classification** is the term used for sorting organisms into groups, and the smallest unit of classification is a species.

What is a species?

Several definitions of a species exist. For example, organisms can be grouped into species based on their appearance, the ecological niche they occupy, and whether they recognise each other as potential mates. The one factor underlying every definition is that members of the same species must possess similar genetic information. Two of the most common definitions are the biological species concept and the phylogenetic species concept.

Biological species concept

The **biological species concept** defines a species as a group of organisms with similar morphological, physiological, biochemical, and ecological features that can interbreed to produce fertile offspring. The ability to reproduce together but be reproductively isolated from other species is an indication that organisms share a high percentage of genetic information. One problem with this definition is that it fails to consider species that only reproduce asexually.

Phylogenetic species concept

The **phylogenetic species concept** defines a species as a group of organisms that share a common ancestor and have the same evolutionary history. This enables the inclusion of organisms that reproduce both sexually and asexually.

The hierarchy of classification

The botanist Carl Linnaeus devised a system for classifying and naming organisms approximately 250 years ago. It remains the basis of our current classification system.

The groups in the classification system are known as **taxa** (singular – taxon). Taxa are organised in a hierarchy as shown in the first column of Table 1.

The number of species in each group increases as you move up the hierarchy from species to domain. For example, a genus (plural – genera) contains at least one species, a family contains several genera, and an order contains several families. The similarity between the organisms in each group increases as you progress down the hierarchy from domain to species.

The largest taxon that scientists use is domains. The domain Eukaryota contains four kingdoms – protoctists, fungi, plants, and animals.

Study tip

Remembering the order of the taxonomic groups can be tricky. Inventing a mnemonic can help. One example is 'Don't Keep Pickled Cucumber Or Fried Gherkin Sauce'.

▼ **Table 1** *The classification of three organisms from different domains*

Taxon	Organism		
	E. coli	Common wheat	Tiger
Domain	Eubacteria	Eukaryota	Eukaryota
Kingdom	Prokaryota	Plantae	Animalia
Phylum	Proteobacteria	Magnoliophyta	Chordata
Class	Proteobacteria	Liliopsida	Mammalia
Order	Enterobacteriales	Poales	Carnivora
Family	Enterobacteriaceae	Poaceae	Felidae
Genus	*Escherichia*	*Triticum*	*Panthera*
Species	*coli*	*aestivum*	*tigris*

Species are named using a binomial system. For example, humans are named *Homo sapiens*. The first word, in this case *Homo*, indicates the genus and the second word, in this case *sapiens*, indicates the species. The binomial system includes a few rules for you to remember:

1. The genus name begins with a capital letter.
2. The species name is always in lower case.
3. You should write both words in italics or, if handwritten, underline both words.

The development of a standard classification system that has been agreed by scientists around the world ensures that everyone is using the same names. This reduces the chance of confusion and ambiguity. The two species in Figure 1, for example, are both known locally as robins. Species (a) would be called a robin in the USA, whereas species (b) is known as a robin in the UK.

It is important to bear in mind that names and classification of organisms can change if new evidence is found.

a *Turdus migratorius*

b *Erithacus rubecula*

▲ **Figure 1** *Two different species that are both called robins*

Hominid classification

The Primates order comprises all species of prosimians (such as lemurs), monkeys, and apes. The term 'apes' includes two families – Hylobatidae and Hominidae. The Hylobatidae family contains several species of gibbons. The Hominidae family includes humans (*Homo sapiens*), chimpanzees (*Pan troglodytes*), bonobos (*Pan paniscus*), two species of gorilla (genus *Gorilla*), two species of orangutan (genus *Pongo*), and many extinct hominid species.

The evidence used in classification

Biologists can use different sources of evidence when classifying organisms into taxa or establishing evolutionary relationships.

Biochemical/molecular evidence

DNA

Species can be classified by comparing their DNA sequences. The more similar the sequences, the more closely related the two species. Scientists have worked out the entire DNA sequence (also known as a genome) of many species. The genomes of over 100 animal species and hundreds of prokaryotes, fungi, and plants are now known. Knowledge of genomes allows scientists to classify species with a high level of accuracy.

The evolutionary relationships between humans and the other extant (living) members of the Hominidae family (gorillas, chimpanzees, bonobos, and orangutans) have been revealed by sequencing the genomes of these species. For example, the chimpanzee genome differs from the human genome by a little over 1%.

Amino acids

The DNA sequence determines the order of amino acids in a protein. Amino acid sequences can therefore be analysed in order to classify organisms. Some proteins are found in virtually all species, but the amino acid sequence of the protein will vary between species. As with DNA, the greater the similarity in sequences, the more closely related the species. Good examples of proteins that can be used to classify species are haemoglobin and a respiratory protein called cytochrome c.

Synoptic link

You learnt how the DNA sequence determines the amino acids in proteins in Topic 4.4, The genetic code.

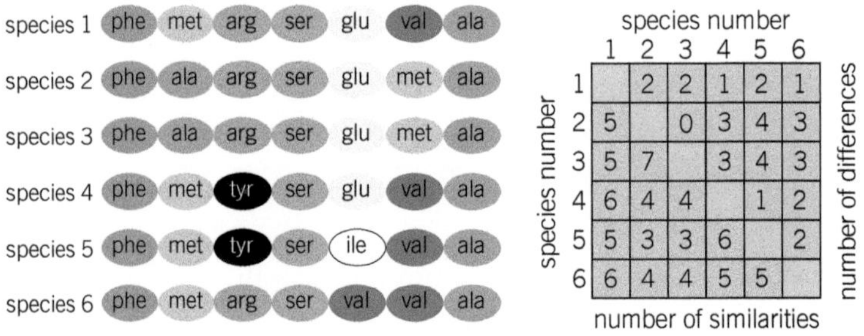

▲ **Figure 2** *Comparison of amino acid sequences from part of the same protein in six species*

▲ **Figure 3** *Ants from the* Eciton burchellii *species. These ants are from the same species but they look very different*

Anatomical evidence

The earliest classification systems were based on appearance and anatomy alone (morphological similarities). This meant many errors were made. For example, the philosopher Aristotle grouped birds and insects together. The invention of microscopes in the 1600s increased the detail seen by scientists.

Even with microscopes, comparing the appearance of species can be an inaccurate method of determining how closely they are related. Many species have evolved similar physical traits even though, in evolutionary terms, they are not closely related. This is known as convergent evolution. For example, the fossils of extinct marine reptiles called ichthyosaurs suggest they resemble dolphins, but dolphins are mammals and have a very different evolutionary history to reptilian ichthyosaurs.

Another problem with basing classification on appearance is that members of the same species can look very different. Ant colonies, for example, comprise several groups of ants with different roles, such as queens, workers, and soldiers. These groups have different appearances even though they are part of the same species.

Fossil evidence

Fossil evidence, however, has been useful in studying the classification of extinct species of hominids. DNA can sometimes be retrieved from more recent fossils, enabling molecular analysis to be carried out. However, DNA is not preserved well in older hominid fossils, such as those of the *Australopithecus* genus. This means evolutionary relationships must be estimated using only the morphological differences (e.g., the length of bones and the shape of skulls) between fossils. Different interpretations of this fossil evidence by different groups of scientists result in competing theories of classification.

Immunological evidence

Another method for comparing the proteins in different species relies on immune responses. Proteins from one species will act as antigens if they are injected into another species. In other words, the proteins will cause an immune response. Closely related species will have similar antigens, which will produce similar immune responses. Figure 5 outlines the method used for these immunological comparisons.

Behavioural evidence

Shared behaviours can provide insights into the relationships between species. Most primates, for example, live in complex social groups and communicate using vocalisations and facial expressions.

Embryological evidence

Early embryos of different species can look very similar. For example, a tail-like structure is visible in all early vertebrate embryos, which indicates they have evolved from a common ancestor as shown in Figure 6.

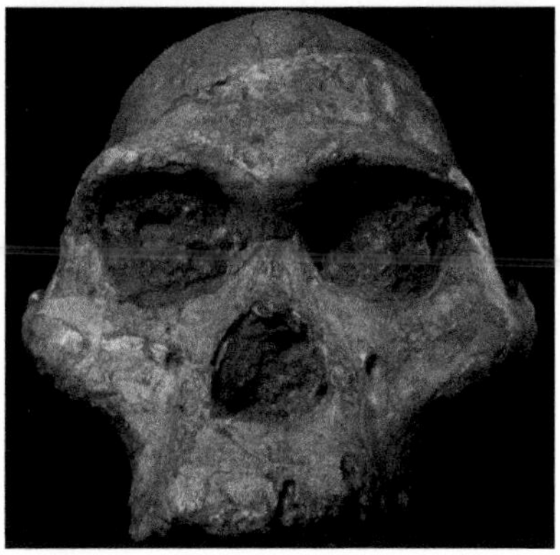

▲ **Figure 4** *A fossil skull of* Australopithecus africanus

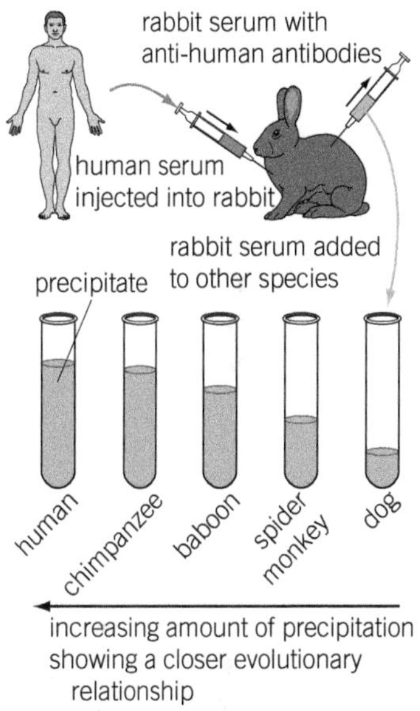

The results show that humans are very closely related to chimpanzees, less so to baboons and even less so to spider monkeys. They are only distantly related to dogs.

▲ **Figure 5** *Immunological comparisons of human blood serum with that of other species*

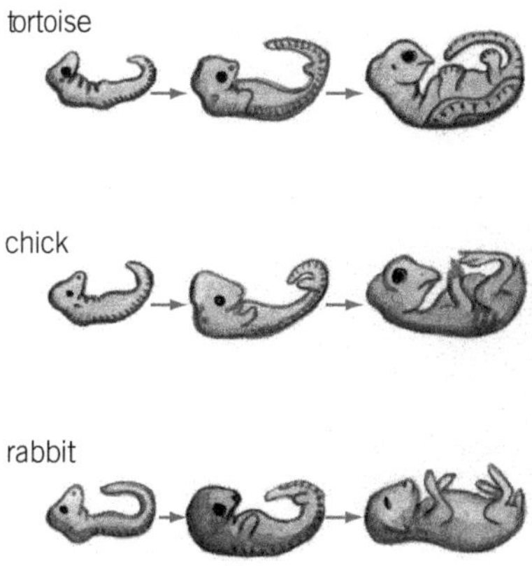

▲ **Figure 6** *Stages in the development of three different vertebrate embryos*

Synoptic link

You will discover more about how the immune system functions in Topic 12.1, The immune system.

DNA barcodes

A **DNA barcode** is a short sequence of DNA that is used to identify a species. A good DNA barcode needs to have the following properties:

1. Universal (i.e., possessed by most taxa).
2. Short, so that it is cheap and quick to analyse.
3. Show a large amount of variation between species.
4. Show little variation within a species.

In eukaryotes, mitochondria contain DNA that has a faster rate of mutation than chromosomal DNA in nuclei. Mitochondrial DNA is therefore a good source of DNA barcode sequences because the relatively fast mutation rate produces significant differences in genetic sequences between species. Cytochrome c oxidase is one of the most widely used barcodes. A short (658 base pairs) sequence of the cytochrome c oxidase gene shows large variation between animal species.

The cytochrome c oxidase gene has a much slower rate of mutation in plants than it does in animals. This means it is less effective as an identifier of species. Instead, a pair of chloroplast genes, rbcL and matK, can be used as a DNA barcode for plant species.

Why domains?

Until recently, kingdoms were the largest taxa in the taxonomic hierarchy. In 1990, Carl Woese proposed a change to the classification system. He suggested dividing the Prokaryotae kingdom into two groups – Bacteria (or Eubacteria) and Archaea (or Archaebacteria). He also placed the other four kingdoms (Protoctists, Fungi, Plants, and Animals) into a single group, the Eukaryota. These groups were named domains.

Woese based his ideas on several observations. Bacteria differ from Archaea by possessing a different cell membrane structure, having no DNA-bound proteins, different enzymes, and a different DNA replication mechanism. In contrast, Archaea share traits with Eukaryotes, such as using a similar DNA replication mechanism and possessing similar enzymes.

1. What similarities do the four eukaryotic kingdoms have that enable them to be classified together in a single domain?
2. Suggest why Woese proposed separating Bacteria and Archaea into two domains.

Summary questions

1 Copy and complete the table to show the classification of humans and chimpanzees, *Pan troglodytes*. *(8 marks)*

Taxon	Human	Chimpanzee
Domain		
Kingdom		
		Chordata
	Mammalia	Mammalia
	Primates	
	Hominidae	Hominidae
Genus		
Species		

2 Use the information in Figure 2 to answer the following:

a Deduce which two species show the closest evolutionary relationship. *(1 mark)*

b Determine which position in the amino acid sequence shows the greatest variation. *(1 mark)*

c Deduce which species are most distantly related to species 5. *(1 mark)*

3 Explain why molecular evidence is more accurate than anatomical evidence when classifying species. *(3 marks)*

10.2 Phylogeny

Specification references: 3.1.3

Learning outcomes

Demonstrate knowledge, understanding, and application of:

→ the construction and interpretation of phylogenetic trees.

In the previous topic you learnt about the methods scientists use to classify organisms into groups that share particular features and evolutionary histories. The study of evolutionary relationships between species is known as **phylogeny**. These evolutionary relationships can be illustrated by constructing tree-like diagrams called evolutionary (or phylogenetic) trees.

Constructing phylogenetic trees

DNA mutations occur randomly and spontaneously over time at a relatively constant rate. The evolutionary relationship between two species can be determined by comparing their DNA. The fewer the differences in their DNA sequences, the more recently two species will have evolved from a common ancestor and the closer their evolutionary relationship. Relationships can be shown in phylogenetic trees, which use the following rules:

- branch length is proportional to time
- nodes at branching points represent common ancestors
- the closer the relationship between two species, the more recently they will have branched from a common ancestor
- branches for extinct species will end before the present day.

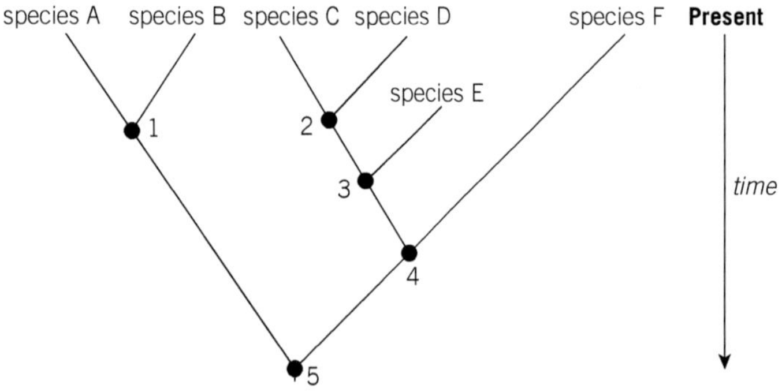

▲ **Figure 1** *A typical phylogenetic tree*

Figure 1 shows a generalised phylogenetic tree, which illustrates these points.

Worked example: Interpreting a phylogenetic tree

The following questions are based on the phylogenetic tree in Figure 1.

Which species is most closely related to species A?

- Species B shares a common ancestor with species A at node 1, which shows that they evolved into separate species relatively recently.

Which species is most closely related to species C?

- Species C, D, E, and F all share a common ancestor at node 5, however species C and D share a common ancestor at node 2 and so are the most closely related.

Which species is extinct?

- Species E is extinct because the branch for species E ends before the present day.

Study tip

Remember that the time scale over which different species evolve from a common ancestor can be many millions of years. A common unit for describing such time scales is 'mya' meaning 'millions of years ago'. Alternative units covering the equivalent time scale include myr, Myr or ma.

Trees can change

Phylogenetic trees are not necessarily fixed. New evidence and different interpretations of evidence can result in the re-evaluation of evolutionary relationships and changes to the arrangement of phylogenetic trees.

For example, a phylogenetic tree could be constructed from fossil evidence, but genetic evidence might be contradictory and indicate an alternative phylogenetic tree. Two sets of scientists might analyse different molecules and draw conflicting conclusions about the evolutionary relationships between species. Figure 2 provides an excellent example of conflicting evidence. It shows two alternative phylogenetic trees for animals based on the analysis of DNA sequences in different sets of genes.

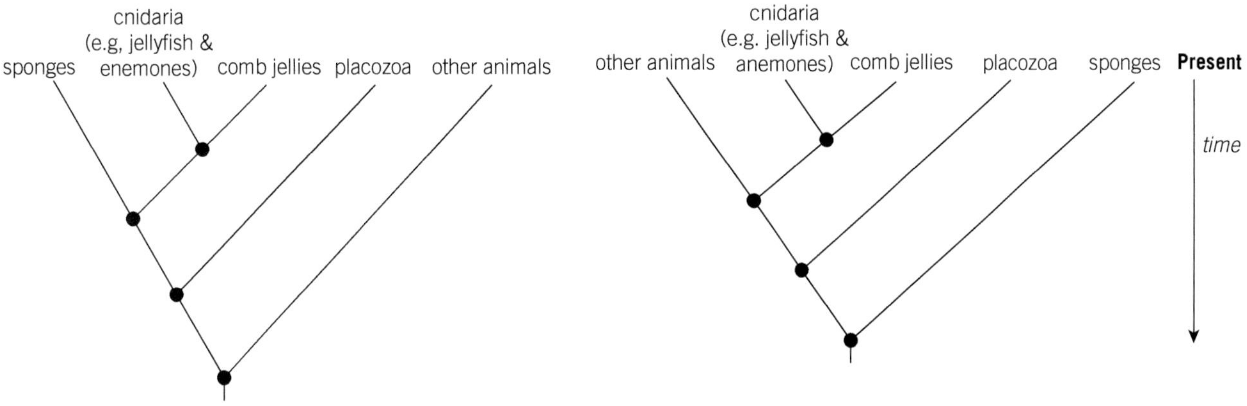

▲ **Figure 2** *Two different phylogenetic trees for animal evolution, based on different evidence*

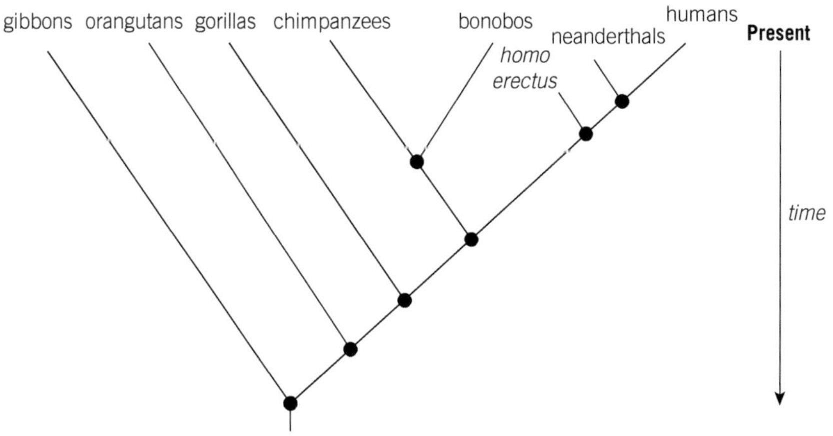

▲ **Figure 3** *Phylogenetic tree for apes*

Summary questions

1 The phylogenetic tree in Figure 3 shows one interpretation of the evolutionary relationships between apes. State which two species shown in the tree are extinct. *(2 marks)*

2 Based on the information provided in Figure 3, determine which living species humans are most closely related to. *(2 marks)*

3 Use the phylogenetic tree to explain whether gorillas are more closely related to gibbons or humans. *(1 mark)*

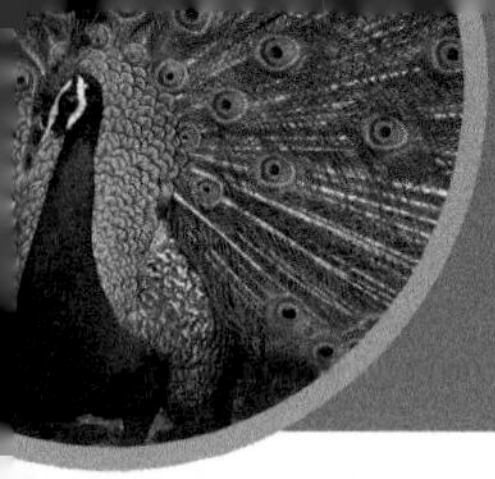

10.3 Adaptation

Specification references: 3.1.3

Learning outcomes

Demonstrate knowledge, understanding, and application of:

- → adaptations to the environment and how they can be behavioural, physiological, or anatomical.
- → adaptations in *Homo sapiens*.

What are adaptations?

So far in this chapter you have learnt about how organisms can be classified based on characteristics that they share or characteristics that differentiate them. Now you will take a closer look at the types of characteristics that can evolve and examine how they benefit organisms.

Members of the same species can exhibit small or significant differences in their appearance, physiology, or behaviour. This is known as **variation**. Any variation that benefits an organism in its environment is an **adaptation**. Adaptations are evolved and maintained by **natural selection**, which you will read about in Topic 10.5, Evolution.

Adaptations help organisms survive in several ways, including:

- obtaining food, nutrients, and water
- avoiding disease and predators
- coping with abiotic factors in the environment, such as temperature, light, and pH levels
- communicating to other members of the species in order to maintain social groupings and to reproduce.

Types of adaptation

The adaptations shown by organisms can be anatomical, behavioural, or physiological.

Anatomical adaptations

Anatomical adaptation involves the evolution of structures that help an organism in its environment.

Behavioural adaptations

Behavioural adaptations range from simple reflex responses, such as an earthworm moving underground to escape predators in response to vibrations, to elaborate courtship displays by birds such as birds of paradise.

Physiological adaptations

Physiological adaptations are biochemical and cellular characteristics. The type of enzymes and hormones an organism produces will have evolved to suit their environment. For example, Antarctic icefish such as *Chaenocephalus aceratus* produce enzymes that are able to function in very cold temperatures.

▲ **Figure 1** *A peacock's tail is an adaptation that helps it attract mates*

Study tip

Adaptation is something that happens to a species over many generations. To speak about an individual organism becoming adapted is incorrect. Individuals *exhibit* adaptations that have evolved over a long period of time.

Human adaptations

Humans have evolved many adaptations to the environment, some of which are described below.

Anatomical adaptations in humans

Bipedalism

Unlike other apes such as chimpanzees and gorillas, humans walk on two feet. This is known as bipedalism. Evidence from a fossil skeleton nicknamed 'Lucy', which belongs to the extinct hominid species *Australopithecus afarensis*, suggests bipedalism evolved approximately 3.5 million years ago (mya). Scientists are not certain why bipedalism evolved. However, one advantage of walking on two feet is endurance – being able to travel long distances. This might have helped early humans to hunt. Bipedalism also means our hands are free, allowing tool use to develop.

Brain size

Brain size has approximately tripled over the course of human evolution, as you can see in Table 1. Larger brains allow more complex information to be processed. Environmental challenges during human evolution, such as social interactions, tool use, and encountering new habitats produced by climate change, would all have required complex information to be processed.

▼ **Table 1** *Brain size and body mass for several species of hominids*

Species	Geological age (mya)	Brain size (cm^3)	Body mass (kg)	
			Males	Females
Australopithecus africanus	3.0–2.4	452	41	30
Australopithecus boisei	2.3–1.4	521	49	34
Homo erectus (early)	1.8–1.5	863	66	54
Homo erectus (late)	0.5–0.3	980	60	55
Homo sapiens	0.4–0.0	1350	58	49

▲ **Figure 2** *Human ancestor skulls. From left to right:* Adapis, Proconsul, Australopithecus boisei, Homo habilis, *and* Homo erectus

Behavioural adaptations in humans

Tool use

The ability of human ancestors to make stone tools would have helped them hunt and obtain high-protein, energy-rich meat. We now use a vast array of tools and technology in all aspects of our lives. The evolution of bipedalism enabled tool use because hands were no longer needed for movement. The development of tool use is very well correlated with an increase in human brain size.

Social adaptations

Primates such as chimpanzees groom each other to maintain social bonds. Some scientists think that human behaviours such as laughter, dancing, and music perform similar functions to grooming. Scientists have hypothesised that the evolution of such behaviours help humans to form and maintain much larger social groups than would be possible with grooming alone.

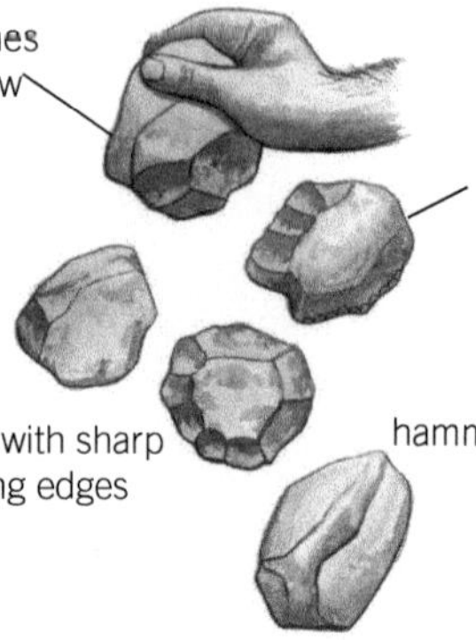

▲ **Figure 3** *Examples of early stone tools*

Physiological adaptations in humans

Lactose tolerance

Infant mammals produce the enzyme lactase. This allows them to digest lactose in their mother's milk. In most mammalian species lactase production is greatly reduced or stopped in adulthood. In some human populations, however, adults continue to produce lactase. This adaptation probably evolved when human populations began farming and obtaining milk from other animals. The ability to digest milk provides humans with a good source of calcium.

Skin pigmentation

Skin pigmentation is thought to have evolved following the loss of body hair in human populations. The production of the pigment melanin in skin cells protects individuals from potentially dangerous UV-B radiation.

Summary questions

1 State whether the following adaptations are behavioural, anatomical, or physiological: opposable digits, colour vision, laughter, sweating, a broader pelvis. *(4 marks)*

2 Examine the data in Table 1.

a Calculate the percentage increase from *A. africanus* to *H. sapiens* in brain size and body mass for both sexes. Give your answers to three significant figures.

b Calculate the rate of increase in hominid brain size over the 3 million year period (use the units cm^3 100 000 $years^{-1}$).

3 Explain the advantages of bipedalism, increased brain size, and tool use for humans, and describe the probable connections in the evolution of these three adaptations. *(4 marks)*

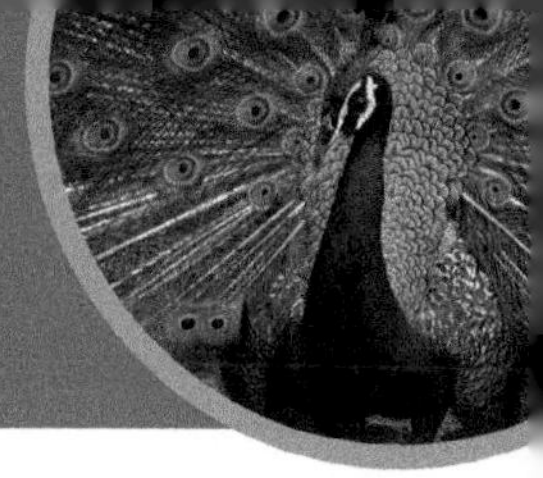

Adaptations to abiotic factors

You discussed the principles of adaptation in Topic 10.3, focusing on examples of behavioural, physiological, and anatomical adaptations in humans. Plant species also show a range of adaptations to the abiotic factors in their habitats.

Temperature

Plant species that live in cold ecosystems have evolved adaptations to maximise the amount of heat energy they receive to prevent frost damage. For example, plants in tundra ecosystems are usually low growing to reduce the risk of freezing, and dark in colour to help them absorb radiation. Some species have evolved dish-shaped flowers, which focus heat from the Sun to the centre of the flower.

Plants that live in ecosystems with high temperatures, such as deserts, have evolved to resist heat damage. For example, the desert paintbrush (*Castilleja linariifolia*) produces heat-resistant seeds that can remain dormant until conditions are suitable for germination. Just as plants in cold ecosystems are often dark, many desert plants possess light-coloured hairs and spines. These act as radiation reflectors to prevent overheating.

Water

Xerophytes are plants that have adapted to survive in habitats with a shortage of water. Two examples of xerophytes are cacti (in deserts) and marram grass.

Typical adaptations in cacti include:

- Their stomata are closed when water is limited to reduce transpiration and conserve water.
- They store water in their swollen stems.
- Their long roots spread out over a wide area to help maximise water absorption.
- Their waxy, needle-like leaves have reduced surface areas, which minimises water loss via evaporation.

A lack of water is, however, not a problem for hydrophytes, which are plants adapted for aquatic habitats. Their adaptations include floating leaves (to maximise sunlight exposure) and seeds, and stems that contain air spaces for buoyancy and are flexible so that they can move with water currents.

Light

The duration, wavelength, and intensity of light to which plants are exposed dictate their rate of photosynthesis. Aquatic plants have adapted to maximise the amount of light energy they absorb. As you

Learning outcomes

Demonstrate knowledge, understanding, and application of:

- → adaptations of plants to their environment
- → practical investigation of the responses of plants to environmental factors.

▲ **Figure 1** *These cacti have needle-like leaves with reduced surface areas to minimise water loss*

Synoptic link

Revisit Topic 6.4, Transport of water in plants if you need to revise transpiration.

Synoptic link

Some plant responses to the environment can be investigated using a potometer as discussed in Topic 6.5, Factors affecting transpiration.

▲ **Figure 2** *A hoverfly captured by a carnivorous sundew plant*

discovered earlier, hydrophytes do this by having leaves that float on the water surface to increase their exposure to sunlight. Plants that live deeper in water have evolved colours to increase their light absorption. For example, some plants (and algal species) are red or brown to absorb the blue wavelengths of light that penetrate deeper water.

Mineral concentration

Some plants live in soil types (e.g., waterlogged bogs) that lack nitrates. Carnivorous plants overcome the shortage of nitrogen-containing compounds in their soils by obtaining nitrogen from animals that they trap and digest. For example, the sundew (*Drosera rotundifolia*) attracts insects with sweet droplets secreted from its leaves. It then traps the lured insect and digests it.

While insectivorous plants obtain some nutrients from animals, they gain most of their energy, like other plant species, through photosynthesis.

Summary questions

1. Describe three ways in which plants can be adapted to ecosystems with a lack of water. *(3 marks)*
2. Suggest how the stomata on hydrophyte leaves might differ from land-based plants. Explain your suggestion. *(2 marks)*
3. Suggest why the results from a potometer experiment are not wholly representative of the transpiration rate of a plant in its natural habitat. *(3 marks)*

10.5 Evolution

Specification references: 3.1.3

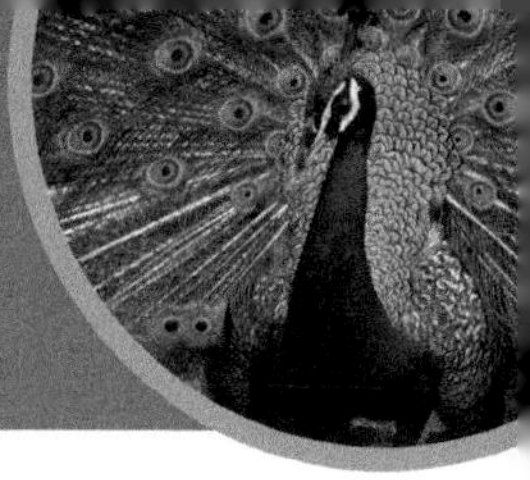

Variation is needed for evolution

In the previous topic, you looked at examples of adaptations that organisms have evolved to benefit them in their environment. You will now take a closer look at how beneficial **traits** evolve.

Although members of a species possess very similar traits, they exhibit variation. This variation can be produced by genetics or the environment. Mutations to DNA can produce genetic variation in a species. This variation can be inherited by offspring from parents. However, even when no differences in genetic information exist, as in identical twins, differences in appearance can develop. This is because the environment can produce variation in a species as well. For example, adult height will be affected by both an individual's genetics and the amount of nutrition they obtain from their environment.

For evolution to occur, genetic variation must exist in a population. Variation caused by the environment cannot be passed on to the next generation.

Learning outcomes

Demonstrate knowledge, understanding, and application of:

- → variation, adaptation, and selection as components of evolution
- → evidence for the competing theories of language evolution.

Synoptic link

You first learnt about mutations in Topic 4.3, Semi-conservative DNA replication.

Selection pressures

A **selection pressure** is an environmental factor that drives evolution in a particular direction. Members of a species must possess traits that enable them to survive the selection pressure. Examples of selection pressures include:

- the threat of predation
- diseases
- a change in food availability
- climate change
- the presence of toxic chemicals – for example, antibiotics act as a selection pressure for bacteria.

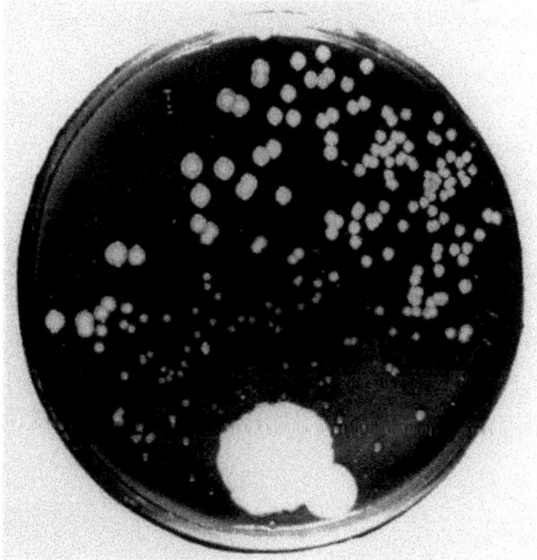

▲ **Figure 1** *The Petri dish in which Alexander Fleming showed that antibiotics can inhibit bacterial growth*

Synoptic link

You will learn more about the evolution of antibiotic resistance in bacteria in Topic 13.3, Antibiotics.

Only individuals possessing traits that enable them to overcome a selection pressure will survive and be able to reproduce. These individuals are better adapted to the environment than individuals lacking the beneficial traits. They will pass on the versions of genes (known as alleles) responsible for the beneficial characteristics that enabled their survival. Other members of the species who lack the beneficial allele will be more likely to die. The individuals that survive the selection pressure are said to have been 'selected'. The process is therefore known as **natural selection**.

Selection pressure operates over many generations. Any beneficial variation that arises will be selected, and the specie's traits will change over time to suit its environment. Often a species can be separated into isolated populations. These populations might experience different

Study tip

Remember that the definitions of evolution and speciation are different. Evolution is a gradual change in the inherited traits within a population. Speciation is the formation of a new species resulting from of the evolution of two reproductively isolated populations.

selection pressures, which can result in the evolution of different traits and the eventual formation of new species.

Sometimes new species can evolve in the same area, without geographical isolation. Groups of individuals within a population might be unable to reproduce with each other because of different behaviours (such as differences in mating season or mating rituals) or mechanical barriers (such as genitalia becoming incompatible). For example, two species of toad (*Bufo fowleri* and *Bufo americanus*) can be mated in a laboratory, but in the wild they do not mate with each other, despite their distributions overlapping. This is because *B. fowleri* mates in late summer and *B. americanus* mates in early summer.

Evidence for evolution

Two scientists, Charles Darwin and Alfred Russell Wallace, first proposed the idea of natural selection as the mechanism driving evolution. They developed the concept independently, but they discussed their ideas and both published articles in 1858 describing natural selection. Darwin published *On the Origin of Species by Means of Natural Selection* in 1859. It is this book that has become famous as the foundation of natural selection theory.

Darwin based his theory of natural selection on several observations. He noticed that members of the same species show variation and most offspring fail to survive to adulthood. This means population sizes remain fairly constant. He concluded that some individuals must be more adapted to their environment than others. These adapted individuals are able to outcompete other members of the species and survive.

Since Darwin proposed his theory of natural selection, much more supporting evidence has been found.

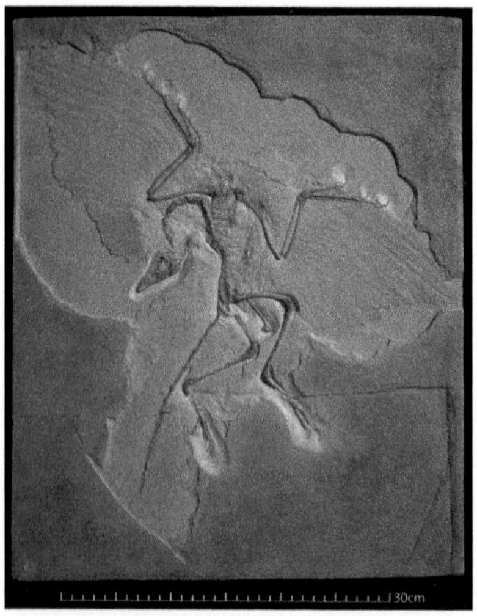

▲ **Figure 2** *An Archeopteryx fossil*

Fossils

The remains of organisms can be preserved in rocks and are known as fossils. The age of rocks, and therefore fossils, can be calculated using a technique called carbon dating. This means the analysis of fossils can show the gradual changes in appearance that occurred in species.

Some fossils indicate evolutionary links between taxa. For example, *Archeopteryx* fossils from 150 million years ago show an intermediate species in the evolution of birds from reptiles. *Archeopteryx* has very basic feather structures but otherwise resembles dinosaurs in having, for example, teeth with serrated edges and a long bony tail.

Molecular evidence

The more recently two species evolved from a common ancestor, the greater the similarity in their DNA and protein sequences. For example, similarities in the protein cytochrome c reflect the closeness of evolutionary relationships between species (Topic 10.1, Classification).

A modern example of natural selection

Deer mice (*Peromyscus maniculatus*) are widespread across the USA. Most deer mice have a dark coat. However, in an area of Nebraska known as Sand Hills, a deer mouse population has evolved lighter fur to match the sandy soils in this habitat. The remarkable part of this story is that Sand Hills only formed 8000–15 000 years ago. The evolution of the deer mouse coat happened within the space of a few thousand years.

Researchers estimate that the allele for a sandy-coloured coat arose 4000 years ago. This produced variation in fur colour within the population. The deer mice would have faced a selection pressure in the form of predation (principally from birds). The mice with sandy-coloured coats would have been better camouflaged and less likely to be eaten. These individuals would have been better adapted than those with dark fur. The mice with sandy-coloured coats would have been more likely to survive, reproduce, and pass on their alleles to offspring.

Over many generations, the frequency of the allele for sandy fur would have increased until the population comprised mainly sandy-coloured mice. The researchers calculated that the sandy coat allele gave these mice a 0.5% survival advantage. This seems small, but over thousands of generations it has a significant effect.

1 Explain why this is an example of natural selection rather than speciation.

▲ **Figure 3** *Two deer mice, one with a dark coat and one with a light coat that evolved within the last 4000 years*

The evolution of language

Language evolution is considered one of the hardest topics to research in science. Scientists want to answer a variety of questions about language. When did it evolve? How did language benefit human ancestors? Which selection pressures caused the evolution of language? Language does not leave a fossil record, however, so relatively little direct evidence exists to answer these questions.

Methods used to study language evolution

A good scientific theory should be testable. This enables evidence to be gathered to either support the theory or reject it. However, it is difficult to perform experiments to research language evolution. Scientists can instead find evidence using other methods.

Genetics

The FOXP2 gene is found in many organisms, but humans have a unique variant of the gene that might be involved in speech production and language. DNA from the fossils of human ancestors enables estimations of when mutations linked to language might have occurred. Genes linked to dyslexia and autism have been identified recently. These genes might provide insights into language development.

Comparisons with other species

Some scientists believed that the evolution of a 'descended larynx' in humans enabled language by making it easier to produce speech. It has now been found that many other species have a 'descended larynx' without having language. This suggests that the evolution of the brain is responsible for human language. Advances in neuroscience may enable comparisons with other species to provide insights into the role played by brain evolution.

Fossils

Language does not fossilise, but the fossil record suggests humans emigrated from Africa approximately 50 000 years ago. All humans have the ability to learn language. This suggests that language probably evolved before this emigration, more than 50 000 years ago.

Computer simulations

Simulations and mathematical models allow theories of language evolution to be tested.

Language evolution theories

Many different theories have been proposed to explain how language evolved. Two of the most prominent ideas are the 'gossip' theory and the 'mother tongues' theory.

The gossip/grooming theory

This theory suggests that 'vocal grooming' replaced physical grooming as a way of maintaining social bonds as human social groups became larger. Vocal grooming would initially have consisted of pleasant sounds, but not what we would think of as a language. This could then have evolved into a more complex language. Some evidence is provided by comparisons of other primate groups in which group size is correlated with brain size, and in particular the size of the neocortex (the part of the brain that controls language in humans). However, the theory struggles to explain how the simple sounds used for vocal grooming could evolve into a complex language.

Mother tongues theory

This theory proposes that words and basic language could have evolved for communication between mothers and offspring. Scientists have suggested that the words spoken by someone would need to be trustworthy for language to be able to evolve initially. It is easy to be dishonest with language. This hypothesis suggests that communication between a mother and her offspring is likely to be honest, in general. However, other scientists say that the 'mother tongues' hypothesis does not answer two important questions:

- Why have other animals not evolved language for communication between mothers and offspring?
- How did language extend to communication between non-relatives?

Summary questions

1 State three examples of a selection pressure. *(3 marks)*

2 a Explain why language evolution is a difficult topic to study. *(2 marks)*

b Describe the methods scientists can use to study language evolution. *(3 marks)*

3 We have developed chemicals called insecticides to kill undesired insect species. Explain how a species of insect could become resistant to an insecticide. *(4 marks)*

10.6 Biodiversity

Specification references: 3.1.3

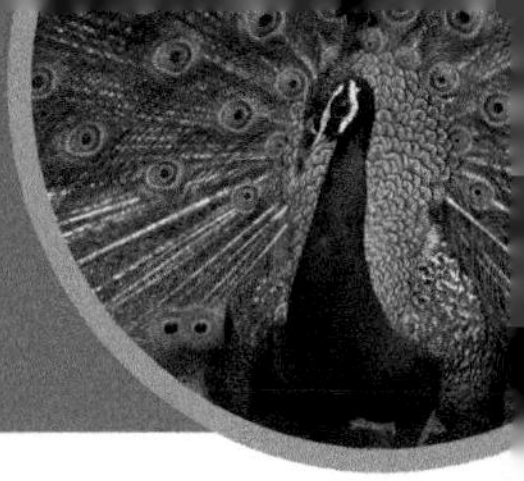

Biodiversity on several levels

Many species live on Earth today, but some regions of the world contain more species than others. Members of a species show variation, as you read in Topic 10.5, but some species exhibit more variation than others. This variation in life is known as **biodiversity**. Biodiversity can be considered on different scales. Scientists use definitions of biodiversity at the genetic, species and ecosystem level.

Learning outcomes

Demonstrate knowledge, understanding, and application of:

→ biodiversity at the genetic, species, and ecosystem level

→ the calculation of genetic diversity within populations.

Genetic diversity

Genetic diversity is a measure of the variety of genes in members of a species. Genetic diversity determines how easily a species is able to adapt to changes in its environment. As you learnt in Topic 10.5, Evolution variation is necessary for evolution to occur. A species with high genetic diversity will contain a wide range of traits on which natural selection can act. Therefore genetic diversity increases the chance of a species adapting to and surviving environmental changes. There are several ways of measuring the genetic diversity of a species. Some of these methods are discussed later in this topic.

Species diversity

Species diversity is a measure of the number of species in a habitat (species richness) and the relative abundance of individuals in each of these species (species evenness).

Ecosystem diversity

An **ecosystem** is formed from a community of species (the biotic component) and their non-living surroundings (the abiotic component such as air, water, and soil) interacting together. Ecosystem diversity measures the range of ecosystems in a particular area. For example, global ecosystem diversity is a measure of the range of different ecosystems on Earth.

Ecosystem diversity is the most complex of the three levels. It is the hardest measure of biodiversity to calculate because the boundaries of ecosystems are often difficult to determine with accuracy.

Measuring genetic diversity

Members of a species have the same genes, but they can have different alleles. A species that has a high percentage of genes with only one possible variant would have low genetic diversity. A species that has a high percentage of genes with several possible variants would have high genetic diversity.

Several methods can be used to calculate genetic diversity:

- the number of alleles per gene
- heterozygosity – the proportion of individuals in a population that have two different alleles for a particular gene

▲ **Figure 1** *A tropical rainforest ecosystem has high species diversity*

- the proportion of genes for which more than one allele exists. A gene that has two or more possible variants/alleles is known as a polymorphic gene. A gene for which only one variant/allele exists is called a monomorphic gene.

The formula used to calculate the proportion of polymorphic genes is:

$$\text{Proportion of polymorphic genes} = \frac{\text{Number of polymorphic genes}}{\text{Total number of genes}}$$

You should note that the 'total number of genes' used in the calculation tends to be a small sample of genes rather than every gene in a specie's genome.

Synoptic link

You will have learnt about the genetic code in 4.4, The genetic code.

Worked example: Calculating genetic diversity

Table 1 shows genetic data for three populations of monkeys.

▼ **Table 1** *Genetic data for three populations of monkeys*

	Blue monkey pop'n	Vervet monkey pop'n 1	Vervet monkey pop'n 2
Sample size (number of individuals)	93	124	364
Number of genes studied	33	23	18
Number of polymorphic genes	5	4	3
Average heterozygosity (%)	4.6	5.6	4.0

Is it possible to conclude which of the populations has the highest genetic diversity?

Step 1 The proportions of polymorphic genes in each population are:

Blue monkeys: $\frac{5}{33} = 0.15$ (or 15%)

Vervet monkey population 1: $\frac{4}{23} = 0.17$ (or 17%)

Vervet monkey population 2: $\frac{3}{18} = 0.17$ (or 17%)

Step 2 Based on average heterozygosity, population 1 of vervet monkeys has the highest genetic diversity. However, based on the proportion of polymorphic genes, the two vervet monkey populations have the same genetic diversity.

Summary questions

1 Outline the difference between species diversity and ecosystem diversity. *(2 marks)*

2 Explain the importance of genetic diversity to species survival. *(2 marks)*

3 Copy and complete the table and discuss which species has the greatest genetic diversity. *(5 marks)*

Species	Number of monomorphic genes studied	Number of polymorphic genes studied	Percentage of polymorphic genes (%)	Average heterozygosity (%)
A	28	8		4.2
B		10	22	7.0
C	8	3		5.4

Practice questions

1 Which of the following is an example of a physiological adaptation?

A tool use

B bipedalism

C lactose tolerance

D opposable digits *(1 mark)*

2 The volume of water taken up by a plant in a potometer is calculated using this formula: volume = length of air bubble × πr^2 (where r = radius of the capillary tube).

In an experiment, the mean distance moved by the air bubble in a potometer was 1.5 cm in one minute. The capillary tube had a diameter of 1.0 mm. Which of the following is the rate of water uptake in $mm^3\ hr^{-1}$?

A 11.8

B 70.7

C 706.5

D 2826.0 *(1 mark)*

3 Which of the following statements is/are true of the evolution of language?

1 Fossil evidence can indicate when language evolved.

2 Both anatomical and neuronal adaptations were probably required for language evolution.

3 Language evolution is unlikely to be associated with particular genes.

A 1, 2 and 3

B Only 1 and 2

C Only 2 and 3

D Only 1 *(1 mark)*

4 The genetic diversity of four populations of a species was being studied. Data for the four populations are shown in the table.

Population	Number of loci studied	Number of monomorphic genes
A	45	37
B	42	38
C	20	16
D	33	22

Based on the data in the table, which population has the highest genetic diversity? *(1 mark)*

5 Which of the following statements is/are true of the possible adaptations exhibited by hydrophytes?

1 Air spaces within stems

2 Rolled leaves to trap water vapour

3 Needle-shaped leaves with reduced surface areas

A 1, 2 and 3

B Only 1 and 2

C Only 2 and 3

D Only 1 *(1 mark)*

6 Describe the evidence that can be used to classify species.

You should illustrate each type of evidence you include in your answer with examples. *(6 marks)*

7 a Define the term classification. *(2 marks)*

b (i) Suggest what criteria a taxonomist may take into account when classifying a new species. *(3 marks)*

(ii) The table shows the main taxonomic groups, but they are not in the correct order.

	Q	R	S	T	U	V	W
Taxonomic group	species	Order	class	phylum	genus	kingom	Family

Place the letters representing the taxonomic groups into the correct order. *(3 marks)*

c Describe the differences between a classification system based on domains and one based on kingdoms. *(4 marks)*

[OCR Q2 from f212 june 2010]

8 a It has been found that 98.4% of chimpanzee DNA is identical to that of a human.

(i) Suggest how the information obtained by DNA analysis can be useful to taxonomists. *(2 marks)*

(ii) State two types of evidence, other than biochemical evidence, that are used by taxonomists when classifying organisms. *(2 marks)*

b The structure of cytochrome C varies between species. However, closely related species have similar cytochrome C.

The diagram shows a possible evolutionary tree for vertebrates. Common ancestors are indicated by the number 1 and various letters.

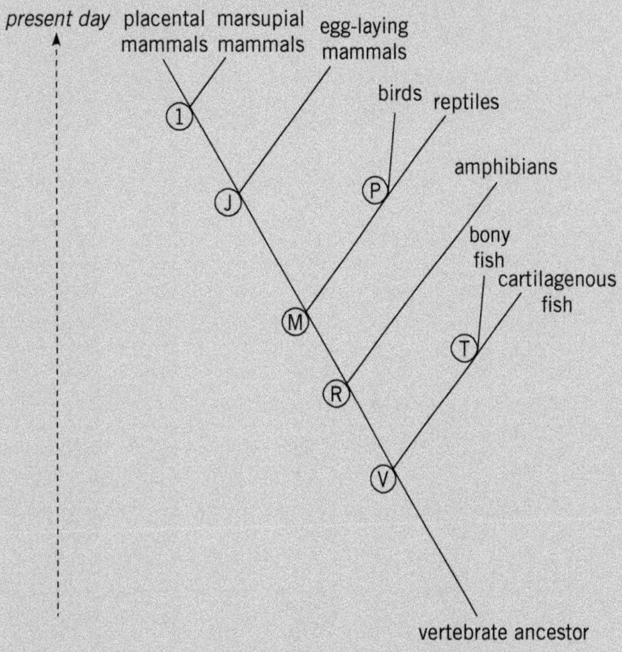

State the letter of the common ancestor that has cytochrome C which will be:

Most similar in structure to common ancestor 1

Least similar in structure to common ancestor 1 *(2 marks)*

[Question 5 OCR Jan 2011 f212]

9 a When studying the biodiversity of a habitat, a student placed their quadrat on areas they considered to have the most biodiversity.

Explain what is wrong with this technique. *(2 marks)*

b The student looked at the abundance of three plants at different distances from the bottom of the slope.

The results table drawn by the student is shown here.

	Percentage cover of each plant species		
Distance from bottom of slope	Cotton grass	Ling	Bracken
0m	76	0	0
10m	68	0	0
20m	0	2	0
30m	0	35	0
40m	0	50	0
50m	0	60	7
60m	0	40	17
70m	0	10	42
80m	0	0	68
90m	0	0	71
100m	0	0	74

The format of the table is incorrect. Suggest one way in which the student could correct the table. *(1 mark)*

[Question 7, Jan 2010 F212 OCR]

10 Bats are the only mammals that can truly fly. Many species of bat hunt flying insects at night. Bats are able to use sound waves (echolocation) in order to help them find their prey in the dark.

a Suggest how the ability to use echolocation may have evolved from an ancestor that did not have that ability. *(4 marks)*

The pipistrelle is the most common species of bat in Europe. It was originally thought that all pipistrelles belonged to the same species, *Pipistrellus pipistrellus*. However, in the 1990s, it was decided that there were two species: the common pipistrelle, *Pipistrellus pipistrellus*, and the soprano pipistrelle, *Pipistrellus pygmaeus*.

Data for both species are provided in the table.

Species	Mean body mass (g)	Mean wingspan (m)	Range of echolocation (kHz)	colour
Common pipistrelle	5.5	0.22	42–47	Medium to dark brown
Soprano pipistrelle	5.5	0.21	52–60	Medium to dark brown

b (i) Name the genus to which the soprano pipistrelle belongs. (*1 mark*)

(ii) Using the data in the table, suggest why pipistrelles were originally classified as one species. (*1 mark*)

(iii) State two pieces of molecular evidence that can be used to identify organisms as belonging to different species. (*2 marks*)

(iv) Describe how it is possible to confirm, over a longer period of time, whether two organisms belong to different species or the same species. (*2 marks*)

[Question 3 June 2012 f212 OCR]

11 The system used by scientists for classifying living things has developed from the original classification system proposed by Carl Linnaeus around 250 years ago.

a Copy and complete the following paragraph by using the most appropriate term(s).

The system of classifying organisms according to their observable features or genetic characteristics is called..... . Organisms are classified into large groups which are then subdivided into increasingly smaller groups. A system such as this is called a .... The term that describes the evolutionary relationship between organisms is .... (*3 marks*)

b New Zealand is made up of two large and many smaller islands and is situated a long distance from any other land mass. In New Zealand there is a large variety of birds not found elsewhere in the world.

Among its many species of the parrot family, Psittacidae, are:

- Kaka (*Nestor meridonalis*)
- Kea (*Nestor notablis*)
- Kakapo (*Strigops habroptila*)

These birds are shown in the photos.

(i) Name the domain to which these parrots belong. (*1 mark*)

(ii) Species that are more closely related in evolutionary terms have more genes in common than species that are less closely related.

Using the information provided, suggest the likely genetic relationship between the three parrot species. (*4 marks*)

c At some point in the past, distinct species of parrot are likely to have arisen from an original ancestral population.

State the name of the process by which new species arise and suggest the mechanisms necessary for this process to occur. (*3 marks*)

[Question 4 Jan 2012 f212 OCR]

11 DISEASES

11.1 Communicable diseases

Specification references: 3.2.1

Learning outcomes

Demonstrate knowledge, understanding, and application of:

- how pathogens cause communicable diseases
- investigating TB and HIV/AIDS
- the structure of HIV.

Disease

Disease is a very difficult thing to define, but it can be thought of as a malfunction of the body or mind, which adversely affects the health of the individual. When we talk about diseases we are really describing the symptoms of the disease that we recognise as being typical of that disease. The **pathogen** must enter the host and penetrate the first line of defence of the body – the interface with the environment. The pathogen must then begin to colonise the tissues – it is only at this point that the infection and its symptoms develop.

Pathogens and infectious diseases

There are many different types of pathogen causing a large range of infectious diseases. An **infectious** or **communicable** disease is one that can be spread from one living organism to another. Some bacteria, fungi, and viruses can act as pathogens. The ability of a pathogen to cause a disease is referred to as the virulence and this can depend on both features of the pathogen and how the body reacts to it.

Pathogens cause disease in a number of different ways:

- Producing **toxins** that damage the cells and produce many of the symptoms such as the rash, spots, and fever typical of that particular disease. For example, some species of the bacteria *Streptococcus* produce several toxins including a 'pyrogenic' toxin, which leads to the rash and high temperature associated with scarlet fever. Typically **bacterial pathogens** create infection and the disease symptoms by toxin production, once they have penetrated the body.
- Damaging the host cells and tissues by entering the cells and preventing the tissue from functioning normally. **Viruses** enter the host cell, inhibit the normal DNA, RNA, and protein synthesis of the cell and use the mechanism to produce new viral particles rather than new host proteins. Once sufficient new viruses are formed, they can either cause the cells to rupture, thereby releasing the viruses ready to infect new cells or the viruses can 'bud off' the host cell, usually taking some of the cell surface membrane with them.
- Secreting enzymes that allow the pathogen to spread through tissues. The bacteria that causes gangrene, *Clostridium perfringens*, uses this form of virulence as do the fungal pathogens that cause ringworm and athlete's foot. The itchiness associated with athlete's foot is an allergic reaction to the organism and its enzymes.

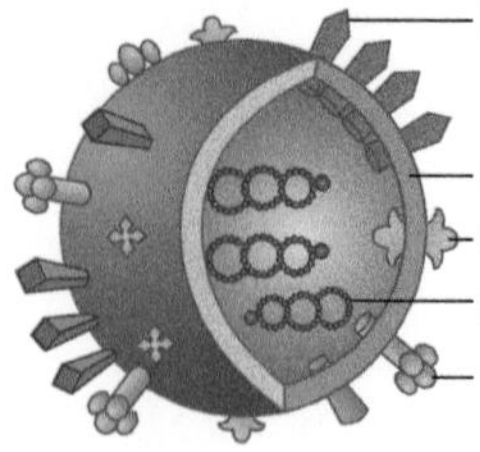

▲ **Figure 1** *The structure of the influenza virus*

Causes and treatment of tuberculosis (TB)

Tuberculosis (TB) is caused by two different species of bacterium, *Mycobacterium tuberculosis* and *Mycobacterium bovis*. TB usually infects the lungs (pulmonary TB) but it may be found in many other parts of the body, including the bones. The virulence of the bacterium is due to its ability to inhibit the action of lysosomes in the phagocytic cells that engulf it. It survives and can multiply inside these cells.

The bacterium is spread through inhaling infected droplets, but it requires long periods of close contact with someone with the disease for the infection to spread. Once caught it is most likely to infect the lungs and causes **primary TB** with fever, weight loss, and tiredness. In a healthy person the immune system may deal with the infection and it may not develop further.

If it does develop further it will enter the stage known as secondary TB, when many phagocytic and other cells accumulate round the infected cells forming a tubercle (a granuloma). The lung tissue is damaged and the sputum is bloodstained. The patient suffers chest pains and night sweats and the condition can be fatal. Infection is more likely in overcrowded conditions and damp housing with poor diet and poor health. Improvement in living conditions has gone a long way to reducing the spread of this disease.

Treatment of TB requires intensive care and extensive use of antibiotics. Pulmonary TB is treated using a six-month course of a combination of two different antibiotics. The usual course of treatment is taking two antibiotics, isoniazid and rifampicin, every day for six months with two additional antibiotics, pyrazinamide and ethambutol, taken every day for the first two months. If TB is in the tissues outside the lungs then the treatment lasts for 12 months with the two additional antibiotics, pyrazinamide and ethambutol, taken every day for the first two months.

One thing that can further complicate the treatment of TB is that the bacteria that cause TB can develop resistance to antibiotics. Multidrug-resistant tuberculosis (MDR-TB) is the term used to describe TB if the TB bacterium is resistant to two antibiotics. Extensively drug-resistant tuberculosis (XDR-TB) describes TB that is resistant to three or more antibiotics. Both these types of resistance require at least 18 months of treatment with four different antibiotics. In all cases a TB Treatment Team is assigned to each patient as part of their care.

Latent TB is where there are no symptoms although the patient is infected. In this case treatment is not recommended for the over 35-year-old age group due to the harm that could result from the drug regime needed, but in these cases monitoring is required to determine whether the infection begins to become active.

Causes and treatment of HIV/AIDS

HIV/AIDS is an acquired immunodeficiency disease caused by human immunodeficiency virus. The virus is a **retrovirus**, which means its genetic information is in the form of RNA not DNA. The

Study tip

Remember it is the pathogen not the disease that is transmitted.

Synoptic link

See Topic 12.1, The immune system to find out more about the action of phagocytic cells in the immune system.

Study tip

Whenever you are asked to name an organism, the full Latin name must be given (e.g., *Mycobacterium tuberculosis*) but when answering a question it is acceptable to use the full name once and then the abbreviated form (e.g., *M. tuberculosis*).

Synoptic link

Transcription is an important part of the incorporation of a retrovirus into the host cells. You first encountered transcription in Topic 4.6, Protein synthesis – transcription.

virus also contains an enzyme, reverse transcriptase, which will create the double-stranded DNA copy of the viral genome once inside the host T helper cell. This viral DNA copy then becomes part of the DNA of the host cell. At this stage the virus is called a provirus. The provirus will be copied before the cell divides but the cell remains normal.

Once the viral DNA becomes active, viral RNA and proteins are synthesised by the cell and more viruses are produced while the T helper cell is destroyed. This prevents the immune system resisting infections within the body. The result is an ever increasing susceptibility to other infections and diseases as the virus destroys more and more T helper cells. These infections are referred to as opportunistic infections, and may include candidiasis (a yeast infection), *Pneumocystis carinii* pneumonia, TB, and cancers such as Kaposi's sarcoma, which are not usually virulent in humans but become so due to the compromised immune system.

There is no cure for HIV/AIDS although antiretroviral treatments can limit the reproduction of the virus and the availability of these and the use of antibiotics can prolong life extensively.

HIV is transmitted only by direct contact with the body fluids of an infected person. It is not able to survive outside the body so will not be transmitted by objects such as cups and handles. Transmissions are known to occur by unprotected sex with an infected person, through infected blood products such as factor VIII, which is given to people with haemophilia (the inherited condition that affects blood clotting), through contaminated needles, or between a baby and infected mother during pregnancy or breast feeding.

The most effective method of managing the spread of HIV/AIDS at the moment is prevention and so a program of education has been embarked upon which includes:

- encouraging the use of condoms and protected sex
- a program of needle exchange to reduce needle sharing in drug users
- screening blood for HIV before use in transfusions or for factor VIII transfusions
- being tested if in an 'at risk' group
- discouraging infected mothers from breastfeeding.

The structure of the HIV virus

Viruses are *acellular particles*, that is, they are not made of cells, and as such they do not carry out any of the normal life processes:

- they cannot reproduce on their own, do not grow, or undergo division
- they do not transform energy
- they lack the machinery for protein synthesis.

Study tip

Be aware of the basic principles of the cell theory. Be prepared to evaluate statements such as 'viruses should not be considered as organisms.'

This has led to the idea that viruses are highly specialised parasites that have lost many of their own functions. However, some scientists believe that viruses are unique structures in their own right.

Viruses all have a central core of genetic material, either DNA or RNA, surrounded by a protein coat called the **capsid** which protects the genetic material within it. The capsid is made up of protein units called **capsomeres**, which may be surrounded by a lipid and protein membrane.

The HIV virus has a spherical shape, a central core of two RNA strands and an enzyme, **reverse transcriptase** within the capsid.

The capsid is shaped like a cone and is surrounded by a lipid and glycoprotein membrane forming the envelope. The glycoproteins form peg-like structures, which bind to receptors on the cell surface membrane of the T helper lymphocyte.

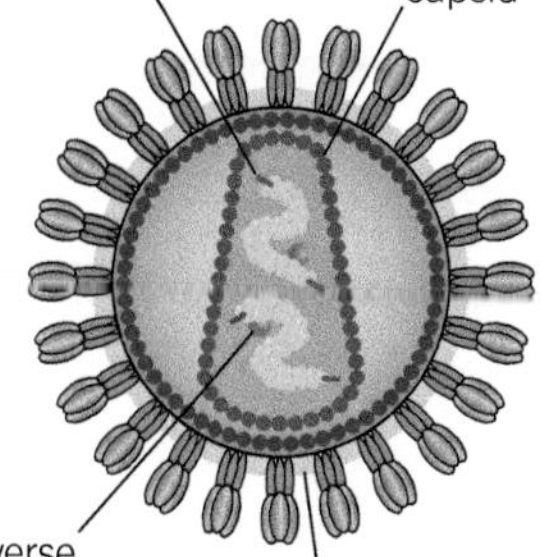

▲ **Figure 2** *The structure of the HIV virus*

▼ **Table 1** *Comparison of TB and HIV/AIDS*

Pathogen	Disease	Features	Symptoms	Transmission
Mycobacterium tuberculosis bacterium	TB	May be present in the lungs but dormant as it is controlled by the immune system, becomes active in weakened patients, destroying lung tissue and infecting after organs.	In the primary infection fever and weight loss, but secondary TB gives a persistent cough, chest pains, blood in sputum, night sweats, and fatigue.	Spreads by droplet infection with close contact.
Human immunodeficiency virus	HIV/AIDS	Infects and enters the T helper cells and may lie dormant for up to 10 years. AIDS develops once the virus is active and then compromises the immune system.	Loss of T helper cells so immunity becomes weakened. Infections such as TB and unusual cancers develop which cause the weight loss and weakness that is typical.	Spreads only by direct contact with bodily fluids.

Summary questions

1 Explain why people living in deprivation, such as refugees, are more likely to:
 a become infected with TB *(1 mark)*
 b develop secondary TB. *(1 mark)*

2 Suggest why the long course of treatment required for TB make it more likely that drug-resistant strains will develop. *(2 marks)*

3 Outline the reasons why some biologists do not consider viruses to be living organisms. *(4 marks)*

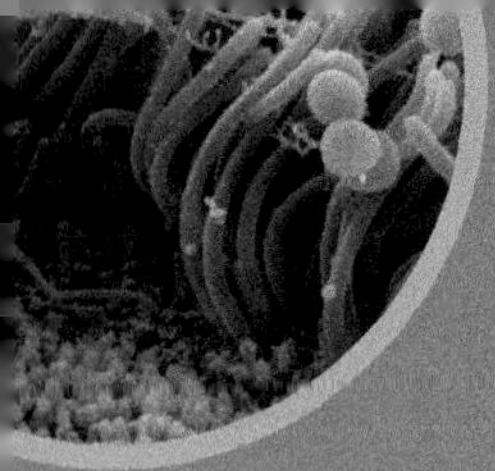

11.2 Identifying and culturing bacteria

Specification references: 3.2.1

Learning outcomes

Demonstrate knowledge, understanding, and application of:

→ the classification of bacteria using Gram staining, cell and colony morphology

→ the culturing of bacteria.

▲ **Figure 1** *Bacteria being cultured in a nutrient broth*

Culturing bacteria

Bacteria can be cultured using either nutrient broth or solid media such as nutrient agar. In both cases the culture medium is prepared using **aseptic techniques** to ensure there is no contamination. The culture medium provides the necessary nutrients required for the bacterial colonies to grow and survive.

- Use sterile containers. For solid cultures use a Petri dish, whereas for nutrient broth use any suitable sealed container of the required size. The broth is usually sterilised in the container.
- For Petri dishes, add sterile nutrient medium using aseptic techniques and store in a refrigerator until needed.
- Inoculate the medium with the bacteria then seal it and incubate at a suitable temperature.
- Use aseptic techniques to dispose of all apparatus used and safely dispose of the cultures after use.

Identifying bacteria using Gram staining

A traditional method of classifying bacteria is to use a stain called **Gram stain**, which contains the chemical crystal violet. Gram staining splits bacteria into two types – Gram-positive and Gram-negative. **Gram-negative bacteria** will stain pink, while **Gram-positive bacteria** will stain purple in the presence of the Gram stain.

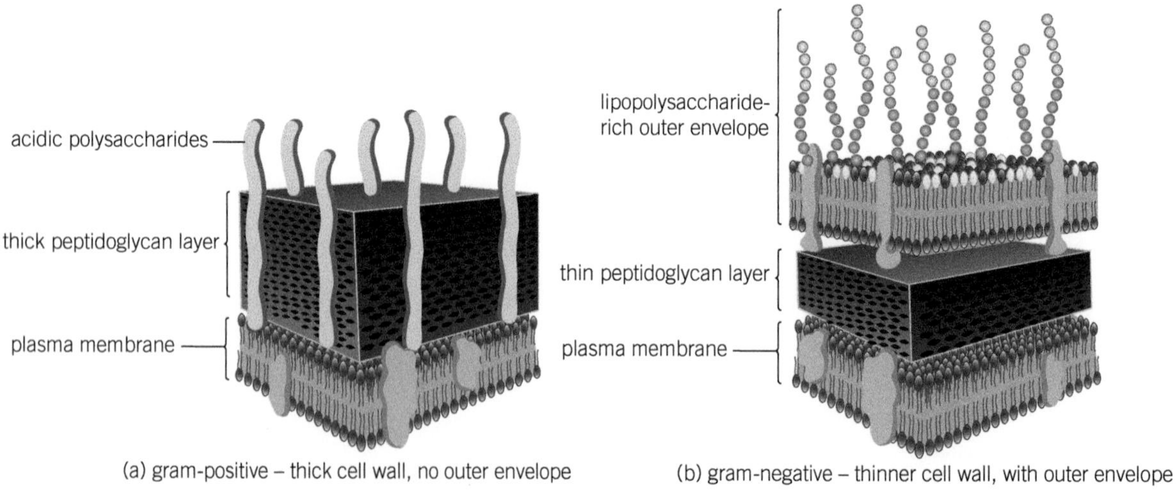

▲ **Figure 2** *Staining in Gram-positive and Gram-negative bacterial cell walls*

Gram staining

1 Prepared bacterial slides are stained with crystal violet for 30 seconds before washing briefly with water.
2 The slide is then flooded with Gram's iodine to bring the crystal violet-iodine complex to the peptidoglycan wall.
3 After one minute the slide is rinsed with water followed by alternate washes of 95% alcohol and water for 30 seconds until no further colour loss can be seen.
4 A counterstain is now used to colour any of the cells that have none of the crystal violet colour left after the washing. A counterstain such as safranin is added and left for 30 seconds. This stains any cells that have not retained the crystal violet.
5 The slide is then blotted, dried, and observed under the microscope. The differences in colour will determine whether Gram-negative or Gram-positive bacteria are present.

Bacteriological loops can be used to spread a small sample of a liquid culture of bacteria onto a slide for gram staining.

What treatment should be carried out a on the loop b on the slide with the culture on prior to gram staining?

The differences in colour are a result of how the stain is taken up by the cells, which is determined by the thickness of the cell wall and the presence or absence of an outer layer:

- Gram-positive bacteria have a thick peptidoglycan wall (consisting of amino acids and sugars) outside the plasma membrane. This thick wall takes up the crystal violet and binds to it. When washed with alcohol, the alcohol does not wash any of this thick layer away and the violet stain remains, leaving the cells with their characteristic purple colour. The counterstain, safranin, will have no effect on the violet colour.
- In Gram-negative bacteria the peptidoglycan wall is much thinner and is covered by a lipopolysaccharide layer. During Gram-staining the crystal violet binds but the alcohol added later washes away the outer layer and so any crystal violet stain is leached out. When the safranin stain is added, the remaining thin peptidoglycan layer becomes stained with this counterstain, giving the characteristic pink colour.

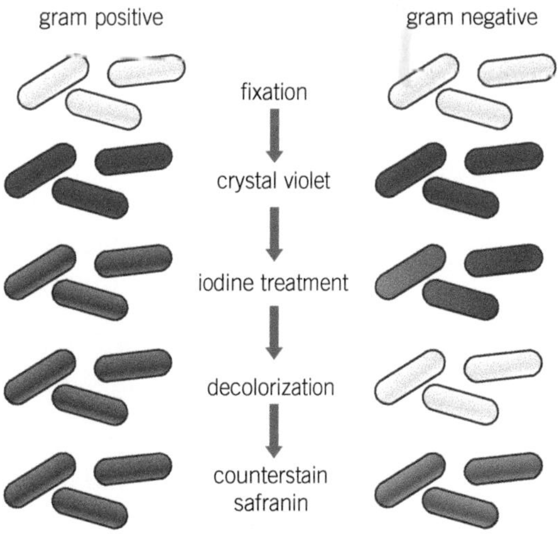

▲ Figure 3 *Flow diagram showing the Gram-staining colours*

Synoptic link

The structure of prokaryotic cells is covered fully in Topic 1.6, Cell ultrastructure.

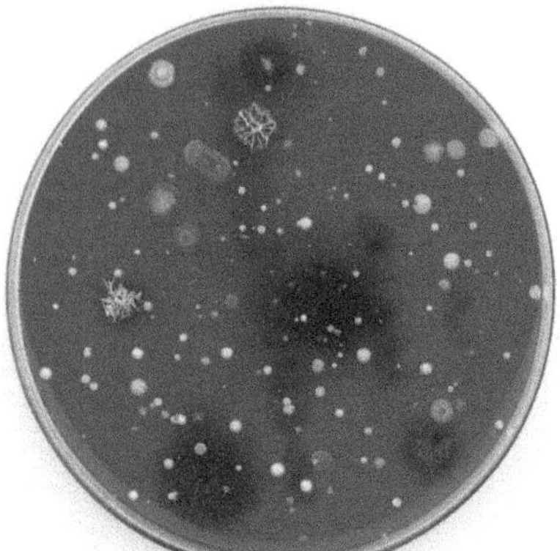

▲ **Figure 4** *Bacterial colonies from a mixed culture*

▲ **Figure 5** *Scanning electron micrograph showing spherical* Staphylococcus aureus *bacteria (yellow) on human nasal epithelial cells, ×7 000 magnification*

Identifying bacteria using colony morphology

In a sample of bacteria there will usually be many different species of bacteria. Usually the first step is to identify them based on the appearance of their cells and colonies. The bacterial **colony** is a small spot growing on a culture medium. The spot is actually composed of millions of identical bacteria that arose through binary fission from a single bacterial cell.

The shape, colour, edges, and the surface features of the colony are all characteristics that differ between different types of bacteria and can be used in identification. These characteristics are called **colony morphology**. Some colonies may be round and smooth, whilst others may be wrinkled with a wavy edge. Many are creamy in colour but there are also grey, yellow, red, or orange colonies.

Although identifying different colonies based on their appearance is a good first step in identification, it must be remembered that many different types of bacteria have similar colony morphology and so this is not a reliable identification method.

Colony morphology is also useful when checking to see if plates of pure culture have become contaminated. If the colonies in a culture are similar in appearance, then it is likely the plate is not contaminated. However, if there are some colonies that have different characteristics from the others then it is likely these colonies have arisen from contamination.

Bacterial cell morphology

Cell morphology relies on the appearance of individual bacterial cells. There are three most commonly recognised types – cocci, bacilli, and spirilla.

- Cocci, such as *Staphylococcus aureus*, are spherical in shape.
- Bacilli, such as *Mycobacterium tuberculosis*, are rod shaped but with a wide variation in length, with some so short it is difficult to differentiate between them and cocci, whilst others are long and thin.
- Spirilla are shaped like a corkscrew, with the number of spirals considerably different from one species to another.

Summary questions

1. Explain why the counterstain used in the Gram staining technique will have no effect on Gram-positive bacteria *(2 marks)*
2. Explain why it is difficult to identify bacteria by the shape of their cells when using a light microscope in a school laboratory. *(2 marks)*
3. In identifying bacteria, a number of methods are used. Describe the different techniques used and state the advantages and disadvantages of each method. *(6 marks)*

11.3 Epidemiology

Specification references: 3.2.1

Epidemiology

Epidemiology is the study of the patterns, distribution, and causes of diseases in a population. Monitoring how and where a disease spreads can help to predict further outbreaks and prepare control measures to minimise the spread of an outbreak. Investigating geographical distribution allows the origin of an outbreak to be determined. Focusing all the control measures available on the origin of an outbreak will help to limit the spread. The data collected during monitoring will include **morbidity** data (people that have the disease) and **mortality** data (people that have died from the disease).

If an infectious disease is always present in a population it is **endemic**. However, when there is a sudden increase in the incidence of an infectious disease in an area, there is said to be an **epidemic**. If that increase is measured in several countries then the disease is said to be a **pandemic**.

Some endemic diseases may vary in terms of how many people have the disease at any one time – this is the **prevalence** of the disease. The number of new cases in a population each year is the **incidence rate** of the disease.

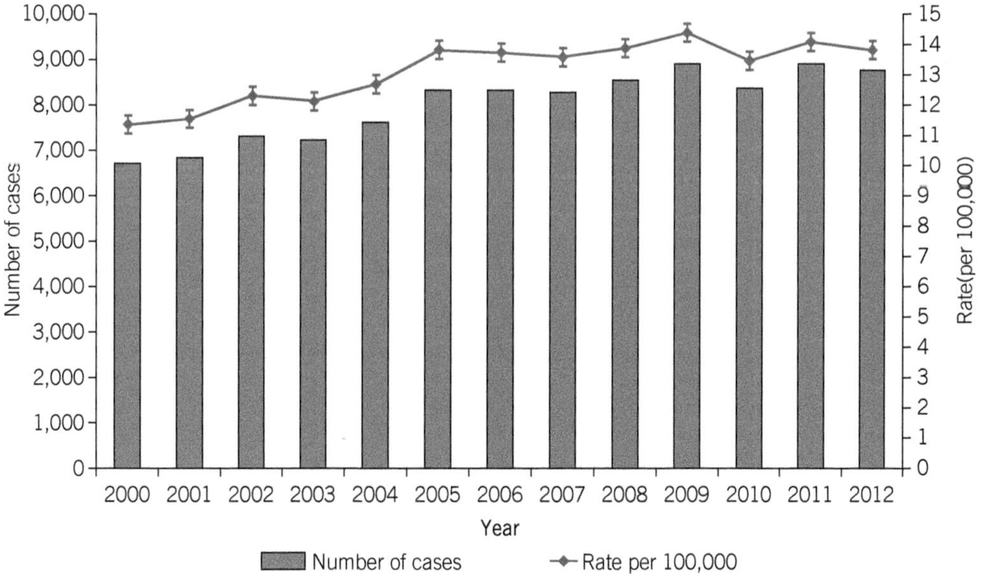

▲ **Figure 1** *The graph shows the number of cases and prevalence rate of TB in the UK between 2000 and 2012*

Learning outcomes

Demonstrate knowledge, understanding, and application of:

- → epidemiology
- → changes in the incidence and prevalence of diseases
- → controlling and preventing diseases.

Study tip

Incidence and prevalence numbers can vary between different sized populations so it is more usual to use incidence rates and prevalence rates to compare different countries or areas.

Study tip

When answering questions on epidemiology data it is important to read figures accurately from graphs. Read the graph to the nearest grid line and quote units correctly from both axes.

Evaluating data to assess the impact of a disease

The table shows some data on the global distribution of both TB and HIV. Study the table and then answer the questions that follow.

Global distribution of HIV (prevalence %)		Global distribution of TB deaths per 100 000	
< 0.1	Middle East and North East Africa	< 2.5	North America, Canada, Australia, and New Zealand
0.1–0.5	Europe incl. UK, Canada, Panama	2.5–10.0	Europe incl. UK, Libya
0.5–1.0	USA, South America, India and North Africa	10.0–35.0	Argentina and Eastern Europe, Middle East, South Africa, Mexico, and Peru
1.0–5.0	Central Africa, Russia and South East Asia	35.0–70.0	West Africa and Central South America, India, and South East Asia
5.0–15.0	East Africa	70.0–100.0	Russia and China, all of Indonesian islands, and Philippines
15.0–34.0	South Africa	>100.0	Peru, South, and Central Africa

1 Look at the epidemiological data for HIV/AIDS and TB. Suggest why the prevalence and mortality rates for the two diseases is higher in some countries than others.

2 Describe what sort of graph or chart you would use to display the data in the table. Explain your answer.

Control measures

There are many different methods for controlling diseases, including vaccination programs, provision of drugs, and education of the population. Improving hygiene, housing, and nutrition will also be highly beneficial and one vital improvement has been the clean water campaign which aims to provide good access to uncontaminated water to all populations. The control of population movements such as isolating affected individuals or preventing access to infected areas may also be necessary in the case of severe infections, or in highly contagious diseases such as Ebola in West Africa.

Examples of infectious diseases

Chicken pox

Chicken pox is a childhood disease that is endemic to the UK, which means that it is always present at some level within the population. It occurs throughout the year, but it is most common in winter and spring. It is a communicable disease but generally it does not cause a severe reaction if it is contracted during childhood. However, there is an increase in the incidence of first infections in older age groups, which is a major concern since the symptoms and reactions in an adult contracting the virus for the first time are more severe.

The prevalence of chicken pox may vary from year to year since in some years there are fewer infected children than in other years. The incidence in those years would also be low, but the basic principle is that it is always present in the population.

Influenza

Influenza is a common disease in the UK, particularly in winter. This means there is a widespread prevalence across the country in most years. However, in some years there is an epidemic, with many new cases being reported. In those years the incidence of the disease is high. In some years the epidemic is restricted to one part of the country and so the incidence rate in that part of the country for influenza will be higher than in other parts.

There are many different strains of influenza as the virus varies in the antigens present on its surface. These differences are a result of mutations. It is the high mutation rate in influenza that causes the different strains to appear. When a new strain appears there is no 'herd immunity' in the population so many people succumb to the infection leading to an epidemic.

Study tip

Herd immunity is the term used to describe how many people in a population have immunity to that pathogen. If herd immunity is high, epidemics are less likely to happen as someone with the disease is more likely to encounter people who are already immune.

H1N1 influenza (swine flu)

Some influenza viruses have a worldwide impact (a pandemic), which is almost certainly a result of a completely new strain developing.

One such pandemic was caused by the influenza A virus H1N1. H and N refer to the haemoglobin and neuraminidase glycoproteins present on the outer surface of the virus. Some strains of H1N1 are endemic in humans and cause some of the seasonal flu infections, whilst other strains of H1N1 caused swine influenza, which actually infected both pigs and birds.

The World Health Organisation declared a new strain of H1N1 a pandemic in June 2009. It was named swine flu and by the start of 2010 it had caused 17 000 deaths worldwide. This type of outbreak only happens when a new strain occurs after mutation. The H1N1 swine flu virus had elements of four different flu viruses and so formed a strain that has not been encountered before. Swine flu antigens are now included in the seasonal flu vaccine offered in the UK.

1 Neuraminidase is responsible for the release of virus particles from the cell. Suggest how neuraminidase inhibitors might be used in the management of H1N1 influenza A.

Severe acute respiratory syndrome (SARS)

SARS or severe acute respiratory syndrome is a highly contagious, serious, and potentially life-threatening form of pneumonia caused by the SARS coronavirus. It is thought to kill one in 10 people infected. It originated in the Guangdong province of southern China in 2002, where it caused an epidemic. The epidemic was contained within South China for almost four months with little spread to other areas, even to neighbouring Hong Kong.

However in 2003, it quickly spread from Southern China to other countries causing a pandemic. The virus resulted in more than 8000 cases and 774 deaths before it was eventually brought under control.

Synoptic link

See Topic 13.1, Controlling communicable diseases which looks more at vaccination and Topic 13.3, Antibiotics to find out more about how they contribute to controlling disease.

Study tip

If a question asks for data to be manipulated then it is expected that you will calculate the size of the increase or decrease (i.e., a 50% increase or the 'rate has doubled').

Notifiable diseases

Doctors and medical practitioners have a duty to inform their local authority of any cases of certain infectious diseases. The local authority in turn must inform Public Health England every week. This weekly data is collated, analysed and published to inform national trends and ensure relevant vaccines are available as required or follow-up measures and investigations are carried out. Chicken pox is not a notifiable disease but many childhood diseases, such as measles, mumps, whooping cough, and rubella are. Viral hepatitis, TB, poliomyelitis, anthrax, malaria, cholera, and food poisoning are also notifiable diseases and will all be treated by the authorities in this way to avoid being spread to others in the population and potentially causing fatalities.

Summary questions

1 Explain why the prevalence rate of a disease in two different areas might be the same but the actual number of cases is different. *(2 marks)*

2 Use the data from the bar chart in Figure 1 to describe the changes in the incidence rate of TB between 2000 and 2012. In your answer you should use manipulated data. *(4 marks)*

3 a State the properties of the new strain of H1N1 that made it ideal for creating a pandemic. *(2 marks)*

b Explain what occurred to ensure the disease reverted to the normal seasonal impact of flu. *(2 mark)*

Practice questions

1 Which of the following statements describes the typical mechanism of bacterial pathogenicity?

A Integration of genetic material into host cell chromosomes.

B Production of toxins.

C Enzyme secretion.

D Activation of host cell genes. *(1 mark)*

2 A nation has a population of 3.4×10^7 people.

In one year, this nation had 2.0×10^5 cases of tuberculosis (TB), of which 40,000 were new cases.

What was the incidence rate (per 100,000) of TB in this nation?

A 118

B 588

C 1176

D 5882 *(1 mark)*

3 The same nation in question **2** had 6.8×10^3 deaths from TB over a one year period.

What was the mortality rate (per 100,000) from TB in this nation?

A 2

B 20

C 200

D 2000 *(1 mark)*

4 Which of the following statements is/are true of notifiable diseases?

Statement 1: these diseases must be reported by doctors to their Local Authority.

Statement 2: the Disease Prevention Agency collates the information on notifiable diseases.

Statement 3: only endemic diseases are included on the list of notifiable diseases.

A 1, 2 and 3

B Only 1 and 2

C Only 2 and 3

D Only 1

5 a State what is meant by the following terms:

(i) endemic *(2 marks)*

(ii) epidemic *(2 marks)*

(iii) pandemic *(1 mark)*

b The World Health Organisation (WHO) coordinates a Global Influenza Programme. The WHO has published guidelines in the form of recommended actions that may be needed to combat the threat of a new influenza pandemic arising from H5N1 virus. This virus is also known as 'Avian Flu' and 'Bird Flu'.

Part of the guidelines will be aimed at reducing morbidity and mortality.

The table shows data for human cases of H5N1 virus in six countries between 2003 and 2007.

Country	Number of cases	Number of deaths
China	25	18
Egypt	38	15
Nigeria	1	1
Thailand	25	17
Turkey	12	4
Vietnam	109	46

(i) What is meant by morbidity and mortality? *(2 marks)*

(ii) Suggest why the information given in the table would not allow a valid comparison of morbidity or mortality rates between different countries. *(3 marks)*

c Suggest three ways in which a country could act to reduce the risk of an H5N1 epidemic developing. *(3 marks)*

[Question 7, OCR June 2009 F222]

6 The diagram here shows the HIV virus.

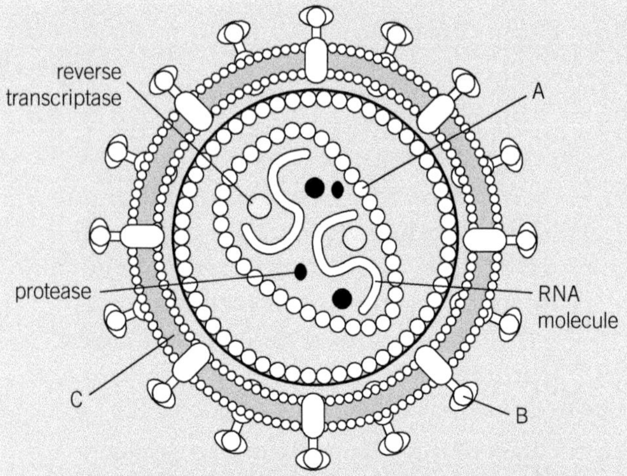

a Identify the structures labelled A, B, and C (*3 marks*)

b Name two features visible in the diagram that identify HIV as a retrovirus. (*2 marks*)

c AIDS is the acronym for Acquired immunodeficiency syndrome.

Suggest what is meant by a syndrome. Credit will be given for the use of examples that show AIDS is a syndrome. (*3 marks*)

d Some people have suggested that HIV might not be the cause of AIDS.

However, at the Durban AIDS conference in South Africa in 2000, the Durban Declaration was signed stating that there is clear-cut evidence that AIDS is caused by the HIV virus.

Some of the evidence is summarised here.

1 If not treated, most people with HIV infection show signs of AIDS within 5–10 years.

2 HIV infection is identified in blood by several reliable tests.

3 People who receive HIV-contaminated blood or blood products develop AIDS, whereas those who receive untainted or screened blood do not.

4 Most children who develop AIDS are born to HIV-infected mothers. The higher the virus count in the mother, the greater the risk of the child becoming infected.

5 In the laboratory, HIV infects the type of leucocyte that reduces in number in people with AIDS.

6 Drugs that block HIV replication in the laboratory also reduced virus numbers in people and delay the onset of aids.

7 Treatment, where available, has reduced AIDS mortality by more than 80%.

Describe the transmission of HIV and discuss the possible reasons for the AIDS pandemic.

In your answer you should refer to the information you have been given from the Durban Declaration. (*6 marks*)

[Question 6, OCR June 2010 F222]

7 HIV and *Mycobacterium tuberculosis* are two pathogens that cause infections that contribute significantly to human disease.

Shown here are two electron micrographs of HIV and *Mycobacterium tuberculosis*.

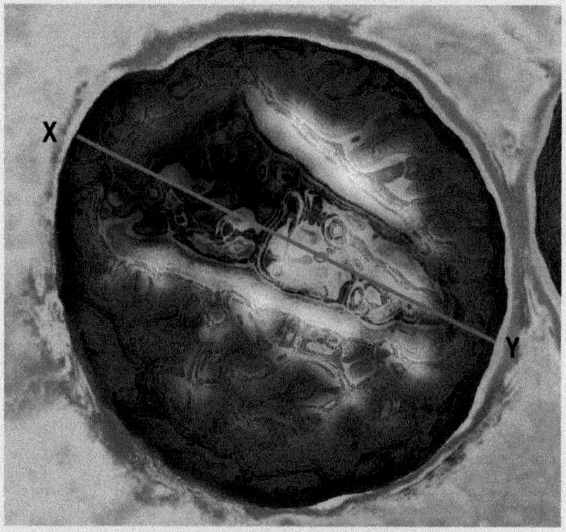

× 500 000

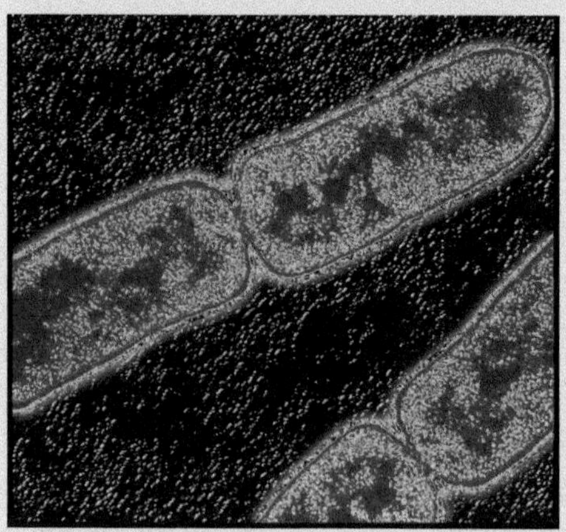

× 50 000

a (i) Calculate the actual diameter (in μm) of the HIV shown in the micrograph between points X and Y. Show your working. Give your answer to two decimal places. (*2 marks*)

(ii) Copy and complete the table by writing each structure listed below into the appropriate column.

Outer membrane; peptidoglycan cell wall; capsid; ribosomes; enzymes; DNA; RNA

Structure found		
Only in HIV	Only in *Mycobacterium tuberculosis*	In **both** *Mycobacterium tuberculosis* and HIV

(*7 marks*)

[Question 4, OCR Jan 2012 F222]

8 In 2009, approximately 1.7 million people worldwide died from tuberculosis (TB). This figure includes 380,000 people with HIV infection.

a Describe how TB is transmitted from one person to another. (*2 marks*)

b One of the symptoms of TB is fatigue. State two other symptoms of TB. (*2 marks*)

c The graph shows the estimated mean percentage change in the incidence of TB for different groups of countries, from 1997 to 2006.

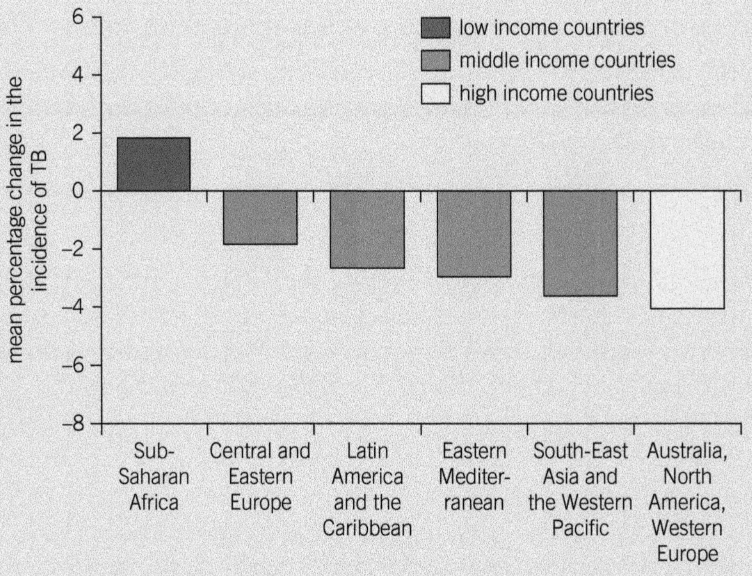

(i) Suggest why the mean percentage change in the incidence of TB is given as an estimate and not a direct measurement. (*1 mark*)

(ii) What conclusions can be drawn about the changes in incidence of TB from the information in the figure? (*3 marks*)

[Question 3, OCR Jan 2013 F222]

12 IMMUNITY

12.1 The immune system

Specification references: 3.2.2

Learning outcomes

Demonstrate knowledge, understanding, and application of:

- → primary defences against pathogens
- → the action of phagocytes
- → the role of B and T lymphocytes
- → long-term immunity and the role of memory cells.

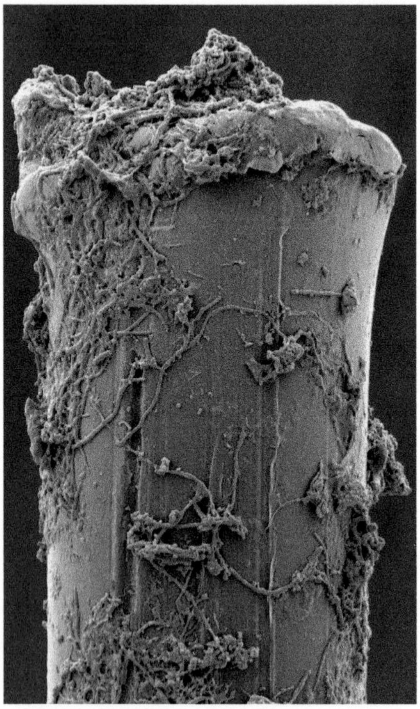

▲ **Figure 2** *Microbial film on a bristle from a tooth brush, ×600 magnification*

Defence mechanisms

We are constantly exposed to the bacteria, fungal spores, viruses, and parasites in our environment, some of which are potentially pathogenic. For example, there is a microbial film that covers most surfaces we interact with in our lives, such as door handles, stair rails, seats, and eating implements.

In order to remain free of disease, the body uses a range of different defence mechanisms to prevent infection.

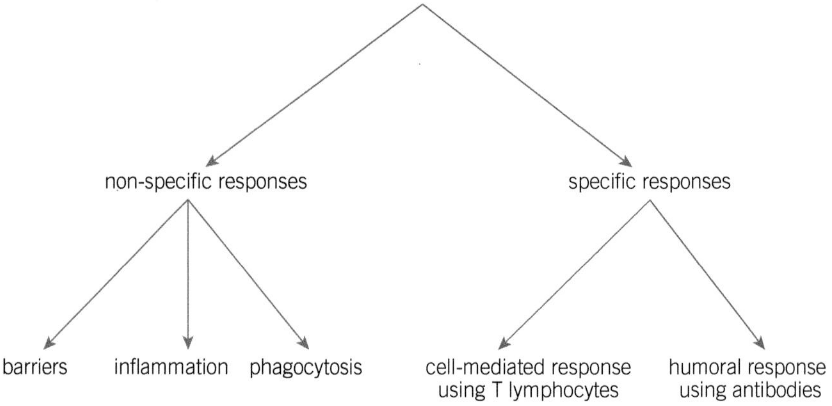

▲ **Figure 1** *The defence mechanisms of the body*

Non-specific defences

The initial or **primary response** to infection by a pathogen is called the **non-specific response**, because the response is the same regardless of what organism is attacking the body or whether it is the first or second time of infection.

Some defences are physical barriers, which actually prevent entry of pathogens:

- The skin is a protective layer covering the surface of the body. The outer layer is made of compacted cells impregnated with keratin (an insoluble protein), which prevents entry of pathogens. A break in the skin is a potential risk but the blood clotting mechanism seals the wound quickly to prevent pathogens entering as well as restricting blood loss.
- The membrane covering the eye, the conjunctiva, is a thin layer protected by secretion from the tear ducts containing an enzyme called lysozyme. This digests bacterial cell walls, destroying the bacteria and keeping the eye free from infections.

- The ciliated epithelial lining of the airways is a protective layer, further protected by a mucus layer secreted by it and covering it. The ciliated cells waft the mucus and any trapped particles and pathogens in the mucus up and out of the airways to the throat, where it is swallowed.
- The stomach lining is also protected by a mucus layer and glands in the wall secrete hydrochloric acid. This acid provides the correct pH for the stomach enzymes to function, but also destroys many of the pathogens that are swallowed or that enter with the food.
- The lining of the vagina is also protected by mucus and an acidic pH. The acidic pH reduces the ability of bacteria to survive since their enzymes do not function at low pH values.

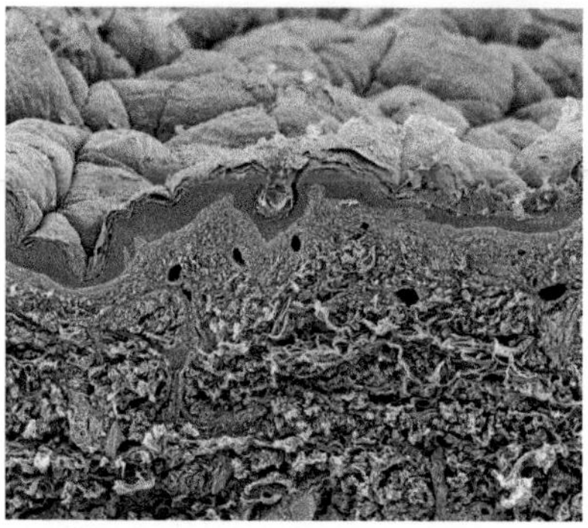

▲ **Figure 3** *The human skin acts as a barrier to the entry of pathogens. The outermost layers are made of dead cells and are tough and water-proofed with keratin*

Inflammation and phagocytosis

If the pathogens breach the barriers and actually enter the body, further non-specific responses will be triggered. One of these responses is the **inflammatory response**. The other non-specific response involves phagocytic cells.

Synoptic link

Look at Topic 3.8, Enzymes in medicine to review how aspirin reduces the symptoms of inflammation.

The inflammatory response

In the inflammatory response, damage from pathogens results in some of the local cells – the mast cells – releasing a range of compounds including histamine, serotonin, and prostaglandins. These dilate arterioles and make the walls of the capillaries in that area more permeable, so phagocytic white blood cells can leave the blood more easily. They also increase the blood flow to the area and increase the pain sensitivity. The result is soreness, swelling, and increased temperature of the area (redness), which are the signs of inflammation.

Synoptic link

You will recall from Topic 2.6, Osmosis in cells that the role of the cell wall in plants is to prevent the cells from bursting when the solution around the cell has a higher water potential than the cytoplasm. You should be able to use this principle to explain why bacterial cells are destroyed if lysozyme weakens the wall.

Phagocytosis

The phagocytic cells in the blood are called neutrophils, and they are effective against microorganisms. The sequence of events in **phagocytosis** is as follows:

1 Pathogens release chemicals and the damaged cells release **cytokines**, which attract the phagocytes.

2 The pathogen becomes attached to receptors on the cell surface membrane of the phagocyte either directly or by 'linking' molecules. Some of these molecules (the **opsonins**) are produced as antibodies by other white blood cells whilst others (complement) are found naturally in the blood plasma.

3 The phagocytes surround the pathogen with a vesicle called a **phagosome**.

4 **Lysosomes** containing hydrolytic enzymes (e.g., lysozyme) fuse with the phagosome. The enzymes digest and destroy the bacteria within the vesicle. Eventually the harmless products are released into the cytoplasm.

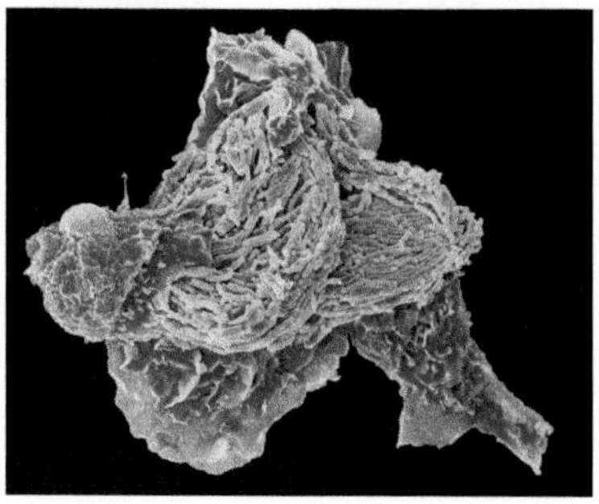

▲ **Figure 4** *Coloured scanning electron micrograph of a phagocyte engulfing a TB bacterium (green), ×1275 magnification*

Monocytes are white blood cells that differentiate to form macrophages. These are large phagocytic cells found in tissues. These release more chemicals that attack bacterial cells, inhibit viral replication, and attract more macrophages. They engulf and destroy pathogens and also damaged cells that are undergoing apoptosis (Topic 8.3, Apoptosis).

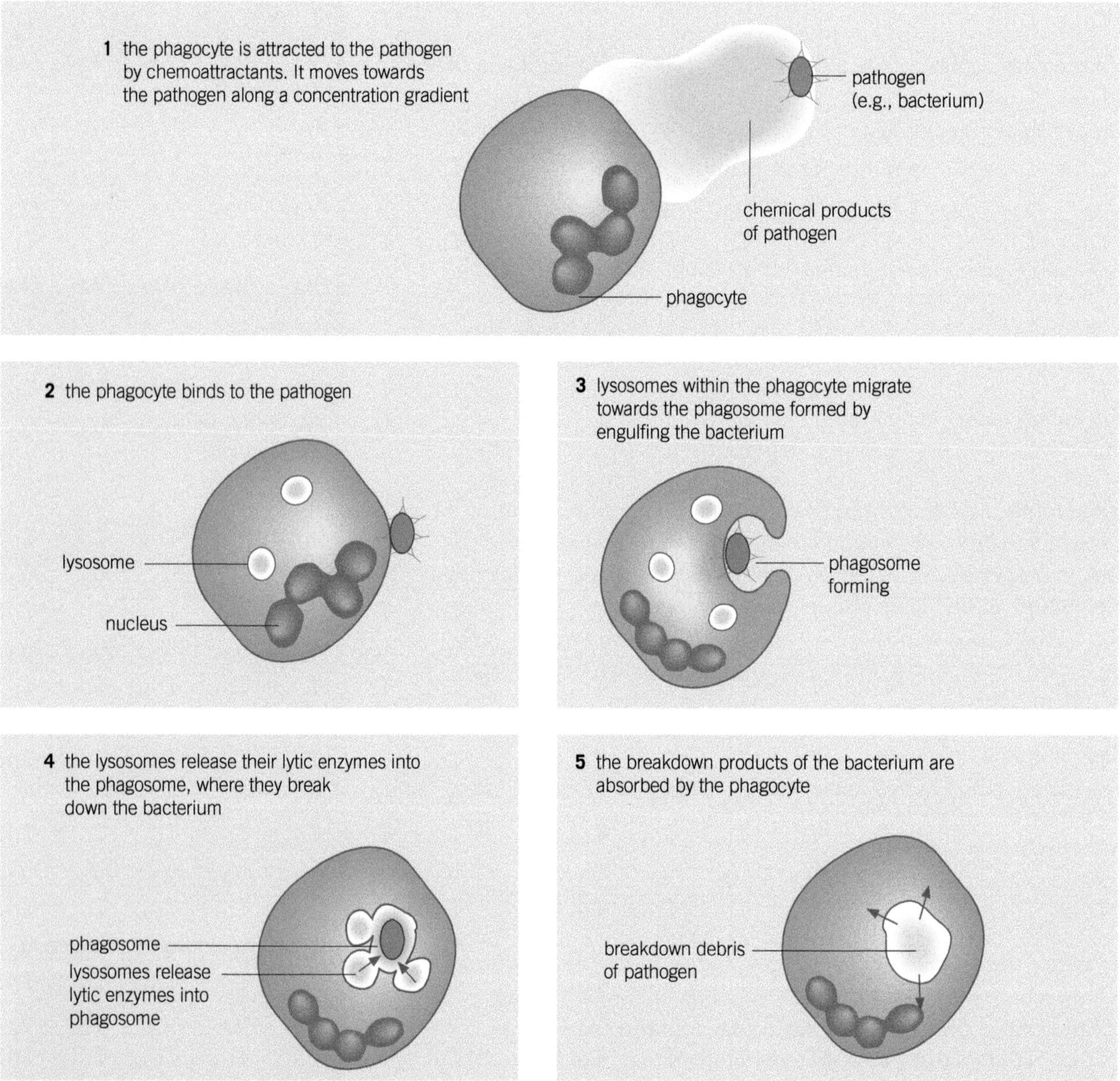
1 the phagocyte is attracted to the pathogen by chemoattractants. It moves towards the pathogen along a concentration gradient
pathogen (e.g., bacterium)
chemical products of pathogen
phagocyte
2 the phagocyte binds to the pathogen
lysosome
nucleus
3 lysosomes within the phagocyte migrate towards the phagosome formed by engulfing the bacterium
phagosome forming
4 the lysosomes release their lytic enzymes into the phagosome, where they break down the bacterium
phagosome
lysosomes release lytic enzymes into phagosome
5 the breakdown products of the bacterium are absorbed by the phagocyte
breakdown debris of pathogen

▲ **Figure 5** *Summary of phagocytosis, showing phagocytes engulfing bacteria, digesting and destroying them*

The specific immune response

The **specific immune response** targets specific types of pathogens invading the body. The response to one pathogen will not provide protection to a different pathogen. The response is triggered by antigens on the outer surfaces of organisms and can distinguish between foreign 'non-self' antigens and the body's own cells with 'self' antigens on the cell surface membrane. Apart from in rare cases of autoimmune diseases, self cells do not trigger an immune response whereas non-self cells are targeted and destroyed. Phagocytes are part of this response, since they are involved in the destruction of the targeted cells. This ability to recognise self cells probably develops in the womb and in the very early years of an individual.

The cells of the specific immune response are the T and B lymphocytes. These are produced in the bone marrow but then 'processed' in the **thymus** (T cells) and the bone marrow (B cells). During processing, lymphocytes acquire receptors on their cell surface membrane. Any cells with receptors exactly complementary to the body's own antigens are destroyed at this processing stage.

Role and action of the T and B lymphocytes

When a pathogen enters the body, it is engulfed by a type of macrophage known as a dendritic cell. This digests the pathogen and displays antigens from the pathogen on its cell surface membrane – becoming an 'antigen presenting cell'. Circulating dendritic cells eventually locate a type of T lymphocyte called a T helper cell – there will be many of these cells but only a few will have the specific receptor that is complementary to this antigen. The macrophage activates the T helper cell causing it to divide rapidly by mitosis to form a clone. The T cells differentiate and different T cells now carry out a number of functions. T helper cells use cytokines to stimulate specific B lymphocytes.

The B cells selected have receptors that are complementary to the pathogen antigens. The selected B cells divide rapidly and repeatedly by mitosis. They differentiate to form plasma cells, which synthesise and secrete **antibodies** with a complementary binding site to the pathogen's antigens (you will learn more about antibodies in Topic 12.2, Antibodies). Some of the B cells differentiate into B memory cells, which circulate for years in the body and provide the immunological memory.

The selection and multiplication of specific T and B lymphocytes described above is known as **clonal selection** and **clonal expansion**.

T lymphocytes also differentiate to produce memory cells. Other T lymphocytes become cytotoxic or T killer cells. They can destroy any cells infected by the pathogen by producing a protein that makes holes in the cell surface membrane of the infected cell. The holes ensure the cell membrane is completely permeable and so the cell will die as a result. Some of the T lymphocytes become **T regulatory cells**. These were previously known as T suppressor cells and as the name suggests they regulate the immune response by maintaining tolerance to self antigens to prevent autoimmune diseases and suppress other T cells, especially after the elimination of the invading organisms. Other T lymphocytes use cytokines to stimulate phagocytic cells to engulf extracellular pathogens.

Synoptic link

Note that the different blood groups result from different antigens on the surface of the red blood cells which trigger an immune response by the phagocytes. You learnt about blood groups in Topic 3.9, Blood donation.

Study tip

Retroviruses such as HIV avoid the cell-mediated response because their RNA is converted into DNA. This becomes incorporated into the host DNA, forming a provirus, so the cell does not appear damaged.

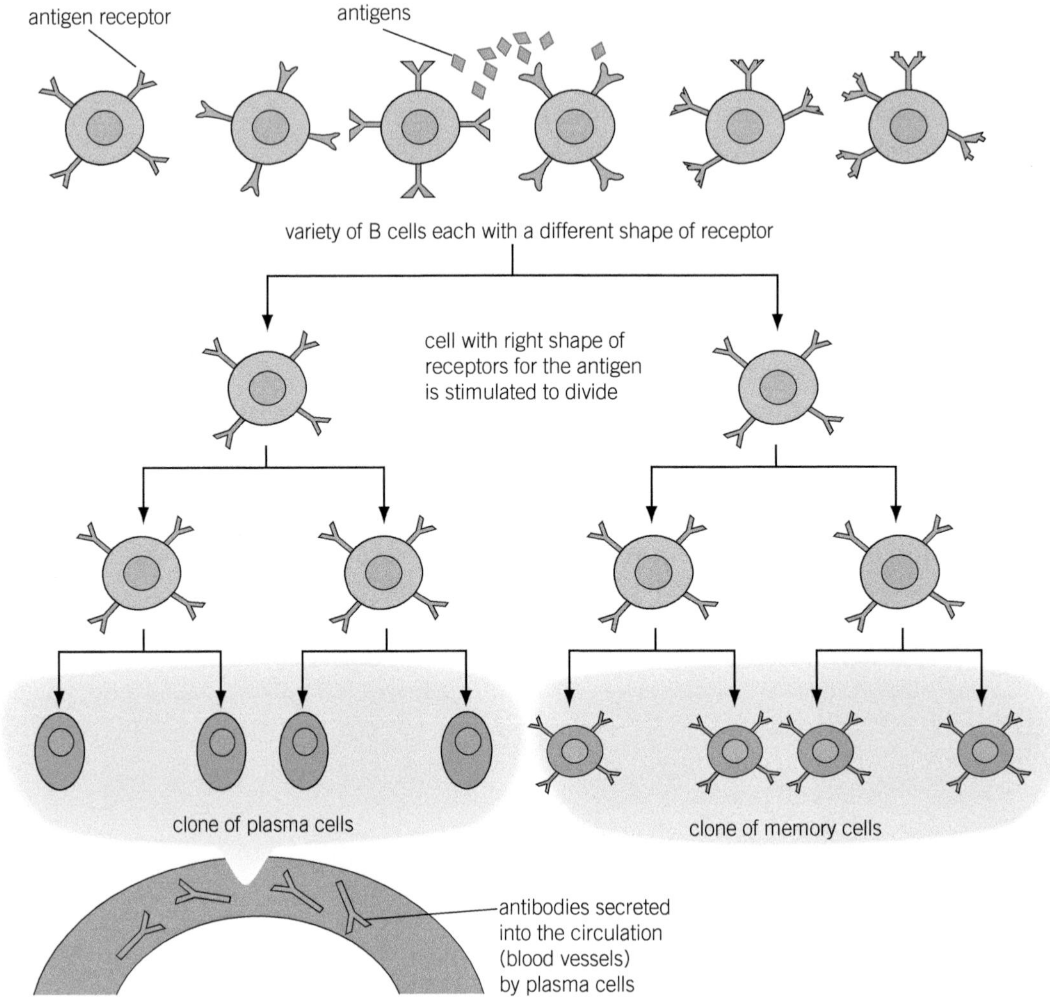

▲ **Figure 6** *How B lymphocytes respond to a foreign antigen*

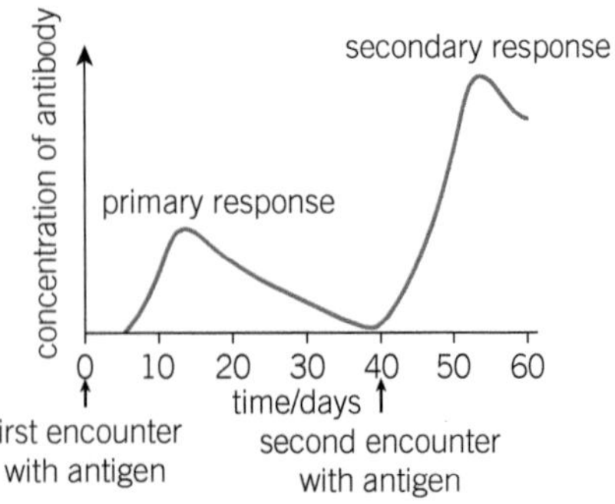

▲ **Figure 7** *Graph of primary and secondary immune response*

Role of memory cells in long-term immunity

Memory B lymphocytes (and memory T lymphocytes) can survive for decades. If a pathogen with the same antigens is encountered again, antigen presentation will be quicker as there will be more of that specific T cell and B cell now in the circulation. Clonal selection and clonal expansion will be more rapid as it is the B memory cells that divide and differentiate to produce more plasma cells. Specific antibodies will be produced faster and in much greater concentrations. The advantage of this is that, on this second encounter, the very rapid response to the pathogen will destroy the organism before it can reproduce and cause the symptoms of the disease. This is the **secondary response**.

Other roles of the T lymphocytes

T lymphocytes attack an organism's own cells that have become infected with other material, for example body cells invaded by viruses – since some of the antigens are presented on the body cell surface and cancer cells – because they present different antigens on their cell surface membranes.

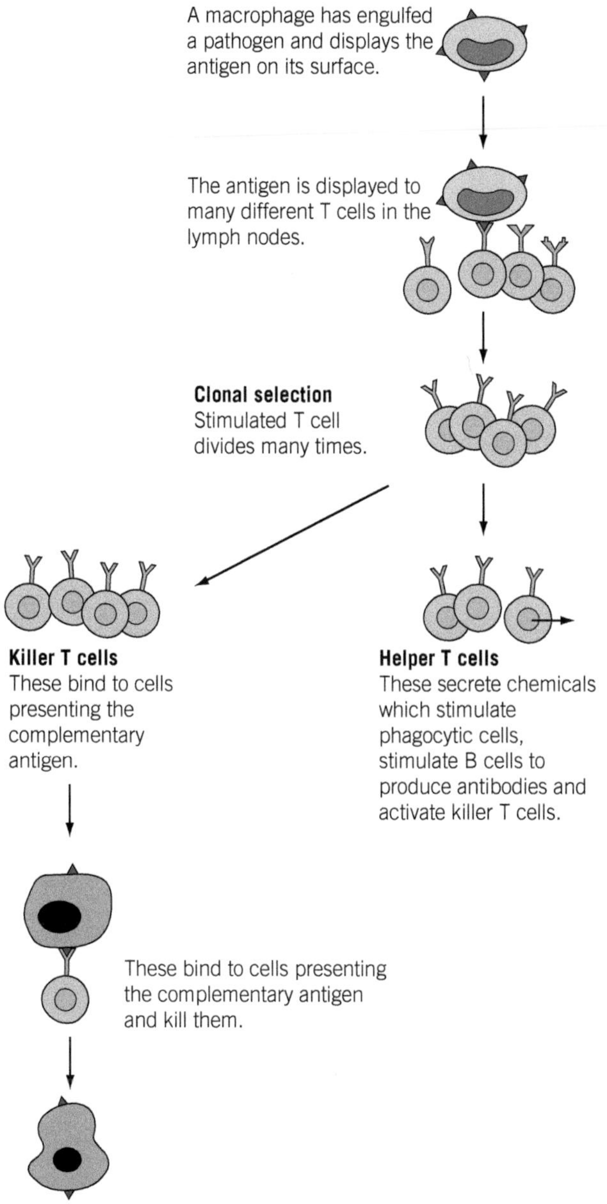

A macrophage has engulfed a pathogen and displays the antigen on its surface.
The antigen is displayed to many different T cells in the lymph nodes.
Clonal selection
Stimulated T cell divides many times.
Killer T cells
These bind to cells presenting the complementary antigen.
Helper T cells
These secrete chemicals which stimulate phagocytic cells, stimulate B cells to produce antibodies and activate killer T cells.
These bind to cells presenting the complementary antigen and kill them.

▲ **Figure 9** *Summary diagram showing the role of B cells and T cells*

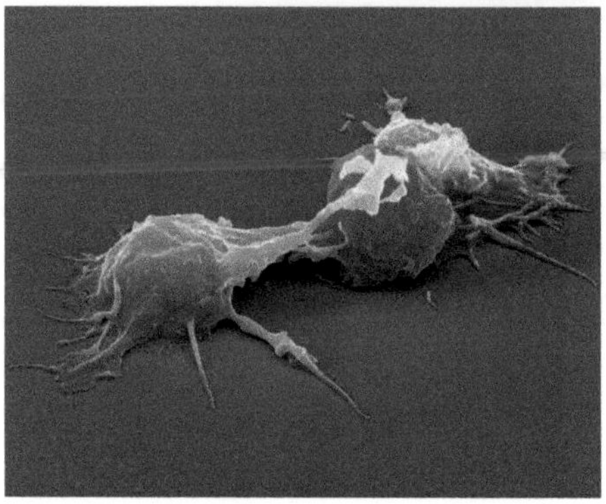

▲ **Figure 8** *False colour SEM of two killer cells (yellow) attacking a cancer cell (red), ×2 800 magnification*

Summary questions

1 Explain what an antigen is and what role it plays in disease and the immune response. *(2 marks)*

2 Describe the role of the B lymphocytes in immunity and explain how the role of the T lymphocytes differs. *(2 marks)*

3 Explain why a person with HIV/AIDS may not exhibit a secondary immune response – even if they have encountered the pathogen previously. *(2 marks)*

12.2 Antibodies and immunity

Specification references: 3.2.2

Learning outcomes

Demonstrate knowledge, understanding, and application of:

- → how antibodies function
- → how individuals can be tested for TB and HIV infection
- → natural and artificial immunity
- → hypersensitivity.

Study tip

Antibodies are synthesised and modified in plasma cells. This explains why plasma cells are much larger and have many more organelles than an undifferentiated B lymphocyte.

Antibodies

Antibodies are soluble glycoprotein molecules or **immunoglobulins** and are produced by plasma cells in response to a specific antigen.

They are Y-shaped molecules with a lower section – the constant region – that is the same for all antibodies that belong to the same class. The variable region is at the end of the two arms. This region is a unique antigen binding site, which is different in all types of antibodies. The region is complementary to the specific antigens of a pathogen. It is the sequence of amino acids in this region that varies between specific antibodies. This in turn varies the tertiary structure giving the variable region its unique three-dimensional shape.

Each antibody consists of four polypeptide chains that are held to each other by disulfide bonds. Most antibodies have a chain of sugar molecules (oligosaccharides) attached to the constant region.

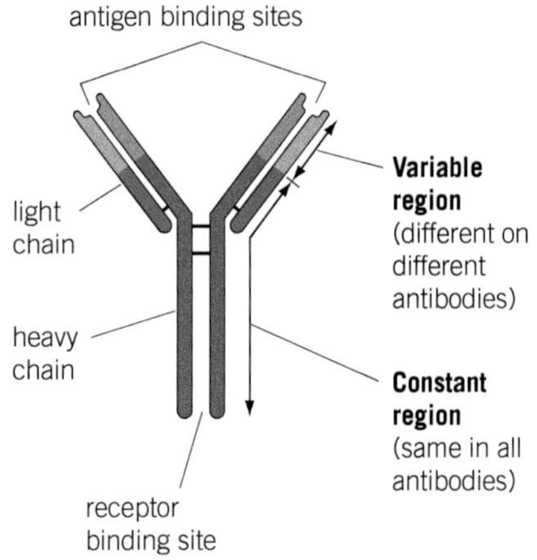

▲ **Figure 1** *Structure of an antibody*

Antibodies destroy pathogens and use a variety of methods to do so:

- agglutination – the antibodies immobilise the pathogens by clumping the bacterial cells together. This stops them entering cells and makes phagocytosis easier
- precipitation of soluble antigens, so again the phagocytes can engulf them easily
- lysis – the antibodies bind to foreign cells and attract *complement* – a collection of proteins in the plasma which, when bound to the antibody, punch 'pores' in the cell surface membrane and destroy the cell
- neutralisation of the toxins produced by bacterial cells so they can do no harm. This is important as the toxins cause many of the symptoms of a disease such as the rashes and fevers

- opsonins – the constant region of the antibodies attach to receptors on the plasma membrane of the phagocyte, while the variable region attaches to antigens on the pathogen or damaged cell, allowing it to be engulfed.

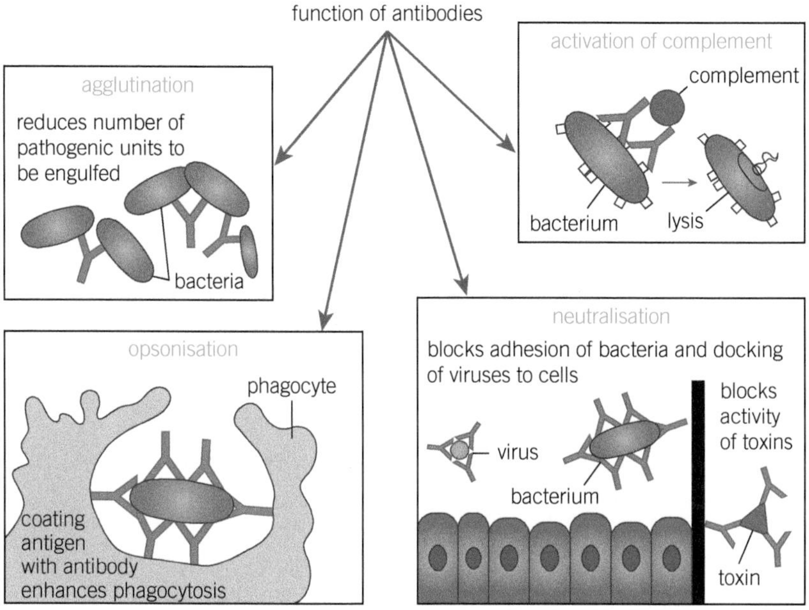

▲ **Figure 2** *Four different methods used by antibodies to destroy bacteria*

Testing for TB and HIV antibodies

A simple blood test can establish if antibodies are present in the blood of a patient for a number of serious diseases. This may be done before any injections are given, since if the antibodies are already present, a vaccination would be unnecessary. It may be done to establish if the patient has had contact with the disease and has become infected.

Before the TB vaccine is given a skin test is performed to see if immunity to TB is already present. This test is called the **Mantoux test**, where a substance called tuberculin is injected just below the skin. This is an extract from *Mycobacterium tuberculosis* containing antigens from the bacterium, but not the actual bacterium itself. If antibodies to the antigen are present, then a skin reaction of inflammation and redness develops around the site of the test. The size of the inflammed area is measured. If the reaction is strong the person is already immune or may have active TB. If the reaction is weak or there is no reaction at all then a TB vaccination is needed.

Screening people who may have contracted HIV can establish if the patient has become infected and allow early treatment. In addition, counselling and education can help prevent the spread of the disease to others.

Measuring antibody concentration using ELISA

The enzyme-linked immunosorbent assay test (ELISA) is used to detect the presence of antibodies to particular pathogens. It can only be used once an immune response has developed and antibodies have been produced.

In the test, capture molecules to a specific antibody are fixed to a surface. Once the test sample is exposed to the fixed capture molecule, any target antibodies present bind to the capture molecules. The next step is to add free capture molecules linked to an enzyme, which will bind to the target antibody if it is present. The complete solution is rinsed, which will wash away the free enzyme-capture molecules if there are no target molecules to bind to. This gives the negative result. If the enzyme-capture molecules do bind to target antibody present they will not be washed away. Substrate molecules are now added and the enzyme will change the colour of the substrate if it is present. The change in colour indicates a positive result. This test can also be used to measure antigens when the capture molecule used is a complementary antibody. This test is outlined in Figure 3.

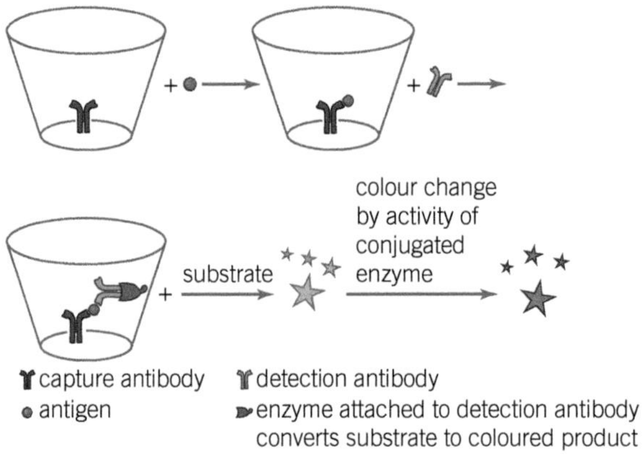

▲ **Figure 3** *The steps in a positive ELISA test for an antigen*

The change in colour can be measured using optical density with a darker colour giving a higher optical density. The antigen concentration and optical density can be used to plot a calibration graph which may be used to read off the antigen concentration in an unknown sample.

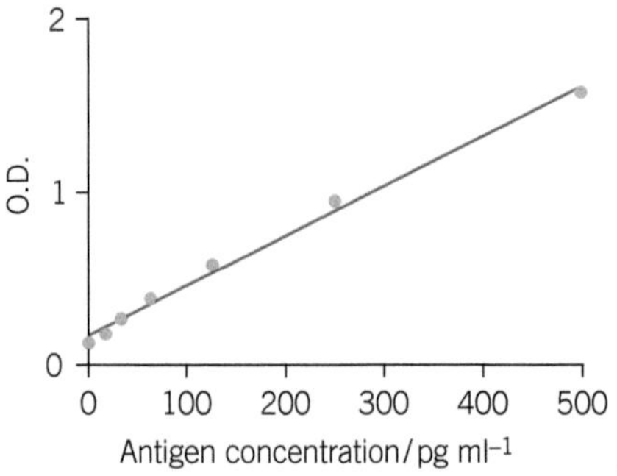

◀ **Figure 4** *A calibration curve for antigen concentration*

1 How would the standard curve be used?
2 Use the curve to determine the concentration of antigen present at an optical density of 1.0.

There are two types of HIV tests recommended by Public Health England in the United Kingdom. These are known as 4th generation tests and test for both HIV antibodies and a specific antigen known as p24. A rapid 'Point of Care' test uses a finger-prick or mouth swab sample. Positive results can be detected between 11 and 28 days after infection and the test takes between 20 and 40 minutes, but there are some issues with the sensitivity. The second method requires a blood sample and again detects the presence of both antibodies and antigens. Again, positive results can be detected within 11–28 days. The results are returned 2–14 days after the test. The test is considered very accurate with fewer false positives and negatives. Where a person is deemed 'at risk' of contracting HIV, a follow-up test is carried out three months after a negative result. Up to three different tests are required before a diagnosis is confirmed.

Study tip

You should know how to evaluate the different tests available for HIV and make clear the advantages and disadvantages of each type of test.

Active and passive immunity

When an individual becomes ill, the immune response to the pathogen activates antibody production and memory cells (B lymphocytes and T lymphocytes) and so immunity to any future invasion of that pathogen results. This is natural **active immunity**. However, a vaccination may provide the same immune response as the antigen. Vaccinations result in artificial active immunity.

In **passive immunity** there is no antibody production by the person themself and so no memory cells are produced. Antibodies provide a temporary immunity but will not protect long term against future infections of this pathogen. Table 1 shows a summary of the different types of immunity.

Synoptic link

In Topic 13.1, Controlling communicable diseases you will find out more about vaccinations.

Synoptic link

Antibodies to the rhesus antigen can result in difficulties in pregnancy if the baby is rhesus positive, but the mother is rhesus negative. You considered this in Topic 9.2, Pregnancy and fetal development.

Table 1 *A summary of the different types of immunity*

Types	Active immunity	Passive immunity
Natural	This type of immunity is from antibody production following infection. The person may become ill and develop symptoms but thereafter is protected from further infection because of the production of memory cells.	Maternal antibodies cross the placenta to the fetus in the uterus or one passed to the baby in breast milk – particularly in colostrum produced in the first few days of lactation. This immunity lasts for the first few months of life, until the baby's own immune system develops.
Artificial	A vaccine of weakened, attenuated, or dead pathogens or a preparation of antigens triggers an immune response in the person and antibodies are produced as a result. Memory cells are also produced so further long-term immunity is conferred.	An injection of antibodies from another source (e.g., serum containing the antibodies). This provides temporary immunity but the antibodies are eventually broken down. One example is tetanus immunoglobulin injections which are advised when there is a risk of tetanus but tetanus vaccinations are not up to date.

Allergies

An **allergic reaction** occurs when the immune system responds inappropriately to an antigen. Antigens which trigger such a response are termed **allergens**. Allergens trigger the inflammatory response even though they are non-pathogenic. Pollen, animal hair, or dust mites may all trigger allergic reactions. These reactions are similar to those made to an invading pathogen and can be serious or even life threatening.

1. Initial contact with the allergen.
2. The allergen triggers a primary immune response, causing IgE antibodies to be produced.
3. The antibodies bind to receptors on mast cells in the tissues lining the airways – this is the sensitisation phase and as yet no symptoms will appear.
4. On the next encounter with the allergen, allergen molecules bind to the variable region of the IgE antibodies attached to the mast cell.
5. The binding of the allergen causes the mast cell to release histamine by exocytosis and the inflammatory response is triggered.

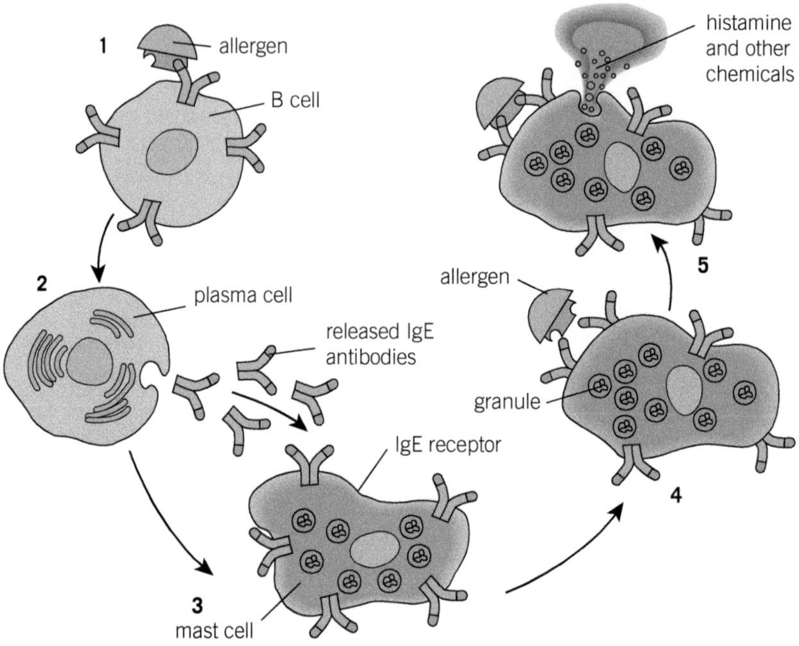

▲ **Figure 5** *The stages in the development of an allergic reaction*

Hypersensitive reaction

A **hypersensitive reaction** is any undesirable reaction produced by the immune system, including allergies and autoimmune responses. These reactions are at least uncomfortable, may be damaging or on occasions fatal. This type of reaction only occurs when the host is already sensitised to the allergen.

Summary questions

1. Explain why a person reacting positively to the Mantoux test for TB does not need to be vaccinated. *(1 mark)*
2. Explain why antibodies produced against one strain of flu virus would not be effective against another flu strain. *(2 marks)*
3. Suggest why the vaccination programme in infants does not begin until the child is approximately two months old. *(2 marks)*

Practice questions

1 Which of the following statements is/are true of a typical allergic response?

Statement 1: antigen-presenting cells detect antigen on the allergen.

Statement 2: plasma cells produce allergen-specific antibodies.

Statement 3: T-helper cells produce histamine.

A 1, 2 and 3

B Only 1 and 2

C Only 2 and 3

D Only 1 *(1 mark)*

2 Which of the following statements represent(s) (a) non-specific immune response?

1 Phagosome formation.

2 Monocytes in plasma leave the capillaries and move to damaged tissue.

3 Macrophages release chemicals that cause inflammation.

A 1, 2 and 3

B Only 1 and 2

C Only 2 and 3

D Only 1 *(1 mark)*

3 The Mantoux test is used to assess whether a person has been exposed to tuberculosis (TB).

Which of the following statements is/are true of the Mantoux test?

Statement 1: a person may have TB if the test results in inflammation within 48-72 hours.

Statement 2: cells of *Mycobacterium tuberculosis* are injected into the person being tested.

Statement 3: the Mantoux test produces a non-specific immune response rather than a specific response.

A 1, 2 and 3

B Only 1 and 2

C Only 2 and 3

D Only 1 *(1 mark)*

4 Which of the following could result in natural passive immunity?

A Vaccination

B Injection of antibodies

C Exposure to pathogenic antigens

D Antibodies passing from a mother to a fetus *(1 mark)*

5 Which of the following is an example of an opsonin molecule?

A Cytokine

B Complement protein

C Lysin

D Interleukin *(1 mark)*

6 In an individual with bronchitis, the mucus contains a large number of pathogenic bacteria. Phagocytic white blood cells destroy the bacteria.

The sequence of events that results in the destruction of bacteria is shown here.

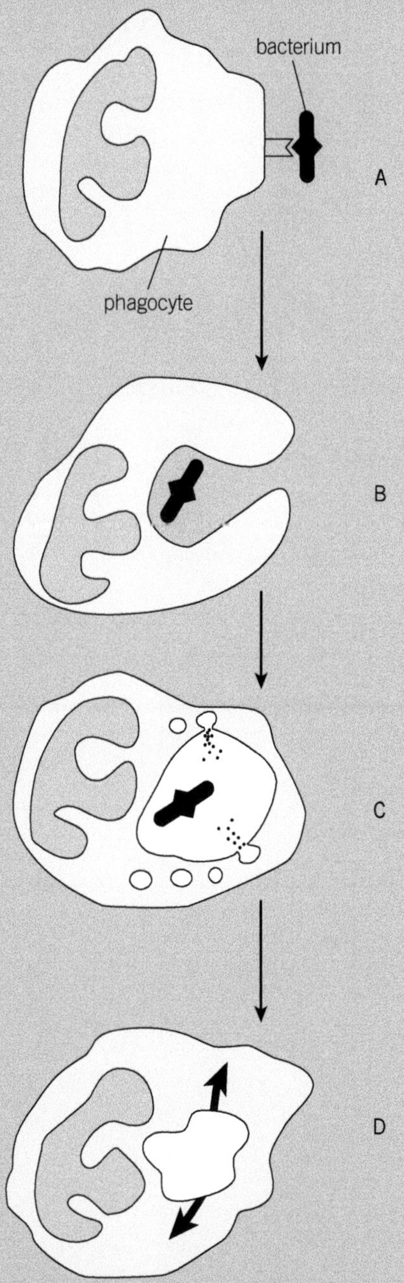

a Describe the events taking place at stages A, B, C, and D in sequence. *(6 marks)*

b The immune system will produce specific antibodies in response to infection.

(i) Name the type of cell that produces antibodies. *(1 mark)*

(ii) Describe how the structure of an antibody molecule is related to its function. *(6 marks)*

(iii) Identify the type of immunity provided by antibodies in breast milk. *(1 mark)*

[question 5 june 2010 F212 OCR]

7 Lymphocytes are important components of the immune system and can be classed as B lymphocytes and T lymphocytes.

a For each of the statements in the table, identify whether the description applies to:

- Only B lymphocytes
- Only T lymphocytes
- Both B and T lymphocytes
- Neither

You may use each response once, more than once, or not at all. The first one has been done for you. *(5 marks)*

Statement	Can be applied to...
Form part of immune response	*Both*
Matures in thymus	
Secrete substances which kill infected cells	
Manufacture antibodies	
Undergo clonal expansion	
Activate other lymphocytes	

b The graph shows the concentration of antibodies in a patient's blood following an initial infection with a pathogen. This is known as the primary response.

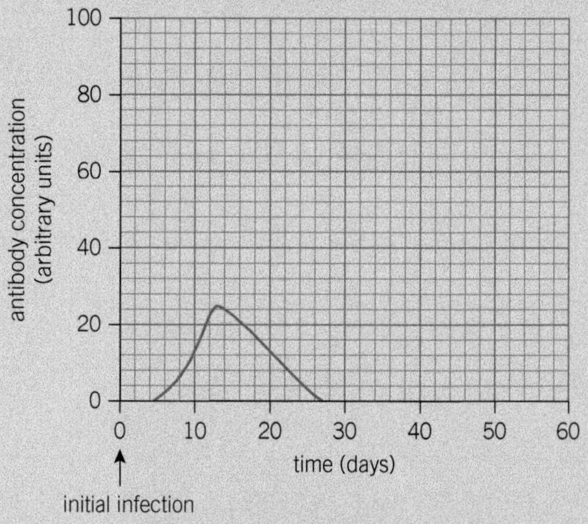

(i) Describe the changes in antibody concentration that occur in the patients blood during the primary response. *(3 marks)*

(ii) The patient was subsequently infected with the same pathogen 30 days after the initial infection.

Copy the graph and draw a line on it to show the likely concentration of antibodies in the patient's blood from 30 days onwards. *(2 marks)* c The diagram shows the structure of an antibody.

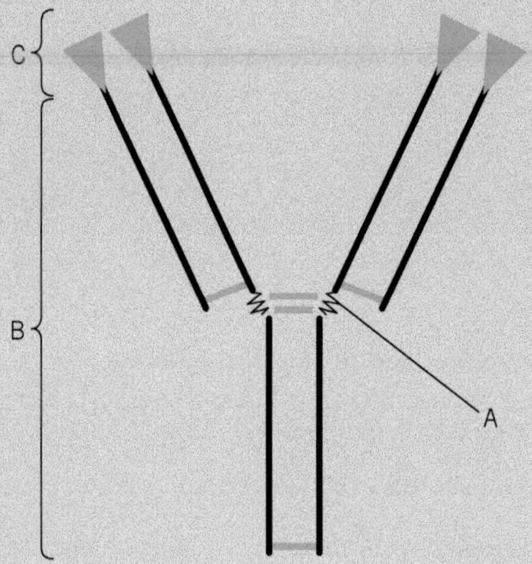

Region	Name	Function
A		
B		
C		

Copy and complete the table by stating the name and function of each of the regions **A**, **B**, and **C** *(3 marks)*

[question 7 jan 2012 F212 OCR]

8 a When an individual is infected with a virus, an immune response is triggered.

(i) Define the term immune response. *(2 marks)*

(ii) One type of cell involved in an immune response is a plasma cell, which releases antibodies.

Plasma cells contain RNA.

Outline the roles of RNA in plasma cells. *(6 marks)*

(iii) Outline two ways in which antibodies reduce the threat from pathogens. *(4 marks)*

[question 2 jan 2013 f212 OCR]

13 COMMUNICABLE DISEASES
13.1 Controlling communicable diseases

Specification references: 3.2.3

Learning outcomes

Demonstrate knowledge, understanding, and application of:

- → the principles of vaccination
- → the role of vaccination programmes in preventing epidemics.

Study tip

The figures given here are both raw data (deaths) and mortality rates. In order to calculate a mortality rate, the size of the population would need to be known. By comparing rates rather than raw data, the fact that the UK population has increased in size between the sets of data is taken into account and a valid comparison can be made.

Synoptic link

See Topic 11.1, Communicable diseases for more detail on the epidemics of influenza.

Vaccinations

The function of a **vaccine** is to afford protection against communicable diseases that cause harm and possible death to humans. These are diseases that can be transmitted easily and quickly and so have the potential to cause epidemics or pandemics.

Diseases such as TB and polio were common and caused widespread deaths in the nineteenth and early twentieth century. In 1913, 36 500 deaths from TB were recorded in the United Kingdom, giving a mortality rate of 99.8 per 100 000. A TB vaccine was introduced into the UK in 1953 and by 1955 the death rate had fallen to 8.8 per 100 000 (3900 deaths) and in 2012 the mortality rate was down to 0.5 per 100 000 (261 deaths).

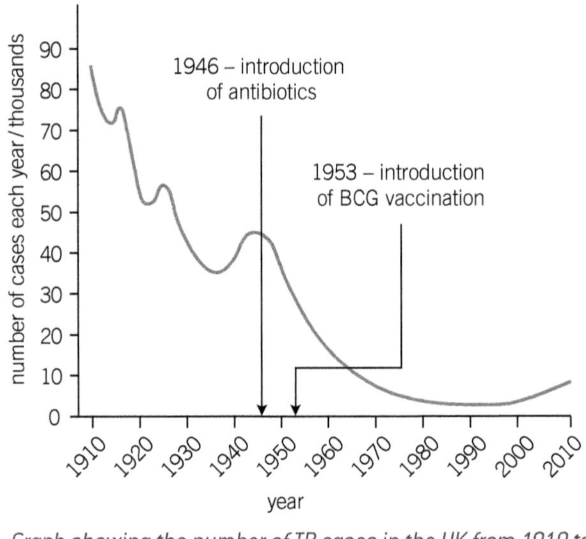

▲ **Figure 1** *Graph showing the number of TB cases in the UK from 1910 to 2006*

Vaccination involves a deliberate exposure of an individual to antigens from a foreign source – usually a pathogen – in order to provoke an immune response in the body and provide immunity to the disease. In this way cycles of disease may be broken.

An individual who is vaccinated and immune will not develop disease symptoms when they encounter the pathogen as the immune system prevents the pathogen multiplying. Hence the person will not be able to pass the pathogen on and the cycle of infection is broken.

Vaccines may be introduced by injection or orally and there are a variety of different forms of the antigen that are used.

▼ **Table 1** *Examples of the different types of vaccine, the diseases they act against, and their advantages and disadvantages*

Type of vaccine	Explanation of what it is	Examples of diseases immunised against	Advantages/disadvantages
Live-attenuated (weakened) vaccines	Modified strains of bacteria or viruses that can multiply but are not pathogenic.	Measles, mumps, polio (oral vaccine), rubella (German measles), TB, typhoid (oral)	Advantages: vaccine gives the strongest response and long lasting immunity. Immunity can be passed on, increasing herd immunity. Disadvantages: cannot be given to anyone with compromised immune systems (e.g., after chemotherapy or people with HIV). The organism may revert and become pathogenic.
Killed inactivated vaccines	Bacteria or viruses are killed by chemical treatment or heat. Antigens are still intact and provoke an immune response.	Influenza, hepatitis A, polio (Salk vaccine), cholera, and whooping cough (also known as pertussis)	Advantages: more stable and safer than live vaccines and doesn't need refrigeration so travels better. Disadvantages: gives a weaker immune response so boosters are necessary.
Toxoids	Toxins are extracted and treated with formaldehyde to prevent the toxin causing the disease symptoms. These are enough to promote an immune response triggering the production of antitoxins.	Diptheria and tetanus	Advantages: safe to use where the toxins, produced by the pathogen, are the reason for its virulence. Disadvantages: may not give the strongest immune response.
Subunit vaccines (isolated extracted antigens) and conjugate vaccines	Only the specific antigens (generally polysaccharides) that promote the immune response are extracted and used in the vaccine. In conjugate vaccines, the antigen is joined to a protein.	Haemophilus influenzae type B (HiB) Meningococcal A and C Pneumococcal vaccine (PCV)	Advantages: can construct vaccines to several different strains. The joining of the antigen to a protein means that both the cellular and humoural responses are triggered.
Artificial antigens or recombinant vector vaccines	Using genetic engineering, genes for the antigens of specific pathogens are transferred to a harmless microorganism. For example, genes for antigens to the rabies virus into a harmless virus creating a 'live vaccine' for rabies.	Rabies and human papilloma virus (HPV) are created using this method.	Advantages: can provide immunity to agents that cannot be easily attenuated or inactivated without destroying antigen activity. Can be used to produce live vaccines.

Vaccination risks

There are some risks with vaccination, but in most cases these risks are very low and are far outweighed by the risk posed by the disease itself. There was wide public concern about the publication of a scientific study in the 1990s on the risks associated with the MMR (Measles, Mumps, and Rubella) triple vaccine. The principal authors of the study were later shown to be biased in favour of single dose vaccines and the research was discredited.

Research published in 2004 indicated that the relative risk of developing autism or other forms of developmental disorder linked to the MMR vaccine was 0.87. Relative risk is a statistical term used to calculate the risk of, in this case, a developmental disorder occurring in MMR-vaccinated children as opposed to unvaccinated children.

▼ **Table 2** *Calculating relative risk*

	Developmental disorder	No developmental disorder
Vaccinated	A	B
Unvaccinated	C	D

The letters refer to the number of children in each category. Relative risk is calculated as:

$$\textit{Relative risk} = \frac{A \div (A + B)}{C \div (C + D)}$$

1 Assuming there are 200 children in each of the vaccinated and the unvaccinated groups, and seven children were diagnosed with a developmental disorder in the vaccinated group and six in the unvaccinated group, calculate the relative risk of a developmental disorder occurring as a result of vaccination.
2 Suggest what a relative risk value of less than one means.

The epidemiology of whooping cough

The bacterium *Bordetella pertussis* can cause an infection of the airways. The early symptoms are similar to a cold but severe bouts of coughing can occur. At the end of each bout a characteristic 'whooping' sound is made and sometimes vomiting follows. This whooping sound gives the disease, whooping cough, its name. The cough can last for two to three months and the disease is more serious in children under the age of one, who can suffer complications such as pneumonia, weight loss, dehydration, and even brain damage due to lack of oxygen.

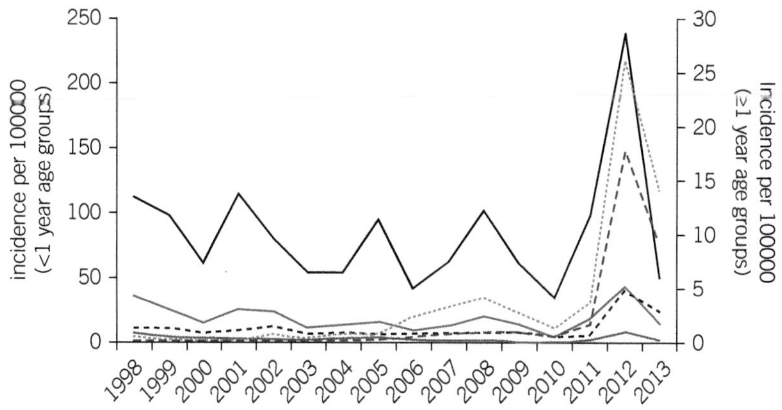

▲ **Figure 2** *Whooping cough incidence rates by age group in England from 1998 to 2013*

Immunity from whooping cough following vaccination or contracting the infection is not life-long. After a time, herd immunity wanes, meaning infections can spread more easily leading to regular epidemics.

Most cases in infants occur before the age when they have fully completed the vaccination programme for whooping cough. In 2012, the Department of Health introduced a temporary programme of whooping cough vaccination in pregnant women. The vaccine is offered between 28 and 38 weeks of pregnancy to new mothers who have never received the vaccine.

1 Using the information in the graphs, what is the evidence that epidemics of whooping cough occur regularly?
2 Explain why this vaccination programme of pregnant mothers will reduce the incidence and mortality of whooping cough in children under the age of one.
3 Why is the vaccine offered to new mothers who have never been vaccinated?

Role of vaccination programmes

The vaccination programme offered in the UK protects the population from a wide range of diseases. Some vaccines require follow up **boosters** – particularly if the first vaccine was not a live vaccine. The boosters increase the number of memory cells and ensure that the secondary response is rapid enough for the pathogen to be destroyed before it can multiply in the body and cause disease.

For a vaccination programme to be successful in preventing epidemics a large proportion of the population must be immune. The chance of someone with the infection being able to pass on the pathogen is reduced if enough are vaccinated, as most people would be immune. This is called **herd immunity**. To establish herd immunity, the percentage of people immune to the disease needs

Study tip

Remember the administration of a vaccine triggers the primary immune response resulting in the production of antibodies and memory cells. A later challenge by the pathogen would trigger the secondary immune response resulting in rapid production of high concentrations of antibodies such that the pathogen is destroyed before it can cause the disease.

to be between 80 and 95% depending on how easily the pathogen is spread. Smallpox needed a herd immunity of 83%, whereas whooping cough requires a herd immunity of between 92 and 94%. In the event of an epidemic **ring vaccination** may be used. This is where all individuals in the immediate area, and any that may have been in contact with those infected by the disease, may be vaccinated.

Summary questions

1 Using the information in Figure 2, calculate the percentage decrease in deaths from TB between 1955 and 2012. *(2 marks)*

2 Evaluate the use of live vaccines. *(3 marks)*

3 Explain why vaccinations against tetanus need frequent boosters. *(2 marks)*

13.2 Problems developing vaccines

Specification references: 3.2.3

Biological problems in the development of vaccines

The **mutation rate** of pathogens can lead to changes in the antigens and protein coats of viruses. Any changes will mean that the antibodies already produced by the body will not be able to bind to the antigen and the effect will be lost. As it takes time to develop a vaccine, there is generally a delay between a new strain appearing, for example a new strain of the influenza virus, and the development, manufacture, and release of the appropriate vaccine.

The variability of the antigens on the pathogen surface means that it can be hard to target all the different forms and this, together with the mutation rate, means an effective vaccine may prove too difficult to develop. This is the case in the search for a vaccine against HIV. There are at least nine different sub-types of the HIV virus. Viruses from different sub-types can recombine to produce new hybrid viruses and the types circulating vary in different geographical locations. A vaccine may be effective in one area but not in another.

Vaccines need to be maintained in sterile conditions throughout the production and packaging process to avoid contamination – this necessity adds to the cost of production and reduces the inclination of pharmaceutical companies to keep producing such vaccines.

Investigators have speculated that a strain of the H1N1 flu virus, which caused a pandemic in Russia in 1977–78, was accidentally released from a laboratory that was working on an earlier 1947–57 version of the virus. This highlights another difficulty in the development of vaccines and the need for strict control measures in laboratories working on vaccines.

Biological problems in implementing vaccination programmes

There are dangers in using live vaccines (**attenuated** pathogens). Future mutations may mean that the pathogen becomes virulent again and disease symptoms develop. The risk varies between different vaccines. Some, like BCG, which has lost 100 genes compared to the original strain, make a reversion less likely. However, cases of vaccine-induced polio have been recorded with the live oral vaccine and some scientists argue that the inactivated polio vaccine (IPV) should be used.

Storage, transport, and distribution of vaccines can prove difficult, especially for live vaccines that require refrigeration at all times before they are administered. In some countries, ready access to such cold storage may prove too difficult.

The nutritional status of the target population may create another problem since any deficiencies in nutrition may impact on the ability of the immune systems to respond strongly. This is especially true

Learning outcomes

Demonstrate knowledge, understanding, and application of:

- → biological problems in the development of vaccines
- → biological problems in implementing vaccination programmes
- → ethical issues related to the development and use of vaccines.

Synoptic link

The issues around protein-energy malnutrition are also dealt with in Topic 2.2, Mammalian and plant biofluids.

with populations suffering from protein-energy malnutrition (PEM), in which amino acids are used as a respiratory substrate, making less available for the manufacture of proteins such as antibodies.

Ethical considerations in the development of vaccines

There are many ethical issues related to the development of new vaccines including the requirement to test on animals and on unaffected individuals before the vaccine can be used in the population.

The cost of developing new vaccines is an important consideration. It is estimated that it costs between 500 to 1 billion US dollars and takes approximately 15 years to bring a vaccine to the market. Approximately 80% of global vaccine production is carried out by five or six multinational pharmaceutical companies. Vaccines, if successful, require a single dose with a limited number of boosters. For example, the search for a vaccine against Ebola has been held back by lack of money. Drugs, on the other hand, may be required three or four times each day. Consequently, there was a period of time when the investment in vaccine research and development was much less than that in drugs as the margin for profit was just too low. With the use of new DNA technology and the development of new approaches to vaccines, there are many, small biotechnology companies now involved and the situation may be changing – particularly with the promise of the development of vaccines against cancer.

Ethical considerations in implementing vaccine programmes

A balance must be struck between an individual's right to refuse their consent to a vaccine and the rights of the entire population to be protected. For serious communicable diseases, there is a strong argument for a compulsory vaccination programme. Of course vaccination itself is controversial and has been debated and discussed many times since the (now disproved) MMR scare in the 1990s.

The question of side effects is also important to consider as for some individuals, the risk and potential impact of these, particularly in childhood vaccinations, outweighs the advantage to the general population.

Some developing countries have become deeply suspicious of vaccination programmes delivered by aid agencies sponsored by governments. The programmes are viewed as 'paternalistic'. Health workers are now viewed with suspicion in some countries – particularly in the light of recent events where a 'fake' vaccination programme was used to collect DNA samples to identify suspected terrorists.

Ethical considerations relating to the HPV vaccine programme

Human Papilloma Virus (HPV) is a small virus with a relatively small genome of 8000 base pairs and only ten genes. Many individuals will have encountered it at some time in their lives (approximately 80% of us) but in most cases the immune system counters it and it is eradicated.

A vaccine has been produced that protects against most of the strains of the HPV virus. It is a subunit vaccine, made of the main proteins on the outside of the virus. There is no DNA within this protein and so infection with HPV will not occur as a result of vaccination. The vaccine should reduce the incidence of cervical cancer significantly, providing it is used to vaccinate girls before they are sexually active. As a result, vaccination of 12 to 14-year-old girls is now included in the vaccination programme in the UK. The decision to have the vaccine or not is taken by the girl and not by a parent or guardian.

Males can also be infected with HPV and infection is linked to cancer of the penis, anus, and mouth, while infected males will spread the virus to their partners. Some strains of HPV cause warts, genital warts, and verrucas, but these are low risk infections and are not the same strain as those causing cancers.

The dilemma now is whether to vaccinate males as well as females and whether to make vaccination compulsory. At present the vaccine is only offered to females and it is not obligatory. This raises several ethical questions:

- Should parents be required to give consent for the HPV vaccination as the young people will be under 16?
- Is it right to ask these young people about any prior sexual activity?
- Side effects from vaccines are an important issue. In some individuals side effects may create unacceptable health risks or long-term harm.
- Can the potential risk to an individual's health caused by use of a vaccine be balanced against the good of the population in controlling a disease?
- Is it acceptable to trial potential vaccines on animals in order to develop the vaccines and ensure they are the most effective for disease control?

Any ethical issue is a moral judgement, which may change over time as new ideas become acceptable and as perceived risks are removed from consideration.

Synoptic link

HPV is sexually transmitted and it is easily transferred in sexually active individuals. The greatest concern with this virus is its link to cervical cancer. Many cases of cervical cancer are caused by strains of HPV. Cervical cancer is the most common cancer in young women with 3000 cases diagnosed each year and about 1500 deaths from the disease annually. Other causes of cancer are discussed in Topic 14.2, Epidemiological evidence and cancer. The principles of clinical trialling are discussed in Topic 15.2, Medicinal drugs and clinical trials.

Summary questions

1 State what is meant by *antigen variability*. (*1 mark*)

2 Stabilisers are substances added to help keep antigens and other vaccine components stable during storage. Gelatin, usually of bovine (cow) or porcine (pig) origin, is added to some vaccines as a stabiliser. Evaluate the advantages and disadvantages of using stabiliser such as gelatin. (*2 marks*)

3 Outline five issues and difficulties in developing new vaccines. (*5 marks*)

13.3 Antibiotics

Specification references: 3.2.3

Learning outcomes

Demonstrate knowledge, understanding, and application of:

- → the use of antibiotics
- → how misuse of antibiotics leads to resistant strains of bacteria
- → the effect of hygiene practices.

Synoptic link

You first read about the differences between prokaryotes and eukaryotes in Topic 1.6, Cell ultrastructure.

Antibiotics

Antibiotics are chemicals that either kill bacteria or slow down their growth. Antibiotics are active against bacteria (prokaryotes) but have no effect on eukaryotic cells or on viral replication. They target the differences in cellular structure and metabolism between eukaryotes and prokaryotes. This makes them invaluable in controlling bacterial infections.

Initially the name 'antibiotic' was restricted to chemical substances produced by microorganisms to inhibit the growth of other microorganisms.

In 1929, Alexander Fleming first showed inhibition of bacterial growth from the chemical (penicillin) produced by the hyphae of the fungus *Penicillium notatum*. His discovery was actually an accident as he had failed to clean up the bacterial plates he had been working on before going away for the weekend. When he returned he noticed the zones of inhibition (areas free of bacteria) around the *Penicillium notatum* fungal growth. Other antibiotics were discovered in the Streptomyces group of bacteria and these now account for most of the antibiotics in current use.

Using antibiotics in disease treatment

There are several types of antibiotics that act in different ways:

- **Bacteriocidal** antibiotics – these kill the bacteria by either preventing cell wall synthesis or by disrupting protein synthesis in bacteria. Penicillin prevents the synthesis of peptidoglycan, which makes up the bacterial cell walls and prevents new cell walls forming and so the bacterial cells burst. Polymyxin B damages the plasma membrane of the bacterial cell so the contents leak out. It does so by targeting phospholipids that are not found in eukaryotic cell membranes.
- **Bacteriostatic** antibiotics – these prevent the growth and reproduction of bacteria, but do not kill them. Tetracycline interferes with bacterial protein synthesis by preventing transfer RNA from binding to the ribosome. The bacteria survives but is unable to synthesise new proteins and so cannot replicate. Tetracycline has no effect on protein synthesis in eukaryotic cells, which have 80S ribosomes rather than the 70S ribosome in prokaryotic cells.
- The sulfonamides are competitive inhibitors of the bacterial cell's metabolism. Mammalian cells can take up folic acid as a dietary vitamin. Bacterial cells have to synthesise folic acid and the sulfonamides inhibit the enzymes in this pathway. In the absence of folic acid, purines and hence DNA cannot be synthesised.

- Erythromycin prevents protein synthesis by blocking one of the sites on the ribosome, which prevents the polypeptide chain elongating.

Many antibiotics available are **broad-spectrum antibiotics**, such as tetracycline, which mean they work well against a wide range of different types of bacteria and can be used to treat many infections from stomach ulcers to gonorrhoea.

Narrow-spectrum antibiotics are only effective against specific groups of bacterial infections, for example penicillin, which works by destroying the bacterial cell wall of Gram-positive bacteria. Polymyxin is only active against Gram-negative bacteria.

The development of penicillin was a major breakthrough in disease treatment and its rapid introduction in the Second World War allowed many injured soldiers to be successfully treated when they would have previously died of wound infections.

Study tip

Remember, viruses cannot be treated with antibiotics because they have a protein coat instead of a peptidoglycan wall and they do not have any metabolism because they rely on the metabolic processes of the cell they invade.

Kirby-Bauer antibiotic testing

Kirby-Bauer antibiotic testing (also known as disc diffusion antibiotic sensitivity testing) uses antibiotic-impregnated discs to test whether certain bacteria are susceptible to specific antibiotics. If a patient has a suspected bacterial infection, a sample of the bacteria is swabbed evenly across an agar plate. The antibiotic diffuses from the paper disc into the agar. The concentration of the antibiotic will be highest next to the disc, and will decrease as the distance from the disc increases. If the antibiotic is effective against the unknown bacteria no colonies will grow where the concentration in the agar is greater than or equal to the effective antibiotic concentration. This is the zone of inhibition.

The size of the zone of inhibition and the rate of antibiotic diffusion are used to estimate the bacterial sensitivity to that particular antibiotic. A larger zone of inhibition correlates with a smaller minimum inhibitory concentration (MIC) of antibiotic for those bacteria. Inhibition data produced by the test is compared with that produced by known concentrations of a reference compound. This information can be used to choose appropriate antibiotics to combat a particular infection.

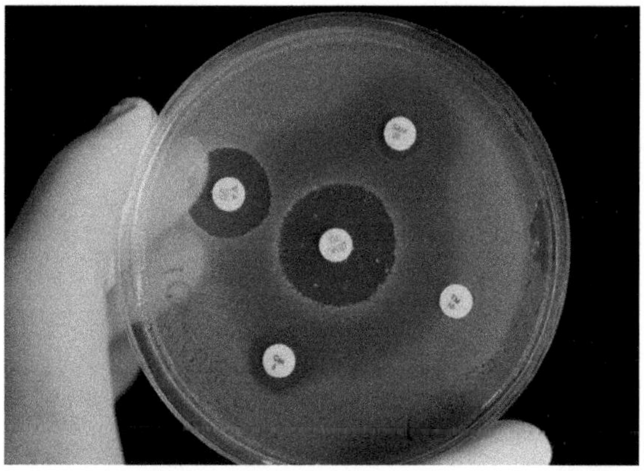

▲ **Figure 1** *Disc diffusion assay showing the effects of different antibiotic drugs on bacteria*

1 Use Figure 1 to determine which discs contain the drug that would be the most and least effective in treating this patient. Explain your decision. *(2 marks)*

2 a How could information in Figure 1 be used to generate quantitative data? *(1 mark)*

b What would an appropriate number of decimal places to give when recording these results in mm? *(1 mark)*

Antibiotic resistance

Mutations in the genes that regulate **resistance** to antibiotics occur spontaneously. This leads to variation within bacterial populations in terms of resistance to antibiotics. So in any population, there will be some cells that can tolerate the antibiotic more than others. If these cells survive, this form of the gene will be passed on and a population of bacteria will develop in which all the cells display increased resistance.

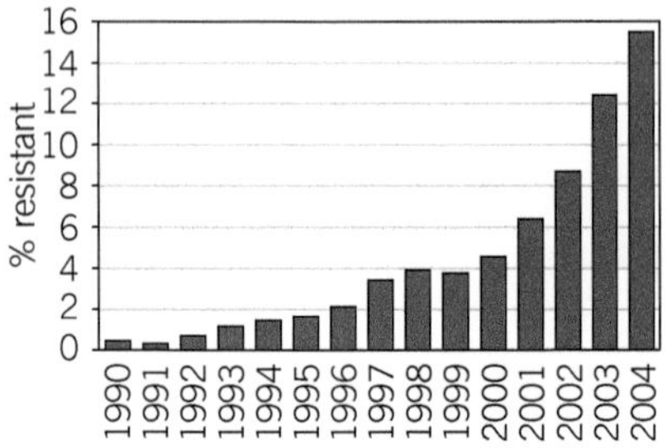

▲ **Figure 2** *Bar chart showing the percentage of resistant strains of bacteria to ciprofloxacin 1990–2004*

The development of strains of bacteria resistant to antibiotics usually begins a few years after the introduction of a new antibiotic. The mechanism is not always the same. The resistance can be due to a mutation on the bacterial genome or it could be due to genes carried on a plasmid. This type of resistance then spreads from cell to cell 'horizontally' to other species of pathogenic bacteria and so the proportion of infections that are caused by resistant strains of bacteria increases, leading to more infections that resist antibiotic treatment.

- **MRSA**, or methicillin-resistant *Staphylococcus aureus*, infections are extremely dangerous because MRSA is resistant to nearly all available antibiotics. It is one of the bacterial infections that may be found in some of our hospitals, where it can be spread from one patient to another unless strict hygiene measures are implemented.
- **MDR-TB**, or multiple drug-resistant strains of the TB bacterium, are now causing major health concerns. These strains will affect the eradication and control of TB worldwide, since any resistant strain will quickly spread to other individuals. TB resistant strains have been isolated in a number of different countries, usually those with poor or incorrect treatment regimes. A big factor is the failure to ensure that all patients given antibiotics complete their course of treatment. If patients do not complete their course, only bacteria that may show slight resistance will be eradicated leaving bacterial populations that are more resistant. There are a number of strains that are resistant to one or more anti-TB drugs, but more worrying are the totally drug-resistant strains or super TB strains.

The issues surrounding bacterial resistance mean it is important to avoid overuse and misuse of antibiotics, so that we are able to slow down, or stop the development of other strains of resistant bacteria. It is also worth remembering that widespread antibiotic use will damage the useful bacteria in our gut and that make up our microbiome.

Reducing the spread of resistance

There are a number of areas that are contributing to the rise in antibiotic-resistant strains of pathogens:

- The failure of patients to complete the antibiotic course, which means the bacterial infection is not completely eliminated.
- The overuse of antibiotics, including unnecessary prescriptions for non-bacterial infections.
- The use of antibiotics in animal feed to prevent infections and allow increased growth in the animals.

Slowing down the spread of resistant bacteria

Listed here are more strategies used by the health care sector to restrict the spread of infections and to slow down the development of more resistant strains:

- reducing contamination in hospitals on door handles, beds, tables, call bells, and all equipment
- improve cleanliness of hands using alcohol rubs, gels, and barrier clothing
- use of aseptic techniques rigorously
- isolation of infected patients with their own dedicated bathroom facilities
- restricting movement of care workers, staff, and visitors, as well as the patient
- emphasis on hygiene outside the ward and at home
- reducing the use of antibacterial hand wash, which removes the non-pathogenic bacteria and allows pathogens to survive.

1 Suggest why the removal of non-pathogenic bacteria by antibacterial hand wash can increase the survival of pathogens.

Summary questions

1 Suggest why tetracycline is a broad-spectrum antibiotic. (*1 mark*)

2 Describe how antibiotic-resistant strains of *Gonococcus* (the bacterium which causes gonorrhoea) developed resistance to sulfonamides within 10 years of their introduction. (*2 marks*)

3 Propose why antibiotics may be damaging to the human body even though it is known that they do not affect human body cells. (*1 mark*)

Practice questions

1 Which of the following statements is/are true of the procedure you would use to differentiate between Gram-positive and Gram-negative bacteria?

1 A primary stain is applied to heat-fixed bacteria.

2 Ethanol can be added to decolourise Gram-negative bacteria.

3 A counterstain such as safranin is applied.

A 1, 2 and 3

B Only 1 and 2

C Only 2 and 3

D Only 1 *(1 mark)*

2 Which sentence is true of bacteriostatic antibiotics?

A They maintain a bacterial cell count at a near constant concentration.

B They prevent bacterial reproduction.

C They kill bacteria by preventing cell wall production.

D They kill bacteria by damaging DNA. *(1 mark)*

3 Which of the following can be used in a vaccine?

1 An inactivated virus

2 Antibodies specific to a bacterial pathogen

3 A protein fragment from a virus

A 1, 2 and 3

B Only 1 and 3

C Only 2 and 3

D Only 1 *(1 mark)*

4 Which of the following statements best describes how herd immunity is usually achieved?

A A large proportion of a population survives a disease epidemic.

B A large proportion of a population is vaccinated.

C A large proportion of a population is given antibodies specific to a pathogen.

D A large proportion of a population receives antibodies specific to a pathogen from their mothers during gestation. *(1 mark)*

5 Which of the following statements can be a mode of antibiotic action?

1 Inhibition of DNA synthesis.

2 Prevention of protein synthesis at rough endoplasmic reticulum.

3 Inhibition of cellulose cell wall synthesis.

A 1, 2 and 3

B Only 1 and 2

C Only 2 and 3

D Only 1 *(1 mark)*

6 Studies have been carried out on the effect of garlic on the growth of bacteria. A summary of the method used in one such study is given here: An extract of garlic is prepared; Agar plates are prepared; a suspension of bacteria is spread over the surface of each agar plate; wells are cut into each agar plate; garlic extract or a known antibiotic is placed in each well; the plates are incubated at 30°C for 24 hours; the size of the zone of inhibition around each well is measured; the diagram shows an agar plate from the above study following incubation.

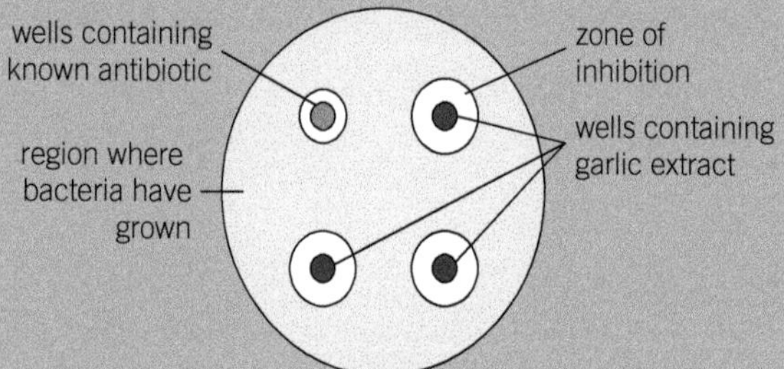

a Explain why a known antibiotic was included in the experiment. *(2 marks)*

b Scientists have used DNA markers to identify species of plant with medicinal properties such as antibiotic activity.

(i) Similar experiments to the one in the diagram have been carried out using onion, *Allium cepa*, instead of garlic, *Allium sativum*.

Suggest what the likely conclusions from such experiments using onion would be. Give a reason for your suggestion. *(2 marks)*

(ii) Suggest two advantages of using DNA markers to identify plants with antibiotic activity rather than using the method in the diagram. *(2 marks)*

[OCR F222 Jan 2010, question 2]

7 Immunisation programmes have resulted in dramatic improvements in health. As a result of immunisation, several infectious diseases which were major causes of ill health are now rare in many countries. The table shows the immunisation programme for young children in the United Kingdom. The vaccines used in the immunisation programme have code names, as shown in the table. Only one of the vaccines and some of the diseases have been omitted from the table – these are indicated by the shaded areas.

Age	Code name of vaccine	Diseases protected against
2 months	DTaP/IPV/HIV (1st dose)	
	PCV (1st dose)	pneumonia
3 months	DTaP/IPV/Hib (2nd dose)	
	MenC (1st dose)	Meningitis C
4 months	DTaP/IPV (3rd dose)	
	PCV (2nd dose)	Pneumonia
12–13 months	MenC (2nd dose)	Meningitis C
	Hib (4th dose) / MenC (3rd dose)	
		Measles, mumps, and rubella
3 years 4 months		Pneumonia
		Measles, mumps, and rubella

a (i) Using the information in the table, name the vaccine that protects children from measles, mumps and rubella. *(1 mark)*

(ii) Name four of the diseases that children are protected against by having the DTaP/IPV/Hib vaccine given at 2, 3, 4 months, and 3 years 4 months. *(4 marks)*

b (i) Outline how a vaccine protects children from infectious disease and explain why more than one dose of the vaccine may be required to give full protection. *(6 marks)*

(ii) Suggest one reason why the routine vaccination programme in the UK starts when children are 2 months old. *(1 mark)*

(iii) Suggest why some parents are worried about their children being vaccinated. *(1 mark)*

c Sometimes a child's immune system does not respond successfully to a vaccine. In the condition known as PEM (protein-energy malnutrition), a child's diet is low in fat and carbohydrates. This results in protein in the diet being used as an energy source. Suggest why a vaccine may not produce an immune response in children with PEM. *(2 marks)*

[OCR F222 June 2013, question 5]

8 a Antibiotic resistance in bacteria is becoming an increasing problem. Describe how an antibiotic-resistant population of bacteria could develop. *(4 marks)*

b Hospitals can check to see if a strain of bacteria causing an infection is resistant to a range of antibiotics by using a **multodisc**. A multodisc contains different antibiotics.

- The bacteria are isolated from a patient
- The bacteria are spread on nutrient agar in a petri dish
- The multodisc is placed on the agar.

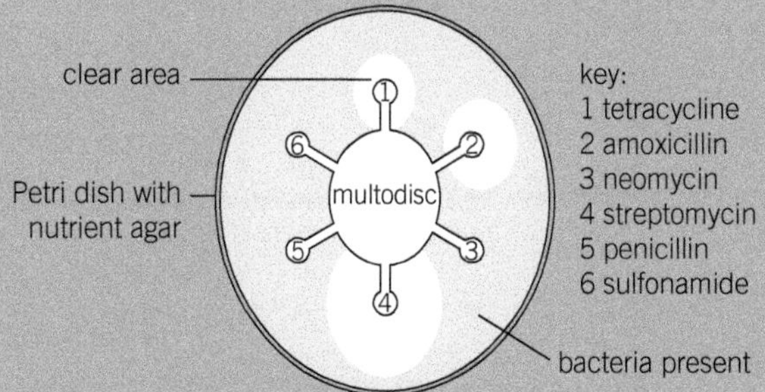

(i) Explain why there are clear discs of agar in the petri dish *(1 mark)*

(ii) Name the antibiotic that is most effective against the bacteria causing the infection. *(1 mark)*

(iii) Suggest three reasons why a hospital might use a multidisc to select the most suitable antibiotic for treating a patient. *(3 marks)*

[question 3 Jan 2010 F212 OCR]

14 NON-COMMUNICABLE DISEASES

14.1 Non-communicable diseases

Specification references: 3.3.1

Learning outcomes

Demonstrate knowledge, understanding, and application of:

→ the factors that may increase the risk of developing non-communicable diseases.

Synoptic link

You met communicable (infectious) diseases in Topic 11.1, Communicable diseases. You will study respiratory disease in detail in Topic 15.1, Pollutants and lung disease.

What are non-communicable diseases?

Non-communicable diseases (NCDs) are diseases which last for a long duration and generally progress slowly (chronic). Unlike communicable diseases, NCDs are not transmitted from person to person via infection. There are four main types of non-communicable diseases:

- cardiovascular diseases (e.g., heart attacks and strokes)
- cancer
- chronic respiratory diseases (e.g., chronic obstructed pulmonary disease and asthma)
- diabetes.

In 2010 the World Health Organisation published the first *Global Status Report on Noncommunicable Diseases*. According to the report, in the year 2008 NCDs were the leading cause of death in the world and accounted for 36 million out of the total 57 million deaths. The report also highlighted that approximately 80% of all NCD deaths occurred in low and middle income countries.

Worked Example: Calculating percentage changes in mortality rates

Example 1

In 2008 there were 57 million deaths recorded globally. 36 million of these were as a result of NCDs. Calculate the percentage of global deaths caused by NCDs in 2008. Record your answer to one decimal place.

$$\frac{36}{57} \times 100 = 63.2\%$$

Example 2

Using the following formula, calculate the expected number of global deaths due to NCDS in 2014.

$$A = P \times (1 + r)^n$$

Where A = expected number, P = initial number, r = the annual increase as a decimal and n = number of years.

$A = 36 \times (1 + 0.075)^6 = 56$ million deaths

(Note that the answer is given to the same number of significant figures as the initial number.)

Risk factors

In epidemiology, a **risk factor** is a variable that is associated with an increased risk of disease or infection. Risk factors can be described in different ways:

- *relative risk* – for example, for each successive 10 years after age 55, the stroke rate more than doubles in both men and women.
- *the fraction of incidences occurring in the group having or being exposed to the risk factor* – for example, 28% of intrahepatic bile duct cancer cases are diagnosed in women.
- *an increase in the incidence in the exposed group* – for example, each daily alcoholic beverage increases the incidence of breast cancer by 11 cases per 1000.

Risk can be evaluated by comparing the number of people exposed to the potential risk factor to those experiencing an event. It can be calculated using the following equation:

$$Risk = \frac{\text{number of people experiencing an event}}{\text{number of people exposed to the risk factor}}$$

Risk factors themselves can often be assigned one of three different groups:

- *conditions* – that is, other medical conditions that increase the risk of a NCD
- *behaviour* – that is, lifestyle choices, for example smoking
- *hereditary* – that is, family history and therefore genetic predisposition.

Table 1 shows some non-communicable diseases and their potential risk factors.

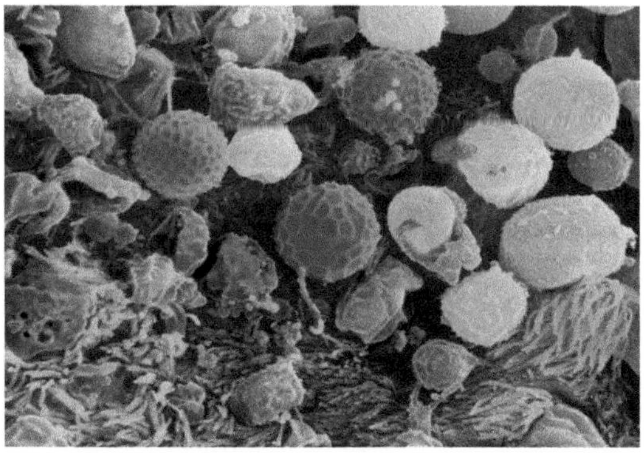

▲ **Figure 1** *Coloured scanning electron micrograph of dust (pale blue) and pollen (pink) on the surface of the trachea. These allergens may bring on the symptoms of asthma, ×1130 magnification*

▲ **Figure 2** *Coloured scanning electron micrograph of dust mites. Millions of dust mites live inside furniture and fabric in the average home. The dead bodies and excrement of dust mites can cause allergic reactions to household dust, ×50 magnification*

Correlation between risk factors and diseases

There can be **correlations** between risk factors and NCDs but this does not necessarily mean the risk factor causes the NCD. For example, being young cannot be said to cause measles, but young people have a higher rate of measles because they are less likely to have developed immunity during a previous epidemic. Correlations can be positive (as variable A increases so does variable B) or negative (as variable A increases, variable B decreases).

▼ **Table 1** *Non-communicable diseases and their risk factors*

Non-communicable disease	Risk factors		
	Conditions	Behaviour	Hereditary
Stroke	• Increased age • High blood pressure • High cholesterol • Diabetes • Previous transient ischaemic attack (TIA)	• Excess salt intake • Physical inactivity • Excess alcohol intake • Exposure to nicotine (direct or passive smoking)	• Sickle cell disease
Lung cancer	• Exposure to radiation • Chronic lung disease, such as emphysema or chronic bronchitis • Increased age	• Smoking or passive smoking • Radon gas in the home • Exposure to ionising radiation • Exposure to asbestos • Medical exposure to radiation to the chest	• Having a first-degree family member (parent, sibling, or child) with lung cancer roughly doubles the risk of developing lung cancer • This risk is more for women and less for men • Having a second-degree relative (an aunt, uncle, niece, or nephew) with lung cancer raises your risk by around 30%
Asthma	• Having another allergic condition, such as atopic dermatitis or allergic rhinitis (hay fever)	• Exposure to allergens, certain parasites – or some types of bacterial or viral infection • Being overweight • Smoking or exposure to second-hand smoke • Having a mother who smoked while pregnant • Exposure to exhaust fumes or other types of pollution • Exposure to occupational triggers, such as chemicals used in farming, hairdressing, and manufacturing	• Having a blood relative (such as a parent or sibling) with asthma
Type 2 diabetes	• Being 45 years of age or older • Having diabetes while pregnant (gestational diabetes), or giving birth to a baby weighing 9 pounds or more	• Being physically active less than three times a week • Being overweight	• Having a parent or sibling with diabetes • Certain ethnicities show a pre-disposition

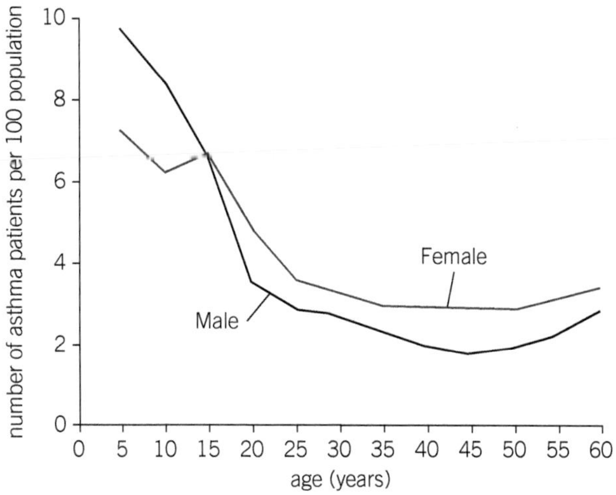

▲ **Figure 3** *Graph to show the gender differences in asthma prevalence according to age*

It is possible to assess the strength of an association between a risk factor and a NCD by applying statistical methods to the data. This method can be used to provide causal evidence (for example in the study of the link between exposure to ionising radiation and leukaemia) and show a causal relationship. An example of this is the evidence that suggests drinking green tea can help reduce the risk of a stroke.

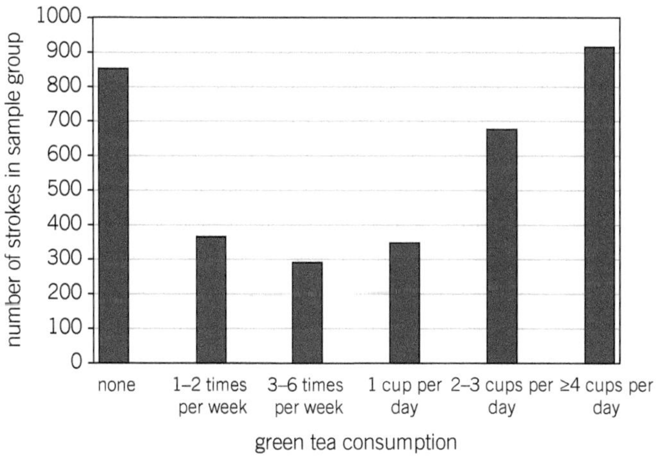

▲ **Figure 5** *A 12-year study in Japan followed over 80 000 volunteers to see if green tea consumption reduced the risk of suffering a stroke*

Study tip

Remember to qualify any descriptions of correlations to indicate if they are positive or negative and (if possible) the strength of the correlation.

Study tip

Remember risk is an assessment of probability and not definitive outcome. For example, some people may smoke for 60 years and never develop lung cancer whereas others may not directly smoke themselves but are exposed to passive smoke and still develop it.

▲ **Figure 4** *Tea is an infusion of leaves from* Camellia sisensis. *Green tea has been shown to have a causal relationship with decreased strokes*

Summary questions

1 Describe the correlation between age and the incidence of asthma for males in Figure 3. *(2 marks)*

2 a It has been estimated that active smoking is responsible for close to 90% of lung cancer cases – radon causes 10%, occupational exposures to carcinogens account for approximately 9 to 15% and outdoor air pollution 1 to 2%. Plot an appropriate graph of this data. *(4 marks)*

b Explain why it is possible that in some cases the combined risk of lung cancer exceeds 100%. *(1 mark)*

3 Suggest why members of a family where grandparents and parents have a history of strokes are described as high risk for having a stroke. *(3 marks)*

14.2 Epidemiological evidence and cancer

Specification references: 3.3.1

Learning outcomes

Demonstrate knowledge, understanding, and application of:

→ evaluating epidemiological evidence linking risk factors and particular forms of cancer.

Synoptic link

You learnt about epidemiology in the context of communicable diseases in Topic 13.1, Controlling communicable diseases.

Epidemiological evidence

Previously you learnt about epidemiology when considering communicable diseases. However, epidemiological studies have probably been more valuable in studies of non-communicable diseases, such as cancer.

Epidemiological evidence and lung cancer

The world's longest and largest epidemiological study started in the UK in 1951. This study collected data related to smoking and cancer. It is as a result of this study, and more recent studies, that scientists have been able to establish a correlation between smoking and diseases such as cancer.

Epidemiological evidence for bowel cancer

There have been many studies looking into the risks associated with bowel cancer, including the quantity of red meat consumed (Figure 2) and dietary fibre intake (Figure 3).

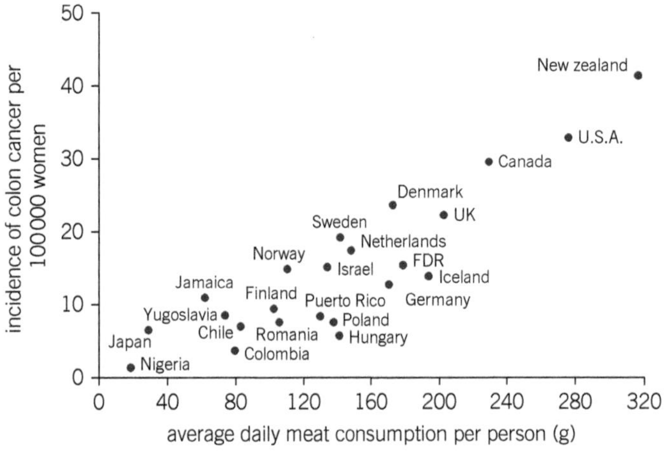

▲ **Figure 1** *Graph to show the incidence of colon cancer in women according to their daily meat intake*

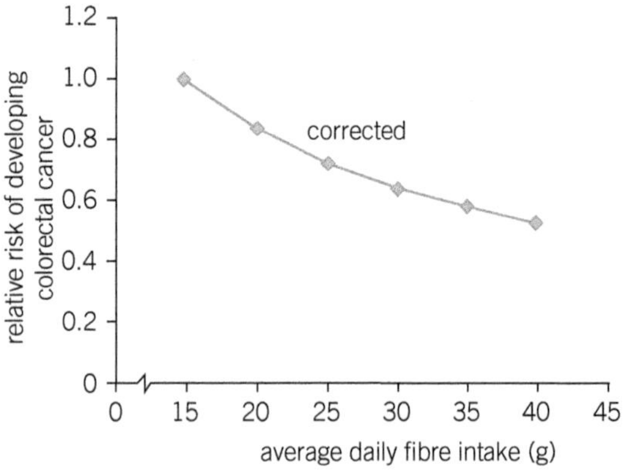

▲ **Figure 2** *Graph to show the relative risk of colorectal cancer according to dietary fibre intake*

Epidemiological evidence for breast cancer

Researchers have identified more than 1000 mutations in the *BRCA1* gene (which is carried on chromosome 17), many of which are associated with an increased risk of cancer (particularly ovarian and breast cancer in women). The *BRCA1* gene is a tumour suppressor gene. You will find out more about tumour suppressor genes in Topic 14.3, Cancer. Like many other tumour suppressors, the protein produced from the *BRCA1* gene helps prevent cells from growing and dividing too rapidly or in an uncontrolled way.

As defects in the genes such as *BRCA1* accumulate, they can allow cells to grow and divide uncontrollably and form a tumour.

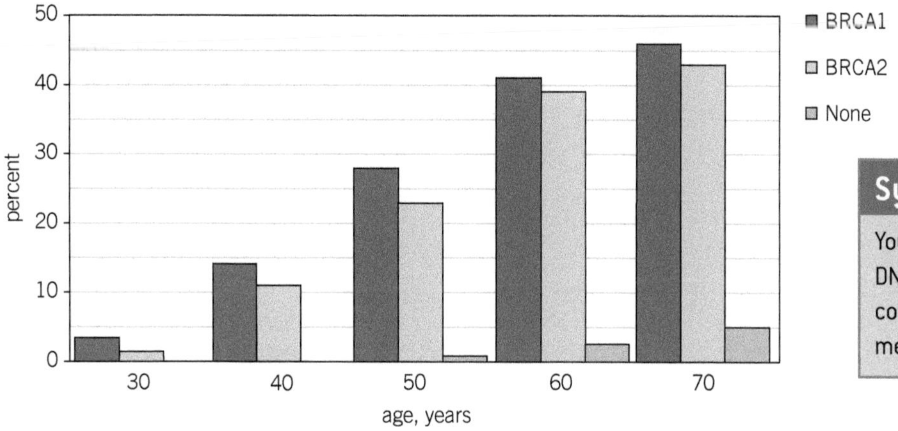

▲ **Figure 3** *Age-specific breast cancer risk associated with* BRCA1 *or* BRCA2 *mutation compared to no* BRCA *mutation*

Synoptic link

You learnt about the process of DNA replication in Topic 4.3, Semi-conservative DNA replication and meiosis in Topic 9.1, Meiosis.

The Bradford Hill criteria for establishing causal relationships

In 1965, the epidemiologist Austin Bradford Hill set out nine criteria for evaluating statistical associations. These continue to be used today. Let's look at this in relation to Figure 1 and to other facts known about lung cancer.

1 Strength of association – the stronger the relationship between the independent variable (smoking) and the dependent variable (lung cancer), the less likely the relationship is due to another variable. Lung cancer is relatively rare in non-smokers.
2 Temporality – the exposure must precede the disease by a reasonable amount of time. Figure 1 shows that there is a time delay between an increase in smoking and an increase in lung cancer.
3 Consistency – multiple observations of an association, with different people under different circumstances. Figure 1 shows similar results for both males and females, and similar graphs can be compiled in countries other than the UK.
4 Theoretical plausibility – it is easier to accept an association as causal when the conclusion is supported by known biological facts. Carcinogens have been identified in tobacco smoke and these have been shown to cause tumours.
5 Coherence – a 'cause and effect' interpretation is possible if all the available data from different types of experiments is consistent and there are no

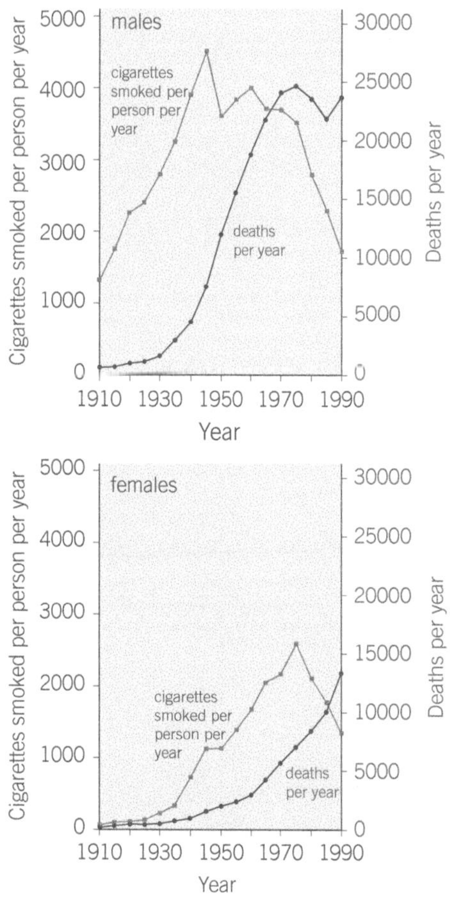

▲ **Figure 4** *Incidence of deaths from lung cancer in the UK correlated to cigarettes smoked per year (1910–1990)*

plausible conflicting theories. The conclusion that smoking causes lung cancer is supported by epidemiologic, laboratory animal, pharmacokinetic, clinical, and other biological data.

6 Specificity in the causes – ideally the effect (lung cancer) only has one cause. You have already seen that lung cancer is relatively rare in non-smokers.

7 Dose-response relationship – there should be a positive correlation between the dose (the number of cigarettes smoked for example) and the disease (lung cancer). Again, Figure 1 supports an increase in lung cancer with an increase in smoking.

8 Experimental evidence – any related research such as animal experiments that supports the epidemiological findings adds to the 'cause-and-effect' conclusion. For example, tar (which is also found in cigarettes) painted on laboratory rabbits' ears was shown to produce cancer in the ear tissue over time.

9 Analogy – Sometimes a commonly accepted phenomenon in one area can be applied to another area. The association between soot and skin cancer in chimney sweeps was first documented over 200 years ago and soot is a product of burning hydrocarbons.

When using these criteria it is important to remember Austin Bradford Hill's own advice:

'None of these nine viewpoints can bring indisputable evidence for or against a cause and effect hypothesis… What they can do, with greater or less strength, is to help answer the fundamental question – is there any other way of explaining the set of facts before us, is there any other answer equally, or more, likely than cause and effect?'

How could the 'strength of association' be established statistically?

Summary questions

1 The Bradford Hill criteria requires both epidemiological and experimental evidence to be considered. Describe one example of each type of evidence used to establish the link between smoking and lung cancer. *(2 marks)*

2 Suggest why it is not correct to say that smoking causes coronary heart disease. *(2 marks)*

3 Currently bowel cancer screening is offered to people aged over 60. Suggest why some people might be offered screening at a younger age. *(2 marks)*

14.3 Cancer

Specification references: 3.3.1

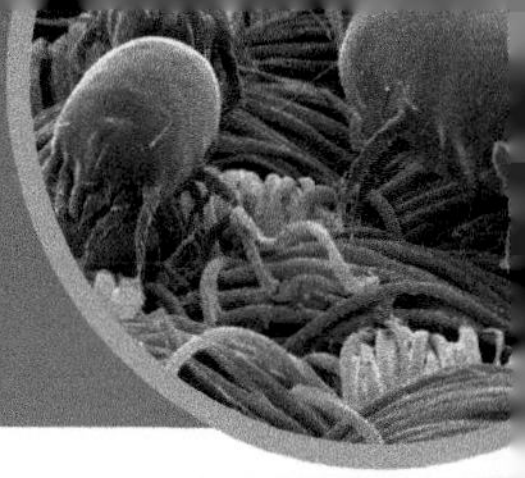

What is cancer?

Worldwide there is estimated to be over 10 million new cases of cancer each year and approximately six million deaths from the disease each year. It is currently estimated that one in three people will suffer from some form of cancer at some point during their lives. However, it is important to remember that certain lifestyle choices can reduce the risk of developing cancer and likewise increased awareness of the symptoms of cancer can lead to earlier diagnosis and consequently more successful treatment.

Learning outcomes

Demonstrate knowledge, understanding, and application of:

→ the cellular basis of cancer

→ the mutations that can cause cancer.

All living cells divide to produce genetically identical daughter cells which are used for growth and repair both in animals and plants. One of the processes involved in the production of these new daughter cells is mitosis, one form of nuclear division. Cells do not divide all the time – their division is carefully monitored and controlled in a healthy person to allow for growth and repair of tissues. The **cell cycle** is normally highly controlled by:

- factors within the cell
- growth factors outside of the cell.

When cell division is not regulated correctly, cells can continue to divide to form a mass of cells at a site where growth or repair is not needed. This mass of abnormal cells is called a **tumour** and it will constantly develop and expand in size.

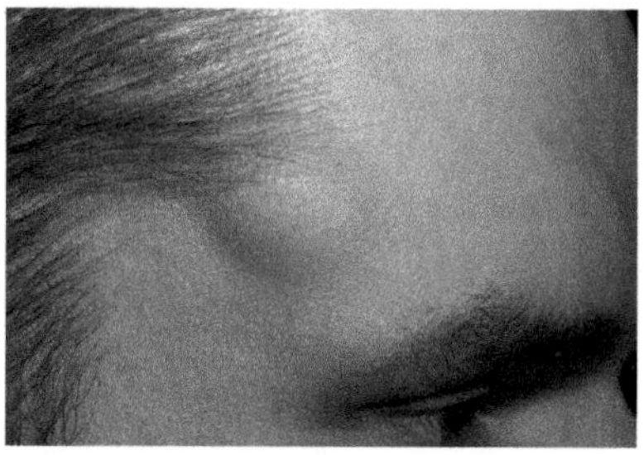

▲ **Figure 1** *Liposarcoma is a malignant tumour that forms in fat cells — the prognosis varies according to the site of origin, depth, and size of the tumour and now close it is to the lymph nodes. Liposarcomas rarely metastasise*

Types of tumour

There are two classes of tumour – benign tumours and malignant tumours.

- **Benign** tumours occur when cells divide too many times and tend to be slow-growing and are located within one specific tissue. Cells do not break off and spread to other parts of the body but remain at their site of origin. As a consequence benign tumours are not normally life-threatening but they can cause serious damage, for example, if the growth of the tumour presses on an important blood vessel or nerve. These tumours are usually removed by surgery and do not usually grow back. Examples of benign tumours include moles and renal adenomas. It is not always possible to safely remove benign tumours by surgery, for example, if the tumour were in the brain.
- **Malignant** tumours usually grow rapidly and if not diagnosed quickly can be very damaging. Some cells can break off from the original tumour (the primary tumour) and spread to other neighbouring tissues via the lymph system or blood plasma. When this occurs the tumour is described as **metastatic** and is now classified as **cancer**. When the primary tumour becomes metastatic and spreads to another location then this new tumour is called a secondary tumour.

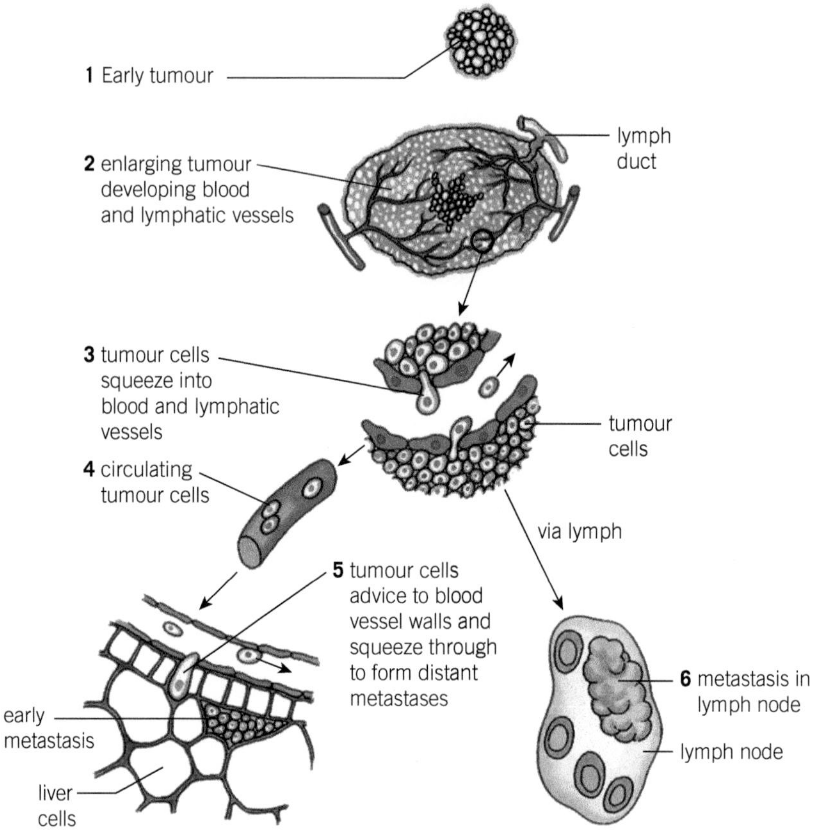
1 Early tumour
2 enlarging tumour developing blood and lymphatic vessels
lymph duct
3 tumour cells squeeze into blood and lymphatic vessels
tumour cells
4 circulating tumour cells
via lymph
5 tumour cells advice to blood vessel walls and squeeze through to form distant metastases
early metastasis
liver cells
6 metastasis in lymph node
lymph node

▲ **Figure 2** *How a primary tumour can develop and spread to form secondary tumours*

Study tip

The classification of cancers has changed drastically over recent years with the focus being more on what changes have occurred in DNA of the cancer cell rather than the organ or tissue in which the cancer is found.

There are about 200 different types of cancer. Cancers can develop in any organ but are most commonly found in the lungs, prostate, breast and ovaries, large intestine, stomach, and pancreas.

How is cell division normally controlled?

In normal cells the rate of cell division is controlled by two types of genes:

- **proto-oncogenes** – these stimulate cell division
- **tumour suppressor genes** – these slow or halt cell division.

Proto-oncogenes

In normal cells, growth factors attach to specific receptor proteins found in the cell surface membrane. When the growth factors bind they activate ('switch-on') genes which control DNA replication and stimulate cell division. An example of one of these types of proto-oncogene is *Ras*.

Other types of proto-oncogenes control the production of growth factors, cyclins, and CDKs (Topic 8.1, The cell cycle) and are regulated by factors within cells.

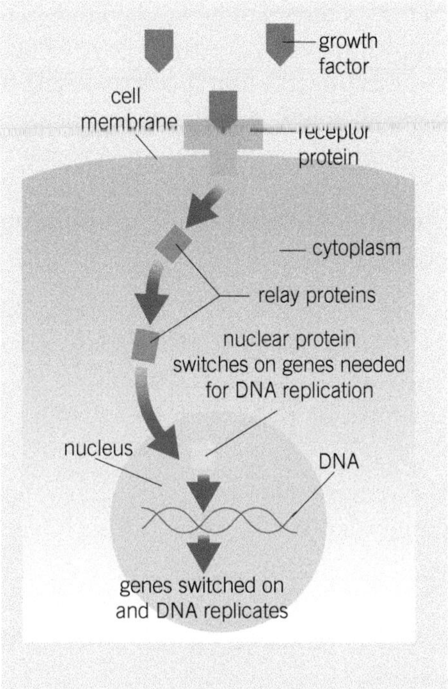

▲ **Figure 3** *Growth factors act as signalling molecules and initiate cell division*

Problems occur when changes occur in the nucleotide base sequence of these proto-oncogenes, that is, when they are mutated. When a proto-oncogene mutates it becomes an **oncogene**. These oncogenes can lead to uncontrolled cell division in one of two ways:

- Some oncogenes produce different forms of receptor proteins which can trigger DNA replication even when the extracellular growth factor is absent.
- Some oncogenes may cause the growth factors to be made in abnormally high quantities.

Synoptic link

You learnt about cell cycle and mitosis in Topic 8.1, The cell cycle and mitosis.

Tumour suppressor genes

Tumour suppressor genes also play a role in the development of tumours. Normally the tumour suppressor genes code for proteins that prevent cells from dividing. They also cause the breakdown of cells with damaged and faulty DNA by the process of apoptosis. If the suppressor genes, therefore, are mutated this 'stop' signal is lost and the cells continue to divide (and replicate their damaged DNA if there is any present).

Synoptic link

You learnt about apoptosis in Topic 8.3, Apoptosis.

Examples of unregulated cell division

There are many different oncogenes and genes which can result in cancer. Three of them are described here.

Ras

In a normal healthy cell, division is only triggered by the *Ras* gene product when the cell is stimulated, for example, by a hormone. However, when the *Ras* gene is mutated and becomes an oncogene, it stimulates cell division continually (even when the hormone is not present) and the cell cycle inhibition is removed.

c-Myc

In the human genome, *c-Myc* is a regulator gene located on chromosome 8 and it is thought that it is responsible for regulating the expression of 15% of all genes. A mutated version of *c-Myc* is found in many cancers. The mutated *c-Myc* causes the *Myc* gene to be constantly expressed, that is, a transcription factor is constantly produced. This leads to the unregulated expression of many genes, some of which are involved in cell proliferation, and results in the formation of cancer.

p53 gene (C)

Human p53 is 393 amino acids long. p53 is another transcription factor which plays a role in apoptosis, genomic stability, and inhibition of uncontrolled cell division. p53 acts in several ways which have an anticancer effect:

- When DNA is damaged p53 can activate the proteins which repair the DNA (an important factor in the process of ageing)
- It can temporarily stop the cell cycle at the G1/S regulation point which allows the DNA repair proteins sufficient time to repair the DNA damage before the cell proceeds through the rest of the cell cycle
- It can also initiate apoptosis.

Study tip

Remember proto-oncogenes are correctly functioning genes which control the cell cycle – oncogenes are mutated forms of proto-oncogenes which can lead to the development of tumours and cancer.

Mutagens and carcinogens

Not all cancer is caused by spontaneous errors in DNA replication, repair, and recombination. A **mutagen** is a physical or chemical agent that changes (mutates) the DNA of an organism and increases the frequency of mutations above the natural background level. As many mutations increase the risk of cancer, mutagens are therefore also likely to be **carcinogens**.

Mutagens may be:

- physical (e.g., ionising or ultraviolet radiation)
- chemical (e.g., benzene, arsenic, asbestos, or alcohol)
- biological (e.g., a virus.)

In healthy cells damaged DNA is normally repaired successfully, so it usually takes a number of exposures to a mutagen before the DNA becomes permanently damaged.

Summary questions

1 Outline the difference between a tumour and cancer. *(3 marks)*

2 Describe precisely which part of the cell cycle is affected by the *Ras* oncogene. *(2 marks)*

3 Suggest why a mass of abnormal cells in a plant would not be called 'cancer'. *(2 marks)*

14.4 Detecting cancer

Specification references: 3.3.1

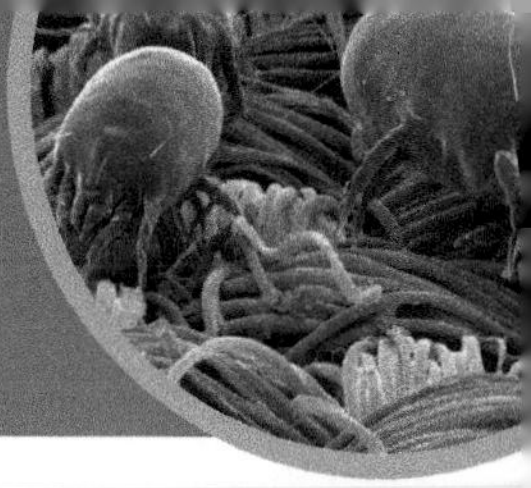

Methods of detection

The success of screening programmes in reducing mortality rates relies on the early detection of cancer and precancerous cells. You will cover screening programmes in Topic 14.5, Screening for cancer. Figure 1 shows the effect of the introduction of the cervical screening programme in the 1980s.

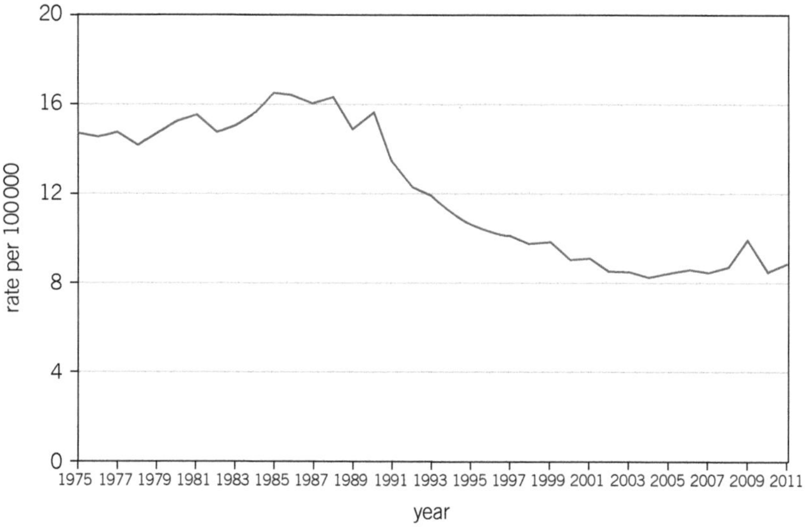

▲ **Figure 1** *Changes in the cervical cancer incidence rate due to the introduction of the screening program*

There are many different techniques which can be used to detect cancers. Each technique is appropriate for the use in detecting different forms of cancer.

Learning outcomes

Demonstrate knowledge, understanding, and application of:

→ how cancers are detected by – MRI scans, X-rays, mammography, CT scans, ultrasound, PET scans, biopsies, and blood tests.

X-rays

X-rays are a form of high energy radiation. They are one of the most established forms of medical imaging. An X-ray is an image of the internal structure of the body, produced by placing the relevant part of the body between a controlled source of X-rays and sensitive photographic film. X-rays can pass easily though soft tissues and the corresponding areas of the film that are exposed to the full amount of X-rays turn black. Dense areas such as bones absorb the X-rays very efficiently and so those areas of the film appear white. Other organs and muscles absorb different levels of X-rays and so appear as different shades of grey.

Mammography

Mammography uses low energy X-rays to examine human breast tissue. It can be used to detect small tumours in the breast. Also in the early stage breast cancer, there may be no tumour, but a mammogram may show small areas of calcium in particular patterns within the breast tissue. These areas of calcium are called calcification. Not all

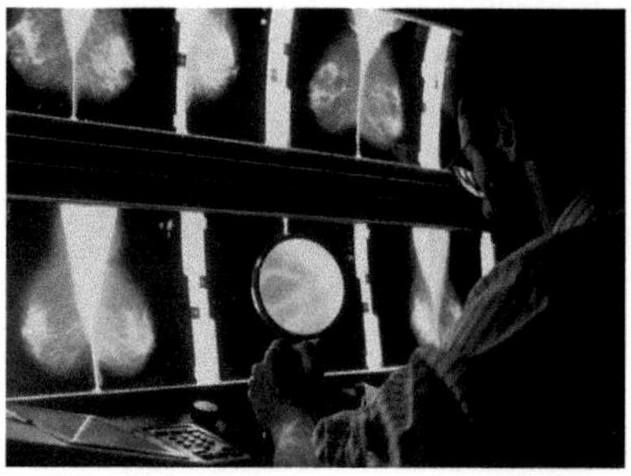

▲ **Figure 2** *A radiographer studying a mammogram*

areas of calcium indicate cancer and this can be established with further diagnostic tests.

During a mammogram a radiographer will position one breast at a time between two small flat plates on the X-ray machine. The plates then press the breast firmly between them for a few moments whilst the X-ray is taken. The compression of the breast helps to give a clear picture.

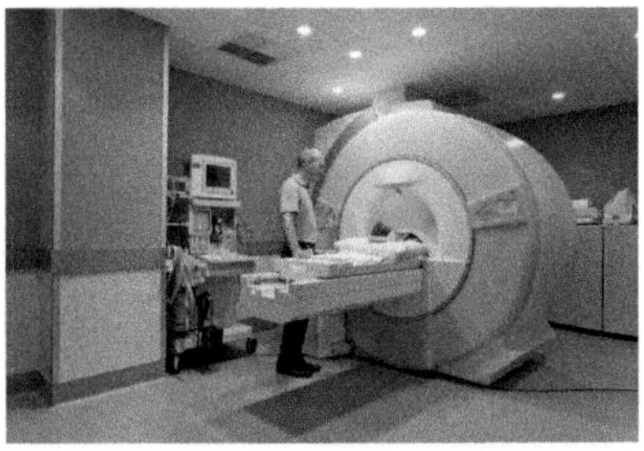

▲ **Figure 3** *A CT scanner*

CT scans

A computerised tomography (CT) scan, also known as computerised axial tomography (CAT), produces detailed images of the internal structure of the body.

During a scan the patient usually lies on their back, strapped onto a flat bed. An X-ray tube rotates around the patient and the patient is continuously moved through this rotating beam. On the opposite side of the body the X-rays are detected and a tomogram (an image of the scan) is produced by a computer.

Tomograms provide more detail than a standard X-ray and can show blood vessels, bones, and tumours. The scan is painless and will usually take between five and ten minutes depending on the part of the body being scanned.

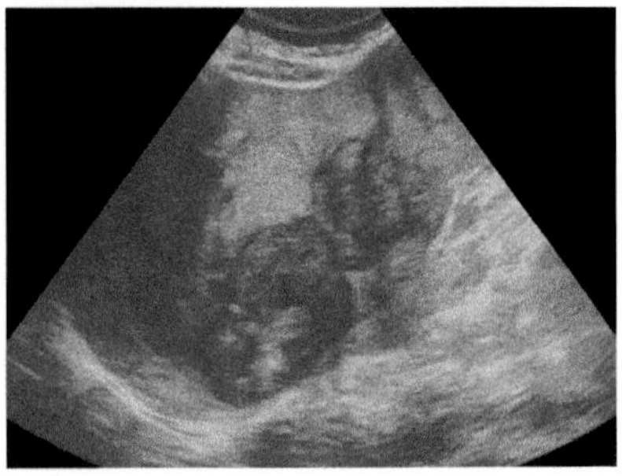

▲ **Figure 4** *An ultrasound scan of cancerous tumours (centre) in a liver. This is a secondary cancer that has metastasised from a primary colon cancer*

Ultrasound

An **ultrasound** scan creates an image of an organ or area inside the body using high frequency sound waves (a frequency beyond what humans can hear). As sound waves are used rather than X-rays the procedure is very safe. A handheld transducer which is moved over the skin's surface delivers the sound waves. These sound waves are then converted into an image by a computer linked to the transducer. The pictures generated are analysed by trained healthcare professionals.

Ultrasound scans can be used to detect tumours in soft tissues of the body. This technique is much cheaper than other techniques and also has the advantage that the machine is relatively portable.

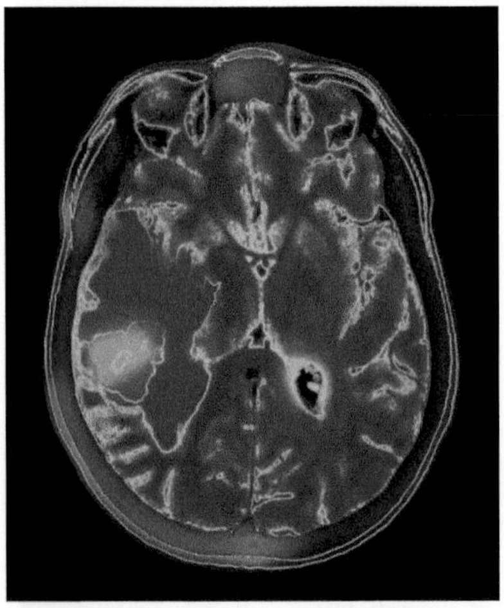

▲ **Figure 5** *A coloured Magnetic Resonance Imaging (MRI) scan of a section through the brain showing a metastatic tumour. At centre left is the tumour (yellow) surrounded by damaged fluid-filled tissue (red)*

MRI scans

Magnetic resonance imaging (MRI) scans can be used show soft tissues across most areas of the body. A single scan can produce multiple pictures from many angles around the body. MRI is especially good for looking at some types of soft tissue sarcomas, primary bone tumours, and brain and spinal cord tumours. MRI scans are less suitable for some types of tumour as the images are affected by movement and won't show up as clearly, for example, swallowing, breathing, or coughing.

Several safety considerations need to be taken into account as the MRI machine requires the patient to lie inside a very large magnet. This means the patient will be asked about existing medical conditions that may include cochlear implants, pins, plates, surgical clips or pacemakers. Metal items such as jewellry and the contraceptive coil

(IUD) need to be removed. The patient may be asked to hold their breath for short periods of time to improve the quality of the images.

The magnet creates a magnetic field that lines up the protons in hydrogen atoms (for example in water molecules). A beam of radiowaves then 'spins' the protons and the signal emitted is converted into a three-dimensional (3-D) detailed image.

PET scans

Positron emission tomograph (PET) is a medical imaging technique used to produce 3-D images of functional processes in the body. The scan uses small amounts of tracer (radioactive drug) to show differences between healthy and diseased tissue. FDG (fluorodeoxyglucose) is the most commonly used tracer. Prior to the scan, the patient is injected with a small amount of FDG. The uptake of FDG is a marker for glucose uptake. Cancerous cells metabolise at a faster rate than healthy tissue and therefore absorb more FDG. The PET scanner then detects gamma radiation which the FDG gives off. This produces a series of coloured images which reveals cancerous and healthy tissue.

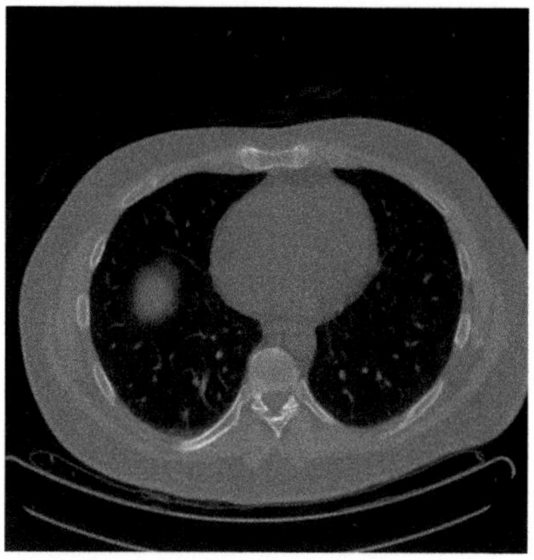

▲ **Figure 6** *A positron emission tomography (PET) scan of a tumour in the left upper lobe of the lungs*

A lot of PET scanners also include a standard CT scanner. Both PET and CT scans can be taken at the same time giving images of both function and anatomy, respectively. As well as cancer detection PET scans can also be used to observe, monitor, and therefore diagnose problems with blood flow to the brain or heart.

Biopsies

A **biopsy** is a procedure that removes a piece of tissue or a sample of cells from your body to be analysed in a laboratory. While imaging tests, such as X-rays, are helpful in detecting masses or areas of abnormality, on their own they can't differentiate cancerous cells from non-cancerous cells. For most cancers, a biopsy is needed to make a definitive diagnosis.

- Bone marrow biopsies – commonly used to diagnose blood cancers, such as leukaemia, lymphoma, and multiple myeloma. This biopsy may also detect cancers that started elsewhere and travelled to the bone marrow. During the biopsy, a doctor draws a sample of bone marrow out of the back of the hipbone under a local anaesthetic.
- Endoscopic biopsies – can be carried out through the mouth, rectum, urinary tract, or a small incision in the skin. Endoscopic biopsy procedures can include cystoscopy to collect tissue from inside of the bladder, bronchoscopy to get tissue from inside a lung, and colonoscopy to collect tissue from inside the colon.
- Needle biopsies – these are often used when a tumour can be felt through the skin, for example in the case of enlarged lymph nodes or unusual breast lumps. If the suspected area can't be felt through the skin, the biopsy may be combined with an imaging technique (e.g., X-ray) to help guide the biopsy.

Smear tests are endoscopic biopsies that take a small sample of cells from the cervix. The cells are checked for abnormalities which could indicate a risk of them becoming cancerous.

Blood tests

Blood tests form an important part of diagnosing cancer and monitoring how patients are responding to treatment. For example, a blood test can be used to detect a protein called CA125 in the blood. CA125 is produced by some ovarian cancer cells and a very high level of CA125 in the blood may mean a person has ovarian cancer. However, CA125 is not specific to ovarian cancer and may also be raised in many benign conditions, so a raised level of CA125 does not definitely mean the person has ovarian cancer.

A prostate specific antigen (PSA) test is a blood test which can be used to help detect prostate cancer in men. Raised prostate concentrations, however, can also be due to:

- A urinary tract infection
- An enlarged prostate
- Infection/inflammation of the prostate (prostatitis).

Biomarkers such as CA 125 and PSA are detected using a technique called ELISA

Synoptic link

You read about ELISA and testing for antibodies in Topic 12.2.

Summary questions

1 Describe the procedure for a mammogram. (*5 marks*)

2 Outline the advantages and disadvantages of MRI and CT scans as techniques to detect cancers. (*6 marks*)

3 Explain why FDG is used to highlight cancerous cells. (*2 marks*)

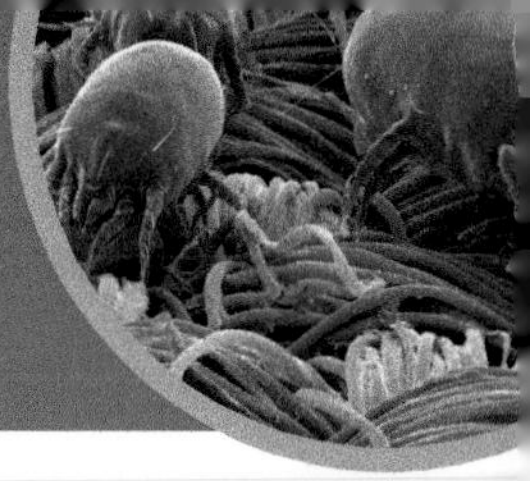

14.5 Screening for cancer

Specification references: 3.3.1

Screening

Screening is used to offer advice to individuals to enable them to make informed decisions and choices about their health. This has important ethical differences from clinical practice. Individuals must be aware that there are risks associated with some screening procedures and that screening programmes will offer limited information which may then need further investigation.

In England some of the screening programmes run by the NHS include:

- breast cancer
- cervical cancer
- bowel cancer
- prostate cancer.

What is screening?

Screening is where healthy individuals with no apparent symptoms are tested for certain diseases or conditions. The individuals may be at increased risk of a certain condition or disease. If this is the case they will be offered further information and tests, as well as appropriate treatment.

The aim of any screening programme is to enable a better outcome for the individual. For screening to make a real difference to their health, testing and intervention must give a better outcome than if they had waited for symptoms or signs and then a diagnosis. Many people are in favour of screening programmes that are carried out soon after birth to determine if a baby has inherited a genetic disorder – however, screening for diseases such as cancer and Alzheimer's which may occur much later in life is more controversial.

An important feature of screening is the context. For example, mammography can be used as part of diagnosis in a patient who has a breast lump, but can also be used as a screening test for a woman who is invited to a routine check-up.

Reliability of screening

Screening programmes have the potential to improve quality of life through early diagnosis, and even to save lives. They are not, however, 100% reliable or accurate. Screening cannot guarantee protection, only to reduce the risk of developing a condition. There is always a chance of false negative (wrongly reported as not having the condition) or false positive results (wrongly reported as having the condition). The UK National Screening Committee (NSC) increasingly presents screening as risk reduction to reiterate this point.

Learning outcomes

Demonstrate knowledge, understanding, and application of:

→ screening programmes

→ the ethical and economical considerations of screening programmes.

- *False negatives* may occur during a mammogram as a result of:
  - A woman having a cancer that doesn't show on an x-ray. This applies in about 7 in 100 breast cancer cases.
  - The limitation of the x-ray as a screening test. In some cases it is possible to look back at the screening mammogram and find the cancer as a tiny speck, but if every tiny speck was recorded as positive then all women would have positive screening results. Therefore, with hindsight it is detectable.
  - Human error. In some rare cases it is possible that the screener should have recorded a positive but failed to do so. This is minimised by the requirement to have a second observer and recording checks.
  - A cancer develops rapidly between one scan and the next.
- *False positives* can occur when mammogram results indicate the presence of cancer, but what is found ends up being benign:
  - The more mammograms a woman has, the greater the probability that she will have a false positive result that will require further tests.
  - Human error can result in a false positive being recorded, but follow-up tests and scans should then lead to the correct diagnosis.

A false positive result can cause fear and worry, but this should not deter people from taking part in screening programmes as it does not outweigh the benefit of mammography for most women. The overall aim is to detect as many cancers as possible, rather than to avoid false positive results.

Ethics of screening

Whether the screening programme is being used to determine the risk of developing cancer or any other disease or disorder, there are many ethical considerations that need to be taken into account:

- *Who should be screened?* The cost and time involved in a screening test can be high and funds are limited. Decisions need to be made to determine who will be offered screening and when, and who it will not be offered to.
- *What happens to the test result data?* There are legal implications regarding the data which is collected from any screening programme. Should all employers and insurance firms have a right to this information?
- *How does treatment influence decisions?* If treatment for a condition can wait until there are symptoms (e.g., for osteoarthritis), then screening gives no better outcome. If effective treatment does not exist (e.g., for motor neurone disease), then screening gives no better outcome.

- *What are the possible outcomes from false positives (or negatives)?* Will women decide to have mastectomies unnecessarily? Will they be exposed to further risk as a consequence, for example, the risk of surgery? What effect will this have emotionally on the woman (and her family)?

Summary questions

1 Discuss the reasons for and against a screening programme for breast cancer. *(4 marks)*

2 a Figure 1 shows data from the US indicating a strong correlation between income and regular mammography for women aged 50–74. Suggest reasons for this finding. *(3 marks)*

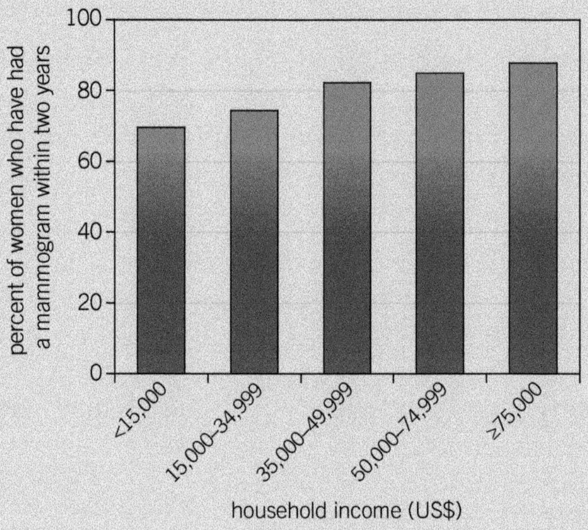

▲ Figure 1

b Consider the data plotted in the graph and suggest one improvement to the method used to collect the data. *(1 mark)*

3 Evaluate the use of HPV testing as a predictor of cervical cancer development. *(4 marks)*

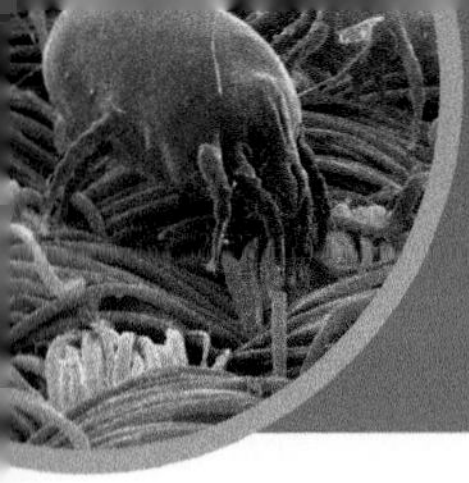

14.6 Genetic tests for cancer

Specification references: 3.3.1

Learning outcomes

Demonstrate knowledge, understanding, and application of:

→ genetic tests for cancer, including:
- BRCA gene testing
- HNPCC gene testing

→ the ethics of genetic tests.

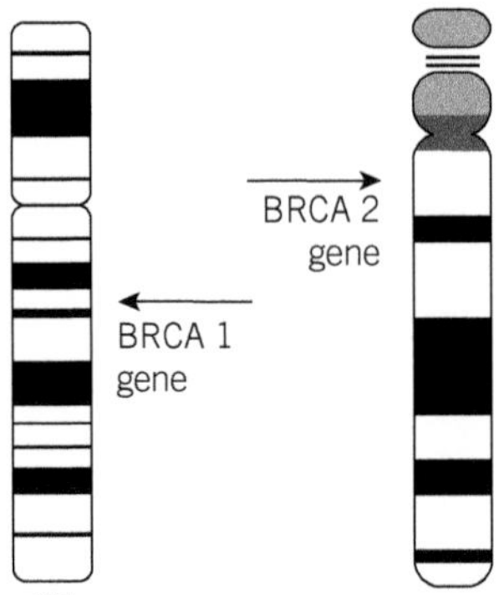

▲ **Figure 1** *BRCA1 and BRCA2 gene loci*

Study tip

BRCA is written in italics for the gene and non-italics for the affected protein.

Synoptic link

You learnt about the cell cycle in Topic 8.1, The cell cycle.

Testing for breast cancer

BRCA1 and *BRCA2* were the first two gene mutations for breast cancer to be identified. The mutations were discovered in the mid-1990s, through studies into families with a strong history of breast and ovarian cancer. It is thought that between 45 and 90% of women carrying *BRCA* mutations will be diagnosed with breast cancer at some point in their lives. There are other genes which have also been found to significantly increase the risk of a woman developing breast cancer, these are *TP53* and *PTEN*. Genetic tests are available to women with a high risk of having changes in these genes.

Genes for breast cancer

BRCA1 and *BRCA2* are genes that produce tumour suppressor proteins. These proteins help repair damaged DNA (Topic 14.3, Cancer). As a result, mutations in these genes will mean DNA damage may not be repaired properly. Affected cells are therefore more likely to develop additional genetic alterations that can lead to cancer.

Research has revealed other genes that when mutated can slightly increase the risk of developing breast cancer, these include – *CASP8, FGFR2, TNRCP, MAP3K1, rs4973768* and *LSP1*. Rare mutations can also increase the risk, for example in a gene called *CHEK2*. As yet, there are no specific genetic tests available for these genes – they may however be detected when the *BRCA* gene is being tested for.

Risks of developing breast cancer

One in eight women in the UK will develop breast cancer during their lifetime. The risk is low in young women but the risk increases with age. Women who have a faulty gene have a higher risk of developing breast cancer than people of the same age. However, the risk of a 30-year-old woman developing breast cancer is much lower than that of as an 80-year-old woman, whether or not she has a faulty gene.

Men who have the BRCA mutation are also at an increased risk of breast cancer. By age 70 the average risks are:

- Around 1% for *BRCA1* mutation carriers
- Around 7% for *BRCA2* mutation carriers.

These risks are still lower than that for women (12.5% for a woman with no *BRCA* mutation). Within the UK population it is estimated that 1 in 1 000 people have inherited either a *BRCA1* or *BRCA2* mutation, therefore, 1 in 500 people carry some form of *BRCA* mutation.

Who can have a genetic test for breast cancer?

Currently in the UK only individuals with a strong family history of breast cancer are eligible to be tested for *BRCA1, BRCA2, TP53*, or *PTEN*. In addition the individual also needs to have a living relative with breast cancer. The relative first needs to be tested to identify which gene may be faulty, this is known as a mutation search and can take many weeks if not months. If a faulty gene is found that same gene is tested for in the individual, this type of testing is known as predictive testing.

Testing for bowel cancer

Hereditary non-polyposis colorectal cancer (HNPCC) is a type of bowel cancer caused by a mutated gene. Approximately 2–5% of all cases of bowel cancer are due to HNPCC. Around 70% of women and 90% of men carrying the HNPCC mutation will develop bowel cancer by the time they are 70 years of age.

The HNPCC gene

HNPCC is caused by an inherited gene mutation or abnormality in genes that normally repair DNA, known as mismatch repair genes. At least five of these have been found to cause HNPCC. If the individual carries the HNPCC gene mutation, the mutation will be present in every cell, meaning other organs are susceptible to developing cancer. Uterine cancer is also very common and can be the main cancer in some families with HNPCC. Cancer in the ovary, urinary system, and gastrointestinal tract can also occur. Despite the fact that there is a high risk of cancer developing with the HNPCC mutation, appropriate check-ups, treatment, and risk awareness can help prevent the cancer and save lives.

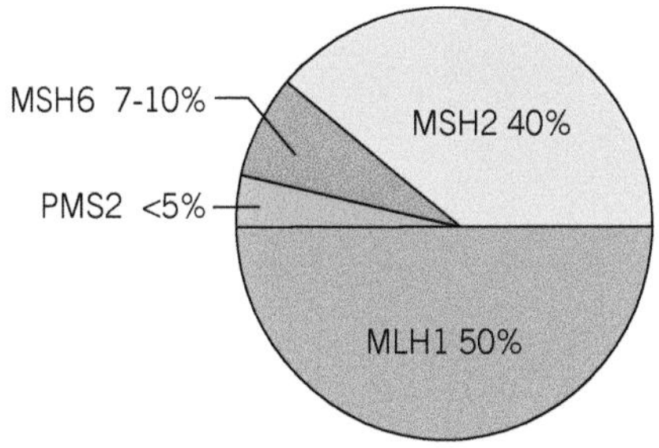

▲ **Figure 2** *Mutations in four mismatch repair (MMR) genes are known to cause HNPCC –* MLH1, MSH2, MSH6, *and* PMS2. *This chart shows the proportions of HNPCC cases caused by each. Recently, a fifth gene, called* EPCAM *or* TACSTD1, *was found to account for a small percentage of HNPCC cases*

Interpreting genetic tests

Results of genetic testing for *BRCA* or HNPCC genes may show:

- a *positive genetic test* – indicates that the gene mutation is found. This does not mean that the person will get cancer, but it does mean they are at increased risk of developing colon cancer during their life. The level of risk depends on the type of gene mutation and whether the individual continues with screening programmes to help reduce their risk. Consultation(s) with a genetic counsellor can help the individual to understand their risks on a personal basis.
- a *negative genetic test* – the absence of the gene mutation has more complicated implications. In cases where other family members have cancer with a specific gene mutation but the individual does not have that gene mutation then there is no increased risk for that individual, it is the same as the general population. If the individual is the first person within the family to be tested a negative result can be misleading as it may be that the individual has a mutation that cannot be detected (or is even known about) for that type of cancer so they could still be at high risk.
- a *gene variation of unknown significance* – the result of a genetic test is not always conclusive, for example it may not give a clear yes or no about the risk of cancer. A gene mutation may be identified which is not fully understood, so it is not possible to know the significance. Genetic counselling is important to help the individual understand the implications of the results.

Benefits and risks of genetic testing

It is generally thought that the direct medical risks of genetic testing are minimal. However, the outcome of test results may have harmful effects on a person's emotions, social relationships, finances, and medical choices. There can be benefits to genetic testing irrespective of whether a person receives a positive or a negative result.

Table 1 *Potential risks and benefits from the results of genetic tests*

	Positive test result	Negative test result
Potential benefits	Relief from uncertainty. It allows individuals to make informed decisions about their own future based on their own risks and test results Equips people to make informed decisions about their future, for example, the appropriate treatment course May be opportunity to take part medical research that could help reduce deaths from cancer in the long run.	A sense of relief. The possibility that special check-ups, tests, or preventive surgeries may not be needed.
Potential risks	Can cause anger, anxiety, and depression. May be difficult to make choices about whether to have preventative surgery or which surgery to have. May be implications in terms of health, employment, and travel and life insurance policies.	Individual may experience 'survivor guilt', caused by the knowledge that they do not have an increased risk of developing a disease that affects one or more loved ones.

The results of genetics tests within a family can cause emotions which can cause tension. Results can affect personal decisions such as whether to get married or have children. If by a small chance the test results are inaccurate people may make important decisions based on incorrect information.

Summary questions

1 What is the chance of a child of someone with a *BRCA* mutation inheriting the same mutation? Explain your answer *(2 marks)*

2 Discuss the statement 'people considering genetic testing for *BRCA1* and *BRCA2* mutations should talk with a genetic counsellor.' *(5 marks)*

3 Some genetic test results are described as a 'true negative'. Suggest what is meant by this term. *(2 marks)*

Cancer treatment

In 2011, more than 28 000 people died from lung cancer in England, which is a significant reduction from the 32 000 who died from lung cancer in 1990. According to figures from the National Cancer Intelligence Network (NCIN), the number of people surviving lung cancer for at least one year has almost doubled over the last 20 years.

Learning outcomes

Demonstrate knowledge, understanding, and application of:

→ Different methods of treating cancer.

Surgery

Surgery can be undertaken to remove tumours in some cases. For example, for breast cancer lumpectomy can be carried out to remove the tumour and a border of tissue surrounding the tumour. In some cases a mastectomy may be needed, which involves the removal of the entire breast. The lymph nodes may also be removed at the same time if it is likely that the breast tumour has spread into the lymph nodes through metastasis. After a mastectomy a woman is often offered reconstructive surgery.

Surgery is also commonly used in the treatment of colon cancer. The type of operation depends on where the tumour is in the colon. The surgeon makes a cut in the abdomen to remove the part of the colon containing the tumour. This operation is called a **colectomy**. How much the surgeon takes away depends on the exact position and size of the cancer. Again, the surgeon will remove the lymph glands closest to the bowel in case any cancer cells have spread there. The ends of the colon are joined back together by an anastomosis. On occasions it can be necessary to bring the end of the bowel out as an opening on the abdomen called a stoma. The stoma is usually temporary and the ends of the bowel are joined back together in another operation a few months later. In between the patient must wear a colostomy bag over the opening of the bowel to collect the faeces.

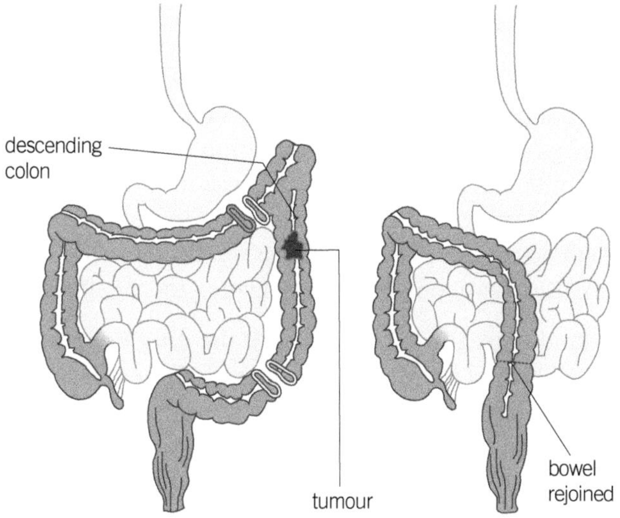

▲ **Figure 1** *A colectomy operation*

Chemotherapy

Chemotherapy involves using chemicals which are toxic to dividing cells. As cancer cells divide much more quickly can normal cells, the chemicals have a greater effect on the cancer cells. Normal cells that divide quickly, however, are also affected. These include cells in the bone marrow, cells lining the mouth and intestines, and hair follicle cells. Side effects depend on the chemicals used, the length of treatment, and the amount taken. Common side effects include:

- loss of appetite
- nausea and vomiting
- increased risk of infection (due to shortage of white blood cells)
- mouth sores
- hair loss
- fatigue (due to low red blood cell count)
- bleeding/bruising after injuries or minor cuts (due to low number of blood platelets)

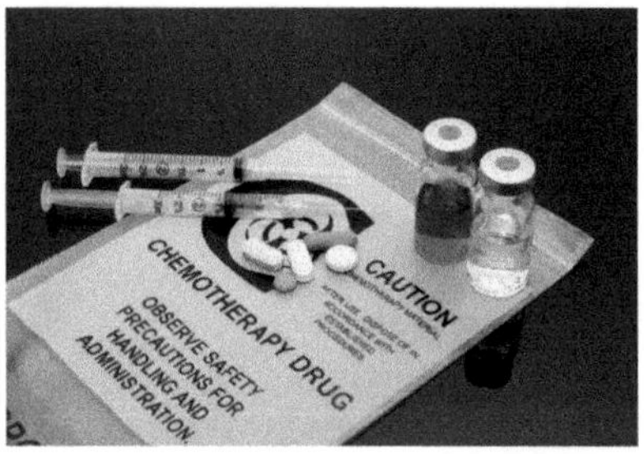

▲ **Figure 2** *Chemotherapy drugs*

Chemotherapy can be used at different stages of cancer treatment:

- before surgery to reduce the size of the tumour
- after surgery to ensure all the tumour cells have been removed
- to treat breast cancer that has spread or returned.

Radiotherapy

Radiotherapy uses ionising radiation to destroy cancer cells. This treatment destroys actively growing cells more than other cells. Ionising radiation damages DNA, particularly during DNA replication, which occurs more frequently in cancer cells. An advantage of this form of treatment is that it can be targeted very accurately to the tumour and therefore reduces the effect on healthy tissue. It can be used before or after chemotherapy.

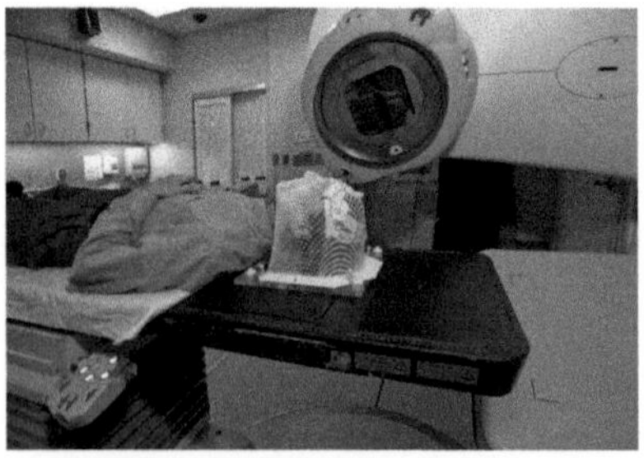

▲ **Figure 3** *A patient undergoing radiotherapy for a brain tumour while wearing a protective mask over their face*

Immunotherapy

It is possible to link anti-cancer drugs to monoclonal antibodies which are attracted to cancer cells. This is called immunotherapy. It is also possible to tag a monoclonal antibody with an enzyme that converts an inactive form of a cytotoxic drug (the prodrug) into the active form. Once injected these antibodies attach to the specific cancerous cells. The prodrug is therefore targeted to just the cancer tissue and can be administered in a high dose since healthy tissue is unaffected. This technique is called ADEPT (antibody-directed enzyme prodrug therapy).

An example of this kind of treatment is the use of the drug Herceptin which is used with 15–20% of breast cancer patients. Herceptin is a specialised antibody that has complementary binding sites for a specific protein receptor found on the surface of breast cancer cells. When Herceptin is bound to these protein receptors it prevents the cell from dividing rapidly.

Complementary therapies

Complementary therapy focuses on making the person feel better during other cancer treatments and to reduce any side effects such as tiredness, nausea, and diarrhoea. Complementary therapies include relaxation therapies, hypnotherapy, reiki, meditation, nutritional therapies (e.g., macrobiotics), acupuncture, stress management, magnetic field therapy, and aromatherapy. There is no evidence that these therapies will cure cancer. However by helping a person feel less anxious, they can have a positive impact, possibly by improving the effectiveness of the immune system.

Synoptic link

You learnt about the action of antibodies and cell surface antigens in Topic 12.2, Antibodies and immunity.

Hormone-related therapies

Tamoxifen is a hormone-related therapy that can be used to treat breast cancer. Oestrogen is a steroid hormone that stimulates gene transcription. Oestrogen diffuses through the cell surface membrane and binds to a specific receptor in the cell cytoplasm. This molecule then binds to another molecule called a coregulator that makes the whole complex bind to chromatin in the nucleus, causing transcription to take place. As a result it increases the growth and division of breast cancer cells.

Tamoxifen works by blocking the binding of the coregulator. The Tamoxifen/oestrogen receptor complex can still bind to the chromatin but in the absence of the coregulator, transcription can't take place. This form of treatment is not suitable for all types of breast cancer but it is very effective in treating some breast cancers in the short term. After 2–3 years the cancer can develop a resistance to the drug and the woman is instead prescribed different drugs that reduce the levels of oestrogen in the circulatory system.

Synoptic link

You first looked at cell surface membranes and surface receptors in Topic 1.9, Cell membranes.

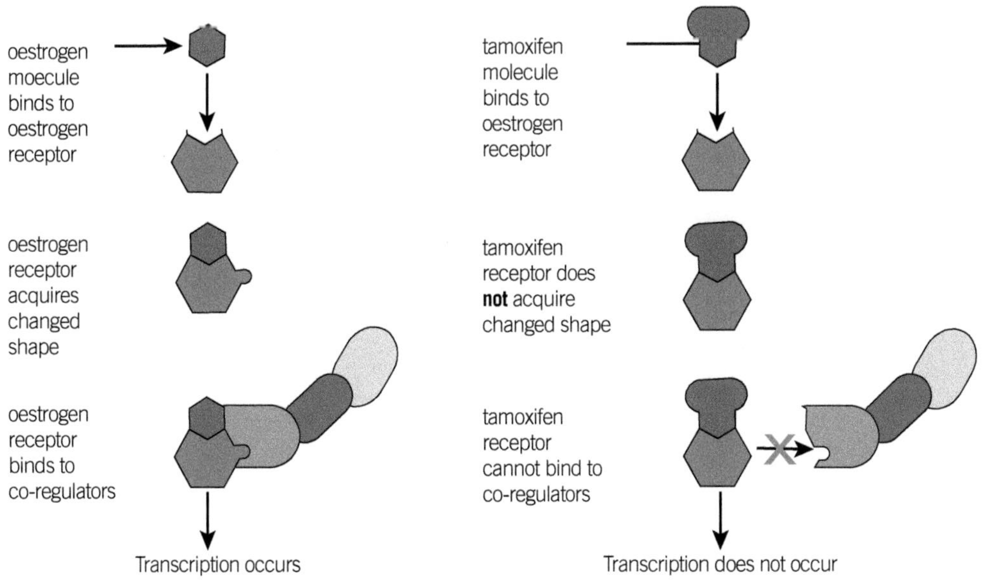
oestrogen moecule binds to oestrogen receptor
oestrogen receptor acquires changed shape
oestrogen receptor binds to co-regulators
Transcription occurs
tamoxifen molecule binds to oestrogen receptor
tamoxifen receptor does **not** acquire changed shape
tamoxifen receptor cannot bind to co-regulators
Transcription does not occur

▲ **Figure 4** *How Tamoxifen blocks transcription in cancer cells*

Factors that can alter the effectiveness of Tamoxifen

The body uses an enzyme called CYP2D6 to convert Tamoxifen into its active form. Two situations can interfere with the body's ability to make this happen:

- Having an abnormal CYP2D6 enzyme – about 10% of people have a CYP2D6 enzyme that doesn't function as well as it should. Having an abnormal CYP2D6 enzyme could stop a person from getting the full benefit of Tamoxifen.
- Taking drugs that interfere with CYP2D6. Some medications block the activity of CYP2D6 to various degrees. These include some antidepressants such as serotonin-specific reuptake inhibitors (SSRIs) and serotonin-norepinephrine reuptake inhibitors (SNRIs). Blocking CYP2D6 activity can interfere with the activation of Tamoxifen – reducing its potency as an anti-cancer treatment.

1 Describe the relationship between the shape of Tamoxifen and oestrogen. Explain your answer.
2 Competitive enzyme inhibitors work by blocking the active site of the enzyme, while non-competitive inhibitors work by binding at a place away from the active site and altering the shape of the active site. Suggest which type of inhibition would lead to the greatest long-term reduction of Tamoxifen effectiveness.

Summary questions

1 Suggest why people who visit a patient receiving radiotherapy are advised to stand further away from the hospital bed. *(2 marks)*

2 Explain why Herceptin is not suitable for all forms of breast cancer. *(2 marks)*

3 Suggest why oestrogen is able to cross the cell surface membrane and why the coregulator complex can enter the nucleus. *(2 marks)*

Practice questions

1 Which of the following statements is/are true of proto-oncogenes?

1 They can be genes that code for growth factors.

2 They can be genes that code for a cell membrane receptor protein.

3 They can be genes that code for proteins that initiate apoptosis.

A 1, 2 and 3

B Only 1 and 2

C Only 2 and 3

D Only 1 *(1 mark)*

2 Which of the following statements is/are true of CT scans?

1 They can be used to build up a three-dimensional image.

2 They use radio waves.

3 They indicate the parts of the body that are warmest.

A 1, 2 and 3

B Only 1 and 2

C Only 2 and 3

D Only 1 *(1 mark)*

3 Which of the following detection methods uses x rays?

A Thermography

B Mammography

C MRI scans

D PET scans *(1 mark)*

4 Which of the following statements is/are true of metastasis?

1 Tumour cells travel through blood vesels.

2 The tumour cells are usually benign.

3 Metastasis occurs during an early stage of cancer (stage 1 or 2).

A 1, 2 and 3

B Only 1 and 2

C Only 2 and 3

D Only 1 *(1 mark)*

5 Which of the following statements is/are true of the potential treatments for cancer?

1 Complementary therapies target the patient as a whole, rather than only the part of the body with cancer.

2 Immunotherapy targets receptors on cancerous cells.

3 Chemotherapy targets cells that are dividing rapidly.

A 1, 2 and 3

B Only 1 and 2

C Only 2 and 3

D Only 1 *(1 mark)*

6 The graph shows the cumulative risk (%) of developing breast cancer in BRCA-1 mutation carriers.

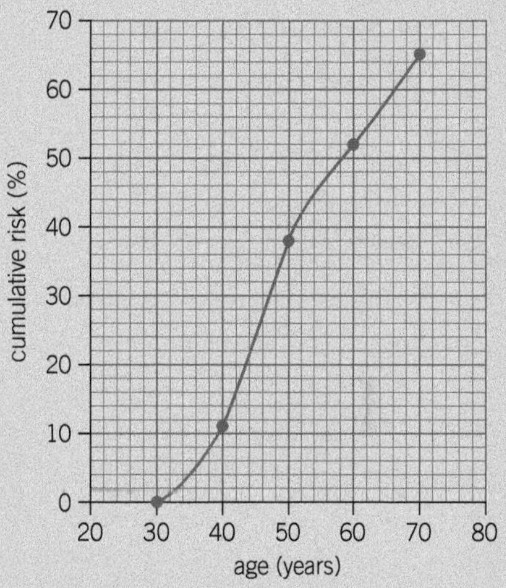

a (i) Suggest what is meant by the 'cumulative risk' of developing breast cancer. *(2 marks)*

(ii) Using the information from the graph describe the effect of age on the cumulative risk of developing breast cancer in BRCA-1 mutation carriers. *(2 marks)*

(iii) Explain the difference between the causes of sporadic breast cancer and familial breast cancer. *(2 marks)*

b Breast cancer mortality in the UK has decreased in all age groups since the 1990s.

The figure below shows the mortality rates from breast cancer in females in the UK between 1988 and 2008.

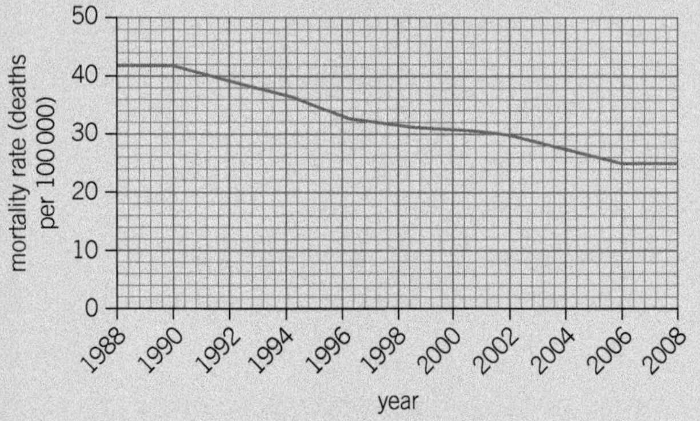

(i) Calculate the percentage decrease in mortality rate between 1990 and 2007. Give your answer to the nearest whole number. *(2 marks)*

(ii) Calculate the rate of decrease in mortality rate between 1990 and 2000. *(2 marks)*

**Level 2 maths*

(iii) Suggest three reasons for this reduction in mortality rate from breast cancer. *(3 marks)*

c For breast cancer to develop, the cells will need to have more than one gene mutation.

Suggest why some carriers of the BRCA-1 gene mutation do not develop breast cancer. *(2 marks)*

[question 1, Jan 2011 f222 OCR]

7 a Recent research suggests that a high intake of soy food reduces the risk of bowel cancer in post-menopausal women.

Suggest two other ways in which women may reduce their risk of bowel cancer. *(2 marks)*

b The figure below shows the incidence and mortality rate for bowel cancer in males and females in England between 1970 and 2005.

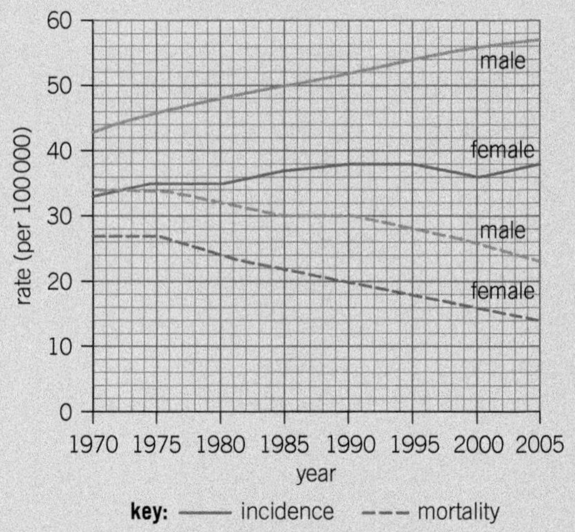

Using the information in the graph, compare the trends in the incidence and mortality rate for bowel cancer in males and females. *(4 marks)*

c The NHS Bowel Cancer Screening Programme in England began in April 2006. In England, people aged 60-69 are now offered screening every two years.

Suggest two reasons for introducing cancer screening programmes, such as the Bowel Cancer Screening programme. *(2 marks)*

d Bowel cancer develops over a period of several years.

Statements **A** to **E** describe the stages in the development of bowel cancer.

A a primary tumour develops

B formation of an oncogene

C cancer cells break away from the primary tumour and form a secondary tumour

D a mutation occurs in a proto-oncogene in cells lining the bowel

E cells lining the bowel continue to divide by mitosis

Use the letters to put the stages into the correct order. *(4 marks)*

e Suggest two factors that would need to be controlled when studying the effect of soy food intake on the risk of bowel cancer. *(2 marks)*

[question 1, June 2011 f222 OCR]

8 The graph shows the incidence of malignant melanoma in males and females, of different ages, in the UK.

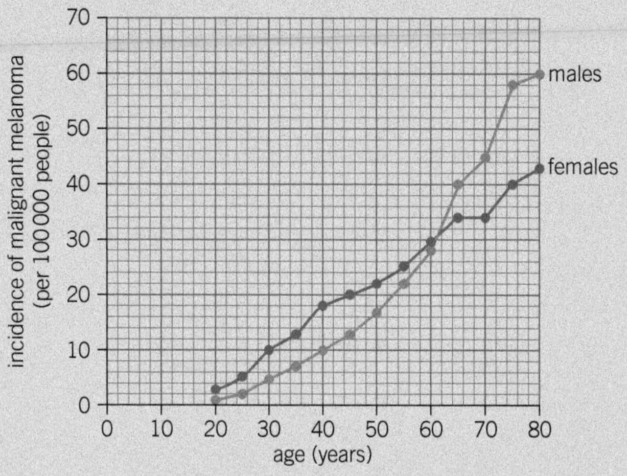

a (i) Using the information in the graph, describe one similarity and one difference in the incidence of malignant melanoma in males and females. *(3 marks)*

(ii) Suggest one reason for the similarity and one reason for the difference described in 1(a)(i). *(2 marks)*

b Cell growth and cell division are controlled by cell signalling pathways.

The figure shows the BRAF protein and other proteins in a signalling pathway that normally controls cell division.

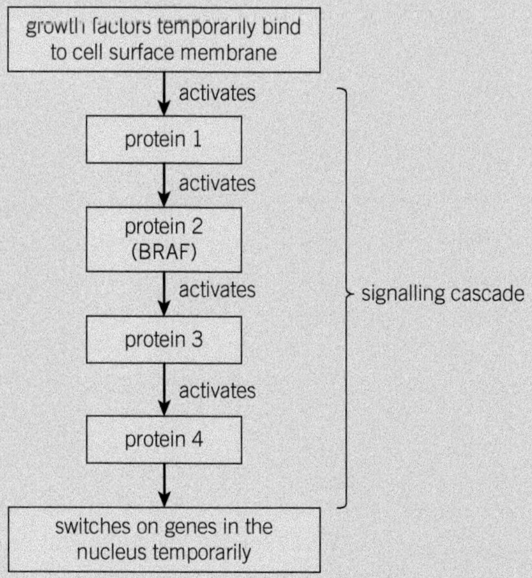

(i) Suggest what feature of the cell surface membrane allows growth factors to bind. *(2 marks)*

(ii) Describe the effect of the BRAF gene mutation on the signalling pathway. *(2 marks)*

c The altered BRAF protein has a significantly different shape to the normal BRAF protein.

(i) Suggest why this difference in shape makes the altered BRAF protein an excellent candidate for drug targeting. *(2 marks)*

(ii) The BRAF protein is an enzyme.

Suggest how a drug might act to reduce or stop the activity of the altered BRAF enzyme. *(2 marks)*

d A research team carried out a phase 3 clinical trial to compare the effects of two drugs in treating malignant melanoma.

(i) Describe what is meant by a phase 3 clinical trial. *(2 marks)*

(ii) Most phase 3 clinical trials are randomised.

Suggest how a randomised trial would be set up to compare the effects of the two drugs. *(2 marks)*

[question 1, June 2013 f222 OCR]

9 Outline the cellular basis of cancer. *(6 marks)*

15 RESPIRATORY DISEASES

15.1 Pollutants and lung disease

Specification references: 3.3.2

Learning outcomes

Demonstrate knowledge, understanding, and application of:

→ the short-term and long-term effects of pollutants on the lungs

→ the causes, symptoms, and treatment of asthma, lung cancer, chronic bronchitis, and emphysema

→ the difference between acute and chronic diseases.

▲ **Figure 1** *The* Aspergillus *fungus. The spores (brown in the picture) can be inhaled and cause lung disease, ×190 magnification*

Synoptic link

You learnt about allergies in Topic 12.2, Antibodies and immunity.

You have seen how diseases can be caused by pathogens (Topic 11.1, Communicable diseases) or genes (Topic 14.1, Non-communicable diseases). Pollutants represent another potential cause of disease. The lungs are the interface between the air and the body. They are exposed to environmental pollutants, some of which have the potential to cause disease. Here some of the lung diseases that can result from inhaling pollutants will be considered.

Pollutants

A pollutant is a substance released into the environment that can harm organisms or the environment itself. For example, tobacco smoke contains many chemical pollutants, including carbon monoxide, tar, and nicotine. Asbestos has been used as a building material because of its tensile strength and heat resistance. However, asbestos fibres pollute the air and, if inhaled, some types can cause lung cancer. Fungal spores can act as air pollutants. Inhalation of spores from the *Aspergillus* mould will produce a condition called aspergillosis, especially in people with weakened immune systems.

Pollutants exert short-term effects such as irritation to tissues or allergic reactions. A substance that causes an allergic reaction is known as an allergen. Here you will examine a common allergic condition – asthma.

Extended periods of exposure to a pollutant can lead to longer term effects, such as **chronic** disease. Smoking regularly increases the risk of lung cancer and other chronic lung diseases.

Diseases can be acute or chronic

An **acute** condition has a rapid onset but a relatively quick recovery time. A chronic condition often develops more slowly but lasts a long time and might be incurable. For example, acute bronchitis can be caused by viral or bacterial infection and tends to last a couple of weeks. Chronic bronchitis, which we will discuss in this topic, is often the result of smoking cigarettes and is a long-term condition.

Asthma

Asthma is a result of inflammation and narrowing of the bronchi in the lungs. Several triggers, such as fur, exercise, cold air, and tobacco smoke, can irritate the bronchi and initiate the symptoms of asthma. Although asthma is a chronic condition, a severe onset of symptoms is considered an 'acute asthma exacerbation' or asthma attack.

Causes

How a person acquires asthma is poorly understood, but both genetic and environmental factors play a role. A person is more likely to have asthma if their parents have the condition, which suggests it is heritable. Dozens of different gene variants have been linked to the risk of developing asthma, but results from studies are not consistent. Epigenetic changes may also cause the disease.

An increased risk of asthma has been linked with a range of environmental factors, such as smoking cigarettes, the inhalation of other air pollutants, high ozone levels, and allergens such as dust mites and mould. The majority of cases are thought to be a result of allergies. Some scientists have hypothesised that a reduced exposure to non-pathogenic bacteria could have increased the number of people with asthma.

Symptoms

Asthma symptoms include coughing and wheezing, difficulty breathing, and chest tightness. The severity of the symptoms will vary widely. In the bronchi and bronchioles, mucus production increases, the smooth muscle in the walls contracts, and inflammation causes swelling in the walls – these events combine to narrow the lumen and restrict air flow.

Treatments

Asthma cannot be cured, but treatments can help to manage the condition. In the UK, one in twelve adults receives treatment for asthma. Treatments can be classed as fast-acting, to deal with acute symptoms, or long-acting. Inhalers are used to deliver the treatments.

Beta-agonists

Beta-agonists tend to be used prior to exercise or during acute asthma attacks. They act as bronchodilators, relaxing the smooth muscle and widening the lumen of the bronchi. Some beta-agonists can be used, in combination with steroids, to provide long-term control of asthma symptoms.

Steroids

Corticosteroids are used to control asthma over the long term. They decrease the probability and severity of an asthma attack by reducing inflammation of the bronchi.

Lung cancer

Causes

As you learnt previously, cancers develop because of specific DNA mutations. These mutations are caused by carcinogens. Many chemical carcinogens are present in tobacco smoke. For example, benzopyrene inactivates *p53*, a tumour suppressor gene. Mutations to the epithelial cells that line the lungs can result in the formation of cancerous tumours.

Synoptic link

You learnt about the development of cancers in Topic 14.3, Cancer.

Symptoms

There are usually no signs or symptoms in the early stages. A tumour might eventually grow large enough to block the airways, causing breathlessness, wheezing, and a persistent cough. Sputum (mucus from the lungs) is likely to be stained with blood if the tumour damages lung tissue. Further symptoms include loss of appetite, fatigue, and unexplained weight loss.

Treatments

You will have learnt about the possible treatments for cancer in Topic 14.7, Treating cancer. Lung cancer can be treated using surgery, radiotherapy, or chemotherapy.

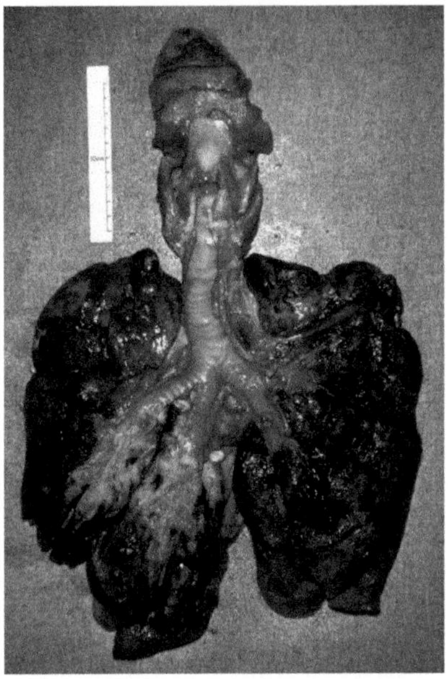

▲ **Figure 2** *A smoker's lungs affected by cancer*

Chronic bronchitis

Causes

Synoptic link

You learnt about some of the harmful chemicals in tobacco smoke in Topic 9.2, Pregnancy and fetal development.

You previously learnt about some of the harmful chemicals in tobacco smoke. Tar, another chemical released from cigarettes, can be deposited in the airways of a smoker. The tar causes inflammation, stimulates mucus production by goblet cells, and paralyses the cilia on ciliated epithelial cells. As a consequence, cilia are no longer able to move mucus up the airways. Mucus containing dirt and microorganisms builds up in the airways trapping bacteria, which can lead to infections.

The extra mucus reduces the diameter of the bronchi and bronchioles. Mucus accumulates in the alveoli and increases the diffusion distance for oxygen and carbon dioxide.

Symptoms

People with chronic bronchitis will experience shortness of breath, wheezing, fatigue, and blood-stained sputum. A persistent cough is one of the most obvious signs of the disease.

Treatments

There is no cure for chronic bronchitis, but, as with asthma, steroids and bronchodilators can be used to treat the condition and relieve the symptoms.

Emphysema

Causes

Large numbers of phagocytic white blood cells are attracted to sites of infection in the alveoli, which result from the presence of bacteria-filled mucus. These phagocytes produce an enzyme called elastase, which digests tissue to enable phagocytes to reach infections. The elastase breaks down proteins such as elastin in the alveoli walls. An inhibitor called alpha-1 antitrypsin (A1AT), which prevents this damage in healthy people, is deactivated in smokers.

The alveoli become enlarged and damaged and can burst, reducing the surface area for gas exchange. Elastin becomes permanently stretched. This prevents the normal elastic recoil of the alveoli. Air becomes harder to remove from the lungs, and stale air remains in the alveoli.

Symptoms

Chronic bronchitis and emphysema tend to occur simultaneously as a result of smoking. Together they are known as chronic obstructive pulmonary disease (COPD). Once again, shortness of breath is a major symptom – breathing will be shallower and exhalation more difficult. The loss of elastic recoil in emphysema means exhalation becomes an effort. Intercostal muscles are required to "pull" the ribcage down. This results in the development of a barrel chest.

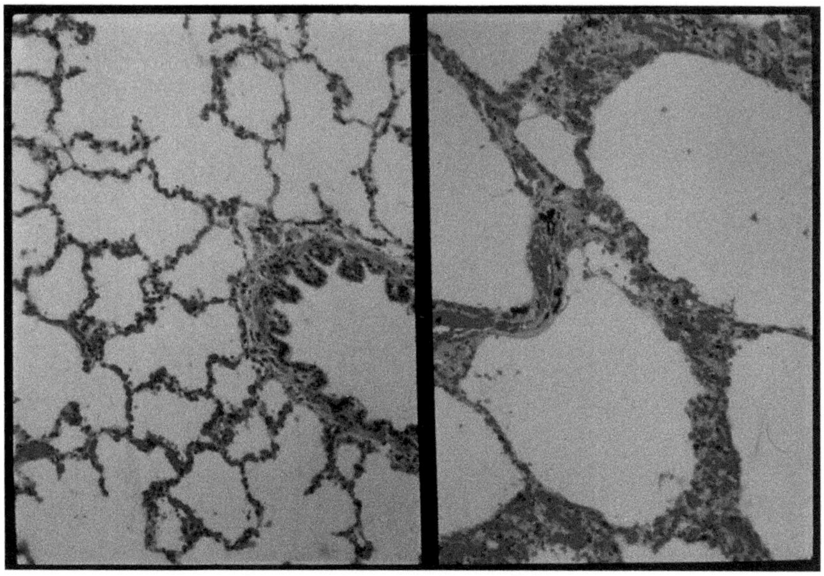

▲ **Figure 3** *Normal lung tissue (left) and lung tissue from a person with emphysema (right). In the case of emphysema the walls of the alveoli have broken down, meaning the spaces are larger, approximately ×40 magnification*

Study tip

Remember, you can be asked questions on learning outcomes you have covered but in a different context. For example, you learnt about protein synthesis in Topic 4.6, and about the role of cell organelles in protein modification in Topic 1.7. You also learnt about enzyme inhibition in Topic 3.5. A1AT is a glycoprotein that inhibits the elastase enzyme. You could be asked to describe how A1AT is produced within cells, how it is released, and how it might inhibit the elastase enzyme.

Treatments

Emphysema and chronic bronchitis are treated together as COPD using steroids and bronchodilators.

Does ozone worsen asthma?

Ozone concentrations are low at ground level. However, interactions between sunlight and air pollutants such as hydrocarbons and nitrogen oxides can produce low level ozone. Therefore ozone is classed as a pollutant at ground level.

Ozone is thought by many scientists to worsen the effects of asthma. Some epidemiological studies have found a correlation between ozone levels and the number of hospital admissions for asthma. However, the results of epidemiological studies are inconsistent.

An experimental result appears to support the theory that ozone worsens asthma. Asthmatic volunteers who were exposed to ozone showed increased airway inflammation and a heightened response to inhaled allergens.

1 Suggest and explain whether you would expect asthma cases to be more severe in urban or rural areas.
2 Suggest other factors that researchers would need to take into account when analysing epidemiological data linking ozone to asthma.

Study tip

Do not confuse 'acute' with 'severe'. Remember that acute diseases are short-lived, but the symptoms can be either relatively mild (e.g., a throat infection) or severe, depending on the disease.

Synoptic link

You can find an overview of the biology underlying allergies (Topic 12.2, Antibodies and immunity) and cancer (Topic 14.3, Cancer) earlier in this book.

Summary questions

1 Explain the difference between an acute disease and a chronic disease. *(2 marks)*

2 Why are smokers more likely than non-smokers to suffer from lung infections? *(4 marks)*

3 Outline the possible causes of asthma. *(3 marks)*

15.2 Medicinal drugs and clinical trials

Specification references: 3.3.2

Discovering potential drugs

You previously learnt about some of the drug treatments available for cancer in Chapter 14 and treatments for lung diseases in Topic 15.1. These medicinal drugs must pass through several stages of testing to ensure they are effective and safe before they can be used to treat patients. In this topic, you will look at how drugs are tested in **clinical trials** and the role of the National Institute for Health and Care Excellence (NICE) in overseeing this process and producing guidance and advice for health and social-care practitioners.

Many drugs we use are based on traditional folk medicine, which relies on plant products. Some estimates suggest that approximately 50% of prescription drugs are derived from chemicals in plants or are synthetic forms of plant products.

Chemicals from plants can produce physiological effects on the body that alleviate the symptoms of a disease. For example, theophylline, from cocoa beans and tea leaves, relaxes the smooth muscle surrounding bronchi and has anti-inflammatory effects. These properties make it useful in the treatment of asthma and COPD. Other plant chemicals exhibit anti-cancer or antimicrobial properties. For example, topotecan is derived from a product of *Camptotheca acuminata*, a tree, and is used to treat lung cancer. A compound from the Ethiopian pepper, *Xylopia aethiopica*, kills a range of bacteria.

Table 1 lists some examples of medicines that are derived from plants.

▼ **Table 1** *Medicines derived from plants*

Medicine	Plant source	Property	Use
Quinine	*Cinchona* sp.	Fever-reducing, painkilling, anti-inflammatory	Antimalarial
Aspirin	*Salix alba* (willow)	Anti-inflammatory, painkilling	Painkiller and anti-thrombotic
Theophylline	*Theobroma cacao* (cocoa)	Muscle relaxant, anti-inflammatory	COPD and asthma treatment
Paclitaxel	*Taxus brevifolia* (yew tree)	Inhibits mitosis	Anti-cancer drug

A small fraction of the world's plant species have been tested for potential therapeutic properties. Medicinal drugs might be present in these species, awaiting discovery. However, with thousands of species under threat of extinction because of deforestation and climate change, these plants might disappear before the chemicals in them are discovered.

Learning outcomes

Demonstrate knowledge, understanding, and application of:

- → the importance of plants as potential sources of medicinal drugs
- → the use of clinical trials in testing new medicinal drugs
- → the role of NICE in providing guidelines for medical treatment.

Synoptic link

You learnt about the treatment of cancer using drugs in Topic 14.7, Treating cancer.

▲ **Figure 1** *A chemical called theophylline, which can be used to treat COPD and asthma, is present in cocoa beans*

Clinical trials

Medicinal drugs must be tested before they can be licensed and made available to patients. Drugs are initially tested on cell cultures and animals in laboratories before they are tested on human patients. This process can last several years. If a drug is considered safe and effective in laboratory trials, only then will it enter clinical trials. Some estimates suggest that only 0.1% of new drugs reach the stage of clinical trials.

Prior to clinical trials, a detailed plan and justification for the trials is submitted to a research ethics committee (in the UK, this is the Medical and Healthcare Products Regulatory Agency). Researchers must prepare an information leaflet about their trials for patients. The research ethics committee checks its accuracy and clarity, and then decides whether to approve the clinical trials.

The information leaflet needs to be in plain language and outline:

- the research questions being asked
- who can take part in the trials and who cannot
- what treatment is being used
- the possible risks and benefits to participants
- who is funding and conducting the research.

Trials consist of four phases, which are intended to confirm that the drug is effective, assess the optimum dose, and check that no side effects exist.

Phase 1

Phase 1 trials usually involve 10–20 people. Patients will initially be given low doses, which increase until the upper limit of a safe dose is established.

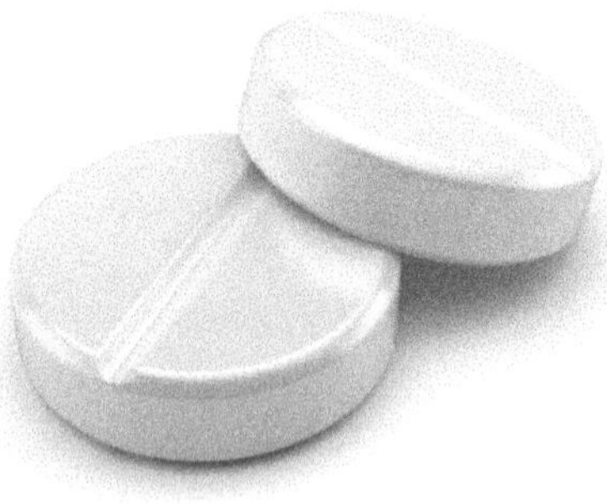

▲ **Figure 2** *Which one is the placebo? In clinical trials, the real drug and the placebo should look identical*

Phase 2

This phase assesses the effectiveness of the drug and whether it works well enough to enter Phase 3 trials. Phase 2 trials often use a **placebo**. A placebo is a tablet, pill, or injection that looks like the real drug but has no effect on the recipient. In these trials, 50% of people will receive the drug and 50% of people will receive the placebo. The effect of the drug can then be compared against the 'dummy' placebo treatment. Phase 2 trials use more participants (approximately 200) than Phase 1.

Phase 3

The new drug is compared against the best current treatment in Phase 3 trials. Many more participants (i.e., thousands) are used than in either of the previous phases. The new drug might produce only slightly better results than a current drug. A large sample size is required to detect a statistically significant difference. Phase 3 trials are conducted on patients with the disease – therefore placebos are unlikely to be used because it would be unethical.

Phase 3 trials are usually randomised. This means patients are randomly assigned to be treated with either the new drug or the established drug. This ensures that the two groups should be similar in terms of the ages and backgrounds of the patients. Alternatively, variables such as patients' ages, genders, and lifestyle can be taken into account when performing statistical tests on the data from the clinical trials.

▼ **Table 2** *What happens between the discovery of a drug and its appearance on the shelf*

	Drug discovery and pre-clinical testing	Phase 1 clinical trials	Phase 2 clinical trials	Phase 3 clinical trials	Licensing approval
What happens?	Laboratory development and testing of the potential drug	Checking for safety (10–20 healthy volunteers)	Checking for efficacy (approximately 200 patients)	Confirm findings with larger sample (>1000 patients)	Clinical results are compiled and submitted to regulatory agencies
Average duration (years)	5.5	7.0	8.5	11.0	12.5
Average cost (millions of £)	533	710	916	1100	1150

Phase 4

Phase 4 trials are conducted once the drug has been licensed and is able to be prescribed by doctors. These trials enable long-term benefits and possible side effects to be monitored.

Blind trials

Most clinical studies are **blind trials**. This means the participants do not know whether they are receiving the new drug, the standard drug, or a placebo. Blind trials prevent the patients' psychology from affecting results and stop the participants showing bias when they answer questions about the treatment they received.

Some studies are **double-blind**. In these trials, neither the patients nor the scientists know which treatment is being issued. This can be done by using computers to randomly assign a code number to each participant. This determines which treatment group the participants enter. Double-blind trials prevent scientists from showing any bias during the procedure.

The role of NICE

NICE is an independent UK organisation that assesses the clinical and cost effectiveness of health technologies such as new drugs, diagnostic agents, procedures, and devices used in health care. Its role is to ensure that all NHS patients have access to the most clinically and cost effective treatments available. Drugs can only be prescribed on the NHS using NICE guidelines.

A new drug could be effective but costly. NICE evaluates whether the effectiveness of the drug is sufficient to justify funding its use. The organisation also issues guidance on the appropriate treatment and procedures for specific diseases or conditions based on the most recent research findings.

Summary questions

1 Describe two essential properties of a placebo. *(2 marks)*

2 Evaluate the role of a research ethics committee in the development of a new drug. *(3 marks)*

3 Describe and explain the differences between Phase 2 and Phase 3 of a clinical trial. *(2 marks)*

Practice questions

1 Which of the following statements is/are always true of an acute disease?

1 Rapid onset
2 Long duration
3 Severe symptoms

A 1, 2 and 3
B Only 1 and 2
C Only 2 and 3
D Only 1 *(1 mark)*

2 Which of the following statements is/are true of asthma?

1 It can affect the trachea, bronchi, and bronchioles.
2 The majority of asthma cases are thought to be caused by allergic reactions.
3 During an asthma attack, muscles of the airways contract.

A 1, 2 and 3
B Only 1 and 2
C Only 2 and 3
D Only 1 *(1 mark)*

3 Which of the following statements is/are usually true of phase 2 of a clinical trial?

1 Placebos are used.
2 Involves thousands of participants.
3 Participants have the disease the drug will be treating.

A 1, 2 and 3
B Only 1 and 2
C Only 2 and 3
D Only 1 *(1 mark)*

4 Which of the following is a cause of chronic bronchitis?

A DNA mutation
B Elastin removal from alveoli walls
C Non-functional cilia
D Alpha Antitrypsin inhibition *(1 mark)*

5 Which of the following is the most accurate description of a blind trial?

A Participants have no knowledge of the drugs they receive.
B Doctors have no knowledge of which drugs participants are receiving.
C Experimental drugs are tested.
D Preliminary results remain unpublished. *(1 mark)*

6 a Give three symptoms that could indicate a student is having an asthma attack. *(3 marks)*

b A peak flow meter can be used to assess asthma.

State two measurements that could be taken using a peak flow meter. *(2 marks)*

c

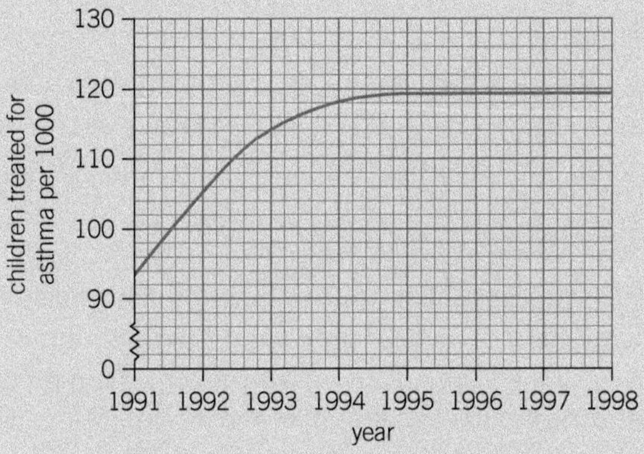

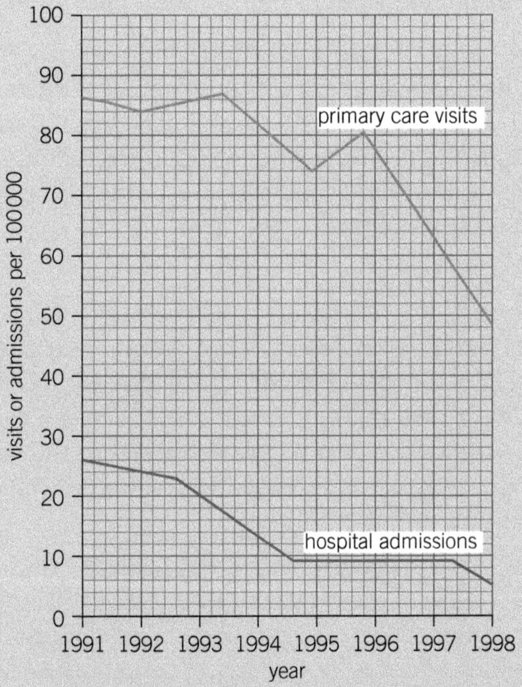

The first graph shows the change in the numbers of children treated for asthma between 1991 and 1998. The second graph shows the change in the numbers of primary care visits, such as visits to GPs, and hospital admissions for the same group of children over the same period.

Using information from the graphs, explain why the treatment given to children for asthma is known to be effective. *(4 marks)*

d Copy and complete the following passage, which explains the role of bronchodilators and corticosteroids in an asthma attack.

An asthma attack can be triggered by exposure to pollen, house mites, or other substances, which are known as..... . In the walls of airways such as the.......... , the .... muscle contracts and the lumen is .... The lining of airways can also become swollen and excess production of ...... can occur. Beta agonists bind to receptors on the muscle cell ..... and cause the muscle to relax and the lumen of the airway dilates. Corticosteroids act to ...... inflammation. *(7 marks)*

[Question 1, OCR June 2010 F222]

7 Chronic obstructive pulmonary disease (COPD) is a chronic disease of the lungs. It affects at least 900 000 people in the UK. The main cause of COPD is smoking.

a (i) Why is COPD described as a chronic disease? *(2 marks)*

(ii) Asthma can be associated with COPD.

Name two other conditions that contribute to COPD. *(2 marks)*

(iii) How is COPD tested for in a health clinic? *(1 mark)*

b Suggest why people with COPD are provided with air supplies enriched with oxygen to relieve their symptoms. *(2 marks)*

c The photomicrograph shows a bronchiole and alveoli in a healthy lung.

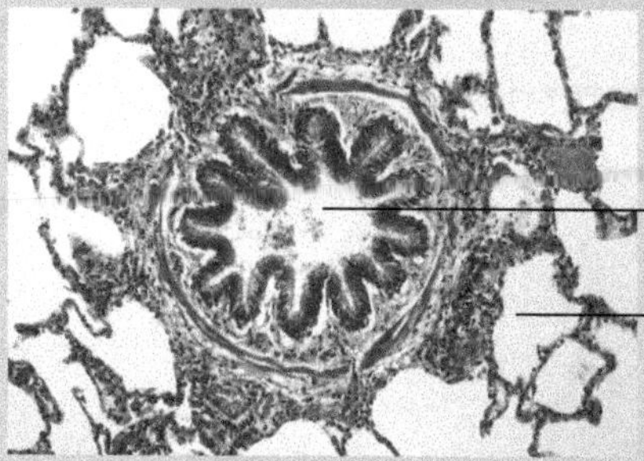

Describe how tobacco smoke causes changes in the lining of the bronchioles and alveoli leading to the development of COPD. *(6 marks)*

[Question 4, OCR Jan 2011 F222]

8 a Asthma is the most common chronic respiratory disease in children.

The diagram shows a cross section through a normal bronchiole and a cross section through a bronchiole during an asthma attack.

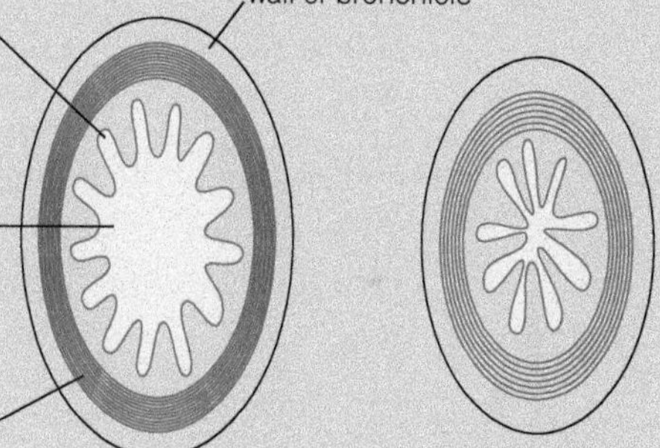

(i) With reference to the diagram, describe three changes that occur in the bronchioles during an asthma attack. *(3 marks)*

(ii) Two types of drug that are used to treat asthma are steroids and beta-agonists.

Describe how these drugs are introduced into the body and explain how they relieve the symptoms of asthma. *(5 marks)*

b The Global Alliance against Chronic Respiratory Diseases (GARD) is trying to reduce the health burden of chronic respiratory diseases.

Suggest two policies that countries could introduce to prevent chronic respiratory diseases. *(2 marks)*

[Question 2c, OCR June 2013 F222]

Breadth of knowledge questions

The diagram shows a generalised plant cell with the organelles labelled P – U. The organelles are not drawn to scale.

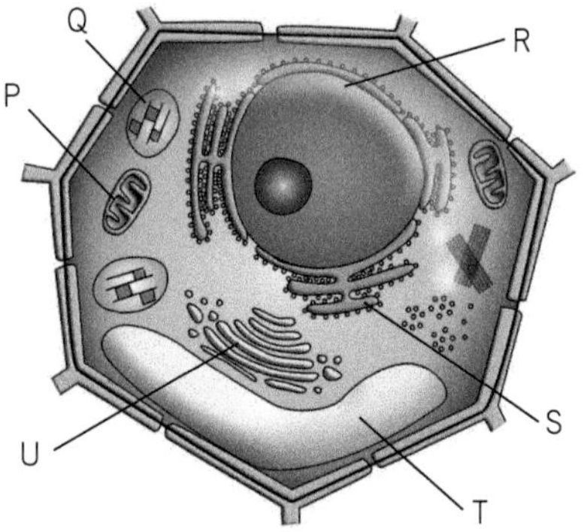

1 Which letter correctly identifies an organelle which is surrounded by the tonoplast?

A P

B Q

C T

D U *(1 mark)*

2 Which of the following in the diagram identifies some of the organelles that would be present in a companion cell in phloem tissue?

A P, Q, R, S

B P, R, S, T

C P and S only

D R and T only *(1 mark)*

3 Which organelles in Fig.1 would be visible using a light microscope?

A P, Q, R, and T

B R and T only

C Q and T only

D Q, R, and T only. *(1 mark)*

4 Which of the following statements could be applied to a phase 1 clinical trial?

A A placebo is used

B The group is randomised as to whether they get the drug or the placebo

C The trial could involve 20 people

D The trial compares the new drug with existing treatments. *(1 mark)*

5 Aspirin has been found to have an effect on the rate of transpiration. Which of the following plants would not be suitable to investigate the effect of aspirin on transpiration?

A *Salix cinerea* (grey willow)

B *Pelargonium* sp. (geranium)

C *Ligustrum ovalifolium* (privet)

D *Solanum tuberosum* (potato) *(1 mark)*

6 The graph shows the changes in volume achieved during one inhalation and one forced exhalation.

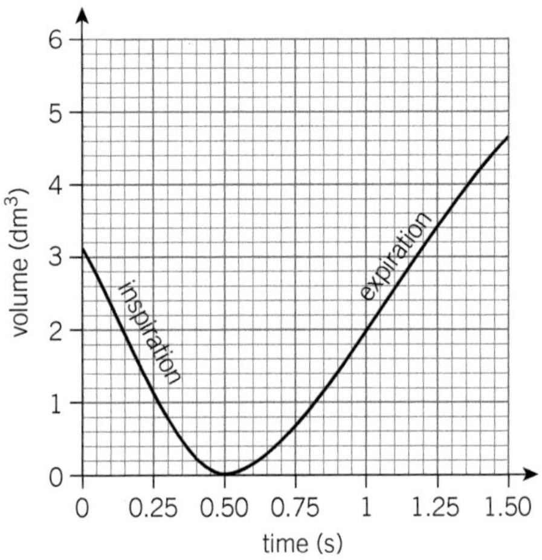

Which of the following statements is a correct interpretation of the graph?

A The tidal volume is 3.5 dm^3

B The vital capacity is 3.5 dm^3

C The FEV_1 is 4.75 dm^3s^{-1}

D The FEV_1 is 4.75 dm^3 *(1 mark)*

7 Measurements of lung volumes can be used in the diagnosis of lung disease. An obstructive lung disease is diagnosed when FEV_1 is less than 70% of the forced vital capacity (FVC).

The table gives data on FEV_1 and FVC for four patients, W, X, Y, and Z.

Which values are likely to be correct for a patient with emphysema?

Patient	FEV_1 (a.u)	FVC (a.u)
W	3.3	4.7
X	4.7	3.2
Y	4.2	4.7
Z	3.2	4.7

A W

B X

C Y

D Z *(1 mark)*

8 The image here shows a breast produced from a routine screening test for breast cancer.

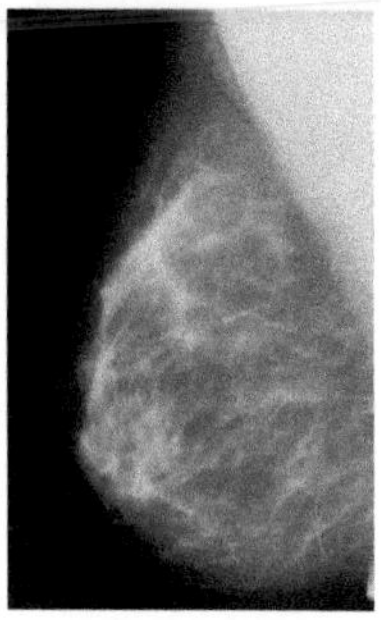

Which of the following methods was used to produce this image?

A X-ray C Biopsy

B MRI D Ultrasound (*1 mark*)

9 The diagram shown here is of the HIV virus. HIV virus particles escape from infected cells by budding off the cell surface membrane.

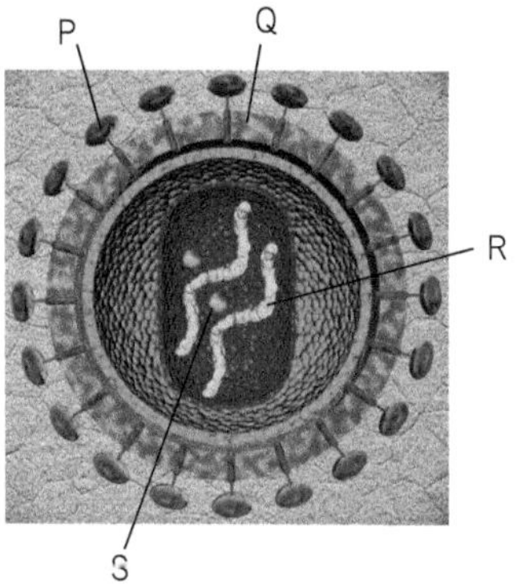

Donated blood is screened for the presence of HIV antibodies. Which letter corresponds to the part of the virus responsible for triggering the production of antibodies?

A P C R

B Q D S (*1 mark*)

10 Using the diagram, which components of the HIV virus consist of molecules which contain phosphorus?

A P and Q

B Q and R

C R and S

D S and P (*1 mark*)

11 The table shows the sequence of amino acids in three different animal species, X, Y, and Z. Each letter represents a different amino acid in part of the haemoglobin molecule.

Species	Amino acid sequence
X	AEEKBBVTALWAKVNVE.....DS...S
Y	AEEKSBVTBLWAKVNVD.....DS...S
Z	BEEKSBVTBLWBKVNVE.....EB....T

Which of the following statements is/are true?

Statement 1: species X is most closely related to species Y.

Statement 2: X, Y, and Z are in the same domain.

Statement 3: the primary sequence of amino acids in species X and Y is the same.

A 1,2 and 3

B Only 1 and 2

C Only 2 and 3

D Only 1 (*1 mark*)

12 The diagram shown here represents the mitotic cell cycle.

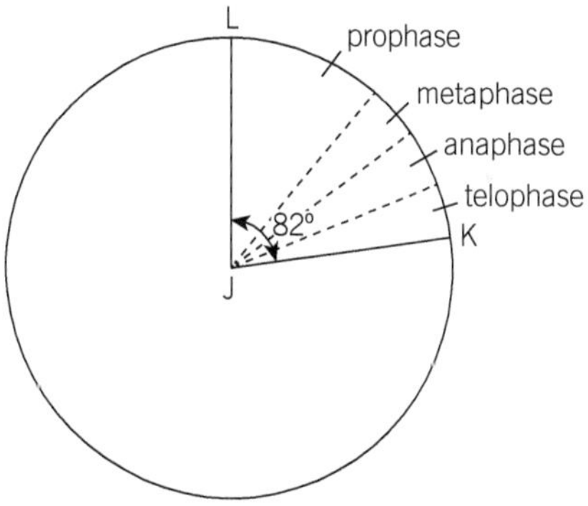

Which of the following statements is/are true?

Statement 1: the cell is a eukaryotic cell.

Statement 2: the cell could be a palisade mesophyll cell.

Statement 3: the DNA content will double in prophase.

A 1,2 and 3

B Only 1 and 2

C Only 2 and 3

D Only 1 (*1 mark*)

13 The time taken each for cell cycle was 48 hours.

Using the information given in the figure, the number of hours spent in interphase was:

A 18.5 C 37

B 11 D 17 (*1 mark*)

14 The image shows a killer T lymphocyte (cell T) attacking a cancer cell (cell V).

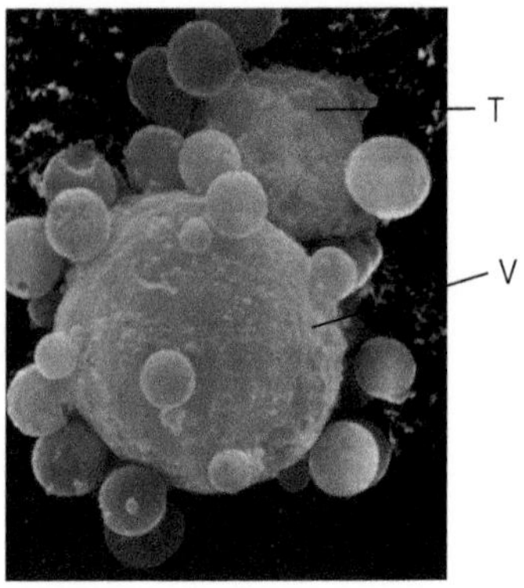

Which of the following is a correct description of the event taking place in the image.

A Phagocytosis triggered by a specific immune response.

B Phagocytosis triggered by a non-specific immune response.

C Apoptosis triggered by a specific immune response.

D Apoptosis triggered by a non-specific immune response. *(1 mark)*

15 One of the factors which affects the rate of entry of substances into cells is the surface area.

a (i) Assuming that a cell is spherical in shape, calculate the surface area of a cell which has a diameter of 7 μm. Give your answer to 1 significant figure. *(2 marks)*

(ii) Human erythrocytes have a diameter of approximately 7 μm. How will the surface area of an erythrocyte compare to the value you calculated in part (a)(i)? Explain your answer. *(1 mark)*

b In blood samples, potassium ions are found both inside erythrocytes and in the blood plasma. The concentration of potassium ions is 20 times greater in the cell cytoplasm than in the plasma.

Suggest how potassium ions enter erythrocytes. Give a reason for your suggestion. *(2 marks)*

c Samples of blood taken for the measurement of potassium ion concentration must be stored correctly.

Suggest why each of the following could result in a **false high** reading of plasma potassium ion concentration:

(i) prolonged storage of blood at low temperatures *(2 marks)*

(ii) the presence of water in the blood collection tube *(2 marks)*

16 All cells require folic acid in order to synthesise nucleic acids and amino acids such as methionine.

a Copy and complete the following table on the similarities and differences between amino acids and nucleic acids by inserting either a tick (✓) or a cross (✗) in the appropriate column.

Feature	Nucleic acid	Amino acid
Polymer		
Contains nitrogen		
Contains phosphorus		

(3 marks)

b In bacterial cells, folic acid is synthesised using an enzyme called DHPS.

- The substrate for this enzyme is PABA.
- Antibiotic drugs known as sulfonamides act as inhibitors of DHPS.

The graph shows the effect of increasing the concentration of PABA on the activity of DHPS.

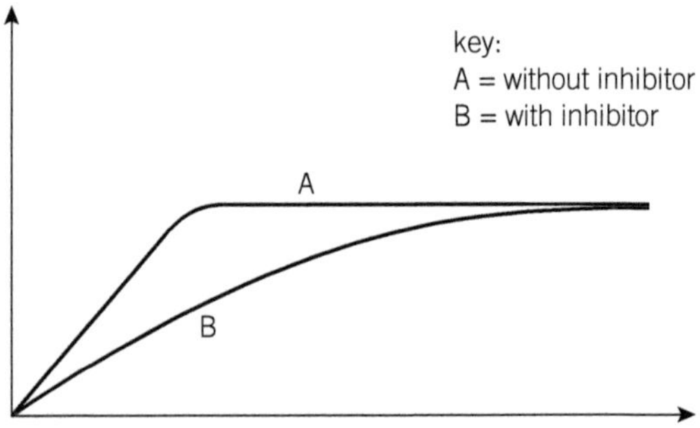

(i) Using the information in the graph, explain how sulfonamides act as inhibitors of DHPS *(3 marks)*

(ii) Suggest why sulfonamides have no effect on human cells *(1 mark)*

c Sulfonamide antibiotics have a bacteriostatic effect on bacterial cultures.

(i) what is meant by the term 'bacteriostatic' *(1 mark)*

(ii) Suggest why sulfonamides are bacteriostatic in their action. *(2 marks)*

17 a The diagram shows a suggested evolutionary relationship between bears, raccoons and two species of panda, the giant panda, *Ailuropoda melanoleuca*, and the red panda *Ailurus fulgens*.

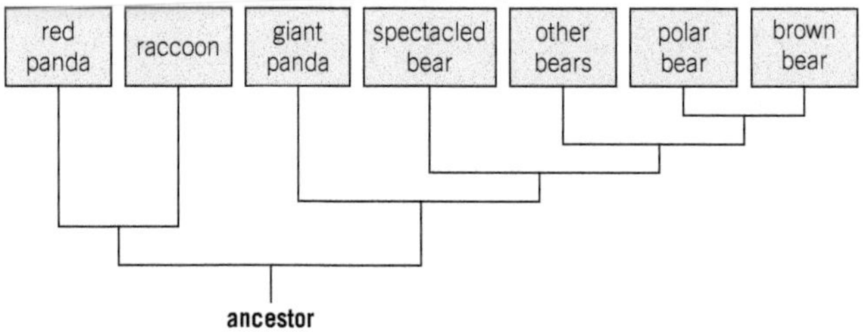

(i) Using the diagram, name the two species that share the most recent common ancestor (*1 mark*)

(ii) State whether pandas form a distinct taxonomic group. Use information from the diagram to justify your answer. (*1 mark*)

b The evolutionary relationship of the giant panda and red panda has been a matter of scientific debate for many years. It was hoped that molecular evidence would provide a definite answer.

Some of the results of scientific studies are listed in the table.

Year	Protein sequenced	Conclusion
1985	albumen	Giant panda is more closely related to bears, and red panda is more closely related to raccoons, than pandas are to each other.
1986	haemoglobin	Giant and red panda are more closely related to each other than the giant panda is to bears or the red panda is to raccoons.
1993	cytochrome c	Giant panda is more closely related to bears, and red panda is more closely related to raccoons, than pandas are to each other.

Comment on what the results in the table show about the nature of scientific knowledge and the role of the scientific community in validating new knowledge. (*2 marks*)

Research on another protein from the giant panda was carried out in 2008. This protein, called crystallin, is found in the lens of the eye, and has a sequence that has been highly conserved in all mammals.

c The panda crystallin protein obtained was 175 amino acids long.

Explain why a protein that is 175 amino acids long is coded for by 528 base pairs of DNA (*3 marks*)

[OCR F215 JUNE 2012]

18 A student was investigating the distribution of stomata on a leaf. The image shows the appearance of stomata on leaf as seen under a light microscope.

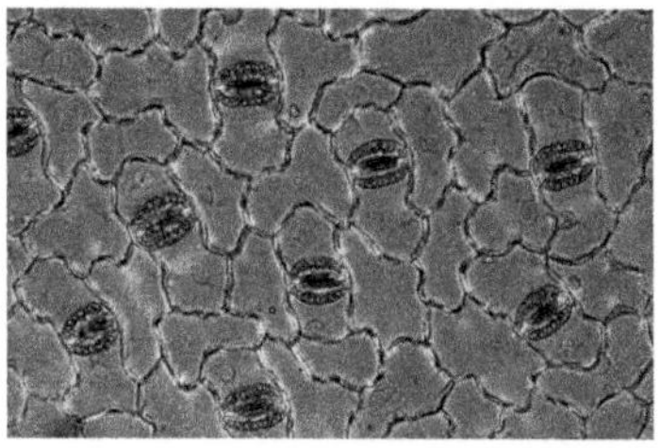

a Explain how a light microscope and graticule could be used to measure the number of stomata per mm^2 on the leaf (the stomatal density). (*3 marks*)

b The student sampled the upper and lower surface of a single leaf and obtained the following results:

Upper surface density (stomata mm^{-2})	Lower surface density (stomata mm^{-2})
43	42
56	7
3	4
56	57
5	51
5	46
45	3
Mean = 30.4	Mean = 30.0
s.d = 24.9	

(i) Calculate the standard deviation for the lower surface density. (*2 marks*)

(ii) Explain how the evidence supports the conclusions that the leaf was taken from a monocotyledon plant. (*1 mark*)

c Describe how changes in the water potential of guard cells bring about the opening of stomata. (*3 marks*)

1 Dietary reference values are given for a number of nutrients. Some of these values change during pregnancy. The table shows how DRV change for different populations of women.

Nutrient	Women (19–50)	Pregnant women
Protein (g)	45.0	51.0
Calcium (mg)	700	700
Iron (mg)	14.8	14.8
Vitamin A(µg)	600	700

a (i) Calculate the percentage increase in the DRV value for protein in pregnancy. *(2 marks)*

(ii) Explain why the DRV value for protein changes in pregnancy. *(2 marks)*

b Both meat and fish are good sources of protein. The table compares some of the nutrients found in one type of fish and one type of meat.

	Mass per 100g	
Nutrient	Mackerel (fish)	Steak (meat)
Protein (g)	18.7	30.9
Calcium (mg)	11.0	15.0
Iron (mg)	0.8	3.0
Vitamin A(µg)	45.0	0.0

(i) Outline how the presence of protein could be detected in a sample of meat or fish. *(3 marks)*

(ii) A pregnant woman decides to include fish rather than meat in her diet as a source of protein.

*Using the information in the Table, outline the roles of vitamins and minerals in pregnancy and evaluate the consequence of including fish rather than meat as a protein source. *(6 marks)*

2 In 2009, approximately 1.7 million people worldwide died from tuberculosis (TB). This figure includes 380 000 people with HIV infection.

a Describe how TB is transmitted from one person to another. *(2 marks)*

b The graph shows the estimated mean percentage change in the incidence of TB for different groups of countries, from 1997 to 2006.

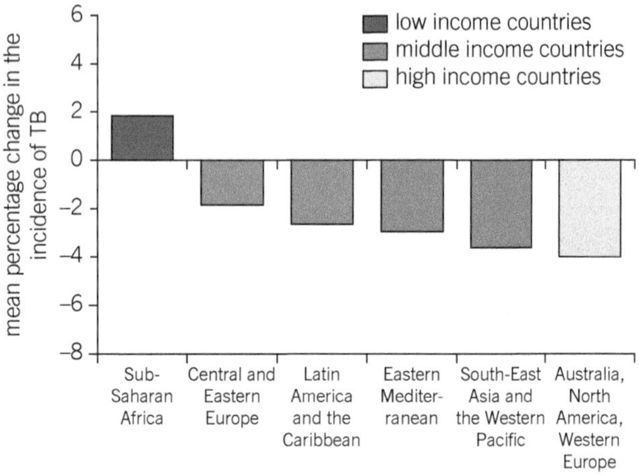

(i) Suggest why the mean percentage change in the incidence of TB is given as an estimate and not as a direct measurement. *(1 mark)*

(ii) What conclusions can be drawn about the changes in incidence of TB from the information in the graph? *(3 marks)*

c The photo shows an agar plate on which colonies of the bacterium responsible for causing TB are growing

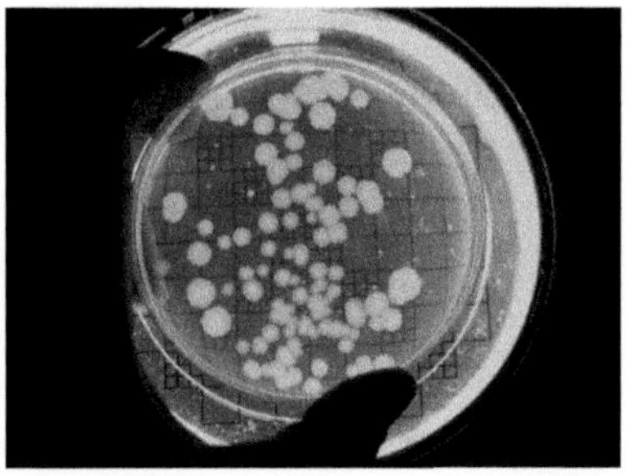

(i) State the genus to which the bacterium which causes TB belongs. *(1 mark)*

(ii) How could you assess that the agar plate shown in the photo contained a pure culture of the TB bacterium? *(2 marks)*

(iii) The agar plate shown in the photo has been produced from a serial dilution of a culture of the TB bacteria.

On a similar plate:

- the dilution used was 10^{-5}.
- 100 µl of the culture was spread onto the surface of the agar plate.
- 90 colonies were found to be growing on the agar plate.

Calculate how many cells were growing in the original culture. Show your working (give your answer in cells cm^{-3}). *(2 marks)*

3 A student used a potometer to investigate the effect of leaf area on the rate of transpiration. This apparatus is shown in Fig. 3.1.

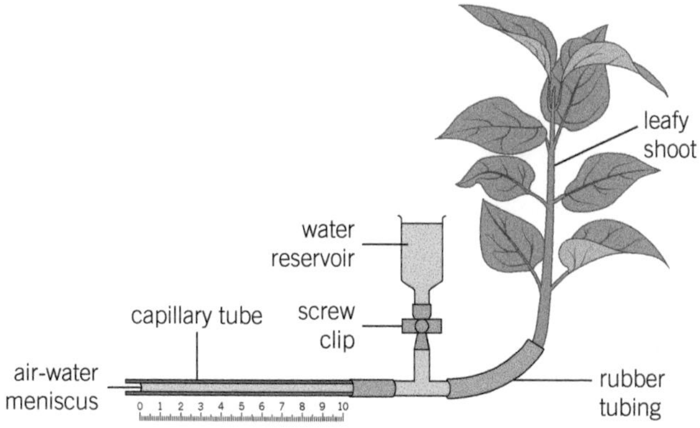

The student presented the results of their investigation in a table as shown.

Number of leaves present on shoot attached to potometer	Mean rate of bubble movement
0	7
2	28
4	49
6	73
8	92

a (i) State what information the student has not included in their table of results *(2 marks)*

(ii) Describe and explain the data shown by the students results. *(3 marks)*

b As part of the evaluation of the experiment, the student wrote the following statements:

1 One limitation is that the leaves were not all the same size

2 I assembled the potometer under water and the leaves got wet.

3 During my investigation the sun came out and the lab warmed up very quickly.

For each statement, explain why this may affect the results **and** suggest how the student could improve the investigation. *(6 marks)*

c The top photomicrograph shows a transverse section through a leaf from a hydrophyte. The bottom photomicrograph shows a transverse section through a leaf from a xerophyte.

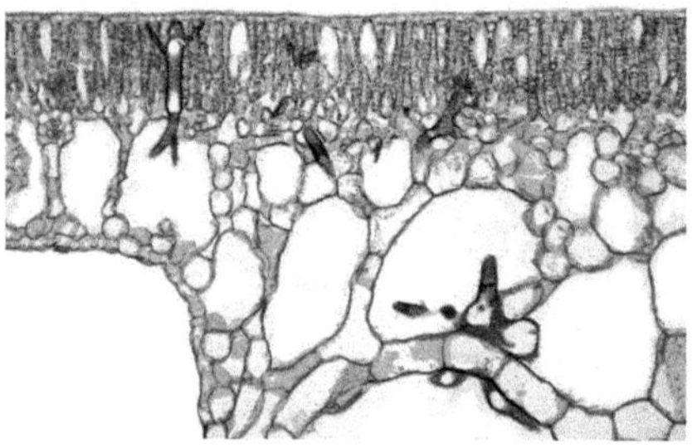

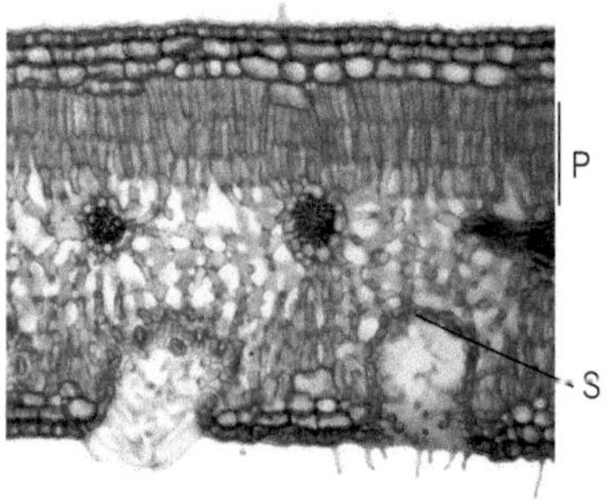

The region labelled P is the palisade mesophyll tissue. S indicates the location of stomata.

*Using the information in both photomicrographs, explain how features of hydrophytes and xerophytes adapt them to survive in extremes of water availability. *(6 marks)*

4 In the United Kingdom, all children are tested shortly after birth for conditions known as inborn error of metabolism (IEMs). IEMs occur due to non-functioning of an enzyme in a metabolic pathway. One of the most common IEMs is Phenylketonuria (PKU).

In PKU, the enzyme which converts the amino acid phenylalanine to the amino acid tyrosine does not function.

The diagram shows a simplification of this pathway.

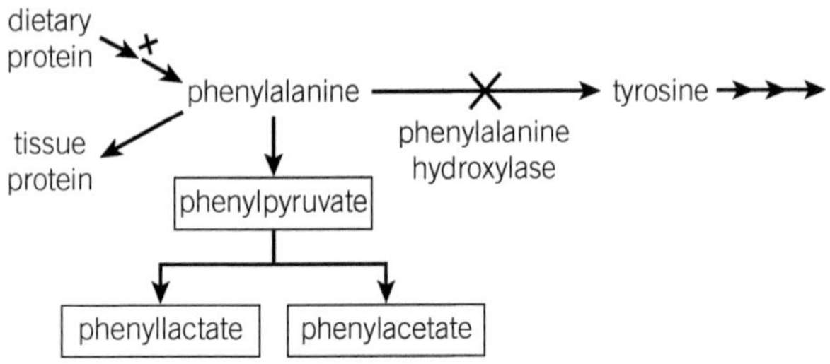

a (i) Suggest how the structure of tyrosine differs from that of phenylalanine. *(1 mark)*

(ii) State one source of tyrosine **other than** from the conversion of phenylalanine. *(1 mark)*

b Phenylalanine is incorporated into tissue proteins on the ribosomes. Explain the role of RNA in the assembly of amino acids such as phenylalanine into proteins on ribosomes. *(4 marks)*

c The presence of PAH (phenylalanine hydroxylase) in human tissue samples can be detected using a test called ELISA.

The diagram outlines how an ELISA detects the presence of PAH.

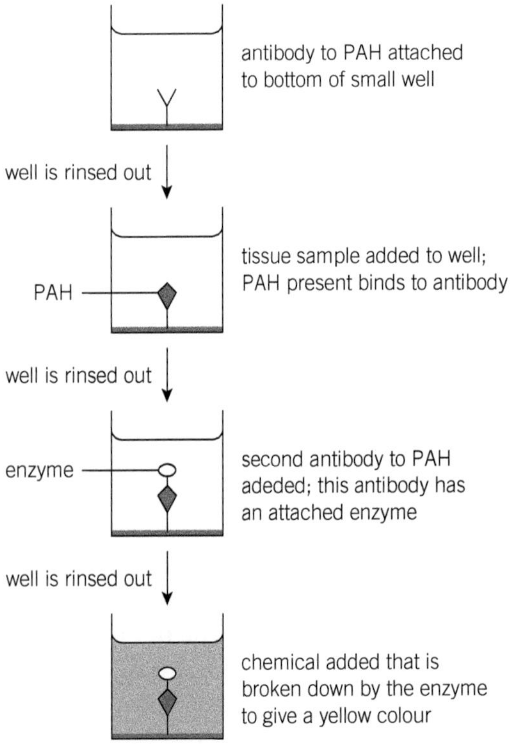

(i) Describe the general structure of antibodies such as the attached antibody. *(3 marks)*

(ii) Explain why both the attached and second antibodies are specific to PAH. *(1 mark)*

d The photo shows the results from an ELISA test. Each circle (well) represents one sample that has been tested. The intensity of the colour can be measured using an optical reader similar to a colorimeter. There is a positive correlation between the colour intensity and the PAH concentration.

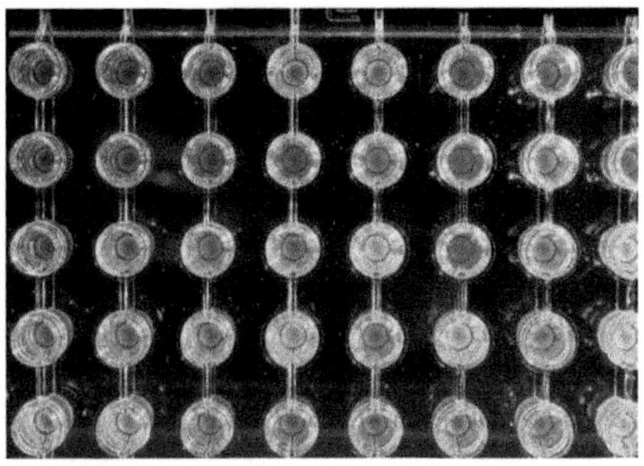

Outline how you would use the results from an ELISA to determine the exact concentration of enzyme present in a sample. *(3 marks)*

e Several different mutations can occur in the gene for PAH which result in a child being born with PKU. A mutation is a change in the sequence of bases in DNA. One mutation replaces the amino acid arginine with the amino acid tryptophan in the PAH enzyme.

(i) Explain why a change in the amino acid sequence of PAH results in an enzyme which is present but does not function. *(3 marks)*

(ii) Suggest why an ELISA test on samples with the mutated enzyme could give a negative result. *(1 mark)*

5 The diagram shows a vertical section of the heart to show the position of certain structures.

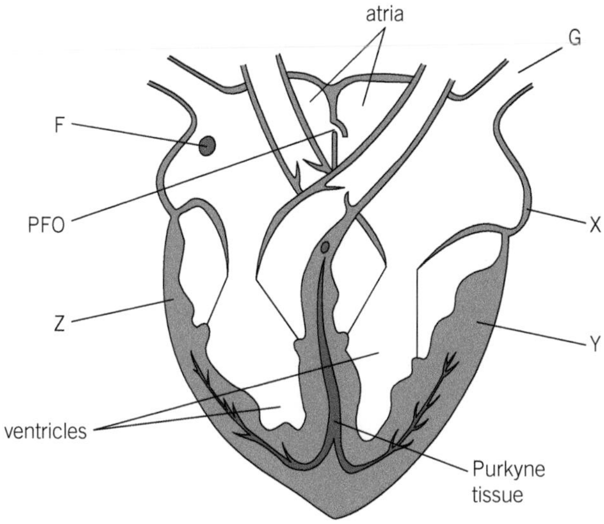

a Name the structure F and G (*2 marks*)

b The statements below were made to a group of students. Explain why each statement is true.

(i) The difference in thickness of the walls of the chambers, as shown by the letters X, Y, and Z is related to the functions of the different chambers. (*3 marks*)

(ii) Without the Purkyne tissue, blood would not be pumped out of the heart efficiently. (*2 marks*)

c Recent research has shown that there may be a link between migraines (severe headaches) and the minor heart defect PFO (patent foramen ovale).

In PFO, the small flap shown in the diagram fails to close completely at birth. Suggest how PFO might lead to a migraine. (*3 marks*)

6 One group of pathogens that can cause diseases in humans are the fungi.

The diagram shows part of a fungal filament (hypha).

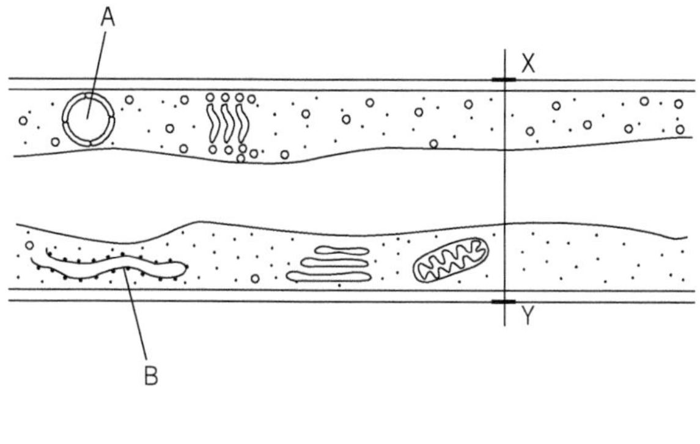

a Name two features visible in the diagram which show that this organism is eukaryotic. Do not include those features labelled A and B. (*2 marks*)

b Assuming fungal filaments are cylindrical in shape, calculate the surface area of a section of a filament 4 mm in length if the distance XY is 12 μm. Show your working (giving your answer in μm^2). (*2 marks*)

c The fungus *Trichophyton* is responsible for the condition known as athletes foot in humans. Protease enzymes are secreted by the filaments. The enzymes break down keratin which is a fibrous protein present in skin.

(i) Give **two** reasons why the protease enzymes synthesised by *Trichophyton* do not break down intracellular proteins in the fungal filament (*2 marks*)

(ii) Outline how the protease enzymes leave the filament and come into contact with human skin. (*2 marks*)

d The protease enzymes produced by *Trichophyton* have been shown to trigger an allergic response.

Complete the passage below which describes how an allergic response occurs.

In the first allergic immune response to the enzyme .... lymphocytes respond by producing .... antibodies. These antibodies attach to receptors on .... cells. This process is called sensitisation.

On a second encounter with the enzyme, the enzyme molecule cross links with the antibodies and stimulates the release of the chemical .... This chemical triggers the inflammation response causing blood vessels near the site to become more.................. leading to the area becoming .... in colour. (*6 marks*)

e Asthma is also an allergic response.

The diagram on the following page shows a drawing made from cross sections of the upper bronchioles of a non-asthmatic (X) and an asthmatic (Y). The sections were drawn using observations made using a light microscope.

Upper bronchioles normally have an epithelium with a few scattered goblet cells.

X – a non-asthmatic

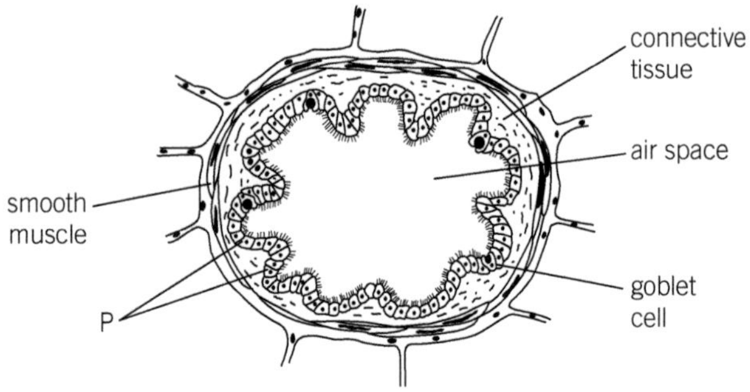

Y – an asthmatic

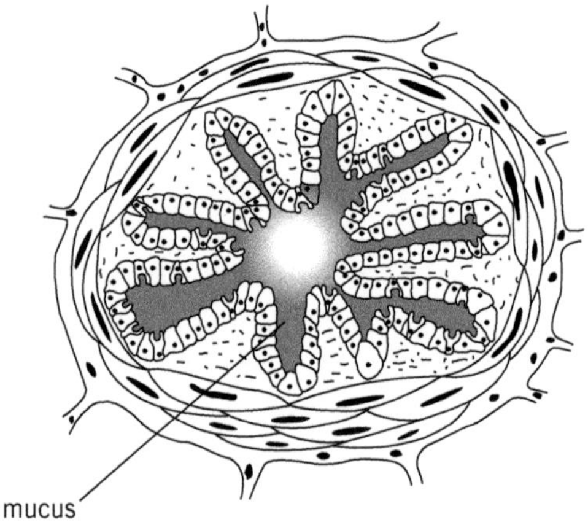

(i) Describe the function of the cells labelled P on the top diagram in the gas exchange system. *(3 marks)*

(ii) Use the information given in the diagrams to explain the following observations made on the bronchioles of an asthmatic during an asthma attack:

the bronchioles fill with mucus ....

the cross sectional area of the air spaces in the bronchioles decreases.................. *(3 marks)*

(iii) Describe the symptoms of asthma and outline the changes that occur in the airways during an asthma attack *(4 marks)*

7 In an experiment to measure the rate of diffusion, a student placed tubes of agar jelly containing an indicator into dilute hydrochloric acid. The indicator changes from pink to colourless in acidic conditions.

The student used cubes of different sizes and recorded the time taken for the pink colour in each cube to disappear completely.

The students results are recorded in the table

Length of side of cube (mm)	Surface area of cube (mm^2)	Volume of cube (mm^3)	Surface area to volume ratio	Time taken for pink colour to disappear (s)	Rate of diffusion ($mm\ s^{-1}$)
2	24	8	3.0:1	50	0.020
5	150	125	1.2:1	120	0.021
10	600	1000		300	0.017
20	2400	8000	0.3:1	700	0.014
30	5400	27000	0.2:1	1200	0.013

a (i) Calculate the surface area to volume ration of the cube with 10mm sides.

Show your working. *(2 marks)*

(ii) **Using the data in the table**, describe the relationship between the rate of diffusion and the surface area to volume ratio. *(2 marks)*

(iii) Explain the significance of the relationship between the rate of diffusion and the surface area to volume ratio for large plants. *(2 marks)*

b Another student used the same raw data obtained in the experiment but calculated a different rate of diffusion for each cube.

This student's results are shown in the Table

Length of side of cube (mm)	Time taken for pink colour to disappear (s)	Rate of diffusion ($mm\ s^{-1}$)
2	50	0.040
5	120	0.042
10	300	0.033
20	700	0.029
30	1200	0.025

In this students table, the calculation of the rate of diffusion is incorrect.

(i) Suggest the method used to calculate the rate of diffusion. *(1 mark)*

(ii) State why the method in **(b)(i)** is not correct. *(1 mark)*

Answers

1.1

1 To enable the light rays to pass through it (1)

2 a × 300 (1 sig. fig) 55 000 ÷ 170 (based on diameter of 55mm)

b 1 mark to show working of (8 × 1000) ÷ 14000

1 mark for answer = 0.57 µm = 0.6 µm

3 Image would be 50 µm × 1500 = 375 000 µm (1) Convert µm to mm = 375 000 ÷ 1000 = 75 mm (1)

1.2

1 A beam of electrons has a shorter wavelength (1)

2

Light bulb	High voltage tungsten filament	High voltage tungsten filament
lenses	electromagnet	electromagnet
200 nm	0.1 nm	0.1 nm
× 1500	× 100 000	× 500 000
living or dead	dead	dead
mm	surface scan	µm
methylene blue	surface coated (e.g. gold)	lead salts
2d	3d	2d
relatively small	very large	very large
can be quick and simple.	can be quick (e.g. freeze fracture) but varies.	time consuming requiring specialist equipment.
cheap to purchase and operate	expensive to purchase and operate	expensive to purchase and operate
no	yes	yes
low	varies	high

(14)

3 Phospholipids which contain hydrogen, oxygen, carbon, and phosphorus (2)

1.3

1 a prevent contamination of the blood sample from the medical practitioner (1); protect the medical practitioner from any pathogens/parasites present (1); latex-free reduces allergies and skin reactions by the medical practitioner (1) *(max 2)*

b to allow the cells to adhere (stick) to the microscope slide (1).

2 a cryosection means freezing the specimen before sectioning the tissue/specimen; used in oncology (studying cancerous tissue) (2)

b *any one from*: freezing may distort the cells/tissue; may damage cell membranes; may introduce ice crystals which rupture cell surface membranes; may introduce artefacts (1)

Transverse radius = 5 mm, Area = 79 mm^2

Oblique section = 5 mm (a) × 8.5 mm (b). Area = 134 mm^2

1.4

1 Carbon dioxide (1)

2 (3) 1 mark per correct row

X	✓	X
X	X	✓
X	X	✓

3 a nucleus is absent (1)

b viruses insert their genes into the host genome in order to use host enzymes to synthesise proteins (1); the host genome is missing from erythrocytes (1)

c Width 70 mm, **i** 5.8 µm (1); **ii** 6 µm (1); **iii** 6 µm (1)

1.5

1 (*any 2 from*): the original mixture is not mixed thoroughly (1); there may have been contamination (1); there may be artefacts present (1)

2 Set up four test tubes labelled A, B, C and D (1); to each test tube add 9.0 cm^3 of distilled water (1); to test tube A add 1.0 cm^3 of the yeast culture and mix thoroughly (1); then transfer 1.0 cm^3 from test tube A into test tube B and mix thoroughly (1); Repeat taking a sample from test tube B and add to test tube C, and then from test tube C into test tube D (1); repeat across all four tubes (1) (max 5)

3 The haemocytometer can also be used to count sperm cells and many other cells e.g., yeast cells and pollen grains. (1)

1.6

1 a RER has ribosomes for protein synthesis (1); antibodies are proteins (1).

b Golgi needed to package antibodies in vesicles (1) for exocytosis (1).

2 a $V = \frac{4}{3}\pi r^3 = \frac{4}{3} \times 3.14 \times 7.5^3 = 1766.25$

$= 1766\ \mu m^3$

1 mark for working, 1 mark for rounding correctly, and 1 mark for correct units (3)

b $\frac{1766}{20\,500} \times 100 = 8.614 = 8.6\%$

1 mark for working and 1 mark for rounding answer correctly (2)

3 Chloroplasts are only found in plant cells (1); they are present in photosynthetic cells, e.g.palisade mesophyll cells (1); they are absent from non-photosynthetic cells, e.g. root hair cells, companion cells (1); hence the statement is not valid (1)

1.7

1 The process converts chemical energy into mechanical energy (1); the chemical energy is released from the hydrolysis of ATP (1); and the mechanical energy is the conformational change of the protein (1)

2 *Any 3 from*: provides stability for the cell (1); determines the shape of the cell (1); moves organelles within the cell (1); make up spindle fibres / centrioles (1); attach to organelles to hold them in place within the cell (1)

3 a line 1 = RER (1); line 2 = Golgi Apparatus (1); line 3 = endosomes (1); line 4 = lysosomes (1)

b there is a finite amount of radioactive amino acid added to the culture (1)

c accept any suitable named tissue with appropriate protein named, e.g. pancreatic tissue & insulin, liver tissue & albumin, kidney tissue & EPO (erythropoietin) (2)

Motor proteins

1 a chemical energy (ATP) into kinetic energy

b Neutrophils – used to extend the cell when pathogens are engulfed.

1.8

1 Area of field of view = $\pi r^2 = \pi \times 1^2 = 3.142$
$= 3.1\ \text{mm}^2$ (to 1 dp)

2 1.5 µm, based on max length of 95 mm

3 0.156 µm (rounding to 3 d.p.)

1.9

1 a phospholipid (1)

b intrinsic protein (1)

2 a membrane would be too fluid (1) and there would be a loss of control entering and exiting the cell (1)

b membrane would not be fluid enough (1) and therefore movement of the cell and substances within it would be restricted (1)

3 a Proteins could 'float' out of position and end up in the wrong section of the membrane (1)

b Protein 'ties' between cells in tissues (tight junctions) (1); prevent the lateral movement of proteins (1)

1.10

1 a uptake of oxygen in alveoli, removal of carbon dioxide in alveoli (2)

b removal of oxygen from stoma (when it is light) (1); uptake of carbon dioxide from stoma (when it is light) (1); diffusion of oxygen and carbon dioxide from cell to cell in plants (remember there is no transport system for respiratory gases (1). (max 3)

2 They are water-soluble molecules (1) and cannot diffuse through the hydrophobic centre of the bilayer easily (1)

3 Increase the surface area of the cell surface membrane (1) and increase the number of protein carriers in the membrane (1)

1.11

1 a secretion of digestive enzymes, peptide hormones (1)

b removal of excess water from unicellular organisms (1)

2 (8 marks, 1 per line)

	Facilitated diffusion	Active transport
Move molecules down a concentration gradient	Yes	No
Move molecules against a concentration gradient	No	Yes
Involves intrinsic proteins	Yes	Yes
Involves carrier proteins	Yes	Yes
Carrier proteins possess at least two binding sites	No	Yes
Carrier protein has a complementary binding site for the molecule to be transported	Yes	Yes
Requires ATP	No	Yes
Involves allostery	No	Yes

3 a *any 6 from*:

Increase temperature (1): this increases the kinetic energy of the molecules (1) therefore more likely to collide with the binding site on the carrier protein (1)

Increase the number of carrier proteins embedded in the membrane (1): this increases the probability of a molecule colliding with the carrier protein (1) and a successful collision between the molecule and the binding site (1)

Increase the rate of respiration (1) to produce more ATP (1): this can be hydrolysed to enable allostery of the carrier proteins

b Carrier proteins used in active transport use energy (ATP) to ensure molecules can only move in one direction (1); carrier proteins used in facilitated diffusion allow the movement of molecules in both directions so the molecules will reach equilibrium either side of the membrane eventually (1)

2.1

1 Due to uneven distribution of charge over the molecule – oxygen atoms being more negative relative to hydrogen atoms (1)

2 Adhesion (1)

The water molecules are attracted to another molecule in the wall (lignin). Cohesion means water molecules are attracted to each other. (1)

3 Amino acids glycine and proline are abundant in collagen (1). Hydrogen bonds form between the NH group of glycine and the C=O group of proline. (1)

Oral rehydration solutions

1 Some bacteria killed by stomach acid
2 Flagellum allows bacterium to swim
3 To increase surface area for absorption
4 Cholera is water born and water used to make up powdered milk.
5 a Source of ATP required for active transport of ions
b To restore the electrolyte balance in body fluids
c Major intracellular ion – some would have been lost in stools.
6 Any reasonable suggestion e.g., contains sucrose rather than glucose, electrolyte balance may not be correct, expensive – relative to ORT powders.

2.2

Fluid retention in the body

1 Not enough amino acids to make blood proteins so COP /OP is too high for fluid to be drawn back into capillaries.

2.2

1 a *Blood* – body's transport medium e.g., from exchange surfaces to the rest of the body and also part of body's defence mechanism (1); *Tissue fluid*– exchange medium between blood and cells (1); *lymph*– recycling 'spent' tissue fluid back to blood.

b To cells: glucose, amino acids, vitamins, from cells: carbon dioxide

2 a 16.5 dm^3 (1) b 5.5 dm^3 (1) c 2.75 dm^3 (1) See Table 1 in Topic 2.2.

3 (5)

Component	Blood	Plasma	Tissue fluid	Lymph
Cells present	Yes	No	Some leucocytes	Some leucocytes
Location	Blood vessels	Blood vessels	Tissues	Lymph vessels
Moved by	Pressure from heart (arteries)and pressure from surrounding muscles (valves in veins)	As blood	HP and OP (formation)	Pressure from surrounding muscles (valves in vessels)
Direction of flow	Systemic and pulmonary circuit		From plasma to tissues with some returned	From tissues to blood
Main components	Water, cells, plasma proteins and dissolved substances	As blood but no cells	Proteins, water, dissolved substances, and some cells but no large plasma proteins	Proteins, water, dissolved substances, and some cells but no large plasma proteins

2.3

1 *any 2 from:* ribose, deoxyribose, fructose, galactose (2)

2 a

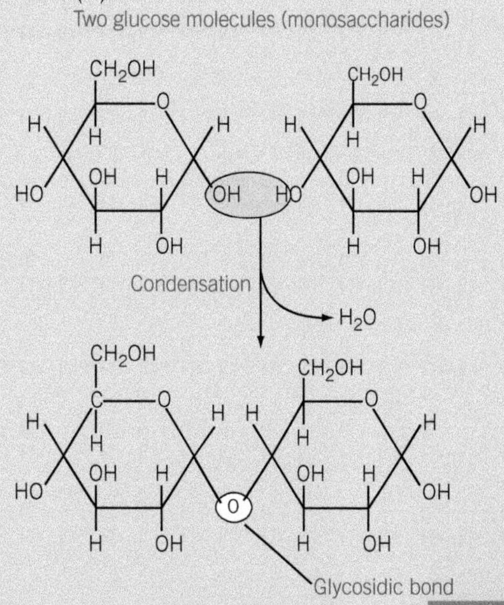

Maltose (a disaccharide)

1 mark for drawing two α-glucose molecules correctly (1)

1 mark for showing where the OH and H groups are removed from (1)

1 mark for showing the maltose molecule with correct glycosidic bond drawn and labelled (1)

b It is a reaction which builds up larger molecules from smaller ones (1) and it requires the input of energy from ATP (1)

3 It is insoluble (1) so it does not affect the water potential of the cells (1); OR it is compact (1) so a large amount of glucose can be stored in a small space (1); it is a branched molecule (1) so there are many terminal ends where glucose can be released quickly (1)

2.4

Benedict's test

The urine sample contained another reducing agent, for example ascorbic acid. People taking vitamin C tablets frequently have high concentrations of the vitamin in their urine and this can lead to 'false positives'.

Testing for a non-reducing sugar

Benedict's solution is alkaline (around pH10). No reaction will occur at low pH.

Using a colorimeter

It is important to choose a filter that produces light that will be absorbed by the solution. A blue solution is blue because it transmits blue light but it will absorb red (630 nm) or orange (600 nm) light. The colour which gives the biggest reading is normally chosen.

1 Benedict's is 'semi-quantitative'. The same colours would be seen in both samples. A 'quantitative' test with excess Benedict's solution could distinguish the two if the samples both fell on the calibration curve (2).

2 Diastix colours are easier to distinguish and a quantitative result is given with a range of concentrations (2).

3 a Any cloudiness on the cuvette and the presence of fingerprints will also increase the absorbance (2).

b (i) x-axis labelled 'glucose concentration' AND y-axis labelled 'absorption' AND curve showing absorption increasing as glucose concentration increases. (1)

(ii) x-axis labelled 'glucose concentration' AND y-axis labelled 'transmission' AND curve showing transmission decreasing as glucose concentration increases. (1)

2.5

Biuret test

The solution would remain pale blue. Biuret tests for the present of peptide bonds not amino acids.

1 Biuret solution detects the presence of peptide bonds present in primary protein structure (1).

Prepare a range of 1 cm^3 protein solutions (e.g. 0 to 5 $mg\,cm^{-3}$ in 1 $mg\,cm^{-3}$ intervals). Add 2 cm^3 Biuret and leave to stand for 10 minutes. Read the absorbance with the filter set at 550 nm. Plot a calibration curve of protein concentration against absorbance. Test the unknown sample using the same volumes (3).

2 This is a large protein and will not be filtered out in the kidney tubules (2).

3 Test strips depend on pH. Alkaline urine will have a high pH which will affect the indicator. This could result in a colour change in the absence of protein (2).

2.6

1 a the cell would increase in size and could burst (1); **b** the cell would shrink (1); **c** the cell would burst quicker than in (a) (1); **d** no change in cell size (1).

2 The presence of aquaporins in the membrane increases the permeability of the membrane to water molecules (1).

3 Sea water has a low(er) water potential than soil water (1); the water potential gradient between the cytoplasm in root hairs and the soil will be less so less water enters (2).

3.1

1 Sulfur <u>and</u> nitrogen (1)

2 Sudden changes in pH can affect enzyme activity (1) and also lead to less oxygen being carried by haemoglobin (1).

3 dipeptide drawn correctly (1); peptide bond correctly labelled (1); hydrolysis reaction annotated OR removal of water shown correctly (1); 2 glycine molecules drawn correctly (1)

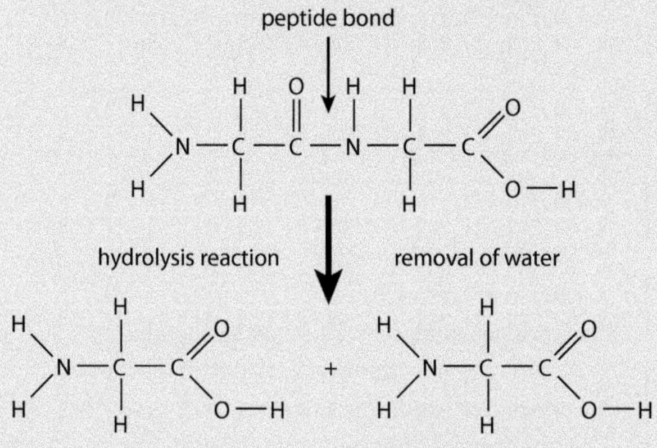

3.2

Separating amino acids using chromatography

The solvent chosen and the R group of the amino acid. The size and nature of the R group (i.e., is it polar, non-polar, acidic, or basic). The R group differs for each of the 20 amino acids.

1 a any two from: lysine, histidine, arginine, aspartatic acid, glutamatic acid, threonine, glutamine, asparagine, serine

b any two from: valine, isoleucine, leucine, methionine, phenylalanine, tryptophan, cysteine, glycine,

2 a pen ink contains dyes and dyes will also dissolve in the solvent and move up the chromatography paper with the substance being separated which would affect experiment; the starting line is needed to enable the R_f value to be calculated; The position of the solvent front won't always be obvious when the paper has dried or after spraying with a staining agent for detection;

b You can get the oils from your hands onto the paper; the solvent may dissolve them and change the rate the solute rises; and this will interfere with the process of the spots forming and travelling;

3 suggestions can include: HAZCHEM safety for the propanol (e.g. highly flammable, risk of serious damage to eyes, vapours may cause drowsiness and dizziness), HAZCHEM safety for ammonia (e.g. irritatnt to eyes, skin and respiratory passages), HAZCHEM safety for the ninhydrin (e.g. causes respiratory tract irritation, harmful if swallowed, may be harmful if absorbed through skin or if inhaled, can cause eye and skin irritation); handling glassware.

3.3

1 Haem group (1)

2 Award one mark per correct row. Do not award a mark if there are no crosses OR if there are any hybrid ticks present.

Level of folding	Disulfide	Hydrogen	Hydrophobic	Ionic	Peptide
Primary	X	X	X	X	✓
Secondary	X	✓	X	X	✓
Tertiary	✓	✓	✓	✓	✓
Quarternary	✓	✓	✓	✓	X

3 *Similarities* (Award a maximum of 4 marks for similarities): *Both* are polymers (1); are made from the monomers called amino acids (1); have a primary, secondary, and tertiary structure (1); contain peptide, ionic and disulphide bonds and hydrophobic interactions (1); have a globular structure (1); have a prosthetic group (iron ions) (1); have 4 polypeptide chains (1); both have binding sites (1); both can bind to four molecules ($4O_2$ for haemoglobin and $4H_2O_2$ for catalase) (1). *Differences*: Polypeptide chains in haemoglobin are of two types, catalase there are four chains all the same (1); binding site in haemoglobin is for oxygen, binding site in catalase is for hydrogen peroxide (1).

3.4

1 substrate and enzyme molecules possess kinetic energy (1); in solution the substrate and enzyme molecules are constantly moving in a random fashion (1); usually the substrate molecules will move faster than the enzyme molecules as they are usually smaller than the enzyme (1); collisions will occur randomly due to the random movement of the molecules (1); a successful collision only occurs when the substrate collides with the active site in the correct orientation (1).

2 both polymers of amino acids (1); both contain α-helices & β-pleated sheets (1); both have a globular shape (1); both consist of a folded polypeptide chain (1).

b State what is represented by
(i) the pink
(ii) the blue
areas of the molecule (2)
a alpha helix (1); b beta-pleated sheet (1)

3 a Concentration of molecule D would decrease (1) less molecules of A means that less molecules of B, and subsequently molecule C and subsequently molecule D would be made (1)

b Concentration of molecule D would remain the same (1) as there is still the same number of A molecules present (1) but the time to reach the concentration of molecule D would be longer (1) as the reaction would proceed slower with less enzyme present (1)

c 3 from: enzyme E3 has an active site with a specific shape (1) molecule B will not be complementary to the shape of the active site (1) and therefore no ESC can form (1) so no product is made (1)

3.5

1

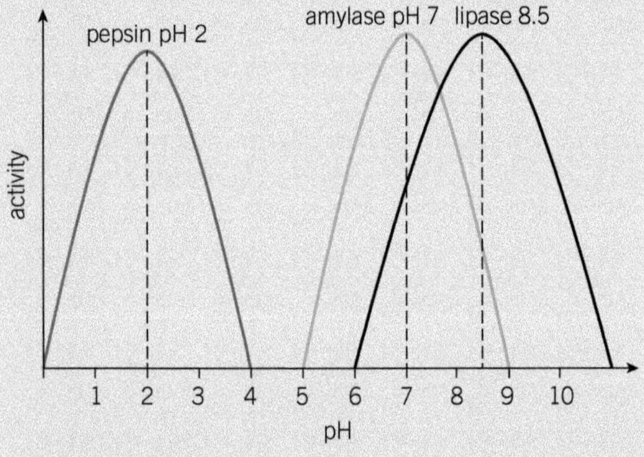

(3)

2 Glucose will be needed to enable the yeast to respire (1); respiration is a series of enzyme-controlled reactions (1); if any products are acidic or alkaline then the pH would be altered which could denature the enzymes (1); which would prevent the collection of valid data (1)

3 a enzymes are needed for DNA replication and protein synthesis (1); without cold-active enzymes these processes would slow/stop (1); as the enzymes would be inactive (1); which would result in the cell cycle stopping/cell death (1).

b enzyme structures are more flexible compared to human enzymes. Enzymes active at low temperatures but become inactive more quickly at higher temperatures and will denature at lower temperatures.More disulfide bonds, fewer hydrogen bonds. (3)

3.6

Tranexamic acid

1 Prevents the breakdown of fibrin, by stopping the action of proteases.
2 Binds to plasmin and competitively inhibits plasminogen – this prevents plasmin breaking down fibrin and clots are stabilised.
3 Clots formed after removal of a tooth are stabilised and excessive blood loss is prevented.

1 Based on diameter of 14 mm. 14 000 ÷ 3000 = 5 µm (2)

2

Enzyme	Cofactors required	Substrate	Product
Thromboplastin	Calcium ions	Prothrombin	Thrombin
Thrombin	Calcium ions	Fibrinogen	Fibrin

(2)

3 *Similarities*. Protease indicates the enzyme works on a protein molecule as is substrate (fibrinogen) (1); the enzyme hydrolyses the end of the polypeptide chain (1); it uses a serine amino acid to perform the cleavage (1) (max 2).
Differences. This may result in a different amino acid being present in the primary structure (1); it could result in shortened polypeptide chain (1). (max 2)

3.7

First-aid procedure

Removing the object could further damage blood vessels resulting in more bleeding

1 To protect the first aider AND the casualty (1); to protect the first aider from any potential infection carried in the casualty's blood (1); and protect the casualty from any bacteria on the first aiders skin entering the wound (1).

2 Slows the rate of blood loss (1); as more resistance to blood flow is increased (1).

3 To prevent dislodging any clot that may have started to form (1); by disrupting any fibrin mesh (1); resulting in the clotting cascade having to start again (1).

3.8

1 *Any 2 from*: Excessive bleeding; fainting or feeling light-headed; hematoma (blood accumulating under the skin); infection (a slight risk any time the skin is broken).

2 *Advantages* (any 2 from): Enzymes are highly specific i.e. perform the exact required reaction (1); conditions caused by a low level/absence of the enzyme can be stopped by replacing the enzyme (1); a lot of research is now being done into gene therapy to give a more permanent solution (1). *Disadvantages* (any 2 from): Production and purification of enzymes is very expensive when compared with drug manufacture (1); they are difficult to store, enzyme activity is lost quickly unless frozen often at –7°C (1); possible side effects due to the species differences between the source enzyme and human form (1); they need to be injected to avoid denaturation in the stomach if taken orally (1).

3 a Catalyse the same chemical reaction (1)

b Slight difference in primary structure/ amino acid sequence (1); different turnover rates (1); different optimum temp (or pH) (1). (2 max)

3.9

1 Freezing damages cells as the ice crystals cause the cell surface membrane to rupture (1) and the membranes of cell organelles (1); whereas plasma and serum do not contain any cells (1)

2 Bar chart drawn with x-axis labelled "acquired condition" and y-axis labelled "number of people" (1); bars drawn of equal width and NOT touching (1); plot area covers 50% of the paper in EACH direction AND equidistant scale used (1); all data plotted to +/– half small square (1); key given or all conditions labelled on x-axis in centre of bar (1).

3 a Leucocytes are likely to promote an immune response (1); and produce antibodies (1); which can cause problems with future transfusions (1). Patients need to boost erythrocyte count (1)

b Whole blood can cause complications during transfusions (1); and using red cell concentrate reduces the risk of the patients' blood volume rising too high (1).

4.1

1 5 (1) Two condensation reactions are needed to make each individual nucleotide molecule (1); one to join the phosphate to the pentose sugar and one to join the pentose sugar to the nitrogen containing base (1); then a 5th condensation reaction is needed to join the two separate nucleotides to form the dinucleotide (1).

2 *Similarities*: both are pyrimidines (1); both contain the same elements (C, H,O and N) (1); both contain a single ring structure (1). *Differences:* thymine is a larger molecule (1) more complex molecule/additional methyl group and H atom (1);

3 *Universal* = found in all living cells (1); *Energy* = molecule can be hydrolysed to release energy or formed by condensation reactions to store energy (1). *Carrier* = ability to release small quantities of energy to link energy releasing reactions to energy requiring (1).

4.2

1 *similarity* = ratio of 1:1 for A:T and G:C is common to both; *difference* = proportion of A and T is significantly lower in the prokaryote (mycobacterium tuberculosis) and the proportion of C and G is much higher (1) i.e. reverse ratios (1).

2 C pairs with G therefore 32.6% will be guanine. Therefore 32.6 + 32.6 = 65.2% of the bases are G or C (1). The remainder will be A and T i.e. 100 – 65.2 = 34.8% (1). This will be split equally between A and T. Therefore $\frac{34.8}{2} = 17.4\%$ will be A (1).

3 Hydrogen bonds are responsible for joining the complementary bases together. Due to the molecular structure of the bases these can't form between A to C and G to T (2)

4 UAC CAG UUA GCC GCA UAG (1)

4.3

1 TTACCGGCATACGCA

2 2 nucleotides drawn correctly each with a single appropriate base (any from bases from A, T, G and C) (1)

Deoxyribose shown (and labelled at least once) within each nucleotide structure (1); phosphate group shown (and labelled at least once) within each nucleotide structure (1); removal of water shown / condensation reaction annotated (1)

3 *Principle A* = semi-conservative replication; as each new DNA helix consists of one parent template strand and one new daughter strand (2) *Principle B* = conservative replication; as the two original template strands of DNA strands remain together in a double helix and a new copy is composed of two new strands containing all of the new DNA base pairs (2) *Principle C* = dispersive replication; more-or-less random interspersion of parental and new parts of parental DNA exist in the daughter DNA molecules (2).

Other specialised enzymes involved in DNA replication

1 DNA polymerase requires a free 3' OH group, due to the shape of its active site.

2 a It removes nucleotides, one at a time from the end (exo) of a polynucleotide chain.

b It works by catalysing a hydrolysis reaction, that breaks phosphodiester bonds, at either the 3' or the 5' end.

3 Quaternary – as there are more than one polypeptide chains present in the functional protein.

4.4

Humulin production

1 Many amino acids joined by peptide bonds, via condensation reactions. Polypeptide chain is then folded in alpha helices and/or beta pleated sheets, and further folded in to globular arrangement, made of more than one polypeptide chain i.e., quaternary structure, correct ref to tertiary and secondary structure.

2 Protein is made of 2 polypeptide chains

3 Linked by disulfide bonds

4 Easy to manipulate the bacterial genetic material, easy to culture, short generation time, less ethical objections

5 Exact match to Humulin insulin (no missing amino acids(s), fewer/no side effects, fewer ethical objections, can be used by religious groups who view avoid contact with materials from pigs.

1 Methionine (ATG) and tryptophan (TGG) (1)

2 a Met, Val, Pro, Ser, Arg, Gly, Asp (1)

b This would code for STOP rather than serine; hence the polypeptide chain would only consist of three amino acids instead of seven, i.e. it would be truncated (2)

3 Advantages: more information can be provided in a given sequence of bases, e.g. a sequence of nine bases could provide the information for seven amino acids instead of only three. (1)

Disadvantages: if one base is changed, e.g. by mutation, it will now potentially affect three amino acids in the polypeptide chain rather then just one. (2)

4.5

1 a 4 (1)

b polypeptide chain (protein primary structure) (1)

c UAC GAG UCC GGA (1)

2 a Both made from many mononucleotides (1); joined by covalent bonds (1); both synthesised by condensation reactions (1); both synthesised in reactions controlled by enzymes (1); both contain phosphate groups (1); both contain pentose sugars (1); both made of the same common elements i.e. C, H, O, P, and N (1); both contain nitrogen-containing bases (1); both contain purine & pyrimidine bases (1); both contain the bases A, G and C (1)

2 b

Feature	DNA	mRNA	tRNA
Number of polynucleotide chains	2	1	1
Relative size	Largest	Middle	Smallest
Overall shape	Double-helix	Single stranded long chain	Single stranded arranged as clover-leaf
Sugar	Deoxyribose	Ribose	Ribose
Organic nitrogen containing bases	A, T, C, G	A, U, C, G	A, U, C, G
Location	Nucleus	Made in nucleus but found throughout the cytosol	Made in nucleus but found throughout the cytosol
Relative quantity in the cell	Fixed quantity	Variable quantity	Variable quantity
Relative stability of the molecule	Very stable	Unstable – easily broken down	More stable than mRNA but less than DNA

3 UGA is a stop codon on the mRNA so there is no amino acid that corresponds to this triplet (2).

4.6

Switching on genes – promoters

1 All have a complementary DNA binding site (and facilitate the binding of RNA polymerase).

1 mRNA leaves via the nuclear pore (1); as the mRNA molecule is too large to diffuse across the nuclear envelope (1).

2 Can be replicated (1); gives stability to the template strand (1); less prone to breakage (1); less prone to getting tangled (1); unlikely to partially pair with itself in an undesirable way, which would make it harder for an RNA polymerase to get to the sequence which should be transcribed (1); more stable (1); one strand serves as a template for repair of small damage to the other (1). *Max 3 marks*

3 Prokaryotes do not have DNA contained in a nucleus, so translation begins as soon as the mRNA has been made, i.e. no delay in time for mRNA to move from nucleoplasm to cytoplasm (2)

4.7

1 a one codon is a stop codon which does not code for an amino acid (1) 1 start codon (for methionine) which has been removed from the chain (1).

b 60 amino acids in the polypeptide chain. 40 amino acids are joined each second. $\frac{60}{40} = 1.5$. Time to produce the polypeptide chain = 1.5 seconds (1)

2 All polypeptides would have an extra amino acid in the primary structure (1); many essential proteins would no longer function (1); as their tertiary structure would change (1).

3 (10)

Feature	Semi-conservative replication	Transcription	Translation
Free nucleotides are activated	yes	yes	yes
How much DNA is involved	whole molecule	cistron / gene	none directly
Number of new molecules per cycle	1	1	1
Hydrogen bonds are broken between complementary base pairs	yes	yes	yes
Cytosine pairs with guanine	yes	yes	yes
Adenine pairs with thymine	yes	no	no
Phosphodiester bonds are made	yes	yes	no
Peptide bonds are made	no	no	yes
Location of process	nucleus	nucleus	cytoplasm
Product	DNA	mRNA	polypeptide chain

5.1

Heart disease

1 The blood flow will be restricted and so the muscle will be deprived of essential nutrients, such as glucose and the gas oxygen and the muscle tissue will die.

1 Large multicellular organisms such as trees have a low metabolic rate (1); and so can survive on relatively low levels of oxygen and produce low levels of carbon dioxide so there is no need for a transport system. (1)

2 Cardiac muscle is unique in being myogenic (1). The cells are branched and interconnected and the muscle will not tire, which is essential to ensure life is maintained (1). The muscle has an intrinsic beat, even if a small piece is dissected out (1). This fundamental difference between cardiac muscle and other muscle tissue is important to understand.

3 Every systole has to be followed by a period of diastole (1); so (unlike skeletal muscle), cardiac muscle cannot go into 'tetanus' (a period of prolonged contraction) (1)

5.2

1 The blood flows through the pulmonary vein to the left atrium, through the left atrium, left ventricle, aorta, then through ONE of the major arteries of the body (carotid, hepatic, mesenteric, renal or another), through a capillary in the corresponding organ or tissue, then returns via the relevant vein to the vena cava and the right atrium, right ventricle and the pulmonary artery to the lungs before returning to the heart again via the pulmonary vein (1); the correct sequence of flow is important (1).

2 The blood is pumped into the aorta under pressure during ventricular systole which causes the artery wall to stretch (1); as the heart relaxes the artery wall recoils to give the characteristic pulse (1); the artery wall contains thick layers of both muscle tissue and elastic tissue to allow this to happen (1)

3 The valves ensure the blood flows in one direction only and back flow is prevented (1); the increase in blood increases the pressure in a chamber and pushes the valves open to allow blood to flow into the next chamber of the heart (1); as this chamber fills with blood the pressure increase here forces the valves to snap shut behind it and prevents the back flow of blood (1)

Atrial kick

10.5 cm^3 and 24.5 cm^3, respectively.

5.3

Testing factors affecting the heart rate

A paired t-test. The same subject is being measured before and after the exercise.

1 age, lifestyle such as exercise /activity levels/ type of exercise / duration of exercise, nutritional health and body hydration. *2 marks – 2 variables required per mark so at least 4 required.* (2)

2 a In A the increase in heart rate is rapid from 2 to 4 minutes (from 65 to 95) and from 4 to 6 minutes the increase slows (from 95 to 105) (1) In B the increase in heart rate is less rapidly increasing from 60 to 80 at 2–4 minutes and 80 to 85 at 4–6 minutes (1)

b A $\frac{40}{65}$ = 0.615 so a 61.5% increase. (1),

B $\frac{25}{60}$ = 0.416 so a 41.6% increase (1)

3 a At the end of ventricular diastole and at the end of ventricular systole. (1)

b Stroke volume = end diastolic volume – end systolic volume (1)

Starling's Law and other factors affecting cardiac output

1 There is less time for the ventricles to fill so stroke volume is lower.

Testing factors affecting the heart rate

1 Primary data is data collected from experiments carried out by the investigator; this gives a small amount of data collected personally and so may be considered accurate with confidence in the control of other variables. Secondary data is data obtained from other sources, i.e. scientific papers or reviews. The advantage is that it provides a much larger sample, so is likely to be reliable, although there may be differences in the methods used and the variables controlled.

5.4

1 The electrical impulse from the SAN is delayed at the AVN and then conducted down to the apex of the heart before ventricular systole happens. (2)

2 550 ms = 0.55 s

$\frac{60}{0.55}$ = 109 bpm (2)

3 The maximum heart rate declines with age (220–age). No systole (P wave) is possible until after a T wave so there will be fewer cardiac cycles per minute and therefore, the maximum heart rate will be lower. (2)

6.1

1 A closed system means that high blood pressure can be maintained throughout the body and there can be differences in pressure in the pulmonary and systemic systems (1); a double circulation ensures an increase in delivery of oxygen and nutrients such as glucose and increase in efficiency in the uptake of these substances (1); oxygenated blood does not mix with deoxygenated blood making oxygen uptake from the exchange site more efficient and lower volumes of fluid are needed than in an open system where the blood would bathe the organs and tissues and so require a high volume of fluid (1); there can be a varied blood supply to different organs depending on their function and need (1). (max 3)

2 Smooth muscle contracts to narrow the lumen and relaxes to allow the lumen to dilate. This regulates blood flow to organs and also maintains the blood pressure (1); elastic fibres stretch during ventricular systole to accommodate the volume of blood ejected into the arteries. They recoil during ventricular diastole which propels the blood forward 'between beats' (1)

3 The skeletal muscles contract around the veins and press on the thin walls of the veins, which increases the pressure on the blood which assists blood flow through the veins (1); the valves prevent backflow and keep the flow moving in one direction back towards the heart, which is necessary given the low pressure (1)

6.2

How to measure blood pressure

A sphygmomanometer is measuring blood pressure of the brachial artery in the arm. The pressure in this artery will be higher for the systolic pressure and lower for the diastolic pressure than the blood pressure in the aorta because the aorta is carrying blood just leaving the heart (under pressure due to contraction of the cardiac muscle of the left ventricle) and as the blood flows further away from the heart the influences of the pressure waves from the blood flowing against the artery walls. It will therefore be showing greater differences between systolic and diastolic pressure.

Risk factors that contribute to hypertension

An ideal life style should include not smoking, a controlled weight (not over weight), a diet low in salt, reduced alcohol consumption, good levels of exercise several times a week, and reduced stress levels. The other two risk factors – gender and age cannot be controlled.

1 Tissue fluid contains no large protein molecules and no blood cells with the exception of some phagocytes if infection is present (1); tissue fluid bathes the cells outside the blood vessels whilst blood plasma is found only within the blood vessels (1)

2 Blood pressure is raised over an extended period of time above the normal level of $\frac{120}{80}$ mmHg (1); consistently over $\frac{140}{90}$ mmHg(1). There are three different stages of hypertension increasing in severity. (2)

3 In cell surface membrane (1); of cells in walls of (peripheral) arterioles (1), and capillaries (1).

6.3

1 Cereals are monocotyledons so xylem and phloem are arranged in a ring with the tissues alternating (1); potatoes are dicotyledons so will have a central area of xylem in an X shape with phloem between the 'arms' of the X. (1)

2 The flow of water is not hindered because there are no end walls and no cell contents (1); the lignin in the walls keeps the cells open and prevents them from collapsing (1); the pits in the side walls allow transport from one tube to the next (1)

3 The organelles in the companion cell carry out the functions (1); communication between sieve tube and companion cell occur by the plasmodesmata or gaps in the walls (1)

6.4

1 The apoplastic route moves in the water-filled spaces between cells and in cell walls (1)

2 The Casparian strip is a waterproof strip around the endodermal cells that ensures water only travels with the symplast route through the endodermis (1); the endodermis selectively takes up minerals such as nitrates by active transport which lower the water potential in the xylem so water moves down the water potential gradient created (1); prevents water returning back to the cortex once it has entered the xylem because the symplast route is blocked (1)

3 Root pressure requires active transport of minerals which use ATP. ATP cannot be produced without respiration. (2)

6.5

Using a potometer to measure transpiration

Weigh the potometer at the beginning and then after a set period of time. The fall in mass will be as a result of water loss.

1 The plant must have the stomata pores open in order to exchange the gases needed for respiration and photosynthesis. As a consequence of this there will inevitably be water loss. In extremely hot conditions the water loss is frequently in excess and the plant wilts and may even die. Only those with specific adaptations survive such conditions. (2)

2 a an increase in temperature increases evaporation and so more water vapour is lost and diffuses out of the stomata. This will increase transpiration rate. (1)

b Windy conditions create an improved diffusion gradient as the wind removes the air with increased humidity. Reduction in wind means there is a build-up of water vapour around the leaf and so the diffusion shell is reduced around the stomata. Less water vapour evaporates and transpiration is reduced. (1)

c In reduced light intensity the stomata close and this reduces evaporation from the mesophyll cells as the humidity builds up in the air spaces. (1)

3 659.4 mm^3 (((3.14 × 0.5) × 14) × 60) (3)

6.6

1 It contains cytoplasm and is destroyed by metabolic poisons or by high temperatures (1); it survives by sharing nucleus, mitochondria, and other organelles with the companion cells adjacent to it (1).

2 The pH of the companion cells is higher than any surrounding cells (1); there is an active process involved as a metabolic poison stops translocation (1); the sucrose concentration in the source is higher than in the sink (1). Movement of solutes is (approximately 10 000 times) faster than possible by mass flow or diffusion alone (1).

3 A multi-bar chart with the different plant parts – **1** leaf blade, **2** vascular bundle in leaf stalk, **3** tissue around leaf stalk, and **4** buds, roots, and tubers on the X axis (1). Mean carbohydrate content ($\mu g g^{-1}$ +/– standard error) on the Y axis (1). Four bars for each tissue type. Each bar to represent one of the four carbohydrates. Range bars added to each bar to represent standard error of mean (1).

4 *Glucose* levels low in all tissues (120 – 624 $\mu g g^{-1}$) as glucose produced (by photosynthesis) is converted to sucrose for transport or storage (1). Glucose has some high levels in the vascular bundles and the tissues immediately surrounding them as some glucose not converted to sucrose or after hydrolysis may be found in these tissues (1). *Fructose* levels are very low in leaf blade and the buds and roots as this carbohydrate is not used directly by the plant so production levels will be low (370 – 494 μg^{-1}) (1) and it is not used for storage as it is not in any high level in the root tubers (1). *Starch* is not in high levels in any of the four plant areas as it is not the main storage molecule in cyclamen (18 – 152 $\mu g g^{-1}$) (1). *Sucrose* is in the highest quantity (1), with highest in vascular bundles (5757 $\mu g g^{-1}$) as this is where sucrose is transported through the plant by translocation and in the buds and tubers (2260 $\mu g g^{-1}$) which is the sink (1). Sucrose will be transported here for plant tissue use or for storage (1) (max 6)

7.1

1 The 'C' shaped rings of cartilage support the trachea and bronchi to prevent them collapsing during exhalation (1); the ring is 'C' shaped to allow flexibility and swallowing movements in the oesophagus (1)

2 Magnification is equal to the size of image / actual size (1) Allow a calculation that shows 0.5 mm / the measured size of the diameter in mm = magnification (1).

3 a Diffusion happens until the gases in the alveoli and blood come into equilibrium. (1)

b The pulmonary artery is carrying blood back to the lung from the tissues via the heart (1); and so will have reduced levels of oxygen and increased carbon dioxide levels due to tissue respiration compared to the alveolar air (1)

7.2

Measuring pulmonary ventilation

1 3.44 litres

How to perform expired air resuscitation

The adult first aider should cover the infants nose and mouth with their mouth

1 Tidal volume is the volume of air moving in and out of the lungs during a normal breath – usually about 0.5 dm^3. Vital capacity is the deepest possible breath in followed by the deepest possible out – about 3.5 dm^3. (2)

2 Pulmonary ventilation is 500 × 9 = 4.5 dm^3. (2)

3 As total lung capacity is about 6 dm^3

This gives $\frac{0.5}{6.0}$ = 8% for tidal volume which will be the normal volume used at rest.
During deepest possible breathe or vital capacity $\frac{3.5}{6.0} \times 100 = 58\%$ (3)

8.1

How can we analyse the cell cycle?

1 9.6 hours

2 Possible health problems include DNA damage, cancer, or malnourishment, such as a shortage of certain nutrients.

1 Cell growth and protein synthesis occur in both phases (1); the second growth phase involves an increase in energy stores (1); whereas this is not such a significant event in the first growth phase. (1)

2 24 hours (to the nearest hour). (1)

3 The timing of each step in the cell cycle is crucial. (1)

Cyclins and CDKs provide the signals to regulate the sequence of events in the cycle. (1)

These molecules will also stop the cell cycle if errors are detected. (1)

8.2

Preparation of plant tissue to view under a microscope

They are fixatives, which stabilise structures in the cells and prevent changes during staining.

1 The use of a sharp razor blade or other cutting tool (1); HCl, which is a strong acid (1); 60°C water (1); excessive pressure on the cover slip could cause it to break and produce sharp edges. (1)

2 Number of cells in mitosis counted (e.g. 14) divided by total cell number (e.g.19) correctly rounded e.g. 0.737 (1)

3 Cancer (1)

The normal control of cell division has broken down. (1)

8.3

1 Destruction of infected cells and damaged cells (1);

Forming the structures of organs and tissues (1);

The formation of connections between neurones in the nervous system (1);

The elimination of useless or harmful immune cells. (1) *max 3 marks*

2 During necrosis, cells rupture and release enzymes. (1)

Necrosis is an uncontrolled process. (1)

Apoptosis is a controlled, regulated process. (1)

Rather than cells rupturing, cell fragments are packaged into vesicles. (1)

3 Phosphatidylserine is positioned on the outside of apoptotic vesicles. (1)

It binds to receptors on macrophages, which engulf the vesicles. (1)

8.4

1 Both totipotent and pluripotent stem cells can differentiate into any type of cell (1); but only totipotent cells can form a whole organism (1)

2 Bone marrow contains multipotent stem cells (haemocytoblasts) (1); that can differentiate into a range of blood cells (1); bone marrow transplants replace the cells lost by the leukaemia patient (1)

3 *Benefits*: The possibility of cell replacement therapy for a vast range of diseases (1); stem cells could also be used to produce various tissues in labs, which would enable drugs to be tested and the development of diseases to be researched *in vitro* (1) *Risks*: Some people worry that stem cell therapy could lead to cloning or the modification of human appearance or behaviour (1); early research raised the risk of stem cell therapy producing tumours (1); some of the ethical issues might be resolved by the use of new techniques such as iPSC production and the removal on ESCs without destroying embryos (1) *max 4 marks*

9.1

Correlations in biology

1 Non-coding DNA is likely to regulate gene expression, controlling which genes are switched on and off in particular cells and the extent to which genes are transcribed. This enables greater fine-tuning in complex organisms.

2 Randomisation; control variables / only a single (independent) variable is changed (i.e. the one being investigated); appropriate statistical testing.

1 Haploid (1); a diploid cell contains pairs of homologous chromosomes, which means the diploid number will be an even number (1)

2 524288 (1)

3 Crossing over (1); during prophase I involves genes being swapped between homologous chromosomes (1); this creates new combinations of alleles (1); independent assortment of chromosomes and chromatids further shuffles the combinations of alleles present in gametes (1).

9.2

What is fetal alcohol syndrome?

1. Additional lifestyle variables, such as diet and smoking, would need to be taken into account when analysing the data and performing statistical tests. The method of data collection would need to be known in order to assess validity (e.g. were women asked to make a daily log, or were they asked to estimate their consumption at the end of pregnancy?).
2. Ethanol crosses the placenta and enters the circulatory system of a fetus. It binds the receptors on cells in the brain (neurones) and interferes with the development of the nervous system. If the patterns of neural growth are altered, this can cause permanent changes in behaviour and psychology.

1 Protein (1); (for growth and synthesis of enzymes and antibodies) (1); vitamin A (1); (for the development of the fetal immune system, gene transcription and the formation of rod cells) (1); vitamin C (1); (for collagen production) (1); folic acid (1); (for the development of the nervous system) (1). (max 4)

2 Regulating alcohol consumption (1); not smoking (1); avoiding raw meat and unpasteurised dairy products (1); maintaining a balanced diet (1).

3 Tobacco smoke contains carbon monoxide and nicotine (1); carbon monoxide binds to haemoglobin (1); reduces the concentration of oxygen received by the fetus (1); reduces the rate of respiration in cells (1); nicotine narrows blood vessels (1); including those in the umbilical cord (1) (which further reduces oxygen supply and lowers respiration rate) (5)

9.3

1 47 chromosomes indicate a chromosomal mutation (e.g. three copies of chromosome 21 would indicate Down syndrome, or two X chromosomes and one Y chromosome would indicate Klinefelter's syndrome). (1)

2 *Advantages*: more cells are collected (1); and the procedure can be performed earlier (1). *Disadvantages*: a higher risk of miscarriage (1); and fetal deformities (1).

3 13.3 mm week^{-1} (2)

10.1

Why domains?

1. Eukaryotes all have a nucleus, membrane-bound organelles and larger ribosomes than prokaryotes (i.e. 80 S).
2. Woese found evidence that the Archaea share more similarities with the Eukaryota than the Eubacteria. This evidence, particularly from DNA and amino acid sequences, suggests that Archaea are more closely related to Eukaryota than Eubacteria. Woese argued that an accurate classification system should reflect this evolutionary relationship.

Taxon	Human	Chimpanzee
Domain	Eukaryota	Eukaryota
Kingdom	Animalia	Animalia
Phylum	Chordata	Chordata
Class	Mammalia	Mammalia
Order	Primates	Primates
Family	Hominidae	Hominidae
Genus	*Homo*	*Pan*
Species	*sapiens*	*troglodytes*

(8, 1 mark per correct row)

2 a 2 and 3 (1) b 5 (1) c 2 and 3 (1)

3 DNA and amino acid sequences accurately reflect the evolutionary relationship between two species (1); the greater the similarity of the sequences, the closer the relationship between two species (1); classification based solely on anatomical evidence is less accurate because members of the same species can show a lot of variation and separate, unrelated species can appear similar because of convergent evolution (1).

10.2

1 *Homo erectus* and Neanderthals. (2)

2 Chimpanzees and bonobos (more genetic data would be needed to know which of these two taxa has a closer evolutionary relationship to humans). (2)

3 Humans. (1)

10.3

1 (A) anatomical (B) physiological (C) behavioural (D) physiological (E) anatomical (4)

2 a Brain size = 199% (1); male body mass = 41.5% (1); female body mass = 63.3%

b 30 cm^3 100 000 years^{-1} (1)

3 Bipedalism enabled early humans to travel longer distances more efficiently and therefore improved hunting capabilities (1); bipedalism also freed the hands to enable tool use (1); making and using tools would have further improved the ability of humans to hunt (1); the evolution of larger brains would have helped humans process the information required to produce tools (1).

10.4

1 Water storage in stems (succulent plants) (1); stomata closure to reduce water loss (1); long roots; reduced leaf surface area (1); rolled leaves and pitted stomata to trap humid air (1). (max 3)

2 Stomata on upper surface (1) – as lower surface is not in contact with the air for gas exchange (1).

3 Lab conditions are likely to be different from those in the wild (1); the plant might be damaged when cut (1); an isolated shoot is smaller than the whole plant (1).

10.5

A modern example of natural selection

1 The frequency of the lighter fur trait has increased in the population so the allele must have conveyed an advantage. The pale and dark coloured mice may still be able to interbreed so may not be separate species.

1 Include predators (1); disease (1); food availability (1); climate change (e.g. temperatures (1); light regimes (1); humidity changes (1)), chemical factors (1). (max 3)

2 a Little direct evidence exists, such as fossils (1); and experiments are very difficult to perform when researching language evolution (1).

b Genetic analysis (1); comparative biology (i.e. comparing brain structure and anatomy with other species (1)), computer simulations (1); archaeology and fossil analysis (1).

3 The insect species has variation between its Members (1); the insecticide acts as a selection Pressure (1); only individuals that have some resistance to the insecticide will survive (1); reproduce and pass on their beneficial alleles to their offspring (1); over generations, the proportion of alleles for resistance will increase and the entire population will be resistant to the insecticide (1) (max 4)

10.6

1 Species diversity measures the number and abundance of species in a specific area/habitat/ecosystem (1); whereas ecosystem diversity measures the number of ecosystems in a specific area/region (1).

2 High genetic diversity indicates a species that has a wide range of traits on which natural selection can act (1); therefore, when a selection pressure is exerted, high genetic diversity increases the chance of some members of the species possessing the traits necessary to adapt and survive (1)

3 (3)

Species	Number of monomorphic genes studied	Number of polymorphic genes studied	Percentage of polymorphic genes (%)	Average heterozygosity (%)
A	28	8	**22**	4.2
B	**35**	10	22	7.0
C	8	3	**27**	5.4

It is not possible to conclude which species has the highest genetic diversity. The data lead to conflicting conclusions (1); average heterozygosity suggests species B, % of polymorphic genes suggests species C (1); however, very few genes have been analysed to determine % of polymorphic genes, especially for species C (1); analysing more genes would improve the validity of the results (1); it is unclear how many genes have been analysed to calculate average heterozygosity. This could be the more accurate of the two measures in this case (1). (max 5)

11.1

1 a Include reference to overcrowded conditions in refugee camps, *any other sensible suggestion* (1)

b Lack of adequate protein (so immune system not producing antibodies OR lack of access to the antibiotics or drugs for treating primary infection). (1)

2 Patients do not finish the course so the more resistant bacterial cells still survive and reproduce – passing on the alleles which make them more resistant leading to populations of drug resistant bacteria. (2)

3 Viruses cannot reproduce on their own; do not grow or undergo division; do not transform energy; lack machinery for protein synthesis; and do not carry out any of the normal functions of life. (4)

11.2

Gram staining

a Loop should be sterilised before and after use by heating in a bunsen flame.
b The slide should be 'fixed' by carefully passing it once through the bunsen flame.

1 The Gram-positive cells have already been stained by the iodine and crystal violet complex and so the counter stain will not be taken up. (2)

2 The resolution of the microscope is not great enough (1); to distinguish between small rods and cocci (1)

3 **Gram staining:** Gram stain contains crystal violet. Two types of bacteria can be identified – Gram-positive and Gram-negative. Gram-negative bacteria will stain pink, while Gram-positive bacteria will stain purple in the presence of the Gram stain. *Advantages* bacteria easily identified and technique easy to use. *Disadvantages*: are only two separate and distinct groups and the cells must be correctly stained. **Colony morphology:** these are the characteristics of shape, colour, edges and the surface features of the colony which all differ in different types of bacteria and are used to identify them. Some colonies are round and smooth edged, others are wrinkled and wavy edged. Many are creamy in colour but others are grey, yellow, red or orange colonies. *Advantages* are there are a number of clear features to identify. *Disadvantage* is that some bacteria share many of the characteristics and so identification cannot be relied upon. **Cell morphology:** cocci are spherical in shape e.g., *Staphylococcus aureus*. Bacilli are rod shaped but there is a lot of variation in size from short and fat to long and thin e.g., *Mycobacterium tuberculosis*. Spirilla are twisted in shape – cork screw like with a number of spirals that differ between different species. *Advantages* identification relatively easy without the need for staining techniques. *Disadvantages* may be many variations especially in bacilli that confuse identification with other groups. (max 2 marks per technique, 1 mark per detail/advantage/disadvantage)

11.3

Evaluating data to assess the impact of a disease:

1 The prevalence and mortality rates may be different for TB and HIV in some countries because of: *Health care availability*; some countries have limited health care and limited finances to fund good health care. This may increase the incidence of TB and increases the chances of death caused by TB. With HIV the increased prevalence is concentrated in South and East Africa where again health care is limited. *A TB vaccination programme* can eliminate or completely reduce TB incident especially if coupled with good health care. With HIV the availability of drugs will also help reduce the incident of developing severe symptoms. *Education* is important in both. TB can be reduced by education on health and preventing the spread. With HIV it is largely education needed to prevent the spread as well as education on the use of condoms and preventing unprotected sex from occurring. *Control measures in place* – which may be a matter of finances available as well as the number of trained professionals able to support the control measures. *Monitoring* – in some countries there is little monitoring and so the actual numbers of individuals counted may be far fewer than actual so reducing the figures for prevalence.

2 A bar chart or a pie chart could be used because the data is categorical – the IV is discrete and so a line graph would be inappropriate. A correlation chart would also be inappropriate because the IV is discrete.

H1N1 Influenza (swine flu)

1 Neuraminidase is one of the markers on the outside of the flu virus. It is an enzyme and is responsible for releasing the virus from host cells. Neuraminidase inhibitors could inactivate the enzyme and so prevent the virus rupturing the host cell and so prevent it being released. This would contain it within the host cell and may allow the host T cells time to attack and destroy the contaminated cells. As a result the spread of the flu virus would be contained and even be eliminated.

1 A prevalence rate looks at the number of cases per 10 000 people (for example). The areas will have different population sizes. (2)

2 There is a general increase from 2000 where the number of cases was 6 500 to 8 500 in 2012 an increase of 31%. However, the data plateaus at 7 000 approximately between 2002 and 2004 and again at 8 000 between 2005 and 2008. (4)

3 a The glycoprotein antigens on the surface were modified by mutation (1). The H1N1 strain had elements of four different flu virus strains and so it had not been encountered before in any great numbers (1).

b Once the flu virus is controlled it will revert to the normal seasonal type of flu. Control measures include ensuring vulnerable people are protected by vaccination programs (1). Control of spread by encouraging people to avoid contact with infected individuals (1) and encouraging those affected from not visiting doctor surgeries or hospitals (1). (max 2)

12.1

1 An antigen is the specific protein that acts as a marker on the cell surface membranes of cells (1); its presence in another body triggers the immune response. A pathogen triggers the immune response as the antigens are seen as foreign and brings about the symptoms of the disease (1)

2 The macrophages are activated by a foreign antigen and cloned to produce many B cells which in turn produce antibodies specific to that antigen / pathogen /disease causing organism. This makes it easier for the phagocytes to attack and destroy the invader. There are different types of T lymphocytes, some of which attack and kill infected host cells, others stimulate the B cells to divide by production of cytokines and yet others regulate the immune attack. (2)

3 The virus destroys T helper lymphocytes (1); Without these to produce cytokines, the specific B memory cells will not be activated (1)

12.2

Measuring antibody concentration using ELISA>

1 Using the Optical density of the unknown antigen concentration read the value on the graph by drawing a ruled line from the OD on the Y axis to the curved plotted line. When the line touches the curve draw a second straight ruled line down from the line to the X axis and read off the value for antigen concentration /pg mL^{-1}

2 Using the technique described in question 1 the value is 300 antigen concentration /pg mL^{-1}

1 A positive reaction to the mantoux test for TB indicates the person already has antibodies to the TB bacterium. (1)

2 Each antibody is specific to only one type of antigen, found on the surface of a particular pathogen, due to the specific shape of the variable region of the antibody (1); different pathogen strains will have different antigens and the antibody receptor will not fit (1)

3 The presence of antibodies in the child's blood due to passive immunity, from the mother, could neutralise the antigens in the vaccine and an immune response would not develop. (2)

13.1

1 93% (2)

2 On the one hand, they give the strongest immune response as the organism can multiply and both the humoral and the cellular response can be triggered. On the other hand, they are more difficult to store and transport and cannot be used on people with compromised immune systems. (3)

3 The bacteria produces a toxin and even small amounts of the toxin can cause paralysis and death. A rapid production of antitoxin antibodies is required so the boosters ensure that the circulating antibodies and memory cells stay in high concentrations in the blood. (2)

The epidemiology of whooping cough

1 There are peaks in the incidence of whooping cough, for example in 2001, 2005 and 2008 with incidences of about 38 cases per 100 000 in 2001 and 30 cases in 2005.

2 Mother will develop active immunity and produce antibodies which will cross the placenta. The baby will be born with some passive immunity to whooping cough.

3 Mothers will develop active immunity and the antibodies can be passed to the baby in colostrum and breast milk.

Vaccination risks

1 $0.035 \div 0.03 = 1.16$

2 It means that having the vaccine reduces the risk of developmental disorders due to disease.

13.2

1 Antigen variability refers to the many different types of antigens found on the pathogen surface, with one pathogen type carrying different types and forms of antigens. For example, there are at least nine different sub-types of the HIV virus each of which carries different antigens. (1)

2 Advantages: antigens more tolerant of temperature changes OR can be stored for longer.

Disadvantages: gelatin could provoke an allergic response; some people may object to the animal origin of the stabiliser for dietary or religious reasons. (2)

3 pathogen mutation, pathogen variability, vaccine induced disease caused by the live pathogen in the vaccine becoming virulent again, storage and security issues, cost. (5)

13.3

Kirby-Bauer antibiotic testing

1 *Most effective* antibiotic is the central one in the petri dish. *Least effective* is the antibiotic which is the one bottom right of the petri dish.

2 a the exclusion zones around each disc could be measured using callipers to give the diameter of the exclusion zone. The larger the zone, the more effective the antibiotic. Or surface area of the exclusion zone could be calculated. (Squared paper may be needed if the zones are irregular in shape).

b this will depend on the resolution of the callipers but maximum 2 decimal places.

Slowing down the spread of resistant bacteria

1 No competition from non-pathogenic bacteria for nutrients or room on a surface.

1 The 70S ribosome which is affected by tetracycline is a feature of all prokaryotic cells. Also, all prokaryotic cell membranes must be permeable to this antibiotic. (1)

2 Prescribed incorrectly or course not finished led to a selection pressure in favour of resistant gonorrhea bacteria which multiplied and passed on the form of the gene which made them resistant. (2)

3 The antibiotics will destroy the normal body gut flora (the microbiome). (1)

14.1

1 A negative correlation (1) between age and the incidence rate of asthma (1)

2 a x-axis labelled 'risk factor' and y-axis labelled 'percentage of cases' (1)

Bar chart plotted with bars of equal width and equal spaces between bars (1)

All data plotted accurately to +/− 1 mm (1)

Graph has an informative title and plot area covers 50% of the available space (1)

b Because of the interactions between exposures (1)

3 Genetic factors likely play some role in high blood pressure, stroke, and other related conditions (1). Inheriting some forms of a gene (certain alleles) can predispose individuals to be at more risk of a stroke (1) (e.g. by increasing blood pressure or the concentration of clotting factors) (1). In addition, it is also possible that people in the same family have environmental factors in common such as diet composition. (1) *Max 3 marks*

14.2

The Bradford Hill criteria for establishing causal relationships

By calculating a correlation coefficient between the dependent and independent variables. The closer the value is to one, the greater the correlation and the 'strength of association'.

1 Epidemiological evidence – evidence such as that shown in Fig. 1 OR the link between number of years having smoked and the development of cancer.

Experimental evidence – evidence such as analysing cigarette smoke and finding known carcinogens (there are other examples). (2)

2 CHD has many other risk factors [1]. Many non-smokers get coronary heart disease. [1]

3 Some people might be at higher risk [1] such as those with a history of bowel cancer in the family. [1]

14.3

1 A tumour is an abnormal mass of cells in a region where no growth or repair was needed as a result of unregulated cell division. Cancer occurs when some cells break off a benign tumour and spread in the lymph nodes or blood to start a secondary tumour in an additional site in the body. (3)

2 G1 stage of the cell cycle (1); gene is permanently switched on so cell moves from G into S phase (1)

3 Plants don't get cancer like animals as the tumours they do get do not metastasise because plant cells don't move around. This is because they are held in place by cell walls. Plant tumours are usually caused by a bacterium, virus or fungus, or may develop as a result of structural damage. (2)

14.4

1 One breast at a time is placed upon the platform on the mammogram unit (1); the breast will be steadily compressed to even out the breast thickness; (1) an x-ray beam will pass through the breast (1); the images taken are recorded (1); this process will be repeated three more times. Multiple images, at least four, are taken (1). The mammogram is examined to identify any small areas of calcium which may indicate changes to the breast tissue (1). *Max 5 marks*

2 (6)

	MRI	CT Scan
Application	Best used for examining soft tissue, showing inflammation, and creating cross-sectional pictures (1)	Best suited for bone injuries, lung and chest imaging or cancer detection (1)
Advantages	Ability to change the contrast of the images. (1) Superior for detection and identification of tumors (1) No known side effects related to radiation. (1) Fewer side effects to MRI contrast. (1)	Very good for imaging bone structures. (1) Time taken to scan is much shorter. (1) Cheaper than MRI (1) Widely available. (1) Less stringent about remaning motionless. (1)
Disadvantages	Patients with claustrophobia may need anesthesia during the scan. (1) Some patients with certian surgical clips or metal fragments, heart monitors or pacemakers may not be able to have an MRI. (1) Cost. (1) Must remain motionless. (1)	Uses ionizing radiation (1) Risk of allergic reaction to the contrast material (1)

3 Cancer cells have a high respiration rate so take up more glucose (high respiration rate)(1); due to high cell division rate (1)

14.5

1 (4)

For	Against
Can result in early detection so more easily treated	Most women will have negative results, so targeted screening would be more cost effective
Early stage cancers can be treated before they spread	Could lead to unnecessary treatment of small benign tumours

2 a Correlation between income and education. (1)

More access to health care if higher earner. (1)

More likely to get screened if they can afford health care to gain treatment if needed. (1)

b Income groups should be of equal increments. (1)

3 Most cases will be due to HPV infections (1), BUT not all forms of the virus lead to cervical cancer. (1)

Microscopic examination of cells is more subjective than testing for the presence of a virus. (1)

HPV testing can be cheaper (1) (as it can be automated – see Topic 12.2).

14.6

1 There is a 1 in 2 chance (1) that the siblings or children of someone with a BRCA mutation will have inherited the same mutation. This is because a child will inherit either their mother or father's copy of each BRCA gene, and this is down to chance. (1)

2 Genetic counselling is generally recommended before and after any genetic test for an inherited cancer syndrome (1); this counselling should be performed by a health care professional who is experienced in cancer genetics (1); genetic counselling usually covers many aspects of the testing process, including: A hereditary cancer risk assessment based on an individual's personal and family medical history (1). Answer should include discussion of: the appropriateness of genetic testing (1); the medical implications of a positive or a negative test result (1); the possibility that a test result might not be informative (see Question 12) (1); the psychological risks and benefits of genetic test results (1); the risk of passing a mutation to children (1); explanation of the specific test(s) that might be used and the technical accuracy of the test(s) (1). (max 5)

3 If a close (first- or second-degree) relative of the tested person is known to carry a harmful BRCA1 or BRCA2 mutation, a negative test result is clear: it means that person does not carry the harmful mutation and cannot pass it on to their children. Such a test result is called a "true negative." (1) A person with such a test result has the same risk of cancer as someone in the general population. (1)

14.7

Factors that can alter the effectiveness of Tamoxifen

1 Tamoxifen has a similar shape to oestrogen, to enable it to bind to the complementary shape of the receptor.

2 Non-competitive inhibitors, such as the allosteric effect on the enzyme, will prevent tamoxifen forming an ESC with the enzyme CYP2D6 permanently. Any increase in tamoxifen dose is unlikely to increase the conversion of tamoxifen into its active form.

1 The further away visitors and staff stand from the patient (and their bed), the less radiation they are exposed to (1). This is because the levels of radiation fall very quickly as the distance away from the radioactive source increases. (1)

2 Herceptin only binds to specific receptors (HER" receptors) (1) so it will not be effective against any kind of cancer that does not have these receptors on their cell surface membranes (1), i.e. it will not work on hormone-receptor-negative breast cancers.

3 It is non-polar so can cross the membrane and the complex can enter through the pores in the nuclear envelope. (2)

15.1

Does ozone worsen asthma?

1 Urban areas are likely to have more air pollution (e.g. ozone), which is liable to exacerbate asthma in some cases.

2 Ozone is more likely to be generated during the summer because there will be more sunlight. However, other allergens linked to asthma are seasonal (e.g. pollen). These and other factors such as respiratory infections would need to be taken into account during the analysis. Is it ozone alone that is increasing the number of hospital admissions?

1 An acute disease has a rapid onset and lasts a short time (1); a chronic disease may have a slower onset and lasts a long time – symptoms often become worse over its duration (1)

2 Tar from tobacco smoke damages cilia in a smoker's airways (1); and causes excess mucus to be produced (1); mucus cannot be moved up the airways by the paralysed cilia and remains in the lungs (1); the mucus can harbour pathogenic microorganisms, resulting in infection (1)

3 Some people have a genetic predisposition to develop asthma (1); allergens such as mould, pollen, dust mites, fur and a range of chemicals can induce asthma attacks and make symptoms worse (1); air pollutants such as ozone and tobacco smoke have also been linked with asthma (1)

15.2

1 Placebos should have the same appearance, smell and taste as the real drug (1); they should not affect the patient (1)

2 To ensure that the planned clinical trials meet ethical requirements and can be justified (1); the committee scrutinises the information that will be presented to potential participants (1); to ensure it clearly outlines all aspects of the trials (1)

3 Phase 3 uses a much larger sample size 91); phase 2 will often test the drug against a placebo, whereas phase 3 tests the drug against the best current treatment (1)

Glossary

activation energy the energy that needs to be put in to cause a reaction. The activation energy is lowered by the presence of an enzyme (biological catalyst).

active immunity a type of resistance developed in an organism through production of specific antibodies in response to an exposure to a pathogen (natural) or to a vaccine (artificial).

active site a group of usually 3–12 amino acid R-groups that makes up a region on the surface of the enzyme into which a complementary substrate temporarily bonds to forming an enzyme-substrate complex.

active transport movement of molecules from an area of low concentration to a region of high concentration gradient against a concentration gradient requiring an input of ATP and involving transport proteins.

acute disease a disease that has a sudden onset and lasts a short time.

adaptation a trait that benefits an organism in its environment and increases its chances of survival and reproduction.

adenosine triphosphate (ATP) a nucleotide composed of a nitrogenous base (adenine), a pentose sugar, and three phosphate groups. The universal energy currency for cells.

adhesion the force of attraction between two different molecules e.g., water and molecules of lignin.

agglutinate clump together.

agglutination the clumping together of antigen-bearing cells, microorganisms, or particles in the presence of specific antibodies.

α helix a right-handed spiral held in place by hydrogen bonds between adjacent C=O and NH groups in a protein.

alveoli small air sacs in the lungs which allow for rapid gaseous exchange.

amino acid an organic compound that has a central carbon atom to which an amine group (NH_2) and carboxyl group (COOH) and variable residual group are attached. They are joined together by condensation reactions to form a polypeptide chain.

amniocentesis a procedure for sampling fetal cells from the amniotic fluid.

anaphase the stage of mitosis in which chromatids separate and are pulled towards opposite poles of the cell by the spindle.

aneurysm a localized bulge of an artery, vein, or the heart wall. The wall of the blood vessel or organ is weakened and may rupture.

antenatal care the care received by a pregnant woman.

antibiotics a substance produced by a living organism that kills or inhibits the growth of microorganisms, has no effect on viruses.

antibodies globular protein molecules (immunoglobulins) produced by plasma cells (B lymphocytes) in response to stimulation by an antigen.

anti-coagulant a substance that prevents blood from clotting e.g., Sodium citrate, heparin.

anticodon a sequence of three bases at the end of a tRNA molecule that determines the specific amino acid carried by the tRNA molecule.

antigen a toxin or other foreign substance which induces an immune response in the body, especially the production of antibodies.

antiparallel a feature of the two strands in a DNA molecule. The 5′ (5 prime) end of one strand is directly opposite the 3′ (3 prime) end of the parallel strand. The two strands run in opposite directions.

antithrombin a protein that inhibits thrombin.

aorta the major artery of the body, supplying oxygenated blood to the circulatory system.

apoplast within a plant, the apoplast is the free diffusional space outside the plasma membrane including the cell wall through which water can travel.

apoplast pathway the transport route taken by water and dissolved substances through the cell walls and intercellular spaces of plants.

apoptosis programmed cell death.

arteries a thick-walled vessel that carries blood away from the heart.

aseptic techniques any techniques/manipulations of equipment or materials that are designed to prevent contamination by microorganisms.

asthma a respiratory condition characterised by the inflammation and narrowing of the bronchi.

atrio-ventricular node (AV Node) a patch of tissue in the septum of the heart that conducts the electrical stimulus from the atria in the heart through to the Purkyne fibres.

atrioventricular valves valves between the atria and ventricles that prevent backflow of blood.

atrium the upper chambers of the heart which receives blood returning from the organs and vessels of the body.

bacteriocidal describes a chemical substance that kills bacteria.

bacteriostatic describes a chemical substance that prevents the reproduction of bacteria.

Benedict's test a biochemical reaction to test for the presence of a reducing sugar, for example glucose. The test can be semi-quantitative or quantitative depending on the procedure used.

benign a tumour that stays in its original location and does not shed any cells into the blood plasma or lymph system.

biodiversity the variety of life, which can be measured on a genetic, species or ecosystem level.

biosensor a device which uses a living organism or biological molecules, especially enzymes or antibodies, to detect the presence of specific chemicals.

biuret test a biochemical reaction to test for the presence of proteins. It is a qualitative test.

blind trials a clinical trial in which participants are unaware whether they are receiving a placebo or a medicinal drug.

blood clot a structure formed from fibrin fibres which traps red blood cells and platelets in response the damage to a blood vessel.

blood group also known as blood type – the classification of blood depending on which antigens are present on the plasma membrane of the erythrocytes.

blood pressure the force exerted by the blood against the walls of the blood vessels.

B lymphocyte a specialised leucocyte that is produced in the bone marrow and matures in the bone marrow. It forms plasma cells after contact with a specific antigen and produces antibodies.

boosters an additional dose of an immunizing agent, such as a vaccine, given at a time after the initial dose to sustain the immune response elicited by the previous dose of the same agent.

bradycardia a slowness of the heartbeat, usually under 60 beats per minute in adults.

broad spectrum antibiotics that are effective against a large variety of organisms.

bronchi the two main branches from the trachea that go into the lungs.

bronchioles one of the smaller subdivisions of the branched bronchial tree that connects the trachea to the alveoli.

buffer a chemical solution which has the ability to absorb or donate hydrogen ions (protons) to maintain the pH of the solution.

bundle of His specialised cardiac muscle fibres that run from the atrioventricular node to the base of the heart.

calibration curve using readings of solutions of known concentrations (e.g., by colourimetry) to construct a calibration curve on a graph to determine the quantity of a substance in a solution of unknown concentration.

catalyst a substance that speeds up the rate of the reaction without itself being altered or used up in the chemical reaction.

cancer a disease usually caused by a mutation that causes uncontrolled cell division and the subsequent formation of a tumour. Some of these (primary tumour) cells may break away and be transported in the plasma or lymph system to form a secondary tumour in a different location.

capillaries very small blood vessels where water, solutes and respiratory gases are exchanged with body tissues.

capsid the outer protein coat of a virus.

capsomere a subunit of the capsid, an outer covering of protein that protects the genetic material of a virus.

carcinogen a chemical or form of radiation that causes cancer.

cardiac monitor a device for the continuous observation of cardiac function.

cardiac muscle the specialised muscle type found in the heart. It has it's own intrinsic heartbeat (it is myogenic).

cardiac output the amount of blood the heart pumps through the circulatory system in a minute.

carrier protein protein found within a cell membrane that carries a specific molecule or ion across the membrane by active transport.

casparian strip a band of impermeable *suberin* found in the walls of endodermal cells in plant roots.

cell cycle the series of events that take place in a cell leading to its division to produce two daughter cells.

cell surface membrane the phospholipid bilayer that forms the membrane surrounding the outside of a cell – sometimes known as the plasma membrane.

cellulose a polysaccharide made from the condensation of many β-glucose molecules to form fibrils. It is used to form plant cell walls.

cell wall a freely permeable structure lying outside of the cell surface membrane of plant fungal and bacterial cells.

centrifugation the process of separating molecules and organelles on a basis of their density by spinning them at different speeds in a centrifuge.

centriole(s) two cylinders composed of microtubules which are involved in the process of mitosis and cell division in some eukaryotic cells.

centromere the region of a chromosome that joins two sister chromatids and attaches to spindle fibres during mitosis.

channel protein a protein pore that spans a cell membrane to enable water soluble molecules and small ions to passively cross the membrane.

chemotherapy destroying cancerous cells using drugs that affect cancerous cells more than other cells in the body.

chiasmata the points at which crossing over occurs between homologous chromosomes.

chordae tendinae also known as the heart strings, they are cord-like tendons that connect the papillary muscles to the tricuspid valve and the mitral valve in the heart.

chorionic villus sampling a procedure for sampling fetal cells from the placenta.

chromatid a DNA molecule; during prophase and metaphase of mitosis, a chromosome consists of two identical chromatids from the replication of DNA in S phase.

chromatography a technique used to separate substances in a mixture according to differences in their solubility.

chromosome a linear DNA molecule, which, during prophase, becomes visible and consists of two chromatids.

chronic disease a disease that lasts a long time and has symptoms that worsen over time.

classification the organisation of organisms into groups based on similarities in biochemistry, anatomy, behaviour and embryology.

clinical trials a series of controlled studies in which a new medicinal drug is tested.

coagulation see blood clotting.

coenzyme an organic non-protein molecule that binds temporarily with the substrate to the active site in order for an enzyme to function.

cofactor a molecule or ion which aids the function of an enzyme – it can be an inorganic ion or a coenzyme.

codon a sequence of three bases on the template strand of the DNA or the mRNA that codes for one amino acid.

cohesion the attraction between water molecules due to hydrogen bonding.

cohesion-tension theory a theory of intermolecular attraction that explains the process of water flow upwards (against the force of gravity) through the xylem of plants.

colony morphology the characteristics of a bacterial colony, in cultures, in terms of shape, colour, edge, and elevation.

colorimeter a device that measures the absorbance of particular wavelengths of light by a specific solution, most commonly used to determine the concentration of a known solute in a given solution.

communicable a disease or infection capable of being communicated or transmitted to another organism.

companion cells a cell in the phloem involved in actively loading sucrose into sieve tube elements. The companion cell is closely associated with the phloem sieve element, to which it is linked by many plasmodesmata.

compartmentalisation the use of intracellular membranes to separate metabolic processes within the cell e.g., the nuclear envelope around the nucleus.

competitive inhibitor a molecule that has a similar shape to the natural substrate which competes for the active site on the enzyme preventing the formation of enzyme substrate complexes and instead forming enzyme-inhibitor complexes.

complementary therapy treatments that involves procedures that are not part of mainstream medicine e.g. acupuncture, aromatherapy.

condensation reaction a chemical process in which two molecules are combined to form a more complex molecule with the removal of a molecule of water to form a covalent bond.

correlation the relationship between two variables e.g. a linear relationship.

co-transporter proteins in the cell surface membrane that allows movement of one molecule when linked to the movement of another molecule by active transport.

cotyledon in seeds, the part of a plant embryo that becomes the first leaf.

crossing over the process in which homologous chromosomes exchange alleles during prophase I of meiosis.

CT scan computer assisted tomography scan (CAT scan): X-rays are used to build up a 3D image of the body using computers.

cytokines cell signalling molecules which are used for communication between cells, allowing some cells to regulate the activity of others.

cytokinesis the division of a cell to form two new cells.

cytoskeleton a network of microtubules and microfilaments that give the cell shape and maintain its structure. They can attach to organelles and move organelles within the cytosol.

defibrillator an apparatus used to control heart fibrillation by application of an electric current to the chest wall or heart.

degenerate code situation where more than one codon codes for the same amino acid.

denaturation usually permanent change to the tertiary structure of a protein resulting in the loss of function. This can be caused by large changes in pH or high temperatures.

dendritic cells (DCs) antigen-presenting cells that act as a key regulator of the immune system. They are capable of activating naive T cells and stimulate the differentiation of B cells.

deoxyribonucleic acid (DNA) the molecule responsible for the storage of genetic information.

diaphragm a sheet of muscular and fibrous tissue separating the chest cavity from the abdominal cavity. Plays an important role in ventilation.

diastole period of relaxation and repolarisation of the cardiac muscle when the chambers fill with blood.

dicotyledons a plant that produces flowers and has two cotyledons (seed leaf) inside the seed, which develops wide leaves with veins.

differential staining staining processes that use more than one chemical stain, for example, multiple stains can be used to distinguish between different microorganisms or structures/cellular components of a single organism.

differentiation the development of unspecialised cells to form specialised cells.

diffusion the movement of molecules down their concentration gradient as a result of random motion. In cells diffusion may occur passively directly across the plasma membrane (simple) or via membrane proteins (facilitated).

diploid cells that have two copies of each chromosome.

disaccharide a dimer made from the condensation of two monosaccharides joined by a glycosidic bond.

disulfide bond a S—S chemical bond between two sulfur atoms in the R groups of two cysteine amino acids.

double circulatory system a type of blood circulation system in which the blood flows through the heart twice for each full circuit of the body.

double-blind trials a clinical trial in which neither participants nor scientists are aware whether a placebo or a medicinal drug has been issued.

ecosystem the organisms and non-living components of a specific area, and their interactions.

electrocardiogram (ECG) a graph showing the electrical activity in the heart during the cardiac cycle (heartbeat).

electrolytes ions such as sodium, potassium, and chloride dissolved in water.

electron microscope a microscope that uses a beam of electrons to view a magnified image of an object giving it greater resolution than a light microscope. There are two main types – scanning and transmission electron microscopes.

endemic describes a disease that is ever present in a population. May also mean a species that is found only in a particular area and nowhere else.

endocytosis the inward transport of large quantities of molecules through the cell surface membrane to form a vesicle within the cytosol. This requires an input of energy in the form of ATP.

endodermis a ring of cells between the cortex of the root and the area housing the xylem and phloem.

endoplasmic reticulum (ER) a series of membranes forming a system of tubes within the cytoplasm which joined to the nuclear envelope and which may have ribosomes attached (Rough ER) or not (smooth ER).

enzyme a globular protein that acts as a biological catalyst.

epidemic a sudden increase in the incidence of a disease that spreads to many people quickly and affects a large proportion of the population.

epidemiology the study of patterns of disease and the factors that influence their spread.

erythrocyte a red blood cell. Red blood cell production is Erythropoiesis and this is regulated by the hormone Erythropoietin.

ester bond a C—O—C chemical bond formed from by a condensation reaction e.g., Between a fatty acid and a glycerol molecule.

eukaryotic cell a cell that has a true nucleus (contained by a nuclear envelope and contains membrane-bound organelles in the cytosol e.g., Golgi apparatus.

exocytosis the outward transport of large quantities of molecules through the cell surface membrane. This requires an input of energy in the form of ATP.

extrinsic proteins proteins which are attached to a monolayer within a cell membrane.

eye piece graticule a square grid that appears imposed in the field of view when inserted into the eye piece of a microscope.

false negative test result indicates that a condition failed while it actually was successful, or indicated that a substance was not present when it was.

false positives test result indicates that a condition was successful while it actually was not successful, or indicated that a substance was present when it actually wasn't.

fetus the unborn offspring of a mammal.

fibrillation a state in which the chamber walls of the heart contract out of rhythm.

fibrin a fibrous protein that forms long fibrils to form a mesh to trap red blood cells and platelets in clot formation.

fibrinogen a soluble protein found in blood plasma that can be converted to insoluble fibrin during the clotting process.

field of view the circular area visible down a light microscope.

flagellum (plural Flagella) whip like structures which function in the movement of both prokaryotic and eukaryotic cells. The structure of a flagellum varies between the two types of cells.

gamete a reproductive (sex) cell that fuses with another gamete during fertilisation.

globular protein proteins that are generally soluble in water and form spherical structures, often having a role in the metabolism of cells.

gene a length of DNA on a chromosome that codes for the production of a specific polypeptide chain.

genome all the DNA that makes up the organism.

glycogen a branched polysaccharide made from the condensation of many α-glucose molecules. It is used as an energy storage molecule in liver and muscle cells.

glycolipid a lipid molecule with a carbohydrate portion (glycocalyx) added to it.

glycoprotein a protein molecule with a carbohydrate portion (glycocalyx) added to it.

glycosidic bond a type of C—O—C covalent bond that joins two saccharides together as a result of a condensation reaction.

Golgi apparatus organelle involved with modification of proteins and production of vesicles (small membrane bound sacks) for storage of enzymes in cells (lysosomes) or exporting molecules from cells (exocytosis).

haemoglobin a globular protein made from four polypeptide chains and four prosthetic haem groups which can combine with four oxygen molecules to form oxyhaemoglobin.

haploid cells that have one copy of each chromosome.

heart rate the number of heart beats per minute. The heart rate is based on the number of contractions of the ventricles per minute. Frequently measured as pulse rate.

herd immunity a form of immunity that occurs when the vaccination of a significant portion of a population (or herd) provides a measure of protection for individuals who have not developed immunity.

histone proteins that are associated with DNA within chromosomes that enable the DNA to condense and coil during nuclear division.

homologous chromosomes a pair of chromosomes, one maternal and one paternal, that have the same gene loci.

hormones chemicals released by endocrine glands which act as cell signalling molecules. They circulate in the blood and regulate the activity of a range of tissues and organs.

hydrogen bond a relatively weak chemical bond formed between the slightly positive charge on a hydrogen atom and the slightly negative charge on another atom (usually oxygen) on an adjacent molecule e.g., between water molecules, found in secondary and tertiary strictures of proteins.

hydrolysis the breaking of a covalent bond within a large molecule to form two smaller molecules by the addition of a molecule of water.

hydrophilic a molecule able to associate with water as it possesses a charge which interacts with the dipoles on water molecules.

hydrophobic a molecule not able to associate with water as it is uncharged i.e., repels water.

hydrophobic interactions tendency of nonpolar substances to aggregate in aqueous solution and exclude water molecules – commonly seen in hydrocarbon R groups which repel water from the centre of the folded polypeptide chain.

hydrostatic pressure pressure created by a fluid pushing against the sides of a vessel.

hypertension a condition in which the resting blood pressure (particularly the diastolic pressure) is raised for prolonged periods.

immunotherapy treatment of disease by inducing, enhancing, or suppressing an immune response.

incidence rate the number of new cases of a disease in a given population, usually in a certain time period.

independent assortment the random assortment of chromosomes during anaphase I of meiosis and chromatids during anaphase II of meiosis.

infectious a disease caused by the entrance into the body of organisms (as bacteria, protozoans, fungi, or viruses) which grow and multiply there.

inflammatory response a tissue reaction to injury or an antigen that may include release of histamine, pain, swelling, itching, redness, heat, and loss of function.

inhibitor a molecule that reduces the rate of an enzyme-controlled reaction by reducing the ability of the enzyme to bind to its substrate(s).

intercostal muscles muscles between the ribs, responsible for moving the rib cage during breathing.

interphase the phase of the cell cycle in which new DNA and organelles are synthesised.

introns portions of DNA within a gene that do not code for a sequence of amino acids within a polypeptide chain. These are removed from the mRNA after transcription.

ionic bond a strong chemical bond caused by the attraction between a negatively charged ion (anion) and a positively charged ion (cation).

karyotype the number and appearance of chromosomes in a cell.

Korotkoff sounds the sounds that medical personnel listen for when they are taking blood pressure using a non-invasive procedure.

leucocyte a white blood cell. There are various types e.g., Neutrophil, T-lymphocyte, phagocyte.

limiting factor a factor that limits the rate of a process. If it is then increase the rate of reaction will increase until a the same (or new) factor becomes limiting.

live-attenuated alive pathogens that have lost their virulence but are still capable of inducing a protective immune response to the virulent forms of the pathogen.

logarithmic scale nonlinear scale used when there is a large range of quantities e.g., pH.

lumpectomy the surgical removal of a discrete portion of the breast tissue, usually as part of the treatment for a malignant tumour or breast cancer.

lymph a colourless fluid containing leucocytes that bathes the tissues and drains through the lymphatic system into the bloodstream.

lysis the disintegration or rupture of the cell membrane, resulting in the release of cell contents or the subsequent death of the cell.

lysosomes membrane bound vesicles made by pinching off from the Golgi body. They contain strong digestive enzyme to break down old cellular components.

magnification the number of times larger an image appears compared to the real specimen.

malignant a tumour that sheds cells that can spread through the body via the blood plasma or lymph system to initiate a secondary tumour in a new location(s).

mammography X-rays of the breast that are usually carried out to detect and screen for the early stages of cancer.

mantoux test a skin test for tuberculosis.

mass flow the movement of fluid in one direction, usually through tube like vessels.

mass transport the transport of molecules in bulk from one part of an organism to another.

mastectomy an operation to remove breast tissue (and sometimes some of the lymph nodes under the arm) to treat breast cancer.

meiosis a type of nuclear division in which the chromosome number is halved.

mesosome an infolding of the cell surface membrane found in prokaryotic cells.

metaphase the stage of mitosis in which chromosomes line up at the equator of the cell.

metastasis process in which cancer cells break from the primary tumour and spread in the plasma or lymph system to initiate a secondary tumour formation at a different location(s).

microtubule protein polymers that form the mitotic spindle and are components of the cytoskeleton.

mitosis a form of nuclear division in which two genetically identical nuclei are formed; occurs in somatic cells.

monocotyledons a flowering plant that produces seeds with only one cotyledon (seed leaf).

monocytes a large phagocytic white blood cell. which can develop into a macrophage.

morbidity the incidence or prevalence of a disease or of all diseases in a population.

mortality the number of deaths in a population.

MRI scan magnetic resonance imaging that builds up images of the inside of the body using different magnetic field strengths.

mRNA a single stranded polynucleotide formed as a result of the transcription of a gene from the template strand of the DNA.

mutagen a chemical or form of radiation that causes a change in the amount or structure of DNA

mutation a change in the structure of DNA, or in the structure or number of chromosomes.

myogenic describes muscle tissue (heart muscle) that generates its own contractions.

natural selection the mechanism for evolution. The best-adapted organisms will be more likely to survive and reproduce, thereby passing on favourable alleles to the next generation.
neutralisation the ability of antibodies to block the site(s) on bacteria or viruses that they use to enter their target cell.
neutrophils a type of granulated phagocytic white blood cell.
non communicable disease a medical condition or disease which is non-infectious and non-transmissible among people.
non competitive inhibitor a molecule that can bind to a binding site separate from the active site on the enzyme. This change the shape of the active site such that it is no longer complementary to the the substrate.
non-polar a substance that contains no permanent dipolar molecules.
non reducing sugar a monosaccharide or disaccharide that cannot donate electrons to other molecules and therefore cannot act as a reducing agent. Sucrose is the most common non-reducing sugar.
non-specific response a resistance manifested innately by a species, it protects you against all possible antigens.
nucleotide monomer made from a nitrogen-containing base, phosphate group and a pentose sugar. Nucleotides can be joined together to form polynucleotides.

objective lens the lens closest to the specimen. Thc magnification can usually be varied to give low, medium, and high power images.
oncogene gene forms when a proto-oncogene mutates which allows cells to divide uncontrollably resulting in cancer.
oncotic pressure also known as colloid osmotic pressure. The force due to the tendency of plasma proteins and other substances to lower the water potential.
opsonin an antibody which binds to a pathogen making them more susceptible to phagocytosis.
organ a collection of tissues that work together to perform a specific overall function or set of functions within a multicellular organism.
organelle a structurally distinct part of a cell that is specialised to carry out specific function(s). Usually they are surrounded by a membrane (true organelle).
osmosis the passive movement of water from an area of high water potential to an area of low water potential down a water potential gradient through a partially permeable membrane.

pandemic describes an epidemic occurring worldwide, or over a very wide area, crossing international boundaries and usually affecting a large number of people.
partially permeable able to let some molecules or ions pass through and not others.
passive immunity immunity acquired by the transfer of antibodies. It may be natural (placental transfer of antibodies during pregnancy and via breast milk) or artificial (injection of antiserum).
passive transport movement of molecules which does not require an input of energy (ATP).
pathogen an organism that has the capacity to cause disease. Its ability to cause disease is called pathogenicity.
pathogen a microorganism that causes disease.
Peptide bond a CO—NH covalent bond that is formed via a condensation reaction between the carboxyl group (COOH) of one amino acid and the amine group (NH_2) of the adjacent amino acid.
PET scan images of the inside of the body built up from positron emission tomography which detects differences in metabolic activity of the tissues.
pH a measure of the number of proteins (hydrogen ions, H^+) in a solution.
phagocyte a specialised leucocyte that destroys bacteria and other foreign material by phagocytosis e.g., neutrophil, macrophage.
phagosome a vacuole inside a phagocyte which is created by an infolding of the plasma membrane to engulf a foreign particle. The foreign particle is held inside the phagosome.
phloem a tissue in plants that is used to transport dissolved sugars and other substances.
phloem sieve tubes one type of cell in phloem tissue joined end-to-end through which nutrients flow.
phosphodiester bond covalent bond found in DNA and RNA joining the 3′ carbon atom of one nucleotide and the 5′ carbon atom of another.
phospholipid specialised lipid molecule containing a phosphate group, two fatty acids, and a glycerol molecule. Consists of a hydrophilic head (phosphate group) and a hydrophobic tail (fatty acids). In water they form bilayers that make up cell membranes.
phylogeny the study of the evolutionary relationships between organisms.
pili a hair-like structure found on prokaryotic cells and used in the exchange of genetic material between bacterial cells.
placebo a 'dummy' pill or injection that resembles the real drug but contains no active ingredient.
plant assimilates the newly formed compounds that result from the incorporation of carbon, from carbon dioxide, into organic substances during photosynthesis.

plasma yellow fluid that transports blood cells and platelets around the body and contains a number of substances, including proteins.
plasmid small circular piece of DNA found in prokaryotic cells.
plasmodesmata microscopic channels which traverse the cell walls of plant cells, enabling transport and communication between them through cytoplasmic connections.
plasmolysed the detachment of the cell surface membrane from the cell wall as a plant cell loses water by osmosis as a result of being placed in a solution with a lower water potential causing the cytoplasm to shrink.
polar having a negative or positive charge.
polymer a large molecule made up of small repeating sub-units or monomers. For example, polysaccharides are made up of monosaccharides joined together.
polypeptide chain more than four amino acids joined together by peptide bonds as a result of a series of condensation reactions.
pressure gradient (heart) the difference in blood pressure across the vessel length or across the valve.
prevalence the proportion of individuals in a population having a disease or characteristic.
primary immune response the response that the immune system displays when first exposed to an antigen – the initial production of antibodies.
primary infection refers to first time exposure to a pathogen.
primary structure the sequence of amino acids in a polypeptide chain that is determined by the sequence of nitrogen-containing bases in a gene.
prokaryotic cell a cell that does not have a true nucleus or any true (membrane-bound) organelles, for example, Blue-green algae, bacteria.
prophase the stage of mitosis in which chromosomes become visible and the spindle forms.
prophylactics a medicine or course of action used to prevent a disease from occurring.
proto-oncogene a gene that helps regulate cell division.
Protoplast a fungal, bacterial, or plant cell which has had its cell wall removed by mechanical means or by enzymes.
pulmonary artery the artery carrying blood from the right ventricle of the heart to the lungs for oxygenation.
pulmonary vein a vein carrying oxygenated blood from the lungs to the left atrium of the heart.
pulmonary ventilation the total volume of gas per minute inspired or expired.
pulse as the heart pushes blood through the arteries, the arteries expand and recoil with the flow of the blood, causing a pulse.
purine nitrogen-containing base with a double ring structure i.e., adenine and guanine.
purkyne tissue (Purkinje tissue) specialised muscle tissue in the septum of the heart that conducts electrical stimulation from the AV node to the ventricles.
pyrimidine nitrogen-containing base with a single ring structure i.e., Thymine, cytosine and, uracil.

qualitative data observations or recordings that are not numerical e.g., presence/absence, colours.
quaternary structure the structure of a protein formed by two or more polypeptide chains and/or the presence of a prosthetic group(s).
quantitative data readings and recordings that are numerical.

radiotherapy treatment that destroys cancerous cells using ionising radiation (radiation that can dislodge electrons from atoms) e.g. X-rays, alpha rays and gamma rays.
reducing sugar a monosaccharide or disaccharide that has the ability to donate electrons to reduce copper sulfate from Cu^{2+} to Cu^{+} and give a positive result in a Benedict's test.
residual volume the volume of air remaining in the lungs after a maximal expiratory effort.
resistance the ability of an organisms to resist the effects of a chemical to which they were once sensitive e.g. bacteria and antibiotics.
resolution the ability to distinguish two distinct objects separately and to see detail.
retrovirus an RNA virus with the ability to transcribe its genetic material into the chromosomal DNA of the host cell, to be expressed there.
reverse transcriptase an enzyme originally derived from retroviruses. The enzyme catalyses the construction of a DNA strand using an mRNA strand as a template.
Rf value ratio of the distance travelled by the centre of a spot to the distance travelled by the solvent front.
rhesus antigen the presence of the immunogenic D antigen on the surface of the erythrocytes (i.e., Rh positive).
ribosomes structures composed of RNA and proteins which may be free in cytoplasm or membrane bound and are the site of protein synthesis.
ring vaccination the vaccination of all susceptible individuals in a prescribed area around an outbreak of an infectious disease.

risk factor an attribute, characteristic or exposure of an individual that increases the likelihood of developing a disease.

rRNA folded polynucleotide chain found in both the small and large subunits of ribsosomes.

screening programme preventive care programs used to determine if a disease/medical condition is present even if a member has not yet experienced symptoms of the problem.

secondary (specific) immune response an integrated bodily response to an antigen, especially one mediated by lymphocytes and involving recognition of antigens by specific antibodies or previously sensitised lymphocytes.

secondary structure the folding of a polypeptide chain into alpha helices and/or beta pleated sheets which are held in place by the presence of many hydrogen bonds.

selection pressure a factor that drives evolution in a particular direction.

semi-conservative replication process by which DNA makes an exact copy of itself. Each DNA molecule will have one parent strand and one newly synthesised strand.

semilunar valves also known as pocket valves and found in veins and between the ventricles and the main arteries. They prevent the backflow of blood.

serum the serum is the component of blood that is neither a blood cell nor a clotting factor – it is the blood plasma not including the fibrinogens.

sink a part of a plant that removes sugars and other solutes from the phloem.

sinoatrial node (SA node) the patch of tissue that initiates the heartbeat by sending waves of excitation over the atria.

solute a solid that dissolves in a liquid (the solvent).

solvent a liquid that dissolves solids (the solute).

source (solutes) a part of a plant that releases sugars and other solutes to the phloem.

species a group of organisms whose members can interbreed to produce fertile offspring and possess similar genetics, physiology, appearance and behaviour.

spindle microtubule fibres that attach to centromeres and separate sister chromatids during eukaryotic mitosis.

sphygmomanometer an instrument used to measure blood pressure.

squamous epithelium a type of thin, flat cell found in layers or sheets covering surfaces such as skin and the linings of blood vessels.

stage micrometer a microscope slide with a finely divided scale marked on the surface.

stem cells undifferentiated cells that are capable of differentiating into a range of cell types.

stroke volume the volume of blood pumped by the left ventricle of the heart in one contraction.

supernatant the liquid component of a mixture left at the top of the centrifuge tube when suspended organelles and molecules have settled to the base of the tube after centrifugation.

surfactant a chemical that can reduce the surface tension of a film of water and is found in the alveoli

symplast the continuous system of protoplasts, linked by plasmodesmata and bound by the cell wall.

symplast pathway the transport route taken by water and dissolved substances through the cytoplasm of plant cells.

systole the contraction of the muscular walls of the atria or ventricles.

T lymphocyte a specialised leucocyte that is produced in the bone marrow and matures in the thymus gland. They coordinate the immune response and destroy infected cells.

T regulatory cells (Tregs) T lymphocytes which suppress the function of other T cells to limit the immune response.

tachycardia an abnormally rapid heart rate, usually defined as greater than 100 beats per minute.

taxon (pl: taxa) one of the groups (e.g. domain, kingdom, phylum, class, order, family, genus or species) used in classification.

telophase the stage of mitosis in which two new nuclear envelopes form around the two sets of daughter chromosomes formed in nuclear division.

tertiary structure the further folding of a polypeptide chain into a fibrous or globular 3D structure – this is held in place by the presence of further hydrogen bonds, ionic bonds, disulfide bonds and/or hydrophobic interactions between R groups.

thrombin a protease enzyme that that converts soluble fibrinogen into insoluble strands of fibrin.

thrombocyte correct name for the platelets involved in blood clotting.

thromboplastin a plasma protein that aids blood coagulation through catalysing the conversion of inactive prothrombin to active thrombin.

thymus an organ that is located in the upper chest in which T lymphocytes mature and differentiate.

tidal volume the volume of air inspired or expired in a single breath during regular breathing.

tissue a group of cells, with a common origin and similar structures, which performs a particular function.

tissue fluid the fluid that surrounds the cells of the body. It has a similar composition to blood plasma except it contains no large proteins and no cells. Its function is to supply the cells with nutrients and remove wastes products.

toxin a poisonous substance produced within living cells or organisms.

trachea the windpipe leading from the back of the mouth to the bronchi.

transcription the formation of mRNA from a section of the template strand of the DNA that corresponds to a gene.

translation the production of a polypeptide gene using the sequence of codons on the mRNA.

tRNA single RNA-polynucleotide chain that carries a specific amino acid to the ribosome during translation.

translocation the movement of sucrose and other substances up and down a plant.

transpiration the loss of water vapour from the aerial parts of a plant due to evaporation.

triplet code a three-nucleotide codon in a nucleic acid sequence codes for a specific single amino acid.

tumour a tumour is a lump or growth in a part of the body and is formed from unregulated cell division by abnormal cells. Benign tumours are not cancerous and are not usually life-threatening.

tumour suppressor genes a gene that protects a cell from becoming potentially cancerous by halting cell division. When this gene mutates to cause a loss or reduction in its function, the cell can become cancerous, usually in combination with other genetic changes.

turgid describes a cell that is full of water as a result of entry of water due to osmosis when the pressure of the cell wall prevents more water entering.

ultrasound scan a procedure that uses high frequency sound waves to create an image of a fetus or some internal organs such as stomach, liver, heart, tendons, muscles, joints, or blood vessels.

vaccination the administration of antigenic material (a vaccine) to stimulate an individual's immune system to produce antibodies to a pathogen.

vacuolar pathway the pathway taken by water in plants as it passes from cell to cell via the cell cytoplasm and vacuole.

vacuole a fluid filled space found within the cytoplasm of some cells which is surrounded by a membrane (the tonoplast).

variation the differences between individuals.

vascular bundles the transport tissue in a plant - usually found as a bundle containing both xylem and phloem.

vasodilate an increase in the internal diameter of an arteriole.

vein a vessel that carries blood towards the heart.

vena cava either of two large veins that carry deoxygenated blood from the body, back to the heart.

venation distribution or arrangement of a system of veins in an organism including in plant leaves.

ventilation rate the volume of air passing into and out of the lungs per minute.

ventricles the muscular lower chambers of the heart, which pump to the organs of the body.

virus a metabolically inert, infectious agent that replicates only within the cells of living hosts: composed of an RNA or DNA core, a protein coat, and a surrounding envelope.

vital capacity the amount of air that can be forcibly expelled from the lungs following breathing in as deeply as possible.

water potential a measure of the ability of water molecules to move freely in solution. Measures the potential for a solution to lose water - water moves from regions of high water potential to one of lower water potential.

xylem a plant tissue containing xylem vessels (and other cells) that are used to transport water in a plant and provide support.

xylem vessels one type of cell found in Xylem tissue – a cylindrical structure forming a tube for transport of water.

zygote a diploid cell formed from two haploid gametes.

Index

Acknowledgements

The authors would like to thank Les Hopper, Rosie Parrish, Amy Johnson, Fran Fuller, Clodagh Burke, Michael Warren, and Ross Laman for their all their hard work and support. In addition the authors would like to thank Sarah Old, Richard Tateson, and Katherine Hands-Taylor at OCR for their help with development of the book.

Cover: SUSUMU NISHINAGA/SCIENCE PHOTO LIBRARY; **p2-3**, **p4-5**: Anna Jurkovska/Shutterstock; **p6**: Giorgiomtb/ Shutterstock; **p7**: Frank Fox/Science Photo Library; **p8**: Natural History Museum, London/Science Photo Library; **p11**(T): Power and Syred/Science Photo Library; **p11**(C): Science Photo Library; **p11**(B): Dr Jeremy Burgess/Science Photo Library; **p12**: Igor Siwanowicz/Science Photo Library; **p14**(T): Dr Keith Wheeler/Science Photo Library; **p14**(B): Zaharia Bogdan Rares/ Shutterstock; **p16**: Biophoto Associates/Science Photo Library; **p18**: Eye of Science/Science Photo Library; **p21**(T): Ed Young/ Science Photo Library; **p21**(B): Lawrence Berkeley National Laboratory/Science Photo Library; **p23**: Steve Gschmeissner/ Science Photo Library; **p24**(T): KR Porter/Science Photo Library; **p24**(B): Microscape/Science Photo Library; **p25**: Science Photo Library; **p26**: Louise Hughes/Science Photo Library; **p32**: Gunilla Elam/Science Photo Library; **p37**: Jacopin, BSIP/ Science Photo Library; **p44**: Science Photo Library; **p46**: Don W Fawcett/Science Photo Library; **p50**: Eye of Science/ Science Photo Library; **p51**: Bikeriderlondon/Shutterstock; **p52**(L): Martin M Rotker/Science Photo Library; **p52**(R): P_Wei/ iStockphoto; **p54**: Mauro Fermariello/Science Photo Library; **p60**: Martyn F Chillmaid/Science Photo Library; **p61**(L): Cordelia Molloy/Science Photo Library; **p61**(R): Saturn Stills/ Science Photo Library; **p62**: Dmitry Lobanov/Shutterstock; **p63**: Andrew Lambert Photography/Science Photo Library; **p64**(T): Czardases/iStockphoto; **p64**(B): Andrew Lambert Photography/ Science Photo Library; **p65**: Science Photo Library; **p81**: Laguna Design/Science Photo Library; **p82**: Laguna Design/Science Photo Library; **p84**: Molekuul.Be/Shutterstock; **p89**: Laguna Design/Science Photo Library; **p90**(T): Dr P Marazzi/Science Photo Library; **p90**(B): James King-Holmes/Science Photo Library; **p91**: Steve Gschmeissner/Science Photo Library; **p92**: Science Photo Library; **p93**: Tek Image/Science Photo Library; **p94**: Alex Bartel/Science Photo Library; **p95**: CNRI/Science Photo Library; **p96**: Laguna Design/Science Photo Library; **p97**(T): Martyn F Chillmaid/Science Photo Library; **p97**(B): Dr P Marazzi/Science Photo Library; **p98**: Eraxion/iStockphoto; **p99**: Revy, ISM/Science Photo Library; **p100**: Jim West/Science Photo Library; **p101**(T): Ria Novosti/Science Photo Library; **p101**(C): Angellodeco/Shutterstock; **p101**(B): Sura Nualpradid/ Shutterstock; **p102**: Ria Novosti/Science Photo Library; **p113**(L): Francis Leroy, Biocosmos/Science Photo Library; **p113**(R): Laguna Design/Science Photo Library; **p115**(T): National Library of Medicine/Science Photo Library; **p115**(B): Catherine Lane/iStockphoto; **p116**: Laguna Design/Science Photo Library; **p117**: Alfred Pasieka/Science Photo Library; **p118**: Hybrid Medical Animation/Science Photo Library; **p121**: Alila Medical Media/Shutterstock; **p123**: Dr Elena Kiseleva/ Science Photo Library; **p126**: Michael Abbey/Science Photo Library; **p127**: GJLP/Science Photo Library; **p141**: Adam Hart-Davis/Science Photo Library; **p145**: CNRI/Science Photo Library; **p147**: Photographee.eu/Shutterstock; **p151**: Biophoto Associates/Science Photo Library; **p152**(T): Garry Delong/ Science Source/Getty Images; **p152**(B): Dr Keith Wheeler/ Science Photo Library; **p152**(C): Dr Keith Wheeler/Science Photo Library; **p153**: Garry Delong/Science Photo Library; **p154**: Dr Richard Kessel & Dr Gene Shih/Visuals Unlimited, Inc/Science Photo Library; **p159**: Sidney Moulds/Science Photo Library; **p164**(L): PIR/CNRI/Science Photo Library; **p164**(R): J C Revy, ISM/Science Photo Library; **p166**: Prof Motta, Correr & Nottola/University "La Sapienza", Rome/Science Photo Library; **p167**: Science Vu, Visuals Unlimited/Science Photo Library; **p168**(T): Biophoto Associates/Science Photo Library; **p168**(B): D Phillips/Science Photo Library; **p179**(TL): Pr G Gimenez-Martin/Science Photo Library; **p179**(BL): Pr G Gimenez-Martin/ Science Photo Library; **p179**(TR): Pr G Gimenez-Martin/Science Photo Library; **p179**(BR): Pr G Gimenez-Martin/Science Photo Library; **p180**: Manfred Kage/Science Photo Library; **p181**: Eye of Science/Science Photo Library; **p183**: Science Picture Co/ Science Photo Library; **p184**: Dr Yorgos Nikas/Science Photo Library; **p188**: Steve Gschmeissner/Science Photo Library; **p193**: Alfred Pasieka/Science Photo Library; **p195**(T): Petarg/ Shutterstock; **p195**(B): Maximshebeko/iStockphoto; **p197**: Monkey Business Images/Shutterstock; **p199**: Dept of Clinical Cytogenetics, Addenbrookes Hospital/Science Photo Library; **p201**: Zephyr/Science Photo Library; **p203**(T): Chatursunil/ Shutterstock; **p203**(B): Denise Allison Coyle/Shutterstock; **p204**: Konrad Wothe/Look/Getty Images; **p205**: John Reader/ Science Photo Library; **p210**: Percom/Shutterstock; **p211**: Pascal Goetgheluck/Science Photo Library; **p213**: Wallenrock/ Shutterstock; **p214**: Dr Jeremy Burgess/Science Photo Library; **p215**: St Mary's Hospital Medical School/Science Photo Library; **p216**: Natural History Museum, London/Science Photo Library; **p217**(L): Stevenrussellsmithphotos/Shutterstock; **p217**(R): Close Encounters Photo/Shutterstock; **p219**: Apiguide/ Shutterstock; **p223**(B): Shaun Jeffers/Shutterstock; **p223**(T): Jt888/Shutterstock; **p223**(C): Tom McHugh/Science Photo Library; **p228**: Javier Larrea/Age Fotostock/Getty Images; **p230**(T): Rudi Gobbo/E+/Getty Images; **p230**(B): Juergen Berger/ Science Photo Library; **p236**(T): Cavallini James/BSIP/Science Photo Library; **p236**(B): Alfred Pasieka/Science Photo Library; **p238**: Steve Gschmeissner/Science Photo Library; **p239**(T): Eye of Science/Science Photo Library; **p239**(B): Science Photo Library; **p243**: Eye of Science/Science Photo Library; **p261**: Hank Morgan/Science Photo Library; **p267**(L): Eddy Gray/ Science Photo Library; **p267**(R): Clouds Hill Imaging Ltd/Science Photo Library; **p269**: Africa Studio/Shutterstock; **p273**: Dr P Marazzi/Science Photo Library; **p277**: Ben Edwards/Getty Images; **p278**(T): Epstock/Shutterstock; **p278**(C): Camal, ISM/ Science Photo Library; **p278**(B): Simon Fraser/Royal Victoria Infirmary, Newcastle Upon Tyne/Science Photo Library; **p279**: Photostock-Israel/Cultura/Science Photo Library; **p288**(T): Fluxfoto/iStockphoto; **p288**(B): Stanley45/iStockphoto; **p294**: David Scharf/Science Photo Library; **p296**: Alamy; **p297**: Biophoto Associates/Science Photo Library; **p299**: Keith Weller/ US Department of Agriculture/Science Photo Library; **p300**: Sashkin/Shutterstock; **p303**: Biophoto Associates/Science Photo Library; **p304**: BSIP/UIG/Getty Images; **p305**(T): Jim Cummins/ The Image Bank/Getty Images; **p305**(B): William Lombardo/The Image Bank/Getty Images; **p306**: Dr Andrejs Liepins/Science Photo Library; **p307**: Marek Mis/Science Photo Library; **p308**: H Raguet/Eurelios/Science Photo Library; **p309**(T): Dr Keith Wheeler/Science Photo Library; **p309**(B): Dr Keith Wheeler/ Science Photo Library; **p310**: Philippe Plailly/Science Photo Library;

Header Photos - 1: Frank Fox/Science Photo Library; **2**: Sashkin/Shutterstock; **3**: Laguna Design/Science Photo Library; **4**: Laguna Design/Science Photo Library; **5**: Michael Abbey/Science Photo Library; **6**: Biophoto Associates/Science Photo Library; **7**: Science Vu, Visuals Unlimited/Science Photo Library; **8**: Eye of Science/Science Photo Library; **9**: Alfred Pasieka/Science Photo Library; **10**: Percom/Shutterstock; **11**: Juergen Berger/Science Photo Library; **12**: Steve Gschmeissner/Science Photo Library; **13**: Hank Morgan/Science Photo Library; **14**: Clouds Hill Imaging Ltd/Science Photo Library; **15**: David Scharf/Science Photo Library;

Artwork by Q2A Media